Construction Accounting and Financial Management

Construction Accounting and Financial Management

William J. Palmer
CPA, Ernst & Young LLP

William E. Coombs, Esq.
Attorney, CPA

Mark A. Smith, Ph.D.
CPA, Ernst & Young LLP

Fifth Edition

McGraw-Hill, Inc.
New York San Francisco Washington, D.C. Auckland Bogotá
Caracas Lisbon London Madrid Mexico City Milan
Montreal New Delhi San Juan Singapore
Sydney Tokyo Toronto

Library of Congress Cataloging-in-Publication Data

Palmer, William J., date.
Construction accounting and financial management / William J. Palmer, William E. Coombs, Mark A. Smith.—5th ed.
p. cm.
William E. Coombs name appears first in previous ed.
Includes index.
ISBN 0-07-012749-2
1. Construction industry—Accounting. 2. Construction industry—Finance. I. Coombs, William E., date. II. Smith, Mark A. III. Title.
HF5686.B7C6 1994
657'.869—dc20 94-32436
CIP

Published previously under the titles *The Handbook of Construction Accounting and Financial Management* (1984) and *Construction Accounting and Financial Management* (1989, 1977, 1958).

16 17 18 DOD/DOD 2 1 0 9

ISBN 978-0-07-135963-4 / 0-07-135963-X

The sponsoring editor for this book was Larry Hager, the editing supervisor was Peggy Lamb, and the production supervisor was Suzanne W. B. Rapcavage. This book was set in Century Schoolbook by McGraw-Hill's Professional Book Group composition unit.

Printed and bound by R. R. Donnelley & Sons Company.

This book is printed on acid-free paper.

Contents

Foreword

As it does to most things, time continues to bring changes to the field of construction accounting and financial management. Since the first edition was published in 1958, major changes have occurred, leading to the second, third, fourth, and now fifth edition of this book. For example, back in 1965, the American Institute of Certified Public Accountants (AICPA) published *Audits of Construction Contractors*, which for years was the Bible for construction accountants. In 1981, this work was updated, expanded and clarified by another AICPA committee of which I had the pleasure of being vice chairman. That important new work is entitled *Audit and Accounting Guide—Construction Contractors*. It is accompanied by an equally important publication of the AICPA *Accounting for Performance of Construction Type and Certain Production Type Contracts (Statement of Position 81-1)*. I had the pleasure of being the chairman of the committee that developed that publication. These developments alone have previously required and continue to require substantial revisions of this book.

Another factor that continues to require new editions of this book is the numerous changes in federal income tax laws that have had a direct effect on the construction industry's accounting for such items as income earned on long-term contracts, depreciation, overhead cost allocation, and cash basis accounting, to name but a few. The introduction of small, inexpensive computers in the early 1980s led to advances in the use of such equipment in the construction industry. This led to the virtual disappearance of bookkeeping machines as a tool for construction accounting. Another important change in the 1980s has been the increased regulation and oversight by governmental, environmental, and other organizations which have caused exterior pressures resulting in new accounting challenges.

When we have so many dynamic changes, there is always the inherent risk that in the rush to adopt new concepts and new procedures, the basic fundamentals of accounting, as applied to the construction industry, can be overlooked. One of our primary objectives with this edition has been to strike a balance between the new and the old.

While accounting principles rarely change, it is an accepted fact that their application to any particular industry, and to any specific business within that industry, usually requires careful analysis and sound judgment. This is certainly true when one looks at the construction industry. This book cannot take the place of detailed information or of the experience with individual problems of any particular construction company. It can, however, survey the problems peculiar to the construction industry and present solutions to deal with them and arrive at sound results.

It is fair to say that no two construction contractors are alike. Operating conditions are so varied and the individual contractor's circumstances so complex that it is often necessary to go back and review basic accounting and business principles in relation to the specific conditions in a given company at a given time. This must be done in order for management to use accounting as a true operational tool. We have prepared this book with the thought that it is devoted more to the "why" than to the "how" of construction accounting. Once the why has been properly determined, almost any good accountant can handle the application of the accounting procedure. Certainly among the techniques described throughout the book, the experienced accountant will find few, if any, that are basically unfamiliar—it is the application which may be somewhat unexpected.

The value of this book should lie in the fact that it provides a review of the reasons for procedures that are too often applied unthinkingly, and the practical reasons why procedures that are standard in other businesses either are not used or are sharply modified. This volume should serve as a checklist of potential problems and opportunities in financial control and management. for the contractor, the banker, the bonding company, the lawyer, and the student, the discussions of how financial data are developed may help to make contractors' business statements more intelligible and useful, and we hope that the overall discussion may offer a worthwhile review of the accounting and financial management aspects of the industry.

The primary purpose in developing editions of this book has been to write it for the individual contractor—"the construction stiff." It is this individual who must buy and use accounting data and who must rely on those figures in making decisions on which the business must stand or fall. Rarely will one find a construction company that is run by an accountant. Yet, if the business is to grow and succeed, the contractor must learn to understand and use the help that accounting and proper management information systems can provide. It is hoped that this book will also help the contractor to better understand and evaluate the recommendation of his financial people and professional advisers.

Finally, the contractor should be able to determine when a given record or procedure is necessary and whether it is worth what it costs.

Time has also brought change to the image of the industry as a purely masculine stronghold. Increasingly, women are occupying an important place in the construction industry. As we write this edition, limitations to the English language continue to exist. Being what they are, masculine nouns and pronouns must be read in the purely generic sense. All such words as used throughout this book are intended to refer to both men and women. We hope that you, our readers, will accept our writings in that context.

In a sense, this book is a compilation of what the authors have learned over the years more by doing than by researching. For years, we have shamelessly picked the brains of everyone we met who had anything to do with construction accounting or financial management. Many of our former associates, clients, and others will recognize some of their thoughts and ideas in the following pages. We have collected forms, graphs, and other data from many of our clients and other sources. To all of the people who have touched upon our careers, we owe a large debt of gratitude.

The most recent effect of time on this book has been the passing of William E. Coombs on January 17, 1994. It was Bill who did the hard work of preparing the very first edition of this publication, and who asked me to coauthor the second edition and to help guide the publication through subsequent editions which have so greatly benefited my career. Bill brought unique perspectives to the subject of construction accounting and financial management. He was a lawyer, Certified Public Accountant, California State Senator, City Attorney, university professor, and a senior accountant, controller, treasurer, and house counsel for a number of construction companies. His contributions to the industry will continue through this and subsequent editions of this book. He will be missed by many who were lucky enough to know him.

With this edition, a new perspective has been added with Mark A. Smith. Mark is a Senior Manager with Ernst & Young LLP, a Ph.D. in Civil Engineering, and a Certified Public Accountant. The chapter on factors affecting project financial results is evidence of the new perspective that he brings to this publication.

Lastly, our special thanks to Ernst & Young LLP for the generous use of their staff and facilities and particularly to David Kirshman and Ross Allen who helped to update Chapter 13, "Tax Considerations

Construction Management." Thanks to Douglas Hughes of the McCosker Corporation for providing Appendix B, "Computerization in the Construction Industry," and to Carone Frazier, who helped with the typing of the manuscript. Our special thanks also to the construction trade associations, whose publications and staff work were of untold value; to the American Institute of Certified Public Accountants; and to all of those who, with exemplary patience and fortitude, put up with our incessantly probing questions.

William J. Palmer

Construction Accounting and Financial Management

Chapter

1

Operating Patterns and Industry Economics

Basic Operating Patterns

The primary objective of accounting in any business is to help that business make the maximum profit after taxes. Unless accounting makes its full contribution to that objective, its cost cannot be justified. In the construction industry, one of the ways accounting pays for itself is to help management to control operations. Another of the ways is to help management utilize its working capital to the greatest possible advantage.

The physical elements entering into normal construction operations are many and varied. The only common denominator is money. Therefore, one of the principal jobs of accounting in the construction industry is to translate operating results into terms of money. To do that, it must follow the basic operational patterns of the industry and of job management.

The job as the basic unit

Essentially, the work of a construction contractor is the improvement of real property. That means the bulk of his work must be done at the site of the property and not, as in manufacturing, at a central factory or plant. Even with the increase in off-site fabrication, the preassembled equipment and materials must be put in place at the jobsite. Accordingly, once a construction contract is made, the contractor's operations necessarily fall into these four major phases of the contract cycle:

1. *Procure* the necessary resources—labor, materials, supplies, and equipment.

2. *Mobilize* the resources to the construction site.
3. *Manage* the resources to complete the work.
4. *Demobilize* the resources from the construction site.

The completion of these four phases in the physical work on any particular project constitutes a "job"—the basic unit of construction operations and so the basic unit of management and of accounting for construction companies.

A job normally consists of the performance of a single contract for a single owner. However, it is not unusual for large projects to be broken down into several contracts. It is somewhat less common to find a single contract breaking down into several jobs, for example, when a single owner lets a single contract covering a wide variety of work at widely separated locations. On certain types of projects, work may be done for a number of owners and the project will still be treated as a single job. One good example is a road contractor who paves numerous driveways for owners along the way. Economy of scale has led to large engineering contractors building two or more refineries or electrical power plants under one contract for a petroleum or utility company or group of companies.

The principal test seems to be one of operating control. If a project can logically be made a unit for purposes of operating control, it will be considered as a single job for purposes of management and accounting. It is true that accounting must take into account the separateness of each individual contract and of each individual owner for billing. It is also necessary to distinguish between owners for the application of lien laws, but operationally a project is looked upon as a single job if it can be controlled and managed as a unit. In many respects a job is comparable to a "profit center" in manufacturing accounting. Chapter 7 also treats the subject of profit centers in more detail.

Types of jobs and contractors

The extent to which any of the four major phases referred to above presents problems to any particular contractor on any one construction job depends on a great many factors. For example, the size, nature, and location of the job in relation to the other jobs which the contractor may have in progress will be important. So will the size and scope of the contractor's organization, its experience with the type of work involved, the necessary financing, and the extent to which specialized aspects of the work can be subcontracted.

The construction industry has its specialists. Some specialize in large process design or construction management known in the industry as "turnkey" or "design-build," which include all aspects of the project from the feasibility study through the final construction. Examples of these jobs or projects are power plants, refineries, hydroelectric facilities, and other such projects requiring extensive management skill and the ability to disperse this skill on a wide geographic basis. Others specialize in earth moving, heavy concrete, and other heavy construction and in turnkey projects. These contractors will generally perform the construction phase, either as a subcontractor to the turnkey contractor or as a co-contractor or joint venture partner with the turnkey contractor. The bulk of the work performed by these heavy construction specialists is done for public agencies and large corporations which do their own designing and engineering. Additionally, there are the general contractors who specialize in building projects such as housing, schools, hospitals, offices, and warehouses, and who perform under the supervision of an architect-designer or engineer-designer.

Cutting across the entire industry are the speciality subcontractors such as electricians, plumbers, roofers, concrete finishers, heating and air-conditioning installers, plasterers, and dozens of others.

Each of these industry divisions has accounting requirements peculiar to its specialty. These accounting requirements are caused primarily by the manner in which the contractor agrees to, or is forced to, accept pricing terms. Pricing terms depend on the degree of competition within the industry's subcategory. For example, because of the extremely large scope and duration of the work in the turnkey segment of the industry, there are relatively few companies with the ability to take on projects of this nature. The contract price is usually cost plus a fee and is not competitively bid, but rather is negotiated. On the other extreme are the general contractors and subcontractors who work in an environment of extreme competition primarily due to the lesser capital requirements. Somewhere in between these two extremes are the heavy construction contractors whose contract price will usually be based on some measure of units performed. This segment of the industry requires large amounts of capital, due in part to the size of the contract jobs and in part to the large amounts of heavy construction equipment required on heavy construction contracts.

Schedule 1-1 is a convenient summary of the broad classifications within the construction industry, their type and scope of work, and the economics of the industry in terms of price arrangement and risk.

Each of the specialist categories has accounting requirements peculiar to the specialty. In general, these accounting requirements are dictated by the nature of the job, scope of the work, pricing arrangements, and risk. For example, a contractor with basically cost-plus work must have an accounting system that generates information to

SCHEDULE 1-1 ECONOMICS OF THE CONSTRUCTION INDUSTRY

		Contract		
General speciality	*Major types of jobs*	*Scope of work*	*Price*	*Major elements of risk*
Design-build contractors (commonly referred to as "turnkey" contractors)	Electric power plants Hydroelectric facilities Petroleum refineries Chemical plants Rapid transit systems Irrigation systems	Engineering Financial services Management Procurement Construction	Cost-plus	Time and cost overruns Warranties
Heavy construction specialists	Roads Bridges Dams Airports Large buildings	Construction and subcontract management	Unit price or lump sum	Unforeseen conditions, cost estimating and omissions, job management failure
General building contractors	Housing Schools Hospitals Office buildings Manufacturing plants Warehouses	Construction and subcontract management	Lump sum	Cost estimating and omissions, job management failures, unforeseen conditions
Subcontractors	All specialty work in all types of construction contracts	Concrete Electrical Mechanical including heating and air conditioning Steel erection Piping Carpentry Dry wall/plastering Flooring	Various	Cost estimating and omissions, job management failures, unforeseen conditions

support the cost-plus billing. Additionally, he must have a network of controls and an information system designed to inform management on a timely basis when there is a danger of time or cost overruns. For this type of contractor, the financial risks generally tend to be geared to penalties for time or cost overruns. On the other hand, the general contractor, whose contracts are of the lump sum or unit price types, needs a system of controls and management information weighted toward emphasizing job management and contract cost control. Because of the competitive nature of general contracting, a low profit margin results in significant financial risk when there are cost overruns. There is no need for the general contractor to have an accounting system designed to provide the documentation necessary for cost-plus billing. What he does need is an accounting system designed to tell him immediately the parts of his jobs where management attention is needed.

In addition, there are certain factors apart from the mechanics of construction that may be significant in operations and hence in accounting for operations. Some examples of such factors are:

1. Job location
2. Procurement of labor and material
3. Use of equipment
4. Control of labor, materials, and equipment
5. Subcontracting
6. Supplying management personnel
7. Financing requirements

Job location. If a contractor confines his operations to one city or to a relatively small local geographic area, problems arising from job location are usually at a minimum. However, when he begins to take jobs greater distances from a central point, he often begins to incur liability for subsistence and travel time and sometimes for the cost of transporting his workers. In addition to these new costs, questions of payroll taxes, income tax withholding, and insurance on the additional payments arise. Also, the contractor faces increased costs in moving materials, supplies, and equipment to the jobsite.

Move the job a little farther away and there are the added costs of moving the families and possessions of key personnel. If the job is far enough away from a city or town, there arises the necessity of providing camp, mess hall, and possibly commissary facilities. If the job is in a foreign country, additional problems arise in connection with banking, letters of credit, multiple currencies, exchange rates, import duties, and taxes levied by the foreign country on the incomes of the company and its employees. The list could be extended indefinitely. The point is that job location is a physical fact which, by itself, can produce a wide variety of operational problems and related accounting problems.

The result is that the smaller prime contracts and a substantial share of the specialty subcontracts tend to be awarded to local contractors because the size of the job does not warrant the added expense incurred by contractors with widely scattered jobs. The larger the job, the farther a large, well-organized, and well-financed firm can afford to go for it, but the farther it goes, the greater the accounting activity needed to deal with the problems created by distance alone. By the same token, when small firms start to branch out and seek work farther from their center of operations, there is a tendency, at least on the first such job, to overlook the problems created by distance in determining the amount of their bids. The increased use of computers in construction accounting has added a new dimension to the problems of job location, but the basic problems remain unchanged.

Procurement of labor and material. The smaller and simpler a job is, the fewer will be the accounting requirements involved in the procurement of labor and materials. The logistics of supply is complicated by quantity and variety as well as by distance. Thus in a simple housebuilding job most, if not all, of the labor and materials will be available locally, and any savings in price which might be made through purchases from distant sources must be weighed against the readiness of local suppliers to serve and the delays and transportation costs inherent in distance. However, as jobs become more complicated, so do the procurement problems and the accounting necessary to control these problems.

In the field of labor, highly skilled specialists like cable splicers or agronomists must often be supplied for relatively short periods of time and very often on short notice. Few organizations have enough work to keep specialists busy for indefinite periods, and when such personnel are required it is necessary to locate, hire, and transport them to the job. They must not arrive on the job too soon or their time, which is usually very expensive, will be wasted. They must not arrive on the job too late because delay is one of the construction industry's most costly bugaboos.

Lead time must be considered in purchasing products which must be fabricated for the job. For example, steel pipe for pipelines, special steel sash and doors, and heavy motors and transformers must often be built to order, and they must be ordered in advance (and followed through with expediting effort) so that they will be ready when needed.

The costs of procuring such specialized labor and materials must be known if they are to be controlled and if funds are to be provided to pay for them. The more complicated procurement operations become, the more important the element of timing becomes, and unless adequate accounting records are kept, control of timing by judicious use of expediting becomes increasingly difficult.

Use of equipment. The extent to which operations of a contractor require the use of equipment depends largely on the type of work being done, and the extent to which those parts of the work requiring special equipment can be subcontracted.

To own construction equipment usually requires a great deal of capital. If a contractor owns equipment he must be constantly alert to avoid having all his working capital represented by idle or worn-out equipment. Many contractors have found it more economical to rent whatever equipment they need rather than tie up the funds required to own it. Those who do own their own equipment must constantly be ready to decide when it is more economical to dispose of any particular item than to keep it in repair.

What equipment is needed is an operational question. Whether to provide it by ownership, rental, or subcontracting is a management question that can be answered intelligently only on the basis of facts that accounting is called upon to supply.

Control of labor, materials, and equipment. Materials (Chapter 4) are subject to two kinds of control—physical and financial. The physical control is, in large measure, an operational fact which must be reflected in the accounts only to the extent necessary to keep management advised of how much is on hand. Management, however, is continually called upon to make financial decisions based on these questions: (1) Do we have it or must we buy it? (2) If we have to buy it, how much are we going to have to pay for it?

Labor and equipment costs are the two most important variables in construction costs. To control their physical application it is necessary to know from day to day how much work is being accomplished in relation to the amount of labor and equipment being used. The tool for doing this is "unit costing"—or cost accounting.

It is not hard to see that the larger and more complicated the job, the greater is the volume of accounting needed to control labor, materials, and equipment costs. There is one important exception. If a large proportion of the work is subcontracted, as in large commercial building projects, the problems of controlling labor, materials, and equipment are considerably lessened. However, as pointed out below in the discussion of subcontracting, this procedure merely exchanges one set of operational and accounting problems for another.

Problems in subcontracting. From the viewpoint of the prime contractor, it is often good business on a large contract to spread the risk, the financing, and the operational problems by subcontracting certain portions of the work. From the viewpoint of the subcontractor, his problems are the same, in both operations and accounting, as those of the prime contractor on those portions of the work not being subcontracted.

Nevertheless, the prime contractor is almost always responsible to the owner for any failure of the subcontractor either to deliver the work he has contracted to perform or to pay his labor and material bills (Chapter 7). For this reason the prime contractor must, to some extent, supervise not only the physical work of the subcontractor but also watch the latter's financial condition. It is usually the responsibility of the accounting department to control payments to subcontractors and take whatever steps are necessary to safeguard the prime contractor and the owner against liens (or comparable claims on government jobs where liens are inapplicable) and against unintentional release of the subcontractor's surety if the subcontractor's work is covered by a performance bond.

The subcontractor who, in turn, subcontracts a portion of his work is, in most respects, in the same position on the subcontracted work as a prime contractor. The main difference is that he is responsible to the prime contractor instead of to the owner.

Both prime contractors and subcontractors must, of course, be alert to see that their contracts are so drawn and their operations and accounting so controlled that they receive payment for work actually done and materials actually supplied. Subcontractors often complain that they are at the mercy of the prime contractor. This is true to some extent, though no more so than the prime contractor's assertion that he is equally at the mercy of the owner or architect. The fact is that it is up to each contractor to take whatever steps are necessary to safeguard his own interests; and before such action can be taken it is necessary to know the facts, which can be supplied by correct accounting.

Supplying management personnel. Job management is essentially an operational function. In other words, the first function of job management, regardless of profit, is to get the work done. In both theory and practice everything else is secondary. If a profit is to be made, the most efficient use of labor, materials, and equipment is required, and this requirement cannot be fulfilled unless management provides adequate supervision and planning.

The need for additional or better job management can be detected in a number of ways. One evidence is failure to meet time schedules, or the presence on the job of idle employees and equipment. Still another is visual observation by an experienced project superintendent of methods and procedures. None of these, however, reveals the inadequacies in job management so promptly as the scrutiny of job costs which have been carefully kept and promptly rendered.

Of course, where the very management whose efficiency is being checked is in a position to control the preparation of the cost figures, the effectiveness of cost keeping can be nullified. Under these conditions, the expense of keeping the records is wasted. Still, properly prepared and used, job cost figures are one of the most useful tools known to the industry for controlling job operations.

Financing requirements. The lifeblood of a construction job, as in any other business undertaking, is money. Progress payments are an established method of financing and it is a rare contractor whose working capital position is so strong that he can be indifferent to them. In fact, through such devices as advances for mobilization, unbalancing of bid and payment schedules, and generous estimates of completion, many contractors not only offset the customary retained percentage but actually recover their initial investments early in the job and continue it entirely on the funds received through progress payments.

However that may be, owners require accounting statements as the basis for payment regardless of whether the contract is lump sum, unit price, time-and-material, or cost-plus-fixed-fee. Regardless of the type of contract, the accounting records must furnish the basis for billings to the owner. This is no less true in those companies where the accounting is done by engineers from informal records. In contracts involving direct reimbursement for cost (such as the fixed-fee or target types discussed below), complete and accurate accounting for expenditures is, of course, vital.

If financing is done through banks, there is no substitute for adequate financial statements in securing maximum financing. Bonding companies rely heavily on the financial position of the company as a whole and of individual jobs in determining bonding capacity (Chapter 10). These are two areas where the “outside” public accounting firm can be of invaluable help in maximizing loan and bonding capacity.

If financing is done through joint ventures it is, of course, axiomatic that the inactive members of the venture will require regular financial reports. The same is true if the owner provides a revolving fund.

Types of construction contracts

It is common to describe construction contracts by the method of payment they provide. When so classified, they fall into three groups.

In the first group are found those contracts under which the contractor agrees to do certain specified things for an agreed price. In this group are lump sum and unit price contracts.

In the second group are those contracts under which the contractor is reimbursed for his costs and receives a fee for his services. In this group are the cost-plus-a-percentage-of-cost, cost-plus-fixed-fee, and force account contracts.

In the third group are those contracts which combine elements of the first and second groups. In this group are the cost-plus-fee contracts with so-called target estimates or with a guaranteed maximum, and the time-and-material contracts in which the contractor's fee or profit is reflected in the rates charged for the time, the materials, and the overhead allowance. Each group has its own peculiar accounting requirements in addition to those general requirements which are common to all.

Regardless of the nature of the contract, it should be possible to answer "Yes" to each of the questions in Schedule 1-2.

Lump sum and unit price jobs will tend to involve greater emphasis on cost accounting and unit costs. Of course, sometimes the jobs are small enough and closely enough supervised to maintain maximum efficiency without cost records. Sometimes the contractor himself does the work of superintendent or foreman. On such lump sum jobs, the cost records will tend to be only what the individual contractor needs to give him estimating data.

The important point is that accounting, whether cost accounting or general accounting, is never an end in itself. Unless it contributes

SCHEDULE 1-2 CONTRACT ACCOUNTING CHECKLIST

1. Do we have the information necessary to submit our progress billings promptly and accurately?
2. Do our permanent records provide complete and accurate support for our progress billings?
3. Are we able to support our claims for extra work done for the client or for others by written orders signed by an authorized representative of the person for whom the work was done?
4. Do we know what we bought, how much it cost, and whether we got it?
5. Are our routines for purchasing, receiving, vouchering, and disbursing arranged so that we can take all discounts, pay bills promptly and accurately enough to keep our credit good, and still avoid duplications and overpayments?
6. Can we tell what we did with the things we bought?
7. Do our timekeeping and payroll procedures let us pay our employees promptly and accurately and in accordance with union rules, wage and hour rules (where applicable), and any special rules in the contract (such as the Copeland Act, the Davis-Bacon Act, and various state and local statutes)?
8. Are our payroll records sufficient for all federal and state regulations as well as for payroll and withholding tax purposes?
9. Can we supply, from the records, the information needed by management?
10. Can we supply our bank and our bonding company the detail they want concerning each job as well as our overall operation?
11. Are there special tax requirements about which we should know and for which we should be compiling data? If so, are the data being compiled?

more than its cost to the final profit it cannot be justified. Thus, in lump sum and unit price contracts, the principal use of cost accounting is to help management save on costs by pointing quickly to the operations where more or better supervision is needed. Its by-product may be better estimating on future work where the unit costs are applicable. To do the job properly, once day-to-day cost figures are developed they must be in the hands of job management within 24 hours.

Cost-plus contracts usually require accounting emphasis on the documentation of costs and expenditures, to the end that nonreimbursable expenses will be held to a minimum and that maximum accountability will be maintained for materials, supplies, labor, and equipment. In this type of contract, it is customary to find agreed rates for equipment rentals with resulting procedures roughly describable as an equipment "payroll." It is also common to see some percentage or lump sum provision for overhead expenses not incurred at the jobsite and special provisions for such items of cost as repairs, maintenance, and the cost of tools and supplies. Unless contracts of this type are written with exceptional care, the job accountant is faced with innumerable problems of interpretation and interminable arguments with the owner's representatives.

One problem which is typical of the cost-plus type of contract is dealing with the owner's representative who tries to substitute his judgment for that of job management. At the accounting level this tendency will usually be manifested in attempts to preaudit financial transactions. If the contract clearly prohibits such action, the problem is usually solved on a policy level. If it does not, the job accountant is usually well advised to resist the first indication of such action.

The line between healthy cooperation and unhealthy interference is sometimes a difficult one to draw. Usually there are considerations of policy beyond the scope of the average job accountant's authority. The temptation is great to hold nonreimbursables to a minimum by having all transactions preaudited. Often the result is delay in operations, damage to credit, and damage to relations with suppliers and even with the client through the activities of an overzealous auditor.

Another accounting problem peculiar to the cost-plus type of contract is the treatment of owner-furnished materials, equipment, or services. If the fee is based on a percentage of cost or the fee increases when extras increase the cost beyond a predetermined percentage of the original estimate, it is usually proper to take the value of owner-furnished items into account to determine the total cost of the work done.

Other contracts combine some of the features of lump sum and cost-plus contracts. The most common of these is the so-called "target" type. Usually the contractor is called upon to submit an initial bid in competition with others. Then the bid, after adjustments arrived at by negotiation, becomes the target that can move upward or downward as a result of change orders, shifts in price and wage rates, and similar factors beyond the contractor's control. There is usually a bonus for doing the job at a cost below the target, and a penalty for running over. Some target-type contracts contain a stop-loss provision limiting penalties to the amount of the fee. Sometimes, too, the target is a guaranteed maximum price and the penalty is 100 percent of the overrun. Obviously, everything that has been said above about both the lump sum and the cost-plus types of contracts would apply to this type.

Time-and-material contracts, on the other hand, are much less demanding. Usually they involve smaller jobs and the principal accounting concern is to see that all time paid for by the contractor is billed to the client, and that all materials withdrawn from stock or purchased directly are charged to a job or are returned to stock. In this type of work it is common practice to add a fixed percentage to the billing for overhead and profit. Otherwise it is necessary to increase the prices charged for labor and for material to cover those items.

Forms of business organization

Construction businesses, like other types of business, are generally organized in one of the traditional forms—corporation, partnership, or individual proprietorship. Insofar as accounting problems peculiar to these forms are concerned, the nature of the business is generally not important since the form of business organization affects principally those accounts which reflect the proprietor's equity in, and not the operations of, the business. One possible exception to this general rule involves corporations. Because of possible personal liability of directors, if there should be a payment of dividends from capital, the method of accounting chosen to account for income from joint ventures and long-term contracts becomes doubly important. Another possible exception occurs when a company needs to make provisions for differences between its financial accounting and its accounting for income tax purposes.

In recent years, favorable corporation tax rates have made it possible to hold down the tax cost of reinvesting earnings in the business with the result that many small and medium-sized construction businesses have been incorporated. However, the changes made by the Tax Reform Act of 1986 have made the corporate form much less attractive in terms of applicable tax rates and of the elimination of cash basis accounting for certain corporations. In some companies, income tax considerations have made the limited partnership a useful alternative. The so-called "S" corporation has also come into wider use.

One common objection offered to the incorporation of a small business is that corporate accounting is much more complicated and costly and that the cost is unjustified. This charge, of course, is simply not true. Accounting for operations is not affected by the form of organization. The principal difference is in the method of accounting for the owner's equity and in formalizing decisions of major importance in the minute book. The accountant who advises against the corporate form solely on the ground that it is too complicated and costly to maintain corporate records is, by his own statement, indicating his lack of qualification to express an opinion. Other factors may, in the opinion of competent legal, tax, and accounting advisers, make the use of the corporate form inadvisable; but the cost of corporate accounting and tax reporting, of itself, is normally not such a factor.

A much more important practical problem, and one which becomes more important and more difficult when the business is incorporated, is that which arises from the failure of the proprietor, partners, or principal officers to realize that firm transactions must be kept separate from their personal accounts, and that no accountant, however skillful, can keep accurate records if all transactions are not disclosed to him.

In this respect, the small one- or two-person businesses are those which create the greatest problem for the accountant and, of course, ultimately for the owners. Checks may be written to "cash." Incoming checks may be cashed and the money spent with no accounting. Equipment may be purchased with no papers to indicate the complete nature of the transaction. Complicated "side deals" may be made with no indication of their existence and may ultimately be liquidated with the proceeds of collections of which the accountant has no knowledge. In other words, the form of organization, as such, presents no serious problem for the construction accountant. His problem, in the average small construction firm, generally is to secure the minimum data necessary to keep adequate books regardless of the form of organization.

One of the most significant developments in comparatively recent years is the widespread use of the joint venture as a form of organization for larger projects. Essentially, a joint venture is a special purpose partnership through which two or more construction firms can pool their finances, their bonding capacity, their know-how, and their equipment in the performance of a large or hazardous job. The joint venture is an effective risk-spreading device and many are formed for that purpose alone. At other times, joint ventures are used to give promoters or key employees participation in a particular job without participation in the operation of the company as a whole.

For all practical purposes, the construction accountant may properly treat a joint venture as a partnership. There are some technical differences in law, and in the past the income tax authorities have sometimes sought to tax them as associations taxable as a corporation. Properly used, however, joint ventures are a safe and well-proved device for pooling abilities and facilities and for spreading construction risks. It is rarely good practice to use the same joint venture for more than one job, and this is true even though the participants and the interests are the same. Some joint ventures even carry this separation feature to the point of securing separate employer's numbers for social security purposes and other licenses and permits on the same basis.

In most, though not all, states a corporation may enter into a joint venture without specific provisions in its articles. So also, generally speaking, may a partnership, a proprietorship, or another joint venture. Whether a trust may do so usually depends upon the provisions of the trust instrument. The safest procedure is always to discuss the formation of a joint venture with legal counsel.

Many joint ventures maintain separate accounting records, complete with general ledger, and it is usually the duty of job management under the supervision of the sponsor (the managing partner) to keep the members of the venture informed on the progress of the work and the financial position of the venture. Sometimes a management committee consisting of representatives from each member organization takes the place of the sponsor, though under some circumstances this procedure might cause the authorities to view the venture as an association taxable as a corporation, and therefore might be undesirable.

Occasionally, tax considerations permitting, a special purpose corporation may be formed instead of a joint venture, and its corporate officers will probably then be chosen from the several member companies.

Basic Accounting Patterns

Once the pattern of operations has been set, the accounting patterns required to reflect operations fall into three major groups, represented by these questions:

1. Where will the accounting be done?
2. What information is needed and why?
3. What is the best way to go about getting it?

At first glance it might appear that the question of what is needed and why should come first. However, it has already been pointed out that many of the functions to be performed by accounting are determined by job location alone. Hence the "where" often has a great deal to do with determining the "what" and the "why."

Centralized and decentralized accounting compared

It has been noted that one of the distinguishing features of the construction industry is that its physical operations cannot be centralized but must, for the most part, take place at the site of the real estate that is being improved. It follows that people at the jobsite have firsthand knowledge of conditions and occurrences there.

One of the most basic of construction accounting problems is how much to do in the general office as opposed to on site. Obviously the answer must vary according to the circumstances of each individual contractor if not with each individual job. It is a fallacy to assume that procedures which work well for one company or on one job will necessarily work equally well for all contractors or on all jobs. The procedures followed in any successful system are flexible enough to meet a wide variety of conditions without sacrificing any basic controls.

It almost goes without saying that the smaller the job, the smaller the proportion of job accounting that will be done at the jobsite. On the smallest, only timekeeping and receiving of materials will be done on the job. Repairmen who carry a truck inventory may add a record of materials used. As the jobs grow larger the question of decentralization becomes more important, and more factors affect the decision. The point here is to recognize the centralization-decentralization decision as one of the basic problems in fitting accounting patterns to operational patterns.

Schedule 1-3 compares centralized and decentralized accounting on the basis of the several control factors involved. It is based on the assumptions that the jobs are large enough to justify decentralization and that the means of communication are good enough to make centralization practical.

It is clear from Schedule 1-3 that where there is a choice between centralized and decentralized job accounting, persuasive arguments exist on both sides of the question. Whether he knows it or not, every contractor must recognize this question and deal with it one way or another the first day he starts his operations.

A basic overview of construction accounting methods

One of the important reasons why construction contractors, like all other business concerns, must keep complete and accurate records is that numerous laws require them. Without such laws many smaller contractors would keep no records at all, and many contractor's records would be less elaborate. Essentially the minimum records required include detailed records of cash receipts and cash disburse-

SCHEDULE 1-3 COMPARISON OF CENTRALIZED AND DECENTRALIZED ACCOUNTING

Centralized	*Decentralized*
Financial Control	
Financial records can be kept by better-trained people under the supervision and control of top accounting personnel, and the need for going to the field for additional facts is held to a minimum. Hence there can be more uniformity and greater accuracy.	Job personnel are closer to the facts and therefore are in a better position to have prompt, accurate information. They also are able to respond quickly to situations as they arise. The general office records can be no better than the data supplied by the field.
Cost Control	
If volume is great enough to permit cost records to be kept by computer equipment, more and better cost statements can be made in the general office, and greater uniformity in presentation of cost statements can be achieved.	Cost statements lose much of their value as tools of job management if they are not completely current. Unless the cost records are kept by data processing equipment, the time required to transmit cost data to the general office, process it, and return it to the field is too great, and cost keeping becomes a useless expense.
Auditing	
The cost of adequate auditing procedures is less when the auditor has all the records in front of him in one place and the responsibilities are centralized.	There can be no adequate audit of operating accounts of a construction company without a thorough understanding of jobsite conditions; this cannot be obtained in the general office.
Reporting	
Financial statements are one of the tools of management and as such must be prompt, accurate, and consistent. Where all accounts are centralized there is no need for delay in securing data from the job; and statements can be prepared by top accounting personnel of the company or by outside accountants without delay and without inconsistencies. If special statements for special purposes are required, they may be taken quickly from records at hand without risk of errors in communication.	Properly organized, there need be no delay or inconsistency in reports from jobs. A good accounting manual, a chart of accounts, and uniform statement presentations are necessary. Special statements usually require data not readily obtainable from the ledgers. Usually such data are more quickly and accurately obtained at the jobsite.
Job Management	
Job management is not a function performed solely at the jobsite. To utilize top management skills effectively, the general office must get prompt, definitive, current job information.	Job management uses accounting records to control job operations. Such records must be completely current to be useful. Usually this precludes centralization because of the necessary transmission time.

SCHEDULE 1-3 COMPARISON OF CENTRALIZED AND DECENTRALIZED ACCOUNTING *(Cont.)*

Centralized	*Decentralized*
Overhead Control	
Centralization permits more specialization, more mechanization, and better supervision. Hence job accounting can be done at lower cost. Also, a more objective view of jobsite overhead is possible to those not directly concerned, and hence fewer personal elements will enter into overhead reduction measures.	The mere fact that a trained accountant may, in a job office, be called upon to do such routine tasks as typing and filing does not necessarily indicate inefficient use of labor. Normally a capable job accountant combines a large amount of internal checking with these operations. If job management is properly informed, it is usually cost-conscious enough to control overhead without help from the general office.

ments, purchases, income, contract revenue (both billed and earned), contract costs, individual earnings records of employees, and a summary of all these, usually in the form of a general ledger.

Since such a record is required by law it would be wasteful not to use it to the greatest possible advantage. Most accountants and all of the more successful contractors consider the legal requirement as secondary to the need of management for adequate records. The fact remains, however, that a considerable number of contractors use their records only as a means of satisfying governmental requirements. This group is losing money by failure to make the greatest possible use of something they must pay for in any event.

A mere historical record is limited in usefulness. Of course, reference to individual accounts in the general or subsidiary ledgers is common, particularly to locate original documents connected with some particular transaction or series of transactions. Thus an account in the general ledger might be used to locate the vouchers and other documents necessary to establish the nature of a transaction for income tax purposes. An account in the accounts receivable ledger might be used to prepare a detailed statement of the account for collection purposes. The cost and commitment ledger might be used to trace the source of certain materials or supplies or to develop data to secure business on a reciprocity basis. However, the most important use for the basic records is to provide the raw materials for financial statements.

Reporting

The reason for the limited usefulness of the basic records is the fact that they are not necessarily designed to make significant groupings or comparisons. That is the function of reporting—in other words, the financial statements. The mere statement that a construction company has $25,000 in cash means very little by itself. However, if it is stated that the company has $25,000 cash and $150,000 in current receivables and other current assets, and that its current liabilities total $100,000, the figures begin to be significant.

To the person in charge of credit at the lumber yard this would normally mean that the company is in a position to pay its bills currently and that, other things being equal, a reasonable line of credit is probably justified. To the banker and to the bonding company, it would indicate the probability that the company could be reasonably expected to handle a backlog of work of, say, $750,000 to $1,000,000, without getting into financial difficulties. To the contractor himself it would indicate that if he has much less than $750,000 backlog his money is not working as hard as it should be, and if he has much more than $1,000,000 backlog he may be overextending himself.

Naturally the significance of the amounts involved will vary with all the circumstances and will not depend solely on current position. It is for the financial statements to bring out all the significant facts essential to a proper evaluation of the company's current financial position and the results of its operations.

Suppose, however, that the company referred to above were a subcontractor working on lump sum contracts, and that included in $150,000 of current receivables was $100,000 in retained percentages due from prime contractors. If the subcontractor could not expect to be paid for these retentions until the prime contractors were paid their retentions by the owners, then the subcontractor's funds could be partially frozen, so that he might have only $50,000 in net working capital available. Naturally, the volume of work that he might take on successfully would be reduced proportionately. On the other hand, the volume of work that the contractor could undertake would not be greatly affected if all his contracts were on a cost-plus-fixed-fee basis using owner-furnished revolving funds.

It is the function of the balance sheet to give, as well as possible, a fair summary of the assets, liabilities, and net worth of the business at a given time. It is the function of the operating statement to give a fair summary of the results of operations during a given period.

In 1971, the American Institute of Certified Public Accountants (AICPA) in the Accounting Principles Board Opinion No. 19 started to require a third major statement called "Statement of Changes in Financial Position." This statement is similar to the old funds statement and is an analysis of changes in the balance sheet from one date to another. The statement of changes in financial position was conceived primarily for use by financial analysts in appraising publicly held companies. As most construction companies are privately or closely held in terms of ownership, the statement is of doubtful value to management. That requirement has been changed. Effective on statements after July 31, 1988, a "Statement of Cash Flow" has been substituted for the "Statement of Changes in Financial Position."

In support of these three basic statements there must be adequate schedules to detail and explain all the important facts and relationships necessary to evaluate the three basic statements correctly.

The questions of what statements are required are covered in Chapter 10. The important facts to be brought out here are these:

1. The information that appears on a financial statement is determined by the use to which the statement will be put.
2. It is a major responsibility of management to utilize its financial reporting to the greatest possible advantage.
3. It must be recognized that financial statements report only past facts; the usefulness of those facts diminishes with time. In other words, conditions change quickly in a construction business, and a statement so old that it may be misleading may be worse than no statement at all.

Forecasting

A balance sheet is, to some extent, itself a forecast. When the person preparing the balance sheet classifies certain assets as "current," he is predicting that they will be converted into cash within a reasonably short time—the normal business cycle, usually one year unless, as in long-term contracting, the cycle is longer. He is predicting, as a rule, that the excess of costs over progress billings represents an asset and not a loss; and when he makes provision for taxes he is predicting that the tax shown will be the amount to be paid on the reported income.

It is on the basis of the balance sheet, the operating statement, and the analysis of jobs in progress that the banker forecasts the need of his construction client for short-term loans and his capacity to repay them. On the basis of the same statements, the bonding company and prospective clients of the contractor predict his ability to finance the work he is undertaking. The contractor himself uses the same statements to forecast how much new work he is in a position to contract for and where he will get the money, equipment, and organization to handle it.

However, since these statements reflect the past only, the next step in the forecasting procedure is for the contractor to adjust his statements to take into account the reasonable expectations for the future. The most common forecast of this type is the cash forecast.

Cash forecast. To make a cash forecast (Schedule 1-4) it is necessary to estimate, usually by months, the times when progress payments may be expected and when any cash from other sources will be available. At this point a forecast of the statement of changes in financial position sometimes proves useful. In making a cash forecast it is necessary to anticipate the cash requirements of known work, overhead, taxes, equipment purchases, and loan repayments. Schedule 1-4 is a brief example of one of several satisfactory forms of cash forecast.

There are contractors who maintain that such forecasts are a waste of time because they become obsolete before they are prepared, due to the fact that the collectibility of payments is not predictable, and because they mislead the reader by anticipating funds not actually on hand. All these objections miss the point. The type of cash forecast illustrated accomplishes two things: (1) it inventories cash resources and balances expected needs against them; (2) it schedules the work that must be done and the collections that must be made if money is to be on hand to pay the bills.

Such cash forecasts, as used by progressive construction companies, are made quickly, because they do not seek to achieve more than approximate accuracy. However, they must be made by someone who knows conditions and who recognizes the forecasts for what they are—a financial plan to be followed as nearly as possible and to be updated as necessary. On the basis of such a forecast it is possible to estimate what short-term loans will be needed and how and when they will be repaid. With this type of forecast to show the banker, it will be easier for a deserving borrower to get the needed credit. If there is reason to believe that the collection of the expected funds will be slow, the forecast will call attention to the need for greater collection effort.

The expected availability of cash may make a great deal of difference in the way new work is estimated. If adequate working capital will be available, the contractor may bid more lump sum work. If working capital is likely to be tied up for some time, he may try to convert some of his lump sum bids to reimbursable cost-plus-fee jobs financed by a client-furnished revolving fund. If work is scarce and jobs in progress are approaching completion, he might bid more competitively and on a lower margin in anticipation of being able to finance the work with funds from jobs closing out.

Some cash forecasts are based on types of work rather than on individual jobs. Thus the captions for receipts might be something like "Contract work," "Maintenance work," "Repair jobs," and so on, and for expenditures, such items as "Payrolls," "Purchases," and "Taxes." In some types of construction businesses this is the only practical

SCHEDULE 1-4 CASH FORECAST

Ecks Construction Co.
Cash Forecast for the Six Months Ending
June 30, 19—
Expected Available Cash

	Jan.	*Feb.*	*Mar.*	*Apr.*	*May*	*June*	*Total*
Cash balance Dec. 31	$30,000						$ 30,000
Monthly balances brought forward		$22,500	$19,500	$ 19,500	$ 22,500	$22,500	
Expected receipts:							
Job A—retention		15,000					15,000
B—fee	3,000	3,000	3,000				9,000
C—progress payments			28,500	36,000	45,000	19,500	129,000
D—progress payments	18,000	9,000					27,000
D—retention					7,500		7,500
E—progress payments		6,000	15,000	30,000	30,000	12,000	93,000
F—fee	1,500	1,500	1,500	1,500	1,500	1,500	9,000
Insurance claim				13,500			13,500
Warehouse scrap sales	1,500						1,500
Total cash before bank loans	$54,000	$57,000	$67,500	$100,500	$106,500	$55,500	$334,500
Bank loans	15,000	30,000	22,500				67,500
TOTAL EXPECTED CASH	$69,000	$87,000	$90,000	$100,500	$106,500	$55,500	$402,000

SCHEDULE 1-4 CASH FORECAST *(Cont.)*

	Expected Cash Requirements						
	Jan.	*Feb.*	*Mar.*	*Apr.*	*May*	*June*	*Total*
Job B—None (revolving fund)							
C—Costs		$12,000	$36,000	$12,000	$ 9,000	$15,000	$ 84,000
D—Costs	$12,000	6,000					18,000
E—Costs	18,000	15,000	21,000	18,000	18,000		90,000
F—None (revolving fund)							
G—Costs					12,000	9,000	21,000
Equipment payments	6,000	6,000	6,000	6,000			24,000
Taxes	3,000			4,500		4,500	12,000
Insurance		21,000					21,000
Overhead	7,500	7,500	7,500	7,500	7,500	7,500	45,000
SUBTOTALS	$46,000	$67,500	$70,500	$48,000	$46,500	$36,000	$315,000
Loan repayments				30,000	37,500		67,500
TOTAL CASH REQUIRED	$46,500	$67,500	$70,500	$78,000	$84,000	$36,000	$382,500
Cash balance carried forward	$22,500	$19,500	$19,500	$22,500	$22,500	$19,500	$ 19,500
TOTAL LOANS DUE TO BANK	$15,000	$45,000	$67,500	$37,500	—	—	—

way. However, where a number of jobs are in progress over several months, the form shown in Schedule 1-4 has the advantage of providing a review of completion schedules at the same time that the forecast is being prepared.

There are three important limitations to the use of such a forecast. First, it is no more accurate than the thinking that goes into it. Therefore it is not a mere juggling of figures to be assigned to anyone who has nothing better to do. It is a monthly review of the immediate past and the best estimate of the course of the business for the immediate future. When it is being prepared, the best combined thinking of management together with its outside accountants and financial advisers is brought to bear on the immediate financial problems of the company. If the cash forecast is not prepared in this way it loses much of the meaning it should have, and may, in fact, become as misleading as its critics charge.

Second, it must be remembered that a forecast guarantees nothing. The money expected will arrive only if the thinking that accompanied the forecast is translated into effective action. There is a strong temptation, particularly when a bank loan is being sought, to indulge in wishful thinking and overly optimistic guessing. This may work once, but if the figures do not prove to be reasonable, the banker will tend to doubt all the contractor's estimates in the future. This could be disastrous to credit, and since the banks and bonding companies usually work very closely together it might well hurt the contractor's bonding capacity. Cash forecasting requires the same hardheaded realism as bidding. Excessive optimism in either can lead to serious trouble. On the other hand, careful, conservative forecasting can be one of the most useful tools of financial management.

Third, a cash forecast is not necessarily indicative of profit or loss. A contractor may have a great deal of highly profitable work in process and still be very short of cash. In fact, he is often in that position. An expanding business works its money very hard, and working funds do not stay any longer in the bank account than working equipment stays in the contractor's yard. For purposes of tax planning and the long-range planning of operations, the proper procedure is to employ the "profit-and-loss" or "operations" forecast.

Profit-and-loss or operations forecast. The forecast of profit and loss is not commonly used by smaller contractors because it is based, essentially, on the concept of a break-even point. In manufacturing and marketing the break-even point is that volume of sales which will produce enough gross profit to pay the fixed overhead expenses. In construction, the volume of job revenue tends to vary more and to be subject to greater fluctuations in the rate of gross profit than the sales of a manufacturing or marketing business. The concept is, therefore, harder to apply to construction. One notable exception is the builder of tract housing whose operations, in many respects, are similar to those of a manufacturer. Another such exception is the plumber or electrician who does a large volume of small maintenance jobs. These operations are similar to a retail marketing business. For the most part, however, construction companies tend to have less predictable fluctuations both in volume of business done and in gross profit.

Nevertheless, in a market where there is a reasonable amount of available business, and assuming a reasonable degree of skill in estimating, the principal factors limiting the volume of a contractor's business are the amount of his capital and the size and ability of his organization. If recognition is given to these two factors, a contractor may, by comparing the results of past operations with a sound estimate of present and future conditions, arrive at a reasonable estimate of the volume of work which he can reasonably expect to complete and the average gross profit which he can reasonably expect that amount of work to produce. In fact, some contractors start their profit-and-loss forecasting with an estimate of expected annual gross profit. Others estimate gross profit in terms of a stated number of dollars per permanent employee.

To illustrate a somewhat more conventional approach, assume a building contractor finds that over the past several years he has been able to earn net job revenue (gross profit) averaging 6 percent on total job revenue (contract price of completed work). Assume further that his estimate of the market for the coming year combined with the capacity of his organization and the amount of available working capital lead him to believe that he can complete $3 million in contract volume and make an average of 6 percent gross profit on it. On the basis of these assumptions, the contractor must keep his overhead under $18,000 if he is to make a profit.

This sort of planning requires a very high degree of management skill and usually the best of outside advice. Once the technique is mastered, however, it enables the contractor to use his available resources to the greatest possible extent and to the greatest possible advantage. To make such estimates, the first requirement is accurate financial statements prepared on a consistent basis over a period of several years. The second, and equally important, requirement is a record of past forecasts and the amounts and reasons for their variations from actual operating results. The third requirement is a sound analysis of existing and future market conditions coupled with sound planning for making the most of them. Combining these three elements, a sound and conservative management can forecast operating results with a high degree of accuracy and in so doing can develop the information necessary to plan the most effective use of its organization, equipment, and capital.

Basic accounting methods

When to report profit on a contract is a problem somewhat unique to construction contractors as well as to other contractors who perform work over a very long period of time. Over the years, four accounting methods have evolved to deal with the timing of profit recognition:

- *Cash Method:* Revenue is the cash received and costs are those that have been paid during the period being accounted for.
- *Accrual Method:* Revenue is what the contractor is entitled to bill under the contract (whether or not it has been billed or collected in cash) and costs are the costs incurred (whether paid for or not) during the period being accounted for.
- *Percentage-of-Completion Method:* Progress of the job is measured in terms of percentage of completion. This derived percentage is applied to the total estimated gross profit which is then added to the cost to date to determine earned revenue. Costs are the costs incurred (whether paid for or not) during the period being accounted for.
- *Completed-Contract Method:* Billings and costs are accumulated on the balance sheet. Revenue, cost, and profit are recognized (transferred to the income statement) when the job has been completed. The Tax Equity and Fiscal Responsibility Act of 1982 and The Tax Reform Act of 1986 now require deferment by some contractors of a number of indirect costs which have heretofore been charged to expense of the year incurred. (See Chapter 13.)

Prior to the Tax Reform Act of 1986 (TRA 86), all four of these methods were acceptable for federal income tax purposes. Effective for taxable years beginning after 1986, TRA 86 made numerous and far-reaching changes in the accounting methods acceptable for federal income tax purposes in the construction industry. Even those methods which nominally survived were drastically changed in much of their application.

One such change was to make hybrid systems unacceptable. Cash basis accounting is no longer acceptable for (1) ordinary business corporations (i.e., the so-called "C" corporations), (2) partnerships in which one or more partners are "C" corporations, (3) tax shelters, and (4) the activities of certain trusts the income from which is taxable as unrelated business income. Accrual basis accounting is given increased recognition for income tax accounting. Completed-contract accounting for long-term contracts has been severely limited, while percentage-of-completion accounting has been given almost complete acceptance.

In addition to the procedural changes, TRA 86 clearly indicates a policy in Congress to give greater acceptance to the long-standing Treasury policy to limit, wherever possible, the acceptability of variations between financial accounting and income tax accounting. (See Chapters 7 and 13.)

Basic accounting methods compared

The cash method. There may be some construction businesses to which the pure cash method is applicable, but they are very rare. The jobs would have to be very small and the receivables, equipment, and inventories minute. Many firms, however, use a modified form of cash method which follows the federal income tax regulations. Under that method most income and expense items are determined on the basis of cash receipts and disbursements. Purchases of equipment and certain other items (such as insurance policies affecting more than one year), however, are not charged to expense but are treated as assets ("capitalized" to use the accounting term), and only the depreciation or "amortization" charged to cost. Equipment purchased on a chattel mortgage or conditional sales contract is considered to be fully owned and the liability is recognized and taken into account on the books. If the jobs are small and are done for cash, if there is little or no inventory, and if equipment is not an important factor, the cash method, as so modified, is simple and effective. However, for the average construction business it is neither economical for tax purposes nor adequate for business purposes.

The accrual method. For construction companies with short-term contracts, the accrual method was at one time considered by many accountants and most taxing authorities to be the only satisfactory method of accounting. In its most recent pronouncement on the subject, Statement of Position (SOP) 81-1, *Accounting for Performance of Construction-Type and Certain Production-Type Contracts,* and a companion work, *Audit and Accounting Guide—Construction Contractors,* the AICPA virtually ignore accrual method. SOP 81-1, under the heading of "Determining a Basic Accounting Policy for Contractors," states: "In accounting for contracts, the basic accounting policy is the choice between two generally accepted methods of accounting for contracts: the percentage method and the completed contract method."

As a practical matter, the AICPA position is not quite as arbitrary as it seems. In terms of actual usage, the completed-contract method, consistently applied by construction contractors doing short-term jobs, yields substantially the same result as a strict accrual method.

Companies with long-term contracts must use one of the recognized variations: percentage-of-completion or completed-contract. Basically the problem is one of matching costs and expenses against the revenue to which they are most closely related.

The courts have occasionally confused this basic concept by holding that income, even though earned, and costs, even though incurred, are not accruable until a legally enforceable claim has arisen. Thus, a retained percentage, though fully earned and accruable by accounting standards, would not be accruable for income tax purposes under the rule of some of the older tax cases until it became due and payable under the contract. By the same reasoning, it would seem to follow that a liability, though fully incurred, would not be accruable until the time for immediate performance had arrived. Obviously, while such standards may be acceptable at law, they would, if applied consistently in accounting, destroy much of the value of a company's financial statements. The accounting test of matching revenues with the related costs is the only practical one from a business viewpoint.

Thus, it is not unusual to see in the current asset section of a construction company's balance sheet such items as "Amounts earned but unbilled." For a contractor doing unit price work this entry might represent the billing price of units completed by the end of the accounting period but not billable until after the end of the period. On a lump sum job it might represent the progress payment which would be due if percentage of completion had been computed to the end of the accounting period instead of at some date prior to the end of the month. Where such practice is followed, the related costs would be included in the cost of operations section of the operating (profit and loss) statement.

Some companies prefer to take into account as job revenue only the amounts actually billable during the accounting period. The balance of costs incurred for which there is no related billing should appear in the current asset section of the balance sheet under a caption such as "Excess of job costs incurred over related progress billings."

Of course if there is a known loss it should be reflected in the accounts. The amount to be taken up should be the total amount of the known loss, even though some of it may not yet have been actually incurred, on the theory that any foreseeable loss should be recognized immediately.

Whatever the detailed mechanics of the accounting may be, the underlying principles are these:

1. Income is to be recognized when earned.
2. Costs and expenses are to be recognized when incurred.
3. Income and the related costs and expenses are to be taken into operating results in the same accounting period.

4. All known losses are to be fully provided for in the period in which they become determinable with reasonable certainty.

Selecting basic accounting policy for contracts. The terms "percentage-of-completion method" and "completed-contract method" are essentially variations of the accrual method of accounting designed to account for operations under long-term contracts of the lump sum or sometimes the unit price type. Their use arises from the fact that any particular job may continue over two or more accounting periods, and it is necessary to determine the period in which the income from the job will be taken into account.

More often than not there is very little relation between amounts billed to owners and actual earnings. For example, it is almost universal practice for contractors to receive progress payments as their jobs progress. Sometimes these payments aggregate less than the costs incurred in order to give the owner protection against defaults and liens in addition to the customary retained percentage. At other times an owner will provide for the cost of mobilization and move-in and also provide the contractor with working capital by permitting progress billings substantially in excess of those which would be allowable strictly on the basis of materials placed or units of work actually completed.

As a practical matter, payment on the basis of physical units actually completed is more fiction than fact. Almost invariably, partially completed units are averaged out to determine the basis for progress payments on a unit price contract. Sometimes, too, mobilization and other preparatory costs incurred on unit price contracts are reimbursed on the basis of billing stated in terms of units though no actual units have been completed. On lump sum contracts the same result is frequently obtained by permitting billings on the basis of percentages of completion not justified by physical progress alone.

Because, as a practical matter, the customary rules of accruing income on the basis of billings to the owner so rarely provide a proper measure of earnings, the two principal variations (completed-contract and percentage-of-completion) have become so generally accepted that they are, in themselves, spoken of as separate accounting methods instead of as mere variations of the accrual method.

The AICPA has given a simple, understandable, and highly authoritative description of the completed-contract method and the percentage-of-completion method of accounting for long-term contracts of the construction type. It is presented here as Schedule 1-5. Although the material in Schedule 1-5 has been expanded and clarified with the publication of SOP 81-1 and the *Audit and Accounting Guide—Construction Contractors,* it is included here because it is the authoritative document on which the subsequent publications are based.

SCHEDULE 1-5 LONG-TERM CONSTRUCTION TYPE CONTRACTS

American Institute of Certified Public Accountants, Accounting Research Bulletin No. 45

1. This bulletin is directed to the accounting problems in relation to construction type contracts in the case of commercial organizations engaged wholly or partly in the contracting business. It does not deal with cost-plus-fixed-fee contracts, which are discussed in Chapter 11, Section A, of *Accounting Research Bulletin No. 43,* other types of cost-plus-fee contracts, or contracts such as those for products or services customarily billed as shipped or rendered. In general the type of contract here under consideration is for construction of a specific project. While such contracts are generally carried on at the jobsite, the bulletin would also be applicable in appropriate cases to the manufacturing or building of special items on a contract basis in a contractor's own plant. The problems in accounting for construction type contracts arise particularly in connection with long-term contracts as compared with those requiring relatively short periods for completion.

2. Considerations other than those acceptable as a basis for the recognition of income frequently enter into the determination of the timing and amounts of interim billings on construction type contracts. For this reason, income to be recognized on such contracts at various stages of performance ordinarily should not be measured by interim billings.

Generally Accepted Methods

3. Two accounting methods commonly followed by contractors are the percentage-of-completion method and the completed-contract method.

Percentage-of-completion method

4. The percentage-of-completion method recognizes income as work on a contract progresses. The committee recommends that the recognized income be that percentage of estimated total income, either:

a. that incurred costs to date bear to estimated total costs after giving effect to estimates of costs to complete based upon most recent information, or
b. that may be indicated by such other measure of progress toward completion as may be appropriate, having due regard to work performed.

Costs as used here might exclude, especially during the early stages of a contract, all or a portion of the cost of such items as materials and subcontracts if it appears that such an exclusion would result in a more meaningful periodic allocation of income.

5. Under this method current assets may include costs and recognized income not yet billed, with respect to certain contracts; and liabilities, in most cases current liabilities, may include billings in excess of costs and recognized income with respect to other contracts.

6. When the current estimate of total contract costs indicates a loss, in most circumstances provision should be made for the loss on the entire contract. If there is a close relationship between profitable and unprofitable contracts, such as in the case of contracts which are parts of the same project, the group may be treated as a unit in determining the necessity for a provision for loss.

7. The principal advantages of the percentage-of-completion method are periodic recognition of income currently rather than irregularly as contracts are completed, and the reflection of the status of the uncompleted contracts provided through the current estimates of costs to complete or of progress toward completion.

SCHEDULE 1-5 LONG-TERM CONSTRUCTION TYPE CONTRACTS *(Cont.)*

8. The principal disadvantage of the percentage-of-completion method is that it is necessarily dependent upon estimates of ultimate costs and consequently of currently accruing income, which are subject to the uncertainties frequently inherent in long-term contracts.

Completed-contract method

9. The completed-contract method recognizes income only when the contract is completed, or substantially so. Accordingly, costs of contracts in process and current billings are accumulated but there are no interim charges or credits to income other than provisions for losses. A contract may be regarded as substantially completed if remaining costs are not significant.
10. When the completed-contract method is used, it may be appropriate to allocate general and administrative expenses to contract costs rather than to periodic income. This may result in a better matching of costs and revenues than would result from treating such expenses as periodic costs, particularly in years when no contracts were completed. It is not so important, however, when the contractor is engaged in numerous projects, and in such circumstances it may be preferable to charge those expenses as incurred to periodic income. In any case there should be no excessive deferring of overhead costs, such as might occur if total overhead were assigned to abnormally few or abnormally small contracts in progress.
11. Although the completed-contract method does not permit the recording of any income prior to completion, provision should be made for expected losses in accordance with the well-established practice of making provision for foreseeable losses. If there is a close relationship between profitable and unprofitable contracts, such as in the case of contracts which are parts of the same project, the group may be treated as a unit in determining the necessity for a provision for losses.
12. When the completed-contract method is used, an excess of accumulated costs over related billings should be shown in the balance sheet as a current asset, and an excess of accumulated billings over related costs should be shown among the liabilities, in most cases as a current liability. If costs exceed billings on some contracts, and billings exceed costs on others, the contracts should ordinarily be segregated so that the figures on the asset side include only those contracts on which costs exceed billings, and those on the liability side include only those on which billings exceed costs. It is suggested that the asset item be described as "Costs of uncompleted contracts in excess of related billings," rather than as "Inventory" or "Work in process," and that the item on the liability side be described as "Billings on uncompleted contracts in excess of related costs."
13. The principal advantage of the completed-contract method is that it is based on results as finally determined, rather than on estimates for unperformed work which may involve unforeseen costs and possible losses.
14. The principal disadvantage of the completed-contract method is that it does not reflect current performance when the period of any contract extends into more than one accounting period and under such circumstances it may result in irregular recognition of income.

Selection of method

15. The committee believes that in general when estimates of cost to complete and extent of progress toward completion of long-term contracts are reasonably dependable, the percentage-of-completion method is preferable. When lack of dependable estimates or inherent hazards cause forecasts to be doubtful, the completed-contract method is preferable. Disclosure of the method followed should be made.

SCHEDULE 1-5 LONG-TERM CONSTRUCTION TYPE CONTRACTS *(Cont.)*

Commitments

16. In special cases disclosures of extraordinary commitments may be required, but generally commitments to complete contracts in process are in the ordinary course of a contractor's business and are not required to be disclosed in a statement of financial position. They partake of the nature of a contractor's business and generally do not represent a prospective drain on his cash resources since they will be financed by current billings.

As set forth in SOP 81-1 the choice between the two generally accepted methods of accounting for contracts represents the basic accounting-policy decision that construction contractors face. The circumstances in which the AICPA recommends the use of the percentage-of-completion and the completed-contract methods are detailed in SOP 81-1 and are summarized in this section.

The percentage-of-completion method. SOP 81-1 recommends the use of the percentage-of-completion method in circumstances in which estimates are reasonably dependable and the following conditions exist:

- Contracts executed by the parties normally include provisions that specify clearly the enforceable rights regarding goods or services to be provided and received by the parties, the consideration to be exchanged, and the manner and terms of settlement.
- The buyer can be expected to satisfy his obligations under the contract.
- The contractor can be expected to perform his contractual obligations.

SOP 81-1 states a presumption that contractors generally have the ability to produce estimates that are sufficiently dependable to justify the use of the percentage-of-completion method of accounting and that persuasive evidence to the contrary is necessary to overcome that presumption. SOP 81-1 recommends that contractors should use the percentage-of-completion method on one or more of the following bases applied to individual contracts or profit centers as appropriate:

a. Normally, a contractor will be able to estimate total contract revenue and total contract cost in single amounts. Those amounts should normally be used as the basis for accounting for contracts under the percentage-of-completion method.

b. For some contracts, a contractor may only be able to estimate total contract revenue and total contract cost in ranges of amounts. If, based on the information arising in estimating the ranges of amounts and all other pertinent data, the contractor can determine the amounts in the ranges that are most likely to occur, those amounts should be used in accounting for the contract under the percentage-of-completion method. If the most likely amounts cannot be determined, the minimum estimated revenue and the maximum estimated cost should be used in accounting for the contract.

c. However, in some circumstances, estimating the final outcome may be impractical except to assure that no loss will be incurred. In those circumstances, a contractor should use a zero estimate of profit; revenue should be recognized equal to costs incurred until results can be estimated more precisely. A change from a zero estimate of profit to a more precise estimate should be accounted for as a change in an accounting estimate. Except for some types of cost-plus contracts, an inability to estimate total contract revenue and total contract costs for most of an entity's contracts either in single amounts or in ranges of amounts is usually a strong indication that a contractor's systems and procedures are weak. A contractor should use this basis only if the bases in (a) or (b) are clearly not appropriate.

The completed-contract method. SOP 81-1 recommends the use of the completed-contract method of accounting in certain situations:

- In circumstances in which a contractor has numerous relatively short-term contracts and in which financial position and results of operations reported on that basis would not vary materially from those that would result from the percentage-of-completion method.
- In circumstances in which estimates cannot meet the criteria for dependability on the basis of estimates in terms of single amounts, estimates in terms of ranges of amounts, or estimates in terms of zero profits, or in circumstances in which there are inherent hazards of the nature of those previously discussed.

SOP 81-1 sets forth in paragraph 47 recommendations on procedures for determining when a contract is substantially completed under the completed-contract method.

As pointed out in considerable detail in Chapter 13, TRA 86 has made substantial changes in the normal accounting concepts as stated in the AICPA's *Accounting Research Bulletin 45.* Practical considerations appear to dictate that in many companies accepted accounting practices will be followed in normal record keeping and that the necessary adjustments will be made at the year end to reflect the required tax practice.

In this connection it should be pointed out that there is no logical reason for requiring that income tax accounting and financial accounting must agree. The purpose of the former is to measure the company's immediate duty to pay income taxes. The purpose of the latter is to make a fair determination of the company's financial position and of the results of its operations as a going concern.

Determining the profit center. In accordance with SOP 81-1, each individual contract is presumed to be the profit center for revenue recognition, cost accumulation, and income measurement unless the criteria for combining or segmenting given in SOP 81-1 can be met.

Measuring the extent of progress toward completion. As set forth in SOP 81-1 the various methods used in practice to determine extent of progress toward completion, such as the cost-to-cost method, efforts-expended methods, and units-of-work-performed method, conform to paragraph 4 of *Accounting Research Bulletin 45.* The objective of all the methods used in practice is to measure the extent of progress in terms of costs, units, or value added. However, a particular method may or may not achieve that result depending on the circumstances of use and the manner in which it is applied. The various measures are identified and classified in SOP 81-1 as input and output measures. Input measures are made in terms of efforts devoted to a contract; output measures are made in terms of results. Both types of measures have drawbacks in some circumstances, and their use requires the exercise of judgment and careful tailoring to circumstances. The results obtained should be evaluated periodically through physical observation by qualified personnel, in the same way that the results of perpetual inventory records are evaluated and adjusted by taking a physical inventory in a manufacturing enterprise.

A good measure of the extent of progress toward completion should give appropriate weight to all elements of a contractor's work and, accordingly, should consider the broad phases of a contractor's operations such as:

- Designing the project (preparing blueprints to meet the owner's specifications)
- Procuring the necessary labor, materials, supplies, and equipment and mobilizing them at the construction site
- Managing the resources to complete the project
- Demobilizing the resources from the construction site

In some instances, not recognizing profits on the procurement and mobilization phases fails to assign profit on some projects to a major portion of a contractor's efforts. For example, projects geographically remote from a contractor may require a major procurement and mobilization effort. The risks related to that portion of the project may be significant, with construction carrying only relatively minor risks once that phase is completed. Construction projects, especially projects in countries foreign to the contractor, may require substantial mobilization and demobilization efforts, which should be included as an element of the measure of extent of progress toward completion.

Cost-plus contracts

Schedule 1-5 does not deal with special accounting problems arising from cost-plus-fixed-fee contracts, and refers specifically to Chapter 14, Section A, of *Accounting Research Bulletin 43* for the principles to be applied to fee work. The applicable portion of this bulletin lists the four most important problems in accounting for cost-plus-fixed-fee contracts:

(a) When should fees under such contracts be included in the contractor's income statement?

(b) What amounts are to be included in sales or revenue accounts?

(c) What is the proper balance sheet classification of unbilled costs and fees?

(d) What is the proper balance sheet treatment of various items, debit and credit, identified with cost-plus-fixed-fee contracts?

The most common practice is to accrue the fees on cost-plus-fixed-fee contracts on the percentage-of-completion basis of accounting in the period when they become earned. The reason is that most cost-plus-fixed-fee contracts are so written that the risk of losing the fee is virtually negligible. However, where there is (as in some target-price type contracts) a chance that the fee may be lost through operation of a penalty clause, then there is very strong authority to support deferring recognition of income from the fee until the contract is completed, or the final outcome can be accurately forecasted.

The second of the four problems in *Bulletin 43* concerns what should be taken into revenue. There would appear, in general, to be three possibilities. First, to take into account only that portion of the fee which will be earned currently. In other words, if the fee were subject to a 10 percent retention, only 90 percent would be accrued. This is not the usual treatment and would normally be used only when there is a good chance that the retained percentage will not be collected. Second, it is the usual practice in maintenance and service contracts where total volume is not important, to accrue the full amount of the fee which is billable even though some portion of it may be retained by the owner until completion or for some guarantee period. Third, and most satisfactory for most construction companies operating on a cost-plus-fee basis, is to take into revenue the entire amount billed or billable including reimbursable costs. The reason for this is obvious. Total volume is significant to the construction contractor in a number of different ways and therefore is the sort of thing which should be directly reflected in the records.

Unbilled costs and fees should appear on the construction contractor's balance sheet as receivables and not as inventory or as an advance. They should be set out separately under a caption such as the following:

Unbilled amounts due on cost-plus-fixed fee contracts:
  Fees earned
  Reimbursable costs

In the current asset and current liability sections of the balance sheet, amounts due to the owner under a cost-plus-fixed fee contract should be shown separately from amounts due from the same owner. Offsetting is so rarely proper that it is preferable to establish a policy of showing the current receivables separately from the current payables and advances. If offsetting is applied, however, the amounts offset and their general nature should be disclosed.

Hybrid methods. Frequently the work of a construction company will include some cost-plus-fixed-fee work, some lump sum and unit price work of a type that justifies the use of percentage-of-completion accounting, and some other work of a type to which, because of risk and uncertainty, the completed-contract method is applicable. So long as the books and the financial statements clearly reflect the facts and the methods used, and so long as the methods are consistently followed, there can be no reasonable objection to such procedure. In fact, the contractor has a legal duty to keep his records so as to reflect income clearly, and certainly one of the requirements for the clear reflection of income is the selection of accounting methods most applicable to the operation to be accounted for. However, indiscriminate hopping from one method to another, merely to gain a temporary tax advantage or to show a better financial statement to the bank or bonding company, is an invitation to serious trouble. Only a very real and substantial change in conditions will justify changing a method once it is adopted, and even then the Commissioner of Internal Revenue may not permit the changed method to be used for tax purposes.

Income tax implications. (See also Chapter 13.) Federal income tax regulations have long required that the accrual method be used if inventories are a substantial income-producing factor. This requirement immediately raises the question of the nature of construction work in progress. Many accountants treat it as an inventory item because it is comparable in many ways to the work-in-process of a manufacturing concern. However, because construction work usually consists in improving the real property of some other person, legal title to the partially completed construction work is generally considered to vest in the owner of the real property. Sometimes, as in U.S. government contracts, passage of title to the work and materials is spelled out in the contract. The technical result of this difference is to change the nature of the contractor's asset from an inventory item to a receivable protected by lien rights (or other comparable statutory rights).

The effect of this distinction on a contractor reporting on the modified cash method permitted for tax purposes is apparent. So long as he has no substantial inventories of materials or supplies he would be in technical compliance with this portion of the regulation. Other factors, such as the effect of individual state laws or the needs of the business for more accurate financial statements must, of course, be considered. Most important of all, in terms of tax effect, is the possibility that a cash basis taxpayer may find income pyramiding in a single year with losses or very low income in the preceding or following years.

Prior to the passage of the Tax Equity and Fiscal Responsibility Act of 1982, the federal income tax regulations and those of most of the states which have income taxes provided that the completed-contract and percentage-of-completion methods of reporting income may be used only for contracts of more than one year's duration. Current tax law and related regulations have not shortened that time. Technically speaking, this precludes most small contractors from using them. If such contractors cannot, or prefer not to, report on the so-called cash method, they are forced to report on the straight accrual method without modification. As a practical matter, the taxing authorities frequently accept the use of the long-term contract methods for contracts running for less than one year if the practice is consistent and no serious distortions result.

If there are no distorting factors such as unbalanced bids, unbalanced payment schedules, mobilization payments, or inflated percentages of completion, it is common to find the results of the ordinary accrual method similar to the accrual method applied with the percentage-of-completion variation. The reason is that many contracts call for the computation of progress payments on the basis of a percentage of completion. It is also common for contractors who report on the accrual method without the long-term contract variations to ignore the legal niceties governing the year in which retained percentages are includable in income and again, as a practical matter, such treatment has been accepted by the taxing authorities if it is consistently followed and no material distortions result.

Taxable income frequently differs from book income, and there is no reason why they must coincide. Nevertheless, where it can properly be done, there are considerable advantages for some construction companies with somewhat less sophisticated financial management techniques in keeping the two as close together as possible. The closer taxable income approaches book income, the easier it is for actual or potential tax liabilities to be checked and for anyone not familiar with the company's detailed tax position to reconcile tax provisions with the income shown.

Generally speaking, among contractors who have the choice, the completed-contract method has been preferred for tax purposes because it defers the payment of the tax until the retained percentage provides the cash for the payment of tax. Apologists who shrink from the idea of accounting methods being chosen for tax purposes alone hasten to point out that wide distortions in income arising from the use of the completed-contract method may be avoided by using the percentage-of-completion method. It may be questioned whether uniformity for its own sake is a desirable goal. However, heads of companies with outside stockholders unfamiliar with the construction business point to an extremely important nontax advantage in keeping the flow of income as uniform as possible. It saves time and trouble in stockholders meetings when lengthy explanations of variations in net income can be avoided.

The problem, of course, lies in the fact that the annual determination of income is virtually a cornerstone in our income tax structure. As applied to a large segment of the construction industry, the concept is totally unrealistic. There are very few construction men who will not admit privately that there is no cash profit until a job is completed. It is true that interim profits on certain types of construction work can be estimated quite accurately provided the individual contractor has the personnel and the records necessary to make such estimates possible. It is equally true, however, that percentage-of-completion reporting for tax purposes can be an invitation to wishful thinking and provide fertile ground for an opportunistic type of so-called tax planning that frequently gets the planners into serious tax trouble.

If differing methods are used for tax and financial accounting and significant amounts of tax are being deferred, the accountant must make sure that management is aware of future cash requirements to cover tax payments.

This still does not deal with the problem of the construction contractor who looks at the bottom line of the operating statement and assumes that any amount he sees there is available in cash. It has been mentioned that for some contractors it is desirable to keep the book income and the income for tax purposes as close together as reasonably possible and that the tax returns be prepared by the same method that is used for the financial statements. If the method chosen is the completed-contract method, there is somewhat less tendency on the part of management to spend, prematurely, profits that are still tied up in jobs.

Probably the most important observation that can be made about the tax effects of accounting decisions in a construction company is that taxation is a highly specialized business and many construction companies do not have people on their staffs competent to make decisions affecting taxes. Even if a company does have a tax specialist on its staff, he usually needs the help of outside specialists to check his thinking. The most costly saving a construction company can make is to skimp on job supervision. The second most costly saving is that made by failure to employ thoroughly competent tax counsel on a retainer basis for year-round consultation.

Prejob Procedures

Staffing the job

Regardless of the size of the contracting firm or the size of the job, somebody has to do the work. An individual contractor doing only one job at a time may put his tools in the back of his pickup truck, call for his helper, and go do the job himself. A contractor who has several small jobs going at one time might send out a journeyman on each job and occasionally call at the job to supervise the work. On larger jobs the contractor himself may act as foreman or superintendent and have journeymen and helpers working directly under him, or he might employ a foreman or superintendent and confine his own jobsite activities to general supervision.

At the other end of the scale, the job might be a multimillion-dollar project operated by a joint venture employing a project manager, a general superintendent, a project engineer, a project office manager, and a full staff of craft or area superintendents, engineers, foremen, journeymen, accountants, clerical personnel, and laborers. On such jobs the number of people on the payroll may run into the thousands.

Whatever the size or the nature of the job, someone must decide, before the job starts, how many people will be required to do the work and who the key people, at least, will be. This responsibility is management's. However, once the size of the force is estimated and the key people are named, it is generally up to the head of the contractor's office (usually the office manager or controller) to see that their time is properly kept, that they are paid the proper amounts at the proper times, and that arrangements are made for payment of their expenses in reaching the jobsite, and often for feeding and housing them after they get to the jobsite. Since modern personnel departments are rare in the construction industry, in all but the very large companies, the controller or office manager is usually the one to decide what accounting and clerical jobs are to be done at the jobsite, and to pick the people to fill them.

Following is a list of a few typical job organization methods with corresponding figures illustrating the flow of accounting data in each situation.

Very small organization. The very small contractor of Figure 1-1 works directly on the job. He keeps the time, buys the materials, and does all the jobsite receiving. He usually writes checks for payroll and purchases, and turns these basic data over to a public accountant or bookkeeping service. His staffing problems are virtually nonexistent unless he is trying to find a helper who is capable of developing into a foreman or superintendent. Subcontractors deal directly with the owner/contractor.[1]

Small organization. The small contractor of Figure 1-2 supervises several journeymen, and usually does his own buying and timekeeping. The journeymen receive materials and supplies delivered at the jobsite and turn the signed delivery receipts over to the owner/contractor when he visits the job. The owner/contractor makes the payrolls, checks the delivery data, writes both the payroll checks and the vendor checks, and turns the basic records of the transactions over to the bookkeeper. Subcontractors deal directly with the owner/contractor.

Small-medium organization. The small-medium owner/contractor of Figure 1-3 works as foreman or superintendent on one large job while

[1]The terms "owner/contractor" and "contractor" are used more or less interchangeably, because in many instances the subject matter could apply equally to either and because, as to certain subcontractors, the owner and the contractor are sometimes the same entity.

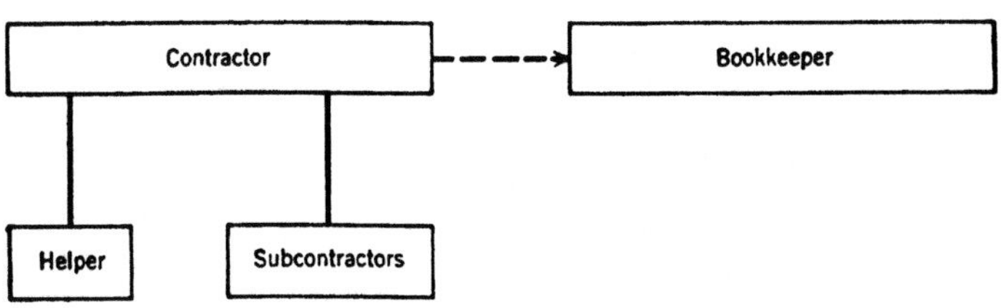

Figure 1-1 Organization Chart for Very Small Organization.

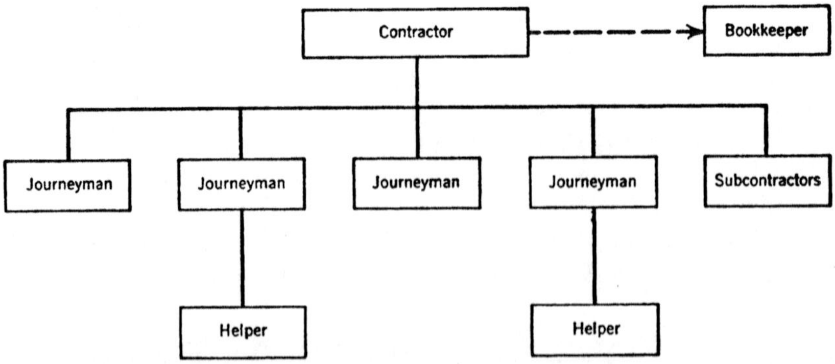

Figure 1-2 **Organization Chart for Small Organization.**

he has other smaller jobs in progress under foremen or journeymen. The flow of accounting data tends to be about the same as in Figure 1-2, but in operations of this size there is usually a company bookkeeper who assembles the data, keeps the books, and writes checks for the contractor's signature in the company office. Usually, too, in this type of operation, the foreman on the large job keeps the time and receives the materials, and turns the time sheets and signed delivery slips over to the contractor for checking. Subcontractors, for the most part, still deal directly with the contractor, though on the big job some of the dealings will be with the foreman.

Medium organization. In the somewhat larger operation of Figure 1-4, the contractor supervises several foremen who keep the time and receive the materials and supplies delivered to the jobsite. Accounting data are picked up and summarized by a field clerk and delivered to

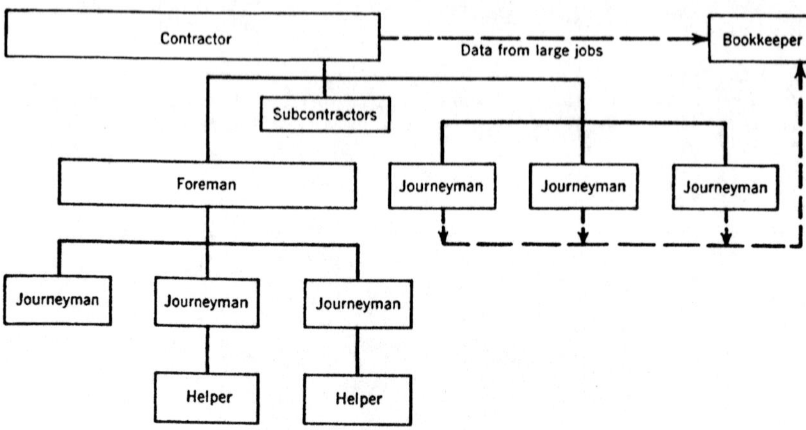

Figure 1-3 **Organization Chart for Small-Medium Organization.**

the office manager for final processing in the general office. It is at about this point that staffing the jobs begins to present special problems. To keep an operation of this size going, the contractor cannot depend so completely on his own personal observations to follow job progress, but must make greater use of reports. In addition to their technical skill, the foremen must be chosen on the basis of cost consciousness, their ability to manage a crew, and their ability to make independent decisions. When an operation reaches this size, the contractor will often have his foremen picked before bidding a job. In the same way, it is quite common for foremen to have a crew that follows them from job to job. In operations of this size, it is usual for subcontractors to make their original deals with the contractor, but supervision of their work is left to the contractor's foremen.

Many larger construction companies, whose jobs normally cost $3 million or less, operate on substantially the same pattern, except that they may substitute a superintendent or a construction or area manager in the place of the contractor shown in the diagram.

As the jobs grow larger—say from $3 million to $20 million—the pattern tends to remain much the same, except that the larger jobs are usually in the charge of a construction manager who may have several foremen and subcontractors under him.

Large organization. As the individual jobs pass the $20 million mark, there is an increasing tendency for each job to have its own organization, headed by a project manager or superintendent (Figure 1-5). With

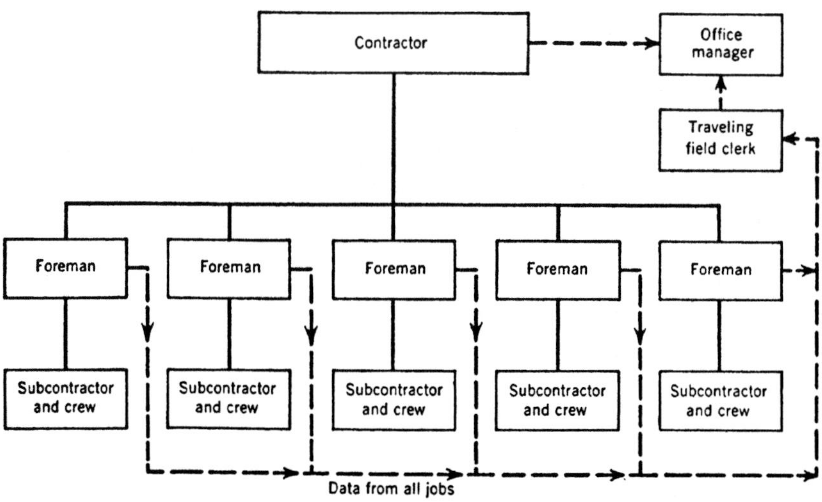

Figure 1-4 Organization Chart for Medium Organization.

centralized accounting, a job clerk and a combination timekeeper-pickup truck driver will usually be enough accounting personnel. With a decentralized system, the job clerk would be replaced by a job accountant, and if there are as many as 100 employees on the payroll, there may be a separate pickup truck driver and a combination timekeeper-paymaster, who would probably then spend at least part of his time operating a small warehouse. If the job is cost-plus-fixed-fee, in addition to an office manager, it may be necessary to have full-time payroll, accounts payable, and warehouse clerks. This pattern tends to apply up to about $20 to $40 million, depending on whether the job is centralized or decentralized.

Very large organization. On jobs costing more than $40 million, the organization will tend to approximate the chart of Figure 1-6. Again, there will be variations in the actual number of key personnel, depending upon the number of employees on the payroll, whether accounting is centralized or decentralized, and upon the availability of living quarters, mess facilities, and first-aid facilities. This organization chart would serve a job in the over $50 million category, for which camp, commissary, and first-aid facilities are needed. Dotted lines show the flow of cost data, which originate in the payroll, timekeeping, accounting, purchasing, and warehouse departments, to the cost engineer. From the cost engineer daily cost reports are sent to the project manager. From the cost engineer also, cost-coded accounting data flow back to the accounting department, where they are incorporated into the general records and the monthly cost reports.

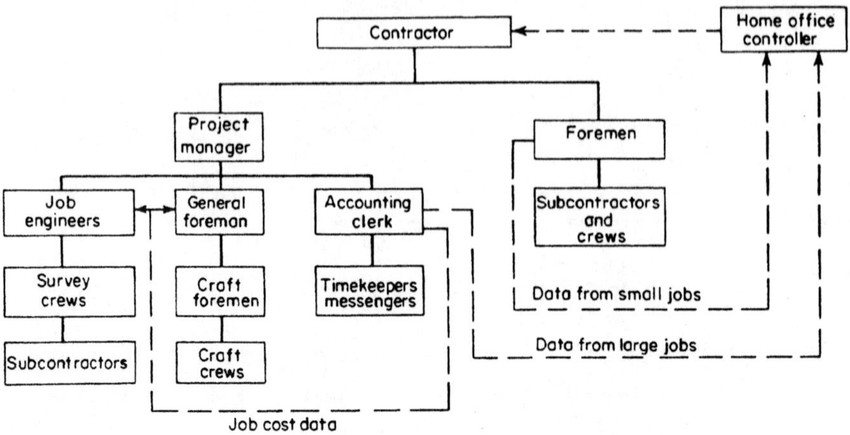

Figure 1-5 Organization Chart for Large Contracts.

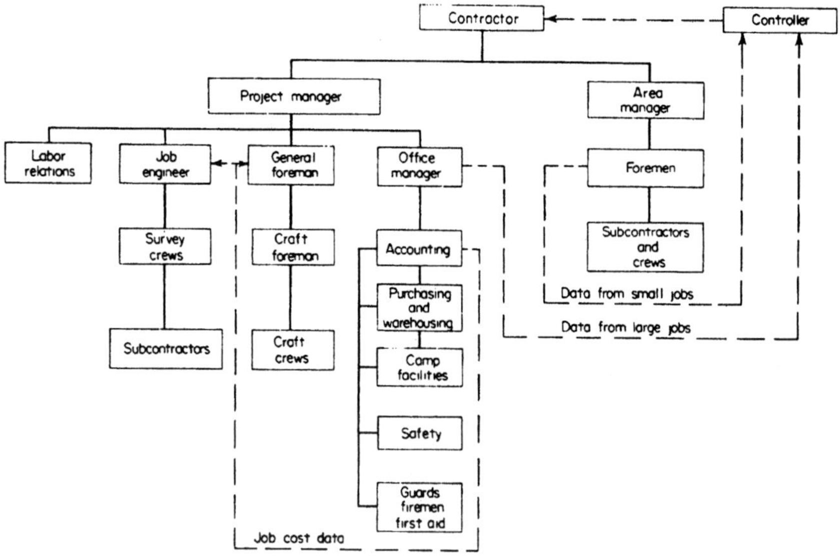

Figure 1-6 Organization Chart for Very Large Contracts.

Picking key personnel

The large decentralized accounting function will vary from project to project depending on the type, size, and nature of the contractual terms. Figure 1-7 depicts the organization of a fairly typical large decentralized accounting department.

The duties of the job office manager (or project controller or business manager, as he is sometimes called on large jobs) will vary widely, depending on the size and nature of the job, and on whether the accounting is centralized or decentralized. Depending on the policies of the company, purchasing is sometimes included in the duties of the project controller. To be project controller on a large job with decentralized accounting, a person should have formal training in accounting equal at least to two years of college plus at least five years of successful accounting experience in a construction field office. At least one of the five years should be as a paymaster on a payroll of 200 or more employees. He must be aggressive, intelligent, and a capable administrator.

On a small job where accounting and payrolls are centralized, the job office manager need be little more than a good timekeeper and clerk. For jobs in this category there are no particular educational requirements, and the individual's experience need only be in the fields of timekeeping, materials checking, and clerical work.

Aside from the office manager, the key accounting jobs in a construction field office are those of payroll and timekeeping. Nothing in

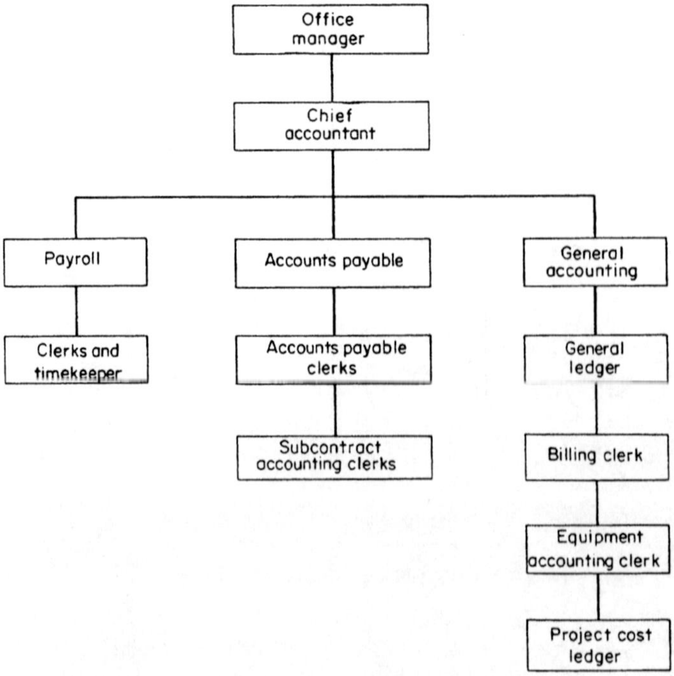

Figure 1-7 Organization Chart for Very Large Decentralized Project Accounting Department.

a construction job office can be more costly than late, messy, inaccurate payrolls. The payroll is the starting point for the paychecks, which must be ready on time and scrupulously accurate if job morale is to be maintained and waiting time penalties avoided. It is also the starting point for the employees' earnings records, the payroll tax returns, and the insurance reports, which must balance with total salaries and wages paid. Most importantly it is the source of key data upon which management relies in monitoring the progress on the job.

Despite the numerous controls and cross-checks on payrolls, there is still a temporary opportunity (and therefore a real temptation) for dishonesty. The fact that ultimate detection is almost certain does not entirely eliminate the possibility of attempts at payroll fraud. Finally, the paymaster is continuously being subjected to pressure. He must detect errors and omissions in timekeeping, and he must fight a running battle with foremen and field timekeepers to keep the flow of time and cost data steady and accurate. He is under pressure to release checks early, make special advances, pay off employees who have been laid off, and make new additions to the payroll. He must watch for garnishments, attachments, and assignments, and he must be sure to deduct for advances and purchases when workers receive termination pay.

If payrolls are centralized in the general office there is no problem in selecting a jobsite paymaster. If payrolls are decentralized, the paymaster will usually be the first or second office employee to be assigned to a new job. Whenever possible he should be an old hand, thoroughly experienced in construction payroll, whose ability to handle the job is certain. When a new paymaster is to be hired, it is a good idea to employ him soon enough to have one of the old hands on another job break him in. If a new paymaster is to be hired and there will not be time to try him out with another paymaster, great care must be taken to see that he personally knows payroll accounting thoroughly and that he fully understands the operation of a payroll department. His first work on the job should be watched very closely. Recommendations from former employers mean very little, and his record on former jobs may be misleading because he might have been getting by on the work of a capable assistant. At the start of a new job things move too fast; there is little time to correct errors in judgment on the selection of a paymaster.

Timekeeping is the raw material for the payrolls and for the labor cost reports. It must be done at the place where the work is being done. If it is inaccurate it leads to one or more of the following consequences:

1. The employees are overpaid and the costs are increased.
2. The employees are underpaid and there is trouble on payday.
3. The paymaster's work is delayed by his having to catch and correct errors.
4. The labor cost figures become meaningless because of inaccurate or careless work designations, and errors in management decisions can occur.

Some companies have the foremen keep the time. The foremen who are chosen to keep time must be willing to take the pains necessary to do it correctly and turn it in promptly. Other companies use timekeepers. Timekeepers need to be familiar enough with construction work to be able to watch a crew at work and know what they are doing. The idea that "anybody can be a timekeeper" is not only false but risky. Nevertheless, it is a job that a person of reasonable intelligence can be trained to do with moderate ease. Some companies break in their young engineers by starting them as timekeepers; their work needs close supervision. Most companies use the job of timekeeper as the first step in learning construction office work.

One of the most important attributes of a timekeeper is personal integrity. He is in a better than average position to soldier on the job without being caught, and he is constantly faced with opportunities to profit by collusion. Therefore, some independent check on the activities of field timekeepers should be provided. Once a person's integrity has been reasonably established, the need for constant supervision no longer exists. On the other hand, it is poor management to place any employee in a position where he is subject to unnecessary temptations. If timekeepers are used, it is better to have more than one and change their assignments from time to time to prevent collusion. If foremen are to keep the time it is well to provide at least one timekeeper to make spot checks and turn in time independently on the crews checked.

Other job office personnel can be chosen principally on the basis of experience in the job for which they are hired. However, in evaluating the experience of a job accountant, it is a good idea to find out if he knows double-entry bookkeeping. If he has worked only for companies with centralized systems, he may never have seen, much less kept, a general ledger. Warehousemen should be chosen largely on the basis of their knowledge of warehousing procedures, and of the various materials, supplies, and spare parts which are to be found on the company's jobs. No amount of accounting knowledge can substitute for these attributes.

Personalities

There are few types of work where pressures and emotional strains are greater than on a construction job. Hard-driving project managers and superintendents often push themselves and the men under them close to the limits of human endurance. The fact that office personnel are generally considered "nonproductive" tends to make the pressure even greater in the job office. It is vital, therefore, that the job office manager be able to get along with the project manager and the project engineer without sacrificing any important matters of principle.

Most of the problems between the office and the operating personnel on a job may be solved or prevented by the attitude of top management. If that attitude is one of mere toleration for "nonproductive overhead," it will be reflected in job management, and the job office will be reduced to a clerical operation, staffed by people who take orders and say nothing. If the attitude of top management is that the office, like every other department, must be productive and contribute its full share to the final job profit, then it is safe to staff the job office with personnel capable of living up to such an assignment. A mixture of the two attitudes, however, can produce either costly dissension or needless expense. If the man in charge of the job considers the job office as "mere nonproductive overhead," it is usually better to centralize all but the most routine clerical functions.

Personnel transfers and transportation

If jobs are small and all the key personnel live nearby, there is no problem of transferring and transporting job personnel. However, when the key people must be brought in from other jobs or from other localities, a number of problems can arise.

In selecting personnel, thought must be given to salary relationships and living-cost differentials. For example, a company which adjusts salaries to make up differences in living costs may find it necessary to reduce some salaries when workers are moved from areas where living costs are high to others where living costs are low. Most employees resent any downward adjustment in the rate of pay, and frequently this results in the loss of capable, experienced workers. The alternative is to raise the salaries of others to maintain proper relationships, and in such cases the contractor may find himself in the unhappy position of disturbing local salary scales and creating an unbalanced relationship between the salaries he pays and those paid by the owner-client or the key subcontractors. This type of problem can usually be met most effectively by setting a uniform salary scale and supplementing it by providing key personnel with housing, transportation, and a cost-of-living allowance based on a differential between living costs at the location of the general office and those at the jobsite. In setting such a differential, the effect of state income taxes on "take home" pay must not be overlooked. Thus, a move from Nevada, which has no personal income tax, to California, which does have such a tax, could have the practical effect of a cut in salary.

When a joint venture is formed to do a job, care must be taken to coordinate the policies of the member companies regarding salaries, expense allowances, bonuses, vacations, and other personnel matters if personnel from more than one company are to be employed. For this reason it is better, if practical, for the sponsoring venturer to staff the job.

It is usually short-sighted economy to pinch pennies in the allowance of family moving costs. It takes only a few costly moves to convince an employee he would be better off to take a steady job at a smaller salary and stay in one place.

Essentially these are management problems, but they touch on the field of accounting in these ways:

1. A centralized accounting system will eliminate most of these problems as they apply to key personnel in the job office.

2. On cost-plus-fixed-fee contracts it is common to find moving and transportation costs and cost-of-living differentials reimbursable to the same extent allowed by the contractor on lump sum work. Also, it is customary to decentralize the accounting on cost-plus jobs because very few owners are willing to pay the cost of centralized accounting even though the cost of decentralized accounting would not be questioned.
3. Administration of expense reimbursement policies is usually the task of the controller's department, and it is therefore up to the controller to see that these problems are considered in setting policy. On large jobs it is of prime importance that these policies be set before the job is started. Changes in such policies made later can easily lead to charges of favoritism.

Delegating responsibility and authority

It is an axiom of business organization (too often ignored) that when an employee is made responsible for the success or failure of some aspect of the operations, he must have the authority necessary to carry out that responsibility. This axiom is just as true in a construction job office as anywhere else, but a construction job moves so fast that the results of failure to observe it are sometimes more dramatic and costly. Fortunately for many contracting companies, construction men are an independent lot, and if something needs to be done able men will do it first and answer questions about their authority later if necessary. Unfortunately, such action is often taken by men whose will to get the job done is exceeded only by their lack of knowledge of how to go about it. As a result, some things that need to be done are often done in a way that is needlessly expensive.

Consider the example of a contractor who had a large United States government contract. The excavation on the job had been subcontracted to a company whose financial position was known to be very weak. One day the accountant in charge of the prime contractor's office received a registered letter from one of the subcontractor's creditors. The letter mentioned, among several other things, the fact that the subcontractor owed the creditor a stated amount of money.

The office manager had never heard of the Miller Act (Chapter 7), so he filed the letter. Then he telephoned the creditor to ask him not to tie up the job, and to assure him that the account was being given personal attention. The accountant also said he would take steps to see that the account was paid. Only after the subcontractor had gone bankrupt and the prime contractor was being sued for his unpaid bills did the accountant's action come to light. Even then its seriousness was not fully appreciated until the prime contractor found himself bound to pay the bill.

So the axiom that authority must be equal to responsibility has a corollary—before delegating responsibility, be sure the employee to whom it is delegated is capable of discharging it.

Perhaps, at this point, a footnote might be in order for those contractors who consider experience as all-important, and who discount the value of education. The accountant in the example just cited had ten years of heavy experience as a construction job office manager. Still he did not know one of the things a course in elementary commercial law would have taught him—that sometimes a promise to guarantee payment of an account does not have to be in writing if the principal purpose of the guarantee is to benefit the party who gives it. A little internal staff training would probably also have made the accountant aware of the existence and general nature of the Miller Act.

General office–job office relationship

One of the most common mistakes in any company where there are branches or subsidiary offices is to imagine that figures compiled in the branch offices cost the company nothing. Construction company executives tend to make the same mistake about job offices, particularly if the accounts are decentralized. Often the effect is to burden the job offices with unnecessary work simply because the cost is charged to job overhead instead of general office overhead. The cost to the company is the same or more, even though charging it to the job may improve the appearance of the general office overhead budget.

Another expensive practice that tends to grow up between general office and job office is that of accepting, without checking, the figures supplied by one to the other. This practice is an open invitation to error, yet it is a rare office where you will not occasionally hear the defense: "It isn't my fault. That's the figure they gave me."

Company policies and job conditions

Because the principal thing a contractor has to sell is his own know-how and that of his organization, the problems of staffing jobs are never ending. Just as each job offers construction problems peculiar to itself, so each has its own staffing problems. It is a rare growing organization that can fill all important jobs from its own ranks. Likewise it is a rare organization that will not find new staffing problems arising from shifts in company policy and job conditions. Therefore, new people are always entering the organization, bringing with them the standards of their former jobs. Each company must necessarily have standards of its own, and it is not practical to attempt to hold employees to a standard unless they are told what the standard is. Therefore, as soon as a contractor reaches the size where authority to make administrative decisions must be delegated to others, a manual of policy and procedure, and a program for making its contents known to key employees, are vital. One of the most important of the prejob procedures is to make sure: (1) that the people who will be assigned to the job are properly briefed on that job; (2) that they are supplied with the proper manuals; and (3) that they know what is in them and how that accounting and policy information is to be used.

Establishing the cost code

Before costs can be kept on a job, it is necessary to determine the cost accounts that will be used and the symbols that will be used to designate those accounts. This procedure must be established before the job starts; otherwise, the expense of keeping job costs will probably be wasted because the figures will come too late to do much good.

Most effective cost codes are set up to follow the bid and to enable the compilation of cost figures on the same units as those used in preparing the initial bid. Only in this way can the effectiveness of the bidding methods be checked against actual operations. In addition, where progress billings are based on percentage of completion, costs kept by bid items are helpful in establishing the percentages.

Standard coding. If the type of work a contractor does is fairly well standardized, it is usually possible to prepare a standard estimating checklist that can be the basis of the cost code. Schedule 1-6 is the bidding checklist used by a builder of larger structures (office buildings, hospitals, public buildings, and so on).

Each of the major subdivisions in Schedule 1-6 will have an internal breakdown. For example, item 14, aluminum and stainless steel work, might well have the subclassifications shown in Schedule 1-7.

It is not hard to see how, by merely adding a series of numbers, this standard bidding analysis can be converted into a ready-made cost code. For example, it is common to see each cost account broken down as follows:

.1 Labor

.2 Incorporated materials

.3 Equipment rentals

.4 Supplies and other direct costs

.5 Subcontracts

.6 Distributed overhead

Thus the hoist rental in hoisting aluminum windows to the upper story of a multistory building would be coded 14B6.3. If it were decid-

SCHEDULE 1-6 CHECKLIST OF STANDARD CODING FOR LARGER BUILDINGS

1 Job overhead
2 Demolition
3 Excavation, grading, and subsurface drainage
4 Stabilization (including piling, shoring, and underpinnings)
5 Paving
6 Structural concrete work (including forming)
7 Scaffolding, hoists, and the like
8 Masonry
9 Caulking and weather stripping
10 Precast concrete, planking, and joists
11 Decking (poured type)
12 Structural steel
13 Miscellaneous and ornamental iron
14 Aluminum and stainless steel work
15 Steel windows
16 Doors, bucks, and frames (metal)
17 Insulation
18 Lumber and carpentry
19 Rough hardware
20 Flooring
21 Millwork, cabinet work, and wood and sash doors
22 Finish hardware
23 Lath and plaster, and interior finish
24 Acoustical work
25 Tile work—ceramic, quarry, slate, marble, and terrazzo
26 Composition tile and linoleum
27 Interior partitions
28 Glass and glazing
29 Roofing and waterproofing
30 Sheet metal
31 Plumbing—internal and outside utilities
32 Heating, ventilating, and air conditioning
33 Elevators and dumb-waiters
34 Fencing, landscaping, seeding, sodding, and planting
35 Painting and decoration
36 Special equipment

ed before the job started that item 14 was not important enough in this job to justify keeping costs on the basis of the full breakdown, the major item might be used alone, in which case the coding of the hoist rental would be coded 14.3. Often house builders confine their cost keeping to the major items in their estimating checklist, such as site preparation, streets and sidewalks, sewers, and concrete foundations and slabs.

Coding by work items. Specifications for large jobs often break the job down into work items. When the job is bid, the estimates will usually follow these work items. Many contractors set up a separate cost code for each job, following the work items in the specifications for

SCHEDULE 1-7 BREAKDOWN OF A MAJOR SUBDIVISION OF A COST ACCOUNT

A Aluminum doors and frames
- A1 Hardware
- A2 Erection and handling

B Aluminum windows
- B1 Projected type
- B2 Double hinge
- B3 Mullions, covers, and hardware
- B4 Stools and sills
- B5 Screens
- B6 Erection and handling

C Stainless steel work
- C1 Erection and handling

D Architectural aluminum
- D1 Spandrels, facia, and trim
- D2 Erection and handling

that job. It is argued in favor of these special cost codes that if the cost figures are used in the preparation of other bids, it is necessary for the estimator to look at the related specifications, and by checking the specifications he avoids the risk of using an inapplicable figure.

There is a certain amount of merit to arguments in favor of setting up separate cost codes for each job. Every job has its own peculiarities, and if a standard code is used to develop average unit costs, those peculiarities become merged in the averages. When the averages are higher than the correct figures, there will be a tendency to lose jobs. When the averages are lower than the correct figures, they will tend to make the contractor low bidder on unprofitable work, or "leave too much on the table." Either alternative is costly.

It is only when the averages fall reasonably close to the correct figures that the bidder would tend to get a reasonable share of the profitable jobs.

Mnemonic or decimal coding. There are three general types of cost codes: mnemonic, decimal, and mixed. In mnemonic codes, letters may be used to give some indication of the item. Thus in the example given, aluminum and stainless steel might have the designation AS instead of 14, and the coding for the hoist rental on aluminum sash handling would be ASB6.3.

A pure decimal system would ignore the nature of the item and would merely code the work item with a number, more or less arbitrarily chosen. If the work item were numbered 75, the erection and handling were designated as 6, and equipment rentals as 3, the coding would be 75.6.3.

The mixed use of letters, figures, and decimals is illustrated by the example first given, in which the item was coded 14B6.3.

Combination coding. Sometimes a company, particularly one doing heavy engineering contracts requiring numerous temporary structures and project shops, will use a standard coding for these distributable costs items, and a decimal system for the actual work items.

Projecting job financial requirements

Every time a new job is undertaken it is necessary to provide financing to operate it, and in order to have money on hand when it is needed, it is important that cash requirements be forecast as soon as possible. This is a simple matter of spreading the estimated costs to the various weeks in which they are expected to occur, and it can usually be done by the estimator as soon as his bid is complete. The spread is usually weekly because the standard pay period is weekly. The estimators should supply this breakdown on any bid that may result in a contract. Thus, on a job estimated for completion in sixteen weeks the breakdown might be similar to the one shown in Schedule 1-8.

Unbalancing the bid. In Schedule 1-8, progress billings are estimated on the basis of a percentage of completion, which is computed by comparing costs on a cash basis to total estimated cost. On this basis the contractor has an investment in the job that at one point reaches $37,000.

One of the methods of decreasing the amount of a contractor's investment in his jobs is known as "unbalancing." The column headed "Estimated progress billings" gives the total price of the job being used as an example as $220,000. Suppose that the job were being broken down by the specifications into ten work items (usually there are more, but ten is enough to illustrate the point), and suppose they are numbered roughly in the order in which the work will be done. If the profit is spread evenly over the job, the payment schedule which would be submitted to the architect would look something like the last column of the payment schedule shown as Schedule 1-9. Of course the schedule given to the architect would show only the "Bid price" column.

Now, suppose that instead of being spread evenly to all work items the profit were spread to the first four work items, and suppose that the $29,600 of cash requirements incurred by the end of the fifth week (Schedule 1-8) were made up as in Schedule 1-10.

The overall percentage of completion, based on cash expenditures alone, would be 14.8 percent (or $29,600/$200,000) but the progress billing, instead of being $29,300 (or $220,000 × .148 = $32,560 less $3,256 retention, or $29,304), would be $32,679, computed as shown in Schedule 1-11.

Sometimes overhead, as well as profit, is used to unbalance the bid figures submitted as a payment schedule. This method has merit

SCHEDULE 1-8 **FORECAST OF CASH REQUIREMENTS ON COST BASIS**

Week	*Direct labor*	*Incorporated materials*	*Equipment rentals*	*Subcontracts*	*Supplies, etc.*	*Distributed overhead*	*Estimated progress billings*	*Cumulative estimated cash investment in job*
1	$ 500					$ 750		$ 1,250
2	750					1,250		3,250
3	1,000					1,000		5,250
4	1,250					800		7,300
5	2,000	$ 2,000	$ 1,500	$15,000	$1,000	800	($ 29,300)	300
6	2,000					700		3,000
7	3,000	14,000	4,500		1,250	700		26,450
8	3,000					700		30,150
9	3,000	4,000	5,000	20,000	1,500	700	(63,400)	950
10	3,500					600		5,050
11	4,000	16,000	5000		1,750	600		32,400
12	4,000					600		37,000
13	2,000	10,000	4,000	20,000	1,500	600	(83,400)	(8,300)
14	1,500					500		(6,300)
15	1,000		2,000		500	500		(2,300)
16	500					400		(1,400)
17				17,000		300		15,900
18							(21,900)	(6,000)
19								
20								
21				8,000			(22,000)	(20,000)
	$33,000	$46,000	$22,000	$80,000	$7,500	$11,500	($220,000)	-0-

NOTE: This schedule is oversimplified for purposes of illustration.

SCHEDULE 1-9 PAYMENT SCHEDULE

Work item	*Estimated cost*	*Estimated profit*	*Bid price*
1	$ 28,000	$ 2,800	$ 30,800
2	22,000	2,200	24,200
3	12,000	1,200	13,200
4	23,000	2,300	25,300
5	10,000	1,000	11,000
6	20,000	2,000	22,000
7	15,000	1,500	16,500
8	25,000	2,500	27,500
9	21,000	2,100	23,100
10	24,000	2,400	26,400
	$200,000	$20,000	$220,000

SCHEDULE 1-10 PERCENTAGE COMPLETION OF WORK ITEMS

Work item	*Estimated total*	*Costs to date*	*Percentage completion by items*
1	$28,000	$21,000	75
2	22,000	5,500	25
3	12,000	1,200	10
4	23,000	1,150	5
5	10,000	750	7.5
	$95,000	$29,600	

SCHEDULE 1-11 COMPUTATION OF PROGRESS BILLINGS ON UNBALANCED BASIS

Work item	*Unbalanced payment schedule*	*Percentage completion by items*	*Progress billings*
1	$ 34,500	75	$25,875
2	27,100	25	6,775
3	14,800	10	1,480
4	28,600	5	1,430
5	10,000	7.5	750
6	20,000	—	—
7	15,000	—	—
8	25,000	—	—
9	21,000	—	—
10	24,000	—	—
TOTAL	$220,000		$36,310
Less	10% Retention		3,631
			$32,679

NOTE: Figures in *italics* reflect estimated cost given in Schedule 1-9 plus entire profits for job, prorated among these four items.

because overhead expenditures tend to be heavier in the earlier stages of the job. Also, by using costs incurred on the accrual basis (instead of on the cash basis as illustrated) it is possible to increase even further the percentages of completion in the earlier stages of the job, and so increase the early progress billings. The result, of course, is to reduce the amount of working capital required to carry the job in its early stages. A more complete illustration of an unbalanced forecast is shown in Schedule 1-12.

Projecting the progress curve. One of the often neglected prejob procedures is to project the progress of the job in terms of percentages of completion. Normally this is done either by the estimator or by the job engineer. However, before any substantial amount of financial or personnel planning can be done, it is necessary to have this information. Sometimes it happens that a progress curve is not prepared by the estimator, and then it is necessary for the accounting department to make its own curve. In some companies the accounting department will prepare a progress curve as a check on the work of the estimator or the engineer.

It is the usual procedure to base such curves on estimates of costs as they will be incurred, and not upon the estimated cash expenditures. Thus payrolls, including all related payroll taxes and insurance, are accrued at the end of the payroll week; purchases of materials, supplies, and small tools are accrued when they are scheduled for delivery on the job; equipment rentals (or prorated shares of equipment ownership expenses) are accrued at the end of each week based on estimated equipment requirements; other estimated costs are accrued at the end of the week in which they are expected to be incurred. A progress curve projected on this basis should approximate very closely the percentage of completion curve customarily prepared by the engineers.

Such a curve usually takes the form of an S. In other words it starts slowly, rises sharply as the job reaches full production, and then tapers off as the job approaches completion. Illustrated is a typical pattern for a job estimated to cost $200,000 and to last 16 weeks (Figure 1-8).

Retained percentages. It is customary for progress payments to be made either monthly or at certain agreed stages of completion. It is also customary for the owner-client to withhold an agreed percentage (usually from 5 to 15 percent) of each payment. To forecast financial requirements, it is necessary to forecast the progress billings and the dates of their collection. It is equally important to forecast the date of collecting retained percentages. Many subcontractors find this problem particularly acute because prime contractors generally require subcontractors to wait for their retentions until the prime contractor receives his at the conclusion of the entire job.

SCHEDULE 1-12 FORECAST OF CASH REQUIREMENTS—UNBALANCED PAYMENT SCHEDULE

Week	*Direct labor*	*Incorporated Materials*	*Equipment rentals*	*Subcontracts*	*Supplies etc.*	*Distributed overhead*	*Estimated progress billings*	*Cumulative estimated cash investment in job*
1	$ 500					$ 750		$ 1,250
2	750					1,250		3,250
3	1,000					1,000		5,250
4	1,250					800		7,300
5	2,000	$ 2,000	$ 1,500	$15,000	$1,000	800	($ 30,800)	(1,200)
6	2,000					700		1,500
7	3,000	14,000	4,500		1,250	700		24,950
8	3,000					700		28,650
9	3,000	4,000	5,000	20,000	1,500	700	(66,600)	(3,750)
10	3,500					600		350
11	4,000	16,000	5,000		1,750	600		27,700
12	4,000					600		32,300
13	2,000	10,000	4,000	20,000	1,500	600	(87,600)	(17,200)
14	1,500					500		(15,200)
15	1,000		2,000		500	500		(11,200)
16	500					400		(10,300)
17				17,000		300		7,000
18							(15,000)	(8,000)
19								
20								
21				8,000				
	$33,000	$46,000	$22,000	$80,000	$7,500	$11,500	($200,000)	
24	On acceptance by owner, retention release						($20,000)	($20,000)
	TOTAL CONTRACT PRICE, CASH RECEIPTS, AND PROFIT						($220,000)	($20,000)

NOTE: The same schedule can be completed for each job, then the calendar weeks added to arrive at total company cash flow.

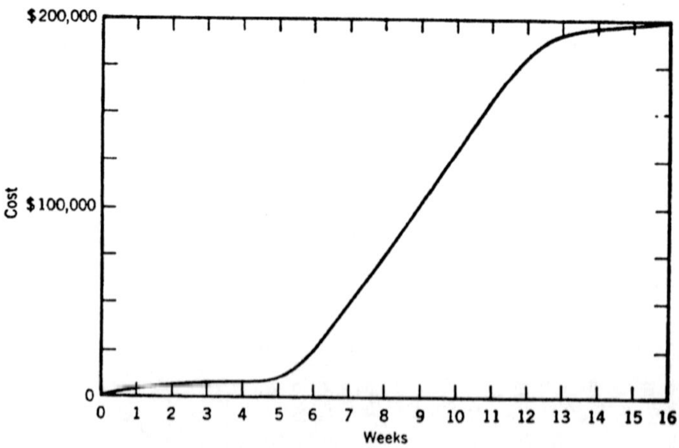

Figure 1-8 **Typical Progress Curve.**

Such a situation must be taken into account in the prejob planning. Often the problem can be avoided by including in the subcontract bid, as a separate bid item, interest on any portion of the subcontractor's retained percentage that remains unpaid after completion of the subcontractor's work.

If the prime contract contains language affecting the percentages to be held back from subcontractors as well as the prime contractor, those provisions must be taken into account in forecasting the cash to be received by the subcontractor on progress billings.

Job bank accounts

When employees are laid off, they must be paid at once. If it is not practical to prepare their final checks in the general office, it is necessary that an imprest account (revolving fund) be kept at the jobsite to handle these payoffs. Then when the regular payroll is prepared a regular paycheck is issued, but instead of being made payable to the employee, it is made payable to the revolving fund account. If all paychecks are written at the jobsite there is no need for the special imprest account.

What bank accounts will be needed and where they will be kept must be determined in advance. It is quite common for the job's general disbursement account to be kept in the bank from which the company borrows money. If this is done, the checks should be of a different size, color, and number series from the checks used for other accounts that are kept in the same bank, in order to prevent confusion.

Payroll accounts are usually kept in some bank near the jobsite because of problems that sometimes arise in cashing payroll checks drawn on out-of-town banks. Provision should be made to keep a large enough average balance in the payroll account to make the account profitable for the bank, so the service will be good. If the payroll is large and the local bank is small, it may be that the demand to cash checks on payday will tax the bank's normal cash reserve. Arrangements should therefore be made in advance to give the bank a reasonable estimate of the size of the payroll so that enough cash can be on hand. Additionally, when a very large project is in a remote location, the small town bank needs to know the projected number of employees so that it will be adequately staffed to handle the volume of checks.

The requirements of each bank regarding signature cards, resolutions, and authorizations should, of course, be determined in advance as far as possible, and complied with before the job is started. If such arrangements cannot be made, it will be necessary to open the accounts on temporary authorization and with temporary signature cards. Most banks will cooperate in these emergencies. If, for some good reason, the bank cannot accept emergency authorizations, the project manager or the job office manager may have to open an account in his own name for a short time until formalities can be complied with.

It is usually a good idea to require two signatures on general disbursement checks, but one signature (sometimes made by a mechanical check signer) usually gives a reasonable degree of safety on paychecks because there are so many controls and cross-checks on the payroll and the amount of each check will be relatively small. Also, the volume is often so great on a payroll that a second signature becomes so automatic that it is virtually meaningless. Sometimes, for example when checks are presigned, the second signature may even be dangerous.

There should be enough authorized signers on the job so that there will never be any reason for presigning checks. There should also be enough authorized signers in the general office so that, if need be, a check may be issued without any signature from the job. If the job is a joint venture, it may be desirable in some circumstances to have a full complement of authorized signers in the office of each venturer. On the other hand, it may be desirable to require one signer from each venturer's office before a check can be issued against the job's general disbursement account.

Planning the job office

If there is to be a job office, there are a number of decisions to be made, before the job starts, about the physical arrangements, supplies, and equipment. Many jobs make use of office trailers. However, the principles of arrangement are substantially the same.

If the job office is to be staffed with only one person, it must be remembered that he will receive all incoming telephone calls and meet all salesmen, new employees, union representatives, and visitors. Therefore, his desk should be located so that he can see people coming on the job and also see the work area. His desk should be behind a counter and the telephone should be outside the reach of persons on the outside of the counter. A public telephone in a protected booth should, whenever possible, be placed outside the office. The window from which the office employee can see the job should be built to serve also as a pay and time window and a place where matters can be discussed with foremen and workers. If the work area is close enough, he needs either a public address system or a bell system to call the superintendent or the engineer to the telephone, and inside the office he should be able to transfer incoming calls to the superintendent, the job engineer, or the office of the owner's representative.

He will need a typewriter (preferably with a 15-inch carriage), and if he is to keep time and check incoming invoices, he will need an eight- or nine-column adding machine, and preferably a small calculator as well. If he is to write paychecks, he will also need a check protector. If he is to have blank paychecks on the job, he needs at least a locked filing cabinet, and preferably a small safe. Illustrated are two typical arrangements which have proved practical under a variety of conditions (Figures 1-9 and 1-10).

If there is more than one man in the office, the same basic floor plan can be extended to form a T, an L, an H, or a U. Thus, in an office for two people, the wall on which the pay window is located could be extended as far as necessary toward the work area to form an I. The superintendent or project manager is centrally located, and the owner's representative can be given adequate privacy close to but not directly connected with the offices of the engineers and the project

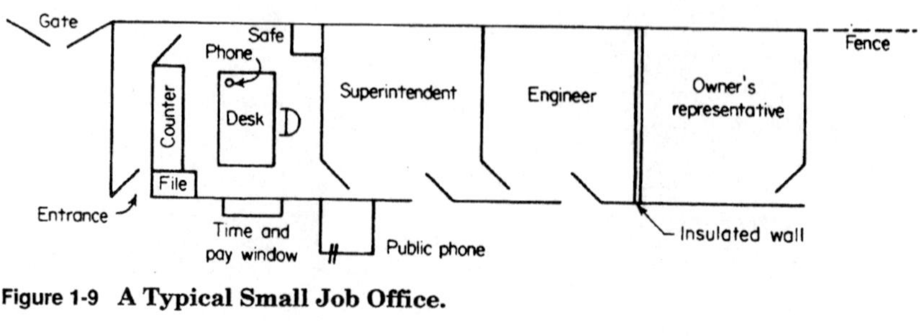

Figure 1-9 A Typical Small Job Office.

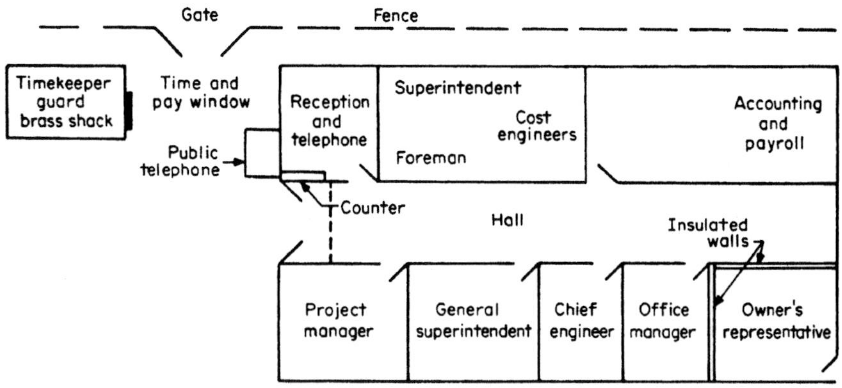

Figure 1-10 **A Typical Job Office for a Larger Job.**

manager. Sometimes it may be desirable to have an especially wide eave on the side toward the work area and, if the weather is likely to be excessively hot or rainy, a covered porch or walkway on that side may be in order.

If job offices are designed on a plain rectangular floor plan like the one in Figure 1-9, it is often possible to build them in 4-foot sections so they can be easily taken down and stored at the end of one job and quickly erected on the next one.

Job stationery. If accounting, payrolls, and accounts payable for a job are decentralized, it is necessary to have checks and stationery printed, and this may take longer than the time available before the job starts. A mailing address for the job may not always be immediately available, and until one is established there can be no stationery printed. However, initially it is easy to print some checks without a bank name and to add the bank name with check stickers which virtually all banks can supply. Usually the business-type checkbooks or the payroll checkbooks issued by the banks are inadequate for a job of the size which would require a separate job office. It is generally better to have check forms that fit the contractor's system and to put the stickers on them. The stationery can easily be overprinted in red, "Please address reply to _________ ," which will normally be satisfactory until any regular job stationery can be printed.

Some of the larger construction companies have a basic initial order of forms and supplies boxed and ready to ship to a job on a moment's notice. This practice can save a great deal of time in the job office. At the start of the job there are always so many interruptions and delays that it is wise to try to avoid the ones caused by shortages of supplies.

Planning a joint venture

When the job is to be done by a joint venture, that fact usually (though not always) is known at the time the job is bid. However, aside from preliminary agreements regarding bidding costs, there is usually little preparatory work done until the bids are in, and it appears that the contract will be awarded to the joint venture.

Some states require that a separate contractor's license be obtained before the joint venture may submit a valid bid. If so, the license must be secured before the bid date. If a license is not required until work starts, one of the first requirements after the award of the contract would be to obtain any required licenses. In addition, there is a joint venture agreement to be prepared and signed; there are powers of attorney in favor of the project manager (and perhaps certain of the other key job personnel) to be prepared; there are signature cards and bank resolutions to be obtained; and the joint venture partners are to be notified by the sponsoring venturer, or by the management committee, how much money they will be asked to furnish and when it will be needed.

It is desirable that there be a complete understanding of what each venturer is to do. The most practical solution to this question is to agree that one of the members is to act as sponsor, that is, managing partner, and to let the sponsor make the decisions after discussions at the policy level.

Under the sponsorship system, complete control would be vested in the sponsor. The nonsponsoring members would normally be entitled to regular (usually monthly) reports on the job finances and job progress. They would also be entitled to share with the sponsoring venturer any decisions on liquidating job-owned equipment and to participate pro rata in job distributions.

Control by a management committee made up of a representative of each member of the venture is a common device to spread the responsibility for management decisions. It is not so popular as the sponsorship method, however, because it is generally less efficient.

There are several accounting matters for a joint venture that should be determined in advance and either written into the joint venture agreement or in supplemental exchanges of correspondence.

First, the accounting method (cash or accrual, and if accrual, whether completed-contract or percentage-of-completion) and fiscal year need to be decided upon and the decision recorded. The point should be considered and determined at the policy level, and not left for someone in the sponsor's office who might bind the joint venturers to a decision without knowing its overall effect.

Policies on depreciation and accounting for ultimate disposal of joint venture equipment should be agreed upon in advance and preferably made a part of the joint venture agreement. In making these decisions it is important to consider the effect of local sales and use tax laws, as well as income taxes.

There seems to be very little doubt that for federal income tax computations, the partnership rules will govern joint ventures. Assets contributed to a joint venture by one of the venturers carry over their "basis," that is, cost for tax purposes, from the books of the venturer to the joint venture. If assets are put into the venture, the joint venture agreement should state clearly how the depreciation and gains or losses from their sale are to be distributed among the venturers. Under the Internal Revenue Code the preparation of a joint venture agreement is a much more important and technical job than it has ever been in the past, and it should not be undertaken without the help of expert tax counsel.

If the joint venture is to have a fully decentralized accounting system, that fact should be covered in the joint venture agreement. So, also, should the extent to which the venturer may be called upon to reimburse traveling expenses of executive personnel of the joint venture partners.

If one of the venturers rents equipment to the job or performs special services for the job (purchasing, payroll accounting, or field auditing), the basis for payment by the joint venture should be covered in the joint venture agreement or by a prejob memorandum supplementing the agreement.

It is not uncommon to find personnel from more than one of the joint venture members on the payroll of the joint venture. This raises a whole series of questions relating to personnel policies, including pay scale, bonus policy, vacations, sick leave, mileage allowances, participation in pension plans, group insurance, and profit sharing trusts. The most satisfactory adjustment of these differences is to let each member of the venture bear the cost of bonuses which exceed those allowed by the other members, but otherwise to pass the costs on to the job. Each member should stand the cost of fringe benefits to his own people, and if vacation or sick leave allowances of one member exceed those agreed upon for the venture, then that member should stand the cost of the excess. In any event, there should be prejob agreement between the venturers on all such personnel policies insofar as they affect the venture.

Permits, licenses, and permissions

One important prejob procedure that is commonly neglected is the securing of required licenses and permits. When a construction company goes into another state to bid or perform work it must:

1. Qualify to do business there, if it is a corporation.
2. Secure any required state and local contractor's licenses.
3. Register as an employer with the state department of employment, and sometimes with the board or commission which deals with employee safety and industrial accidents.
4. Secure the necessary state and local licenses for its automobiles, trucks, and other equipment.
5. Determine whether it is required to hold a state or local sales tax or gross receipts tax permit, to withhold state income taxes from its employees, or to file a property declaration for property tax purposes; and comply with any other applicable laws and regulations.

Even in its own state a construction company must frequently obtain local contractor's licenses and special vehicle licenses as well as local sales tax permits. Sales and use taxes and transportation taxes are growing in importance, both as a source of revenue to governmental bodies and as a tax cost and filing requirement to be reckoned with. To a contractor who is constantly going into new communities to do business, they are important items on his checklist of things to consider in bidding and in carrying out his contracts.

The federal government and many state and local governments prescribe minimum rates of pay and require that special permission be obtained for any deductions from employees' paychecks other than required tax deductions, union dues, group insurance, garnishments, and so on. If pay rates are mentioned in the contract they should be checked before the job starts, to be sure that the prescribed minimum wages do not upset existing pay scales and area differentials. Also, if deductions for meals, lodging, commissary purchases, charitable contributions, or other similar purposes are to be made, the necessity for permission should be investigated.

Another important prejob procedure is a careful consideration of local union requirements with particular reference to payroll and hiring procedures and to contributions to union health and welfare funds.

Since almost all of these requirements are local, detailed instructions are not possible. The best procedure is to obtain local advice wherever a jobsite is to be located.

Planning as a management tool—job planning

There is one aspect of prejob planning that deserves special treatment. Modern construction management involves a sophisticated integration of design and construction methods. Any proper prejob plans for accounting and financial management must include plans for providing the necessary budgetary controls, cost controls, quality controls, and time controls needed by top management personnel to so manage the jobs that they can complete on time and at a profit.

However, once having recognized that principle, it is equally important to recognize that it is a costly mistake to provide for records that management cannot (or will not) use. For example, critical path method (CPM) or program evaluation and review techniques (PERT) may produce excellent results in jobs sufficiently complicated to justify them and where the savings through proper management can justify their cost. On other jobs they may cost more than they are worth. The function of prejob planning is to select the controls, and the accounting methods needed to achieve them, with due consideration for the nature of the job and the needs of the people who will manage it.

The starting point for job planning is the bid. By the time it is complete, the estimator will have built the entire job in his mind. In so doing he will have analyzed the job by operations and will have determined when each operation is to start and finish in relation to the others. The estimator will also have determined how much each operation is to require in terms of worker-hours, equipment hours, and working days. Each of these factors is expressed in both time and money, and the relationship and timing between the various work items will have been determined.

When the job is awarded, the person who is to manage the job will go over each aspect of the bid, refining it where needed. The work items to be subcontracted will have been analyzed and examined, and, it is hoped, all gaps and overlaps in the work to be subcontracted will have been eliminated. In its completed form the bid is, for all practical purposes, a budget for the job. At this time the cost coding will be established, and each work item and the cost breakdowns within the work items will be assigned the account numbers that are to be used in the cost accounting on the job. If computers are to be used, the programming will be worked out for the individual job at this time.

Just where in the overall planning process the accounting and financial management people start to participate depends on the individual company. In many of the more modern operations they are in it from the start. In others they do not enter the picture until the bid is complete and the construction and estimating departments present them with the completed job estimate and cost coding.

At this point, the actual letting of subcontracts begins, and the financial personnel start to fit the cash requirements of the job into the cash requirements of the company as a whole. The same thing happens with the personnel requirements, the equipment requirements, the accounting requirements, and, if computers are to be used, the computer time requirements.

Planning as a management tool—companywide planning

It cannot be too strongly stressed that fitting all these job requirements into the general requirements of the company is a planning task of major importance. How well it is done is a crucial test of the ability of top personnel to manage the company instead of being managed by it. The detailed accounting procedures needed to provide the required operational information are discussed elsewhere in this book. The job of using the information they disclose to further the company's major objectives varies constantly and cannot readily be reduced to a formula.

Before job planning can be fitted into the overall plan of the company, however, the company must have an overall plan. If there is one place where the construction industry's management techniques need to be improved, it is in the area of formulating (and formalizing) the general planning for company operations as a whole. The larger the company, the better the chances are that it has a formalized plan. One reason for this slant in favor of the larger companies is that having a single plan toward which everyone in management is working is one of the things which makes a company grow. Another reason is that the larger a company becomes, the greater is the need to formalize its planning so that everyone who is involved in decision making will know what the plan is and make all decisions accordingly.

The first three steps are relatively simple. First, determine the volume of work the company can handle without overextending its financial capacity, its personnel, and its equipment. Second, determine whether the company wants to grow beyond the level of its present capacity and, if so, how much. Third, make an informed estimate of the facilities the company has at its disposal to help it reach whatever objective is set and how those facilities can be used to the best advantage.

The basic tools for making these determinations are budgeting; controls on costs, billings, and overhead; and standardized procedures for comparing actual operating results with the budget. Then by determining the reasons for variations, it is possible to sharpen the whole estimating, managing, and budgeting procedure to achieve maximum effectiveness. However, these procedures alone are not enough to provide all that is needed to plan the future course of the business as a whole. To get a viable overall plan, it is necessary to estimate future volume. Especially since the Tax Reform Act of 1986, overall planning must include federal income tax and related accounting considerations because these will affect the future growth and development of the company.

Sources of business need to be explored, and the best methods of getting each type of business must be considered. For example, in an expanding economy, it may be possible to achieve the predetermined goal merely by expanding with the economy. In a shrinking economy, it may be necessary to plan for what is necessary to get a larger share of the available business.

It may prove desirable to so build up community contacts that a larger proportion of negotiated work can be obtained. Or it may prove desirable to reduce the amount of private work and concentrate on public works jobs. Or the reverse may be true and negotiated private work may be the desired goal.

If the objective is more public works jobs, it may be necessary to devise ways to reduce overhead and improve construction methods so lower bids may be submitted without sacrificing the necessary margin of profit. If negotiated private work is the goal it may be necessary to devise a special cost-plus-fee arrangement with a guaranteed maximum and a participation in savings. The point is that the potential market must be analyzed and procedures devised to capture whatever share of it is necessary to determine future volume goals realistically and then to devise practical ways of reaching those goals. In so doing it must be borne in mind that the company's competitors are probably doing the same thing.

One word of warning is in order at this point. It is easy to set goals for the total volume of construction work to be done. Also it is not too difficult to achieve those goals. The true objective has to be stated in terms of net profit. The objective is not to achieve a predetermined volume of gross business, but to achieve a predetermined profit after taxes.

After the desired volume of business has been determined and the most desirable "mix" (i.e., bid versus negotiated; public versus private; large versus small; etc.) has been estimated, it is then necessary to explore the credit and bonding capacity necessary to handle the estimated volume made up of the types of jobs included in the mix. A comparable estimate must be made regarding equipment, personnel, and the geographical area in which jobs will be sought.

Finally, all the preliminary thinking will tend to boil down to, say, three or four alternative plans. The next step is to line these plans up side by side and compare them with the business as it is presently operating. This can be done by a simple chart which would shape up somewhat along the lines illustrated in Schedule 1-13.

The construction industry being what it is, the combined timing of all jobs and operations within jobs usually cannot be estimated in more than the most general terms before the jobs are awarded. The reason is obvious. There is no predictable flow of jobs coming up for bid or negotiation. Nevertheless, when there is a choice between jobs to bid, it helps to have an estimate of the times when funds, equip-

SCHEDULE 1-13 A SIMPLE AND EFFECTIVE MEANS OF COMPARING PROJECTED RESULTS OF ALTERNATIVE PLANS

	Last year	*Expected results next year under*			*Year after next under Plan selected*
		Plan 1	*Plan 2*	*Plan 3*	
Gross billings	$	$	$	$	$
Job costs					
Job profits	$				
Home office overhead	$				
Expected profit before taxes	$				
Federal and state taxes	$				
Expected profit after taxes	$				
Expected net increase in retained earnings	$				

ment, and personnel will be available to do a particular type of work.

To do this, the progress charts of the jobs presently in progress can be combined to see when each of various phases of those jobs will be complete so that, by having another job ready to go, maximum use may be made of both financial and construction capabilities. Thus, if for example, by combining the expected progress charts on the jobs in progress, it can be determined that a substantial amount of excavating equipment and personnel will be available by, say, August 1, the people who decide what jobs are to be bid will be looking, back in March or April, for jobs that will utilize that type of equipment and personnel.

Substantially the same thinking must go into planning for the use of credit and bonding capacity, but here another factor enters into the planning. Where there is substantial borrowing, the lenders frequently insist on an annual "clean-up" of borrowing. In other words, some banks like to see all their loans paid off at least once each year. If the company is using bank financing, this factor may well have to be taken into account in the financial planning.

To summarize, the final objective of all construction company planning, whether at the job level or the company level, is to keep the company's money, its equipment, and its personnel profitably employed 100 percent of the time. One of the most effective approaches to that problem is to state all the factors involved in terms of units of personnel and personnel productivity. It has been said elsewhere in this book (and it will bear repeating) that the only thing a contractor has to sell is the know-how of his organization and his ability to finance his work. That fact is at the base of all successful planning.

Chapter

2

Procurement and Subcontracting*

Purchasing Policies and Procedures

Contractors who have no separate job offices must necessarily do their purchasing from the home office. In fact, in very small firms the purchasing functions are centralized in the contractor himself, except for occasional small items bought on the job to fill an immediate need. As jobs spread out over more territory, and as they become larger, there is increasing pressure for more jobsite buying.

The advantages of jobsite buying over centralized buying are that:

1. Local merchants will tend to give better service on emergency requirements if major purchases are made locally.
2. Relationships in local communities where jobs are located are improved.

The advantages of centralized buying over jobsite buying are that:

1. Local buying tends to be in smaller lots, and so the price advantages of volume buying are lost.
2. When shortages develop, suppliers give preference to long-term, large-volume customers.
3. Local jobsite buyers have no way of knowing when surpluses of certain materials and supplies develop at other jobs, and so one job may be buying at market price the same items that another job is liquidating at a loss.

---

*Computers are rapidly taking over much of the record keeping involved in procurement. However, hand-kept forms are used for illustration in this chapter since they point up the factual data needed when the system is computerized (see Chapter 12).

4. Better financial control can be achieved there by helping to eliminate the problem of kickbacks and related irregularities which are prevalent in the construction industry.

Purchasing objectives

There are five major factors to be considered in setting purchasing policies for a construction company:

1. Price
2. Quality assurance
3. Delivery schedule
4. Vendor responsibility
5. Quantity control

The relative importance of these items is not necessarily indicated by the order in which they are named.

Price, of course, is of prime importance, but when lower purchase prices are offset by losses due to delays, additional expediting costs, and questionable quality, they can usually result in a higher final cost. It is common to hear purchasing agents talk about making a "profit on buying." That is a fallacy. A contractor makes a profit only on completed work, and it is only when buying contributes to the final job profit that it contributes to earnings. In other words, it is easy to make a paper profit on buying and then lose it trying to complete the job under adverse conditions that ill-considered "price alone" buying creates.

Quality is controlled by the specifications, and the client-owner can expect only what he has specified and is willing to pay for. Sometimes specifications will call for inferior materials but will set standards which, if met, would require materials of top grade. The estimator must watch for such inconsistencies.

In many respects, however, quality does vary, and when a vendor delivers substandard materials he has cost the contractor time and money through handling costs and delays, even though he may ultimately deliver goods that meet specifications. The cost is even greater when inferior or defective materials are installed and have to be removed. Uniformity and dependability are the most important things to think about when evaluating the quality of a vendor's products.

Delivery schedules can sometimes be the controlling factor in determining what to buy and where to buy it. Because timing is vital on any construction job, delays and errors in deliveries can often be expensive out of all proportion to the value of the items purchased. Delivery ahead of schedule, on the other hand, can result in loss through spoilage or theft (Chapter 4) or disrupt cash forecasting by moving payment and discount dates ahead. Therefore, in evaluating a vendor's proposal, the vendor's reputation for delivering exactly what is ordered on the exact day for which delivery is scheduled is a matter of prime importance.

Vendor responsibility is a term that has a number of meanings, and all of them are important. It is directly related to the supplier's ability to provide goods of uniform quality, make accurate deliveries, keep precise delivery schedules, and still sell at a price which meets comparable competition. Therefore, the vendor's financial stability can be an important factor to consider when selecting a supplier. There are a number of credit sources the contractor can use to make this evaluation—Dun & Bradstreet, bankers, trade associations, and other contractors. Many contractors keep a list of vendors whose responsibility and ability to perform are a matter of record. It also means the willingness of the supplier to make fair and reasonable adjustments when errors do occur, and the extent to which a supplier will back up its salesmen in their efforts to help their contractor-customers solve difficult supply problems. In summary, it means the extent to which a vendor may be relied upon to live up to the spirit as well as to the letter of its obligations, and its willingness to share its customer's problems. These factors are intangibles, but their effect over a period of time can be very real.

Quantity control is an internal problem, but it is still an important one for anyone responsible for the purchases of a construction company. Because money invested in surplus materials and supplies is idle money, it is vital that the purchasing department of any construction firm know what is on hand, and when and if it is going to be used in the foreseeable future. Often the temptation is very great to make large bargain purchases of standard materials or equipment. Sometimes such purchases are good business. Certainly they are when use can be foreseen in the near future. The test is this: Is the amount saved on this purchase as good a return on the money tied up as might be earned elsewhere in the business?

The problem of pickup purchases

On every construction job there will arise unforeseen needs for small items that must be purchased locally. When job needs are properly anticipated, these items will be minor. Still, failure to anticipate job needs cannot ordinarily justify delaying the job until routine purchasing procedures (see "The mechanics of buying," below) can be followed.

There are several alternatives. Purchases can be made from petty cash and supported by paid bills. A variation of this procedure is to allow the individual to make the purchases out of his own pocket and to receive reimbursement on the basis of paid bills. Another procedure is to issue blanket purchase orders to certain local suppliers covering requirements for a month. These devices are all subject to the criticism that there is no control before the purchase is made, and there is no way to compare the invoice or cash register slip with items actually authorized for purchase.

One of the more satisfactory answers to the problem of "pickup purchases" is the field purchase order (Figure 2-1). This is a simple,

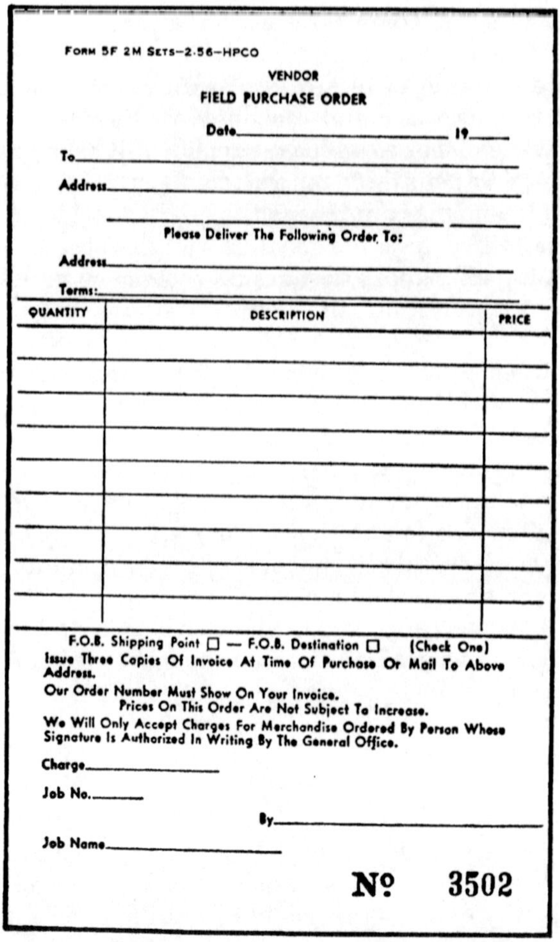

FORM 5F 2M SETS–2.56–HPCO

VENDOR

FIELD PURCHASE ORDER

Date__________ 19____

To__________

Address__________

Please Deliver The Following Order To:

Address__________

Terms:__________

QUANTITY	DESCRIPTION	PRICE

F.O.B. Shipping Point ☐ — F.O.B. Destination ☐ (Check One)

Issue Three Copies Of Invoice At Time Of Purchase Or Mail To Above Address.

Our Order Number Must Show On Your Invoice.

Prices On This Order Are Not Subject To Increase.

We Will Only Accept Charges For Merchandise Ordered By Person Whose Signature Is Authorized In Writing By The General Office.

Charge__________

Job No.______

By__________

Job Name__________

№ 3502

Figure 2-1 **Field Purchase Order in Quadruplicate.** Copies for vendor, jobsite office, central purchasing, and accounting.

handwritten form small enough to be carried in the superintendent's pocket or in the glove compartment of his pickup truck. When the need arises to make a pickup purchase the superintendent can write out the order in a few seconds. The record is kept clear and all purchases are still authorized in advance. At least once each week copies of the field purchase orders are sent to the general office so that the person in charge of purchasing can keep control of the miscellaneous jobsite purchases.

Purchasing and expediting

There are two major steps in the procurement function—purchasing and expediting. Purchasing, as previously discussed, includes finding required materials, supplies, and equipment, and buying them on the most favorable terms that can be arranged. Expediting is the job of following up the purchase to see that the items ordered (and not substitutes) are delivered, and that the delivery is made at the place and time required by the contractor's operations.

Both buyers and expediters need a specialist's knowledge of construction materials, supplies, and equipment. Both need to be able to read plans well enough to take off materials lists. Both need a sufficiently detailed knowledge of construction to know, at least in a general way, where and how the things they buy will be used on the job for which they are buying. Without this knowledge they are not in a position to exercise the judgment necessary to a proper performance of their jobs. Because of these requirements, there has been a tendency in construction companies to place the procurement functions in the category of engineering or operations, and certainly such a classification makes good sense.

On the other hand, there is a tendency on the part of buyers and expediters who operate as a part of the engineering or operations departments to overlook the important accounting aspects of their jobs, and this has led many construction controllers to seek jurisdiction over the procurement functions. For example, it is sometimes possible to make all major purchases of materials and supplies at the start of a job. If this is done and the personnel on the job place orders for still more supplies and materials, it means that one or more of the following is true:

1. The original takeoff was wrong.
2. Materials and supplies are being lost or stolen, either on the way to the job or after they get there.
3. There may be some items which are being paid for twice.

These are all matters that merit careful attention, yet a purchasing department completely controlled by engineering or operations personnel is often inclined to disregard them in the interests of job progress.

Another example of the tendency of operating personnel to disregard the accounting aspects of procurement functions is to be found in the use of the purchase orders (see "The mechanics of buying," below) as a part of the vouchers to support cash disbursements. Prices on purchase orders are used by the accounting department primarily to check the prices on vendors' invoices. Pricing the items purchased on the purchase order is a source of continuous friction between the accounting department and buyers who work primarily for the engineering or construction departments, for reasons that are easy to understand. It takes additional time and effort on the part of the purchasing agent to secure the prices, and it raises arguments between the purchasing agent and the supplier when discrepancies (which would never be found at all unless prices were listed on the purchase orders) are found by the accounting department and referred back to the signer of the purchase order.

However, in spite of vigorous contentions to the contrary by many buyers, pricing the purchase orders is definitely important and practical. It simply means that when negotiating the purchase (usually by telephone) the buyer will have to insist on price quotations even over objections on the part of vendors and their salesmen. Once it has been made clear that prices are required, they will invariably be supplied without question. In fact, insistence by management on the pricing of purchase orders has been known to discourage questionable side deals (kickbacks) between buyers and salesmen.

Actually, provided the engineering and operations departments are as accounting minded as they should be, and the accounting department as operations minded as it should be, it makes little difference who controls the purchasing and expediting.

In very small contracting firms there is no question of control. Almost invariably it is the owner who supervises purchasing. This does not mean, however, that the foregoing discussion is not significant to the small contractor. No one should pay bills without checking the prices and extensions on the invoices. Unless the contractor can remember the exact price of everything he buys, he has nothing against which to check prices. It takes only seconds to write down the price quoted when a purchase is being negotiated, and then there can be no question when it is time to pay the bills. In areas where the building material business is highly competitive, certain unscrupulous dealers have been known to quote prices at cost and make their profit through "errors" in pricing on the invoices.

Items to be bought

Materials and supplies are usually bought on the basis of a materials estimate for use on a specific job, regardless of the size of the contractor's business. The estimate may be a takeoff from the plans or, as often happens on very small jobs, it may be based on the estimator's best guess as to what is needed (see "The materials budget," Chapter 4).

If, as is done by plumbing, electrical, or similar specialty subcontractors or repairmen, an inventory is carried and purchases are made for inventory, it is usually good practice to keep "bin cards" (Figures 4-4 and 4-5). When withdrawals from a particular rack or bin bring the amount on hand below a predetermined quantity, the item is placed on the "want list" to be ordered the next time an order is placed.

Specialty purchases made to meet emergency jobsite needs have already been discussed in this chapter (see "The problem of pickup purchases").

The title of the person responsible for purchasing is not nearly so important as seeing that the responsibility for purchasing is clearly and definitely allocated.

The mechanics of buying

The buying process starts when someone with authority to spend the company's money decides that a purchase is necessary. It may be the owner of the business who goes to the building material dealer's yard and signs an order for whatever is needed. More often, even in very small concerns, the first step is the preparation of a "want list" or a materials estimate covering the materials and supplies needed for work in process or jobs about to start. When operations are a little larger, these want lists and rough materials estimates are formalized into a purchase requisition. Whether it be in the mind of the contractor himself or on a formal requisition, a list of the needed items must be accumulated somewhere.

The items must then be ordered from the appropriate supplier. It is common, among smaller contractors, for these orders to be placed by telephone, or over the counter of the lumber yard, the hardware store, or the materials dealer. So long as operations remain small enough, this informal procedure is satisfactory. There are few enough items on few enough jobs so that the details of each transaction can be remembered by the contractor, who in addition to doing the buying, checks the invoices and pays the bills.

A purchase order, under the Uniform Commercial Code, is a contract. Accordingly, there are both advantages to be gained and risks to be avoided when purchase orders are printed. For that reason, when purchase orders are printed, an attorney familiar with the contractor's particular branch of the construction industry should be consulted.

For example, a company doing work under United States government contracts will sometimes want to incorporate in its purchase orders the specific language of the Code of Federal Regulations dealing with such subjects as the "equal opportunity clause" (see Executive Order 11,246 of September 24, 1965), the "nonsegregated facilities certificate," the "employer information report" (EEO-1 Standard form 100), and the guidelines of the Title 41, Code of Federal Regulations dealing with the "affirmative action compliance" program.

Another example would be company rules regarding back orders, delivery dates, shipping and billing instructions, material guarantees, and similar provisions.

Purchase orders (Figures 2-1 and 2-2) are needed as soon as the operations become large enough so that the person who checks the incoming invoices and pays the bills needs some way to check on what was bought, who ordered it, and the price agreed upon.

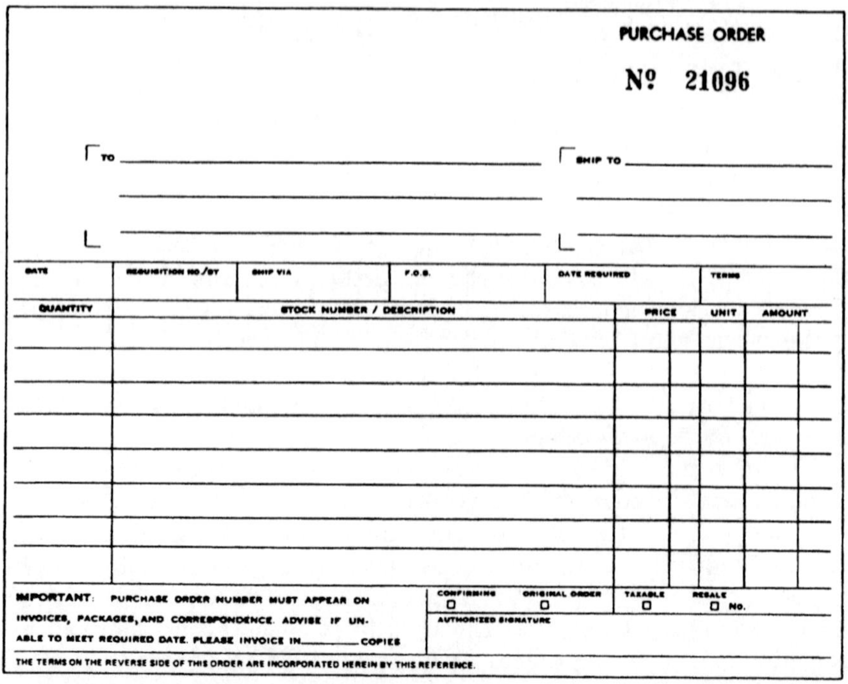

Figure 2-2 **Purchase Order.** As indicated elsewhere, purchase orders can, and probably should, contain the terms of purchase to take full advantage of the Uniform Commercial Code. Where back ordering can create problems such as those referred to in Chapter 4, a provision prohibiting back ordering would be desirable.

The actual buying starts with getting in touch with suppliers, securing quotations, and arranging terms for the purchase of requested items. When the terms are arranged, purchase orders are written and sent to the appropriate suppliers.

One copy of the purchase order is sent to the person who prepares the vouchers for payment. This copy will be compared with the invoice when it arrives, to check authorization, quantities, items, and prices. Another copy goes to the place where the items will be received, and serves as a checklist on incoming shipments. Another copy goes to a numerical file, and the last copy is filed in a tickler file under the date when delivery is due. This last copy is ultimately filed alphabetically by vendor's name after the goods have been delivered. Until delivery, however, it is kept in the tickler file for follow-up. If there are indications that shipment may be delayed, an expediter, who may or may not be the buyer himself, works on the order to keep deliveries on schedule.

Purchase orders are designed principally to do four jobs:

1. Establish the responsibility for buying the things that are needed and only those things.
2. Tell the supplier in writing exactly what has been ordered; how, when, and where it is to be shipped and billed; how much the company expects to pay for it; and any other information necessary to fill the order promptly and accurately.
3. Provide an internal check on purchasing, on receiving, and on the payment of money for items purchased.
4. Under the Uniform Commercial Code, fix certain of the terms of the contract of purchase and sale.

Purchase orders should be prenumbered, and the numbers strictly controlled. If these two checks are not maintained, a good deal of the effectiveness of the purchase order is lost. The same is true of the authority to sign purchase orders. Unless limited to a comparatively small number of individuals, responsibility for purchasing becomes scattered, and purchasing becomes loose and wasteful. In some ways the responsibility which goes with the authority to sign purchase orders is even greater than that which goes with the authority to sign checks. Delegation of such authority deserves careful and serious consideration.

The mechanics of buying machinery and equipment are much the same as those of buying materials and supplies. However, the management decisions that lead to such purchases are somewhat broader in scope. The normal purchase of materials and supplies is made to meet the needs of a specific job or to maintain a complete and well-balanced inventory. Equipment, on the other hand, may be bought for a particular job, but at the end of that job either it must be sold or there must be another job to keep it busy. For tax purposes, unless it is sold at the end of the job, only a pro rata share of its cost may be charged to job costs. Often equipment sitting idle in the contractor's yard is too expensive to keep and too valuable to sell at the prices that can be realized for it. Therefore, only those in the organization who understand the problems created by the ownership of equipment should have authority to sign orders for its purchase.

Expediting

Almost everything a construction company purchases must fit somewhere into a production schedule. Delivery too early creates problems (and generates costs) of storage and protection. Delivery too late means costly delays in completion. It is the job of the expediter to see that the things bought arrive on the job, as nearly as possible, at the exact time they are needed.

In larger concerns there are people who do nothing but expedite. In smaller concerns expediting is often another job for the buyer. Often it is one of the many tasks assigned to the job office manager.

Regardless of who does the expediter's job, it is important that he know what he is trying to accomplish, and that he is not so loaded down with other more urgent duties that he has no time to expedite purchases.

Too often it is the erroneous belief of contractors that expediting and timekeeping jobs are to be assigned to people who lack the ability to make good anywhere else in the organization. When that idea prevails, expediting suffers and the job costs go up.

To do his job properly, the expediter must know exactly what has been purchased, when it is due to arrive on the job, and what will happen if it is not there on time. He needs to have day-to-day knowledge of progress on all jobs, and of shifts in production schedules. If materials are to be fabricated he must be able to go into the supplier's factory and identify the items on their way through production. He must be enough of a diplomat to maintain good supplier relations, enough of a salesman to keep the supplier sold on giving as much priority as necessary to the company's orders, and enough of a slave driver to apply as much pressure as needed to get suppliers and carriers to meet the company's production schedules. It is not essential that he be an engineer, but the training helps.

In order to anticipate job requirements properly, the expediter needs to study the production schedules, materials takeoffs, equipment estimates, and purchase orders. He must catch and correct errors in scheduling the delivery of these items. He must maintain, or have available, a numerical register of purchase orders and a tickler file of unfilled purchase orders, arranged in the order of the "lead time" required to get the items delivered on time. These he must check systematically until he receives notice (usually a receiving report or similar memorandum) from the job that an order has been filled. He then moves the purchase order out of the tickler file into an alphabetical file, by vendors' names, of filled purchase orders.

There will be many purchase orders and purchase contracts that require no expediting effort, but those should still go through the expediter's hands so that his picture of the flow of materials will be complete.

Treatment of back orders

No discussion of purchasing or expediting would be complete without some comment on back orders. Often when a supplier is unable to supply one or two items on a purchase order, the remainder of the order will be filled and the missing items will be left open to be filled later. If the buyer has been careful to find out if the items ordered are available, back ordering will be held to a minimum, but even if there are only a few back orders it may often be necessary to buy the missing items elsewhere or to delay the job. Accordingly, as indicated in Chapter 4, it is better to prohibit all back orders and write a new purchase order for the items to come than to permit any back orders at all. Too often they are forgotten and the materials or supplies arrive after the need has long passed. If that happens they normally must be paid for, just as if they had been shipped on time. Even if they can be returned there are handling costs and shipping charges.

Subcontracting Policies and Procedures

There are few, if any, construction jobs on which some part of the work is not done by a subcontractor. Sometimes, on larger projects, the larger subcontractors will subcontract portions of their own work. This practice results in several "tiers" of subcontractors. The owner holds the prime contractor responsible for the entire job, even though some parts of it may be done by firms of which neither owner nor prime contractor has ever heard. In the same way, the prime contractor holds the first-tier subcontractors responsible for their portion of the work. Where some of the first-tier subcontractors have subcontracted a portion of their job to the second-tier subcontractors, they look to the second-tier subcontractors for a complete job, and so on down the line. Sometimes owners insist on approving the subcontractors used on their jobs, but it is unusual to find this degree of control extending past the first tier.

There are both practical and legal reasons why an owner rarely deals directly with a subcontractor, or a prime contractor with a second- or third-tier subcontractor. Legally, a subcontractor in any tier owes his duties under the subcontract only to the firm with which he contracted. From a practical viewpoint, dealings between the owner and a subcontractor would tend to relieve the prime contractor of responsibility for the subject matter of such dealings, and the same would be true of dealings between the prime contractor and lower-tier subcontractors. Therefore, the chain of responsibility and authority in subcontract matters is observed with great care.

Reasons for subcontracting

There are a number of reasons why so many prime contractors subcontract portions of their work. First, in the plumbing, electrical, and painting trades, the subcontractors maintain an organization of skilled specialists, making it possible to do certain parts of the work better and more economically than might be done by someone less well specialized. In some regions, labor union contracts are so written that, except in emergencies, only specialty contractors can employ specialists in those trades. Also, relatively few general contractors have enough specialty work to keep a crew busy on a continuous basis.

Second, many specialists require equipment peculiar to the trade, and it is often more economical to subcontract the work than to buy or rent the equipment. Thus, many building contractors regularly subcontract their excavation and site preparation work rather than invest in very expensive earth-moving equipment.

Third, since contracting is a risky business and can require large amounts of capital, subcontracting is often used as a device to spread both the risk and the financial burdens of a large job.

From the subcontractor's point of view there are also advantages. It takes less capital to build up and equip a small crew of specialists than a large, diversified organization. The subcontractor receives many jobs that would be beyond his financial capacity were he to attempt to handle them as a prime contractor. Often, too, by associating himself closely with a few active general contractors he can keep his flow of work regular, even when some of the general contractors find themselves in a work slump.

In summary, then, it is advantageous to the prime contractor to subcontract the specialized parts of his work because the subcontractors, as specialists, can perform better, faster, and as cheaply as the contractor could do it himself. Besides, from the prime contractor's point of view, the risk and the financial burden are spread. From the subcontractor's point of view, it generally requires a smaller investment and a smaller full-time organization to be a subcontractor.

Disadvantages of subcontracting

There are, of course, disadvantages on both sides, too. As soon as a prime contractor subcontracts a portion of his work, he sacrifices a certain amount of control over that part of the project. If things start to go wrong it takes longer to correct them, and the risks of delay and of disorganizing other parts of the job are increased. While some of the financial risk of the prime contract is passed to the subcontractors, the potential loss is greater and harder to guard against if the subcontractor proves to be financially unstable. Such a subcontractor can leave the prime contractor with a liability for unpaid bills relating to work for which the subcontractor has already been paid, and subcontractor's performance and payment bonds can only partially guard against this risk.

From the subcontractor's viewpoint, the disadvantages are numerous and potentially serious. He must be able to recognize and protect himself against the all too common practice by certain owners and prime contractors of specifying inferior materials while demanding top quality results. This type of risk faces prime contractors too, and sometimes the prime contractor tries to avoid it by passing portions of the work on to subcontractors.

The subcontractor must be able to move on and off jobs without loss of time or money, because he has to schedule his own operations to coordinate with those of the prime contractor and other subcontractors. He must be constantly alert to avoid situations in which an occasional unscrupulous prime contractor seeks to finance his entire job at the expense of his subcontractors. He must be prepared to combat attempts by the prime contractor to hedge against loss by holding the subcontractor's progress payments to a minimum. He must also exercise extreme care to see that the limits of his work are clearly defined, both in the initial subcontract and in connection with subsequent change orders. Otherwise he may find himself doing work he had not contemplated and for which he had made no allowance in his bid. Finally, if the prime contractor assumes a risk when a subcontractor fails financially, the subcontractor is also assuming a risk in connection with financial failure of the prime contractor. To hedge against such risks, the subcontractor must know (and watch) his lien rights on private work, and his statutory rights under the various bond acts on public works. The wide prevalence of subcontracting indicates that most subcontractors are fully aware of these hazards and have learned how to deal with them. Since time is of the essence on a construction project, the subcontractor generally has considerable leverage in negotiations with the prime contractor.

Subcontract coordination

Coordinating the work of the contractor and the various subcontractors on a job is essentially a job management function. Most written subcontracts give the prime contractor complete power to control the timing of the work done by subcontractors, and it is usually the prime contractor's project manager or superintendent who makes the decisions.

Faulty coordination can create accounting, financial, and credit problems for the prime contractor. When, because of faulty planning, there is excessive delay in making final change orders, allowance must be made for the loss of job momentum. Otherwise unit costs will be distorted, and their usefulness will become extremely limited. When progress is delayed, so also are progress payments. Unit costs, on the other hand, tend to be higher than normal, so that financial planning and credit arrangements require reexamination. If the subcontract requires the subcontractor to wait until the completion of the whole job for the retention money, the delay in completing a poorly coordinated prime contract will require further changes in the subcontractor's financial planning.

One final problem in connection with subcontractor coordination is that which faces subcontractors when they bid. If a prime contractor's jobs are usually well coordinated, the subcontractor is reasonably safe in assuming that he can plan his work. If the prime contractor's jobs are frequently disorganized or poorly coordinated, it is good financial planning to add enough to the bid to cover both the excess unit costs and the excess finishing costs that may reasonably be expected. The idea that a subcontractor is morally bound to submit the same bid to all potential prime contractors on a job is not correct, if only for the reason given above.

Taking subcontract bids

From the prime contractor's viewpoint, there are relatively few matters connected with the taking of subcontract bids to concern the accounting or financial personnel. They should, however, determine, as far as they are able, whether the subcontractor will be financially acceptable, before issuing invitations to bid. It is wise practice to confirm telephone bids in writing immediately. It is up to the office manager to see that subcontractors are supplied with working area in which to make their takeoffs from the plans in the prime contractor's possession, but dealings with the subcontractors on behalf of the prime contractor are usually carried on by the individuals who are preparing the prime contractor's bid estimate.

From the subcontractor's viewpoint, there are several matters which those responsible for accounting and finance should consider at the time bids are submitted. The first and most important consideration is the timing of progress payments. It is quite common for a subcontract to contain a provision that the subcontractor is not to receive payment until the prime contractor has been paid. If such a provision is present, then the payments to any individual subcontractor may be delayed by matters entirely beyond his control. In evaluating this risk, the reputation for prompt payments of the owner, the architect, and the prime contractor must be taken into consideration. If any one of the three has the reputation of delaying payments or holding them up on account of minor questions or arguments, that fact should be reflected in the bid price.

Second is a related consideration—payment of retained percentages. If the subcontract form requires the subcontractor to wait for his retention until completion of the entire job, the cost of providing the additional financing should be considered in the bid price.

Third, it must be determined whether performance or payment bonds will be required and, if so, at whose expense.

Fourth, the insurance requirements must be determined and, if additional coverage will be required, its estimated cost should be taken into the bid price. Particularly important are special risks that might affect compensation insurance premiums or that might require builder's risk insurance. Occupational Safety and Health Act (OSHA) regulations also may be such as to require special insurance on any individual job.

Fifth, if a job office is to be required, its staffing and equipment are matters of concern to accounting and financial personnel.

Sixth, while legal problems are usually not the direct concern of the accounting or financial personnel, it is the responsibility of these individuals to call the attention of the legal department to such problems when they arise. At the bidding stage, it may be remembered, a bid is an offer that, barring such considerations as "promissory estoppel," expires after a reasonable time and may normally be withdrawn at any time before acceptance. Moreover, an acceptance, to create a binding contract, must conform exactly to the bid, and attempts to bargain on any part of the bid will usually result in a counteroffer, which must then be accepted before there is a contract. If negotiations indicate that neither party expects to be bound until an integrated contract document is signed, there will ordinarily be no contract until the document is executed and delivered.

Seventh, if the products of a particular manufacturer are specified, for example, in door hardware, care must be taken to see that the supplier is in a position to provide such items. Often, "protected territory" arrangements between manufacturers and their local representatives give these representatives a virtual monopoly on certain types of items.

Eighth, in bids involving the supplying of materials or transportation of property, the persons taking the bids need to understand the application of state and local sales and use taxes, and state transportation taxes.

One of the most valuable contributions that a subcontractor's financial manager may make to bidding is that of objective analysis of the business aspects of each bid. There is a tendency for estimators, in their eagerness to secure new work, to accept unduly burdensome terms. Often a little informed pessimism on the part of the business manager produces profitable results.

Analysis of subcontract bids

When a large contract is being bid, prime contractors usually receive subcontract bids right up to the last minute. Once a prime contractor receives some indication that he may be awarded the contract, he immediately starts to analyze the various subcontract bids and bidders.

The evaluation of the bid and of the availability and technical qualifications of the bidders, their organizations, and their equipment, is a job for the estimators. The evaluation of the bidders' credit and the effect of their existing workloads on their ability to finance the job are the responsibility of the prime contractor's business manager or controller. It is at this time that the forecasting done by the subcontractors' accounting departments at the time that the bid was submitted can pay for itself.

A realistic and sensible financial forecast by the subcontractor can often convince the prime contractor that the subcontractor operates on a businesslike basis, and can often secure subcontract awards even against lower bidders. If the competitor's bid is so low that he is sure to lose money on the job, there is a good chance that his performance will be unsatisfactory. Besides, overly competitive bidding sometimes indicates financial instability. One subcontractor who is not performing properly can upset planning, create delays, and sometimes turn a closely bid job from a profit to a loss. Even the limited protection of a performance and payment bond cannot prevent a subcontractor's default from being costly to the prime contractor. For these reasons a prime contractor may pass up the lowest bidder and award the subcontract to another bidder whose bid is higher but who shows clearly that he knows how to plan his work well and is able to demonstrate that he is in a financial position to complete the job. This phase of the analysis of subcontractors' bids is usually up to the business manager or the controller of the prime contractor.

In analyzing bids it is important to distinguish between a subcontractor and a materials supplier. If the bid is merely to supply materials, a state sales or use tax may have to be paid in addition to the bid price, rather than being a part of it. For this reason, as well as a number of others, it may be desirable to use a contract form written especially for materials contracts instead of using the form written for subcontractors. Requirements will tend to differ from one state to another, and for that reason the contract form should be checked by legal counsel. Transportation tax is another matter that should be of concern. Some states have a transportation tax on hauling done for a prime contractor.

Many prime contractors, after taking subbids and analyzing them, will isolate those work items that have the lowest unit prices and award a subcontract which covers those items alone. Of course, the subcontractor is not bound to accept such a counteroffer, but there is always a temptation to do so, especially when work is scarce. This is not to say that subbids should always be made on an all-or-nothing basis. It simply calls for a thorough reconsideration of the bid as revised. Failure to do so can be very expensive if the subcontractor has unbalanced his bid. If a subcontractor intends that his bid must be accepted as a whole or not at all, the bid should say so. If a subcontractor is going to unbalance his bid, the price should, wherever possible, be stated as a lump sum bid so that there can be no breakdown on a counteroffer.

When a large building job is broken down and allocated to a number of subcontractors, questions of gaps and overlaps of responsibility inevitably arise. For example, if the excavation and site preparation is sublet to one subcontractor, and the sewers and drainage to another, and the paving to a third, who is responsible if the paving subsides and breaks due to faulty compaction in the sewer line trenches? Is it up to the plumber or the roofer to provide the sheet metal flashings around the roof vents?

Obviously, this sort of problem can be covered by careful wording of the "scope" paragraphs of the subcontracts, so that laps and gaps are eliminated.

The prime contractor's accounting and financial departments become involved when the inevitable disputes arise over who is to do the work and how much is to be paid for it. When an overlap occurs, and the parties cannot agree on which subcontractor is responsible, it may be necessary for the prime contractor to do the work in order to avoid delaying the job. If the prime contractor intends to charge the cost of doing the work to one of the subcontractors, he should allow enough time for the subcontractor to undertake the work himself. Otherwise the prime contractor might find that legally he is a mere volunteer and unable to collect for the work done.

The subcontractor is saved untold cost and difficulty if the limits of his work are precisely defined. For that reason the subcontractor should take the initiative, if need be, to see that laps and gaps are eliminated. If a gap develops and the prime contractor demands that the subcontractor proceed with the work, the subcontractor may be better off if he does the work and at the same time takes the necessary legal steps to save his right to payment for an "extra" or "scope change." In such situations, the costs incurred in performing the disputed work must be recorded with the greatest care, and must be fully documented. The best procedure is to set up a separate work order to cover the disputed work and to document charges to the work order as if it were a separate, cost-plus job.

If a lap develops and one of the overlapping subcontractors does the work, the other subcontractor must be careful to see that any resulting change orders issued by the prime contractor do not make excessive deductions in price. In this connection, unit costs on comparable work from other jobs may prove useful. If the work involved in the overlap is profitable, it may be worthwhile for the subcontractor from whose contract the work is to be deleted to insist on the right to proceed with the disputed work, and to agree to its elimination only if adequate legal safeguards preserve his rights under the contract to the profit which he would otherwise have made on the deleted work.

Awarding the subcontracts

The first step in awarding a subcontract is a decision by the prime contractor on just what work is to be subcontracted. Because the extent and scope of the work may vary according to which subcontractor is chosen, it may be necessary to select the subcontractor at the same time the scope of the work is determined.

The most important consideration in choosing a subcontractor is whether he can do work of the required nature and quality at a competitive price. Usually a prime contractor is reasonably justified in taking this basic know-how for granted. Still, a mistake in judgment in selecting a subcontractor can be extremely expensive, and this fact alone justifies making the selection with more than ordinary care.

Next in importance after technical ability is financial ability. It has been noted before that the two major limitations on a contractor's ability to undertake new work are his organization and his available working capital. If a subcontractor is lacking in either respect, the prime contractor may have to share the former's difficulties. How much total money or bonding capacity the subcontractor may have is immaterial. It is the amount he has available for the particular job that counts. It becomes the job of the prime contractor's accounting department, therefore, to analyze the subcontractor's credit reports and his accounting or financial statements in considerable detail to determine his financial responsibility.

In making such analyses, net working capital is a key figure. According to some bonding companies, a building contractor can carry between $10 and $20 of uncompleted work for each dollar of net working capital, or $4 and $6 of uncompleted work for each dollar of net worth. Naturally, the ratios vary according to individual circumstances. For example, compare the position of Contractor A with that of Contractor B, as shown by their balance sheets in Schedule 2-1.

On the statements alone, Contractor A is clearly in the better position of the two. His ratio of uncompleted work ($1,000,000) to net

SCHEDULE 2-1 SUBCONTRACTOR FINANCIAL CONDITION

	Contractor A	*Contractor B*
Uncompleted work on hand	$1,000,000	$1,000,000
Cash	$ 30,000	$ 50,000
Progress billings currently due	60,000	100,000
Retained percentages	200,000	50,000
Inventory, at cost		
Materials, supplies	70,000	40,000
Equipment parts	40,000	10,000
Prepaid expenses	40,000	30,000
Total current assets	$ 440,000	$ 280,000
Notes payable	$ 100,000	$ 90,000
Trade accounts payable	90,000	70,000
Payroll taxes, insurance, etc.	60,000	50,000
Equipment installments	65,000	40,000
Total current liabilities	$ 315,000	$ 250,000
Net working capital	$ 125,000	$ 50,000
Equipment, at cost net of depreciation	$ 500,000	$ 500,000
Less: Long-term equipment contracts	(450,000)	(450,000)
Real estate, at cost	500,000	100,000
Less: Long-term notes payable	(400,000)	—
Other assets (net)	$ 150,000	$ 150,000
Owner's equity	$ 275,000	$ 200,000

working capital ($125,000) is only 8 to 1, compared to 20 to 1 ($1,000,000/$50,000) for Contractor B. Contractor A's ratio of uncompleted work ($1,000,000) to net owner's equity ($275,000) is something under 3.7 to 1, while B's comparable ratio is 5 to 1.

However, suppose that on analysis the following facts came to light:

1. A has, in his backlog of work, a job that will lose as much as he makes on all his others, so that his overhead for the next few months will be carried from his working capital instead of from job profits. Assume that this loss of working capital can be valued at about $75,000.
2. Of the $200,000 of retained percentages shown as a current asset for A, there is one item of $100,000 tied up in a dispute that will probably not be settled for a year.
3. The unencumbered real estate on B's balance sheet is worth $300,000 and the bank will loan $200,000 on it without question.

Such facts would change the picture completely. Contractor A's $125,000 net working capital is overstated for this purpose by some $175,000 leaving him with a negative net working capital ($125,000 − $175,000 equals a minus $50,000). Contractor B, on the other hand, has a hidden resource of $200,000 so that, if the need arose, his net working capital of $50,000 could be increased to $250,000.

Under such circumstances, and with other things approximately equal, it would probably be safer financially to award any proposed work to B.

A third factor to be considered in selecting a subcontractor is the subcontractor's reputation for maintaining a work schedule. Primarily, the purpose for this requirement is to prevent delay on the jobs. In addition, however, a subcontractor who is behind in his work is also, as a rule, behind in his payment of bills, thus creating a lien risk for the prime contractor that can be avoided by careful selection.

Negotiating the subcontract

Usually, subcontracts are negotiated by the principals or by the operating or engineering personnel of the prime contractor and subcontractors. Operating men and engineers are certainly capable enough so far as the scope of the work and the prices are concerned. Too often, however, they are inclined to brush off a good many important details as "fine print." This attitude they will justify with the sound observation that, "if an individual's word is no good, neither is his contract." Still, disagreements do arise, even between honest people. Sometimes, too, a contractor may think he is dealing with an honest person when, in fact, he is not. When things go wrong, both parties will go to the so-called "fine print." When the point at issue is covered in the contract, the questions are usually settled. It is when there has been no coverage of a point in the contract that the parties find themselves in court.

It is often up to the business manager or controller to see that the final contract documents are complete and that all important matters have been covered. Schedule 2-2 provides a list of questions which should prove helpful in such an analysis.

In all cases, the prime contractor ought to assure himself that the subcontractor fully understands the provisions of his subcontract, and be sure that he is able to establish that the subcontractor's understanding is complete. The temptation is strong to slide rather gently over touchy points in the hope that they may not arise or, if they do, that they can be ironed out. The prime contractor's representative needs to remember that contract documents are usually strictly construed against the party that prepares them. If feasible, the subcontractor should be given a copy of the general conditions at the time he makes his takeoff from the plans. Then when the contract is being negotiated, these general conditions should be reviewed to assure complete understanding.

The form and content of the subcontract document will vary as widely as the work to be covered. The checklist of 39 points given in Schedule 2-2 indicates some of the issues to be covered, but no universally accepted form could be drawn because of the wide variation in actual circumstances. If a prime contractor is preparing a form, it should be reviewed by his attorney before being used. The standard subcontract forms sold by stationers are rarely complete, applicable, or entirely satisfactory. The most satisfactory procedure is for each prime contractor to have his own subcontract forms prepared to fit his subcontracting policies. Each subcontractor, in turn, should have in mind (or better yet, on paper) the basic terms to which he will agree, and then place a value in money on any important deviation from those terms.

It is generally good policy to have the subcontract document signed before work commences. If that is not practical, then the subcontractor is entitled to a "letter of intent" or a "notice to proceed" signed by the prime contractor or someone authorized to act for him.

From the prime contractor's point of view, particular care should be taken to see that the contract is signed by, or on behalf of, the person or company to which the prime contractor is looking for performance. It is fairly common for a manufacturer of national reputation and unquestioned credit to appear as a bidder and then return the subcontract signed by some local distributor whose capacity is so limited

SCHEDULE 2-2 CHECKLIST FOR SUBCONTRACTS

1. Are the names and addresses of the parties correctly stated?

2. Is the form of business organization of each party correctly stated? For example: "a sole proprietorship," or "a partnership consisting of X and Y," or "a limited partnership of which X and Y are the general partners," or "a California corporation."

3. Are all important terms either defined or used in such a way that their meaning is entirely clear?

4. Is the work to be done described completely and accurately? If copies of documents, plans, drawings, and so on are to be furnished, are they actually incorporated by reference? Who is to pay for them?

5. Who is responsible for obtaining and submitting shop drawings?

6. Who owns the plans, drawings, and models furnished?

7. If samples are to be submitted, to whom do they belong?

8. If patented processes or devices are to be used, who pays the royalty?

9. Does the subcontractor pay for lights, water, power, transportation, and similar facilities? If any of these are to be furnished by the prime contractor, will he bill the subcontractor, and if so, on what basis?

10. Who is responsible for lines, grades, and surveys?

11. If special licenses or permits are required for the subcontractor's work, does the subcontractor have the duty of seeing that these are obtained?

12. To what extent does the subcontractor assume the responsibility to protect the work, the public, and the owners of adjacent property from loss or damage?

13. Has the subcontractor agreed to the same standards of supervision and inspection as those imposed on the prime contractor? To whom must the work be satisfactory?

14. Has the subcontractor agreed to the same provisions regarding change orders as those in the prime contract? Are there satisfactory provisions for negotiating the price of the changes? Is provision made for the subcontractor's claims for the cost of excessive delay in issuing change orders or for the cost of complying with special instructions requiring costs beyond those originally contemplated?

15. Have satisfactory provisions been made for correcting unsatisfactory work?

16. Are satisfactory provisions made for delays and extensions of time?

17. Are the provisions for notice to the subcontractor's bonding company of change orders, time extensions, and payment of funds liberal enough to prevent unintentional release of the surety?

18. Do the guarantee provisions of the subcontract have the same coverage as those in the prime contract?

19. Does the subcontract contain the same provisions as the prime contract regarding the owner's privilege to do work which the contractor has done unsatisfactorily, and does the prime contract have a similar privilege with respect to the subcontractor's work?

20. Does the prime contractor have the same power to terminate the subcontract as the owner has to terminate the prime contract?

21. When, and on what basis, does the subcontractor receive progress payments? Is there any relationship between the owner's payments to the prime contractor and the prime contractor's payments to the subcontractor? Would acceleration of the subcontractor's payments release his bondsmen? Is he required by the terms of the contract to apply such payments first to payment of bills incurred in connection with the job? Is the subcontractor entitled to require that progress payments be made even though some of his accounts with his suppliers are delinquent?

22. When, and on what basis, does the subcontractor receive his retained percentage? Must he wait until the entire job is completed and accepted? If so, would payment in advance release his bonding company?

SCHEDULE 2-2 CHECKLIST FOR SUBCONTRACTS *(Cont.)*

23. To what extent must the subcontractor carry:
    a. Workers' compensation insurance
    b. Public liability and property damage insurance
    c. Fire, storm, earthquake, builder's risk, or other similar types of insurance?

24. Is the subcontractor required to furnish a performance and payment bond? If so, who pays for it? If not, is there adequate protection against liens or comparable statutory liabilities?

25. Does the subcontractor agree to hold the prime contractor harmless for damages caused by the subcontractor's operations, and vice versa?

26. Is the prime contractor's duty to pay the final payment and the retained percentage conditional upon his receiving a release of liens from the subcontractor (Figure 2-4)?

27. Are the agreements regarding assignments of rights and delegations of duties under the subcontract satisfactory?

28. Does the subcontractor agree in general terms to be bound by, and to bind his subcontractors to, all applicable terms of the prime contract?

29. Does the prime contractor reserve the right to approve or disapprove any subcontracts made by the subcontractor covering the work on this job?

30. Does the subcontract contain the same provisions regarding arbitration as the prime contract? Should it contain such provisions?

31. Does the subcontract state in so many words that there are no outside understandings, representations, or agreements?

32. Are there adequate and satisfactory provisions covering cleanup at the jobsite?

33. Who pays lawyers' fees and legal expenses in the event of litigation caused by breach on the part of either party?

34. In the event of default by the subcontractor, does the prime contractor, or his nominee, have the right to take over and use for the purposes of completing the subcontractor's work, all materials, supplies, equipment, and facilities which the subcontractor has on the job?

35. Does the subcontract require the subcontractor to include in his subcontracts and purchase orders language permitting the prime contractor to take them over for purposes of completion should the subcontractor default?

36. If either the prime contractor or the subcontractor abandons the project, are there adequate provisions for the protection of the other party?

37. If, through action of one party in legal or labor matters, or for other reasons unrelated to the other party, the job is shut down to the detriment of the other party, are there satisfactory arrangements for dealing with losses thus caused?

38. Does the subcontractor have the right to joint-venture or subcontract significant portions of his work without approval from the prime contractor or owner? If so, to what scope or extent?

39. Does the prime contractor or owner or their representatives have the right to audit, especially if work is done under "cost-plus" or "unit price" contractual terms?

as to invite suspicion. In such circumstances, the prime contractor should either insist on the guarantee of the manufacturer or, as a minimum procedure, require a performance and payment bond.

It is customary for subcontracts to incorporate, by reference, the plans, specifications, drawings, addenda, and so on mentioned in the prime contract. Such comprehensive reference ensures that the subcontract is complete and covers all the requirements contained in documents that may not have been made available to the subcontractor. Therefore, if a subcontractor is to be held responsible for knowing the contents of all the plans, specifications, drawings, addenda, and other documents, he should be careful to see that he has studied them for all matters that might affect his work. It is equally important that the prime contractor be very sure he is able to establish that they were available to the subcontractor for such study.

Once the subcontract is awarded, it is up to the prime contractor to see to it that proper notice is given to the subcontractor and to appropriate members of the prime contractor's organization. When a prime contractor uses the same subcontractors regularly, many of the foregoing details can be eliminated, especially those associated with giving notice. However, it is easy to let long association lead to unbusinesslike practices. To avoid such a situation, even the smallest operators might well review their practices in the light of the suggestions of this chapter. Mere smallness does not necessarily justify failure to take reasonable and businesslike precautions in dealings between prime contractors and subcontractors, nor does it justify a blind reliance by the subcontractor on the forms and procedures of the prime contractor.

Payments to subcontractors

It is customary for the subcontract to call for progress payments to be made to subcontractors as soon as the prime contractor receives payment for the work, and in the same proportion. The subcontractor's partial payment estimate (Figure 2-3) is a useful form, for inclusion in payment vouchers, that summarizes the payment position of each subcontractor.

Payment to the subcontractor only after the prime contractor has been paid has the disadvantage to the subcontractor of requiring him to wait until the whole job is completed before receiving his retention. If the subcontractor wants to collect his retention before final completion and acceptance, he should see that his right to payment at an earlier date is spelled out in the subcontract.

Often the prime contractor's field engineer will try to pay subcontractors less than the amounts requested and received by the prime contractor from the owner for the subcontractor's work. If the subcontractor has been alert at the time of negotiation, the subcontract itself will require that payments to the subcontractor be based on the same percentage of completion as that agreed upon by the prime contractor and the owner.

FORM 74—5M—5-58 CPCO

Date ________

SUB-CONTRACT PARTIAL PAYMENT ESTIMATE NO. ______

JOB NAME ________________________

JOB NO. ________________________

FOR PERIOD FROM ____________ 19___ TO ____________ 19___

SUB-CONTRACTOR ________________________ SUB-CONTRACT NO. ______

DESCRIPTION OF WORK ________________________ ACCOUNT NO. ________

ORIGINAL CONTRACT PRICE	CHANGE ORDERS TO DATE	NET CONTRACT PRICE	% COMPLETED TO DATE	VALUE OF WORK COMPLETED TO DATE
$	$	$	%	$

CHANGE ORDER RECORD

C. O. NO.	DATE	INCREASE	DECREASE	
		$	$	
		$	$	
		$	$	
		$	$	
		$	$	$

PARTIAL PAYMENT RECORD

	PREVIOUS	THIS PERIOD	TO DATE
TOTAL AMOUNT ESTIMATED	$	$	$
RETAINED AMOUNT (%)	$	$	$
NET PAYMENT DUE	$	$	$

ESTIMATE PREPARED BY ________________ APPROVED BY ________________

APPROVED FOR PAYMENT BY ________________

Figure 2-3 **Summary of Subcontractor's Partial Payment Estimate.** This form is designed for separate computation of percentage of completion on change orders. If this procedure is not followed, changes can be lumped into one amount or the work items can be adjusted to take the changes into account.

For the prime contractor to attempt to underpay the subcontractor may be shortsighted. It is true, of course, that if subcontractors are consistently underpaid there will be a greater margin of safety in case of default. On the other hand, a policy of underpaying subcontractors could bring about defaults that otherwise might be avoided. Also, this practice sometimes misleads the prime contractor in his cash forecasting. Certainly such a policy cannot have a healthy effect on future bids received from the subcontractors affected. If a prime contractor is going to unbalance a payment request, it is better, if possible, to avoid doing it with subcontracted work items.

The subcontract should provide that when the retained percentage is paid, a release (Figure 2-4) will be given by the subcontractor. When final payment is made to a bonded subcontractor, the prime contractor should notify the subcontractor's bonding company, and similar notice should be sent by the subcontractor when the payment is received.

Breach of a subcontract

Breach of contract is a legal matter and should be taken up immediately with the company's attorney. In fact, if a breach appears possible or probable, the attorney should be consulted beforehand. If the breach involves a stoppage of work by the subcontractor, regardless of who may be guilty of the breach, the following steps should immediately be taken by both parties:

1. Complete and detailed progress photographs should be taken by an independent professional photographer, preferably one familiar with construction and with the use of photographs as evidence in court. The photographer and the attorney should work together closely on this project, and it is desirable that the attorney be present while the photographs are being taken.
2. Both the contractor and the subcontractor should prepare a complete and detailed inventory of the materials, supplies, and equipment that the subcontractor has on the job.
3. Both the contractor and the subcontractor should make careful and detailed engineering estimates and measurements of those parts of the subcontractor's work that remain to be done. Since questions of faulty materials and workmanship are almost certain to arise, the work that may be questioned should be inspected with extreme care and detailed notes made of its exact extent and condition. If some independent expert can make the examination and render the report, so much the better.
4. If the terms of the subcontract require arbitration of disputes, and if there is no way to avoid the requirement, it is a good idea for the individual contractor to analyze his position carefully to see where

Individual Proprietorship

Los Angeles, California
..............19....

RELEASE

KNOW ALL MEN BY THESE PRESENTS, that..................
of..
an Individual, for valuable consideration, the receipt whereof is hereby acknowledged does for himself and his successors in interest and assigns, hereby release and forever discharge the Prime Company, a California Corporation, Prime Contractor, and their successors and assigns, for the construction and completion of..................................
....
at..
from all actions, debts, accounts, bonds, covenants, contracts, agreements, damages, claims, and demands whatsoever, in law or equity, which have arisen or may arise out of or by reason of, or which are based on that certain contract executed on the ....day of............19...,
by and between the said Prime Company and the undersigned, or which have arisen from or by reason of any cause, matter or thing whatsoever.

The undersigned further releases the owner and the owner's property from any claim of any kind on account of work performed or materials furnished in connection with the above referred to contract, or in connection with construction on above referred to project.

The undersigned further warrants that he has paid and satisfied all claims for labor and materials used by him for or in connection with the work under the said contract. The undersigned further hereby agrees to indemnify and hold the aforesaid persons, firms and corporations free and harmless from any and all such claims and any and all expenses including attorney's fees, which may be incurred in contesting or adjusting any thereof.

IN WITNESS WHEREOF, the said............................
has caused these presents to be executed the day and year first above written.

....
....

Subscribed and sworn to before me this ....day of............19....

....
Notary Public in and for the county of
.......................State of.......................

Figure 2-4 **Release Form.** This form is for use by an individual proprietorship, but with slight modifications in language it could be used by any other form of business organization, such as partnership, corporation, and so on.

his best interests lie. If his legal position is strong, he should seek an arbitration board of lawyers, but if, on the other hand, his legal position is weak and his case depends on the equities of the construction facts, he should try to have construction men on the panel. Because the combination of legal and construction knowledge is rare, it is often true in practice that arbitration in the construction industry is unsatisfactory. Usually the interests of the prime contractor do not lend themselves to arbitration whereas the interests of the subcontractor often do.

When the subcontract is breached by failure of the prime contractor to pay amounts due under the subcontract, the subcontractor needs to move promptly to protect his rights. If the owner is a private individual or business firm, the remedy lies with the state lien laws. If the job is being done for a state or local government body, the remedy is usually a "stop notice" under the state bond laws, followed by a timely suit to enforce the claim. If the job is being done for the federal government, the remedy is under the Miller Act—the Federal Bond Act of 1935.

Lien laws and bond acts are for the protection of contractors, workers, materials suppliers, and others who contribute directly to the work on a construction job. These laws are highly technical in their application, and, generally speaking, they are strictly interpreted. Therefore, as soon as any payment, and particularly a final payment or retention, becomes delinquent, the contractor should consult his attorney at once, so that there will be time to act before the relatively short limitation periods run out. One other fact must be mentioned. In a dispute on a construction contract there is no substitute for a complete, well-kept job diary, maintained by the person responsible for the work, and written in his own handwriting. All project managers, superintendents, and job engineers should be required to maintain such a record daily.

Prime contractors frequently require that subcontractors furnish a performance and payment bond, and there is a tendency to feel that such a bond constitutes full protection against default by the bonded subcontractor. However, this view is only partially true. If the subcontractor goes bankrupt or makes an assignment for the benefit of creditors, the bonding company will normally pay the loss (though usually not the attorney's fees) without much question. If, on the other hand, the subcontractor merely abandons the job and fails to pay his suppliers, it is the prime contractor who must complete the job and pay the bills or suffer the consequences. If the prime contractor sues and obtains judgment against the subcontractor, the bonding company will reimburse the costs up to the limits of its liability, but the bonding company will not act before the subcontractor's legal liability is established lest it lose its right to be reimbursed by the subcontractor. This restraint can put the prime contractor to the trouble and expense of a lengthy lawsuit in addition to the trouble and expense that are the normal result of a default. Often it is better for the prime contractor, under such circumstances, to accept a substantial loss and settle, rather than to pursue either arbitration or litigation.

Accounting for Subcontract Costs

It might appear that, once the subcontract is let, there are few, if any, problems in accounting for the costs. In large measure, this statement is true. On very small jobs the subcontractor renders his bill at the end of the job and the cost is accounted for in that period. If subcontracting is on a cost-plus basis the subcontractor submits periodic billings for his costs and fee, and again the costs are accounted for in the period when billing takes place. If jobs cover a longer period and involve progress billings, retained percentages, change orders, and special work or services, or if there are second- and third-tier subcontractors, jobsite security regulations, or labor problems, accounting procedures must be broadened to meet the broader needs. The larger and more numerous the subcontracts become, the greater are the problems of controlling them. Some contractors with very large and complicated jobs and numerous subcontractors are now turning to computers to help them achieve this control. A more complete discussion of computers is to be found in Chapter 12.

Progress billings

If subcontractor progress billings are based on a percentage of completion or units performed, it is usual for the prime contractor's job engineer and the subcontractor or his superintendent jointly to estimate the percentage of completion or units performed and agree on the amount to be billed by the subcontractor. Such progress estimates may be summarized on special forms (Figure 2-5).

Sometimes, instead of periodic estimates based on percentage of completion or units of work performed, the subcontract will call for estimates to be submitted at a particular stage of completion. For example, a plumber might bill one-third of his contract price when the rough plumbing is in, one-third when the finish plumbing is in and the fixtures are set, and the balance 35 days after final inspection and test. This is just another way of arriving at percentage of completion, and when the job is not too long it is reasonably satisfactory. On the longer jobs, however, periodic billings based on progress are generally preferable. In either event, the same general type of summary is needed to support the entry debiting the general ledger account with "Work in progress" and crediting a liability account for "Amounts due subcontractors" on the prime contractor's books.

The subcontractor's ledger sheet (Figure 2-6) is used in a subsidiary ledger controlled by the "Subcontract cost" column in the "Work in progress" account and by the account "Amounts due subcontractors—Retained percentages." Other columns in the subcontractor's ledger are set up to give a summary of the subcontracts let and the amounts

SUB-CONTRACTOR'S PARTIAL PAYMENT ESTIMATE NO.________

For Period Ending________________

	Total Amount	% Completed to Date	Amount to Date
Sub-Contractor:			
Description of Work:			
Original Sub-Contract Amount:	$		
Change Orders:			
Total Amount of Contract:			
Less Retained Percentage (%)			
Less Previous Payments			
Net Amount to Date:			
Less:			
Payment Now Due			

Figure 2-5 **Payment Estimate.** Figure 2-3 shows one form of subcontractor payment estimate. This is another form, designed to be either typed or handwritten.

remaining to be paid to them. This last information can be useful at two different points. First, when the engineer is making his periodic estimate of cost to complete, he can refer quickly to the relative stage of completion of each subcontractor. Second, if either the subcontractor or the prime contractor is "unbalancing" the subcontract payment requests, any serious discrepancies between physical progress and dollar payments are more easily noticed. Normally, there will be a

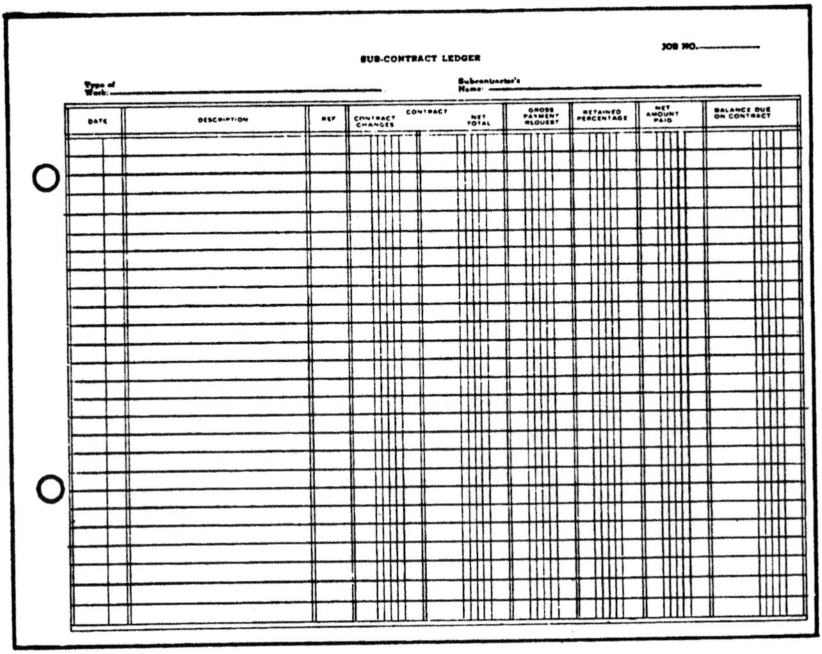

JOB NO.

SUB-CONTRACT LEDGER

Type of Work:

Subcontractor's Name:

DATE	DESCRIPTION	REF	CONTRACT: CONTRACT CHANGES	CONTRACT: NET TOTAL	GROSS PAYMENT REQUEST	RETAINED PERCENTAGE	NET AMOUNT PAID	BALANCE DUE ON CONTRACT

Figure 2-6 Subcontractor's Ledger. A special variety of accounts payable ledger, this form is particularly useful: when there are numerous subcontracts; when these constitute an important part of job costs; and when the amounts owed are important in the contractor's balance sheet.

tendency for the subcontractor to unbalance in favor of overpayment while the tendency of the prime contractor will be to unbalance in favor of underpayment. Usually the two just about offset each other, but when the payments get out of proportion to physical progress, the records should call attention to that fact. Another factor for which both the partial payment estimate form (Figure 2-5) and the subcontractor's ledger form (Figure 2-6) make provision is the change order. Even on relatively small jobs change orders frequently change the contract amount and must be taken into consideration as the job progresses. Occasionally, change orders are taken into account as separate work orders, but the generally preferable procedure is to make them a part of the subcontract itself.

Retained percentages

When a subcontractor bills the prime contractor for a progress payment, construction cost is charged for the full amount. The contra credit is divided between an account called "Amounts due to subcontractors—Current" and another called "Amounts due to subcontractors—Retained percentages." It is the latter account that controls the "Retained percentages" column in the subcontractor's ledger.

The payment of the retained percentages must be made strictly in accordance with the subcontract agreement. Otherwise there is a possibility that the subcontractor's surety (if he has one) may be released.

A subcontractor, quite naturally, wants his retained percentage released as soon as possible after he has finished his work and, wherever possible, he will seek to have provision made in the subcontract agreement to that effect.

The prime contractor, on the other hand, will try to keep the subcontractor's retention until the whole job is completed. There are several reasons for this effort. First, the prime contractor will not, as a rule, receive his retention until the end of the job. Second, many owners and architects reserve the right of final inspection at the completion of the whole job, and if something goes wrong with the subcontractor's work between the end of the subcontract job and the final completion inspection, there is no security for correcting the work. If the prime contractor's policy calls for early release of subcontractor retentions, it might well be necessary, from a conservative accounting viewpoint, to provide a reserve for such losses.

Back charges

One of the most persistent problems of accounting for subcontract costs is that of controlling back charges. A subcontractor needs a piece of equipment for an hour or so. The prime contractor's superintendent supplies it. In theory, when the equipment time is reported, the time worked for the subcontractor will show up. The contractor will charge the subcontractor's account and it will be deducted from his next progress payment.

In practice it may be handled differently. The contractor's superintendent may say to the subcontractor:

"Okay! I'll send the equipment over, but how about having your gang repair those places where the carpenters damaged your work, and we'll call it even?"

If this sort of exchange goes on very much there is a better-than-even chance that after a few months there will be an argument about disputed back charges. Regardless of the outcome each company will feel that it has been cheated.

This observation definitely does not mean that prime contractors should furnish nothing to subcontractors or vice versa. It simply means that nothing should be done without written authorization. In the example above the subcontractor's representative should have given the contractor a written order for the equipment and the contractor's superintendent should have given a like order to the subcontractor.

The excuse for failing to observe these formalities is that by the time a formal order is written up and signed the work is done and forgotten about. That is usually the trouble. It is forgotten about until the time is recorded for the subcontractor's men and the contractor's equipment. Then the books of both companies show an open charge. The prime contractor is in a position to collect by offsetting the charge against the progress payment, and when he does, the argument starts.

There is really no excuse for such a result. The formalities are extremely simple. The contractor's superintendent can easily carry in his pocket, or in the glove compartment of his pickup truck, a pad of field work order forms (Figure 2-7). The forms are in triplicate and padded with one-time carbons. When an agreement is reached or the order is given it takes only a few seconds to record it, give one copy to the superintendent, one to the subcontractor's representative, and leave one in the book for audit. If subcontractors clearly understand that no work will be paid for without such an authorization and if superintendents and project managers are held strictly accountable for work done without authority, there should be no problems on chargebacks. In other words, there should be no recognition of oral instructions or oral deals.

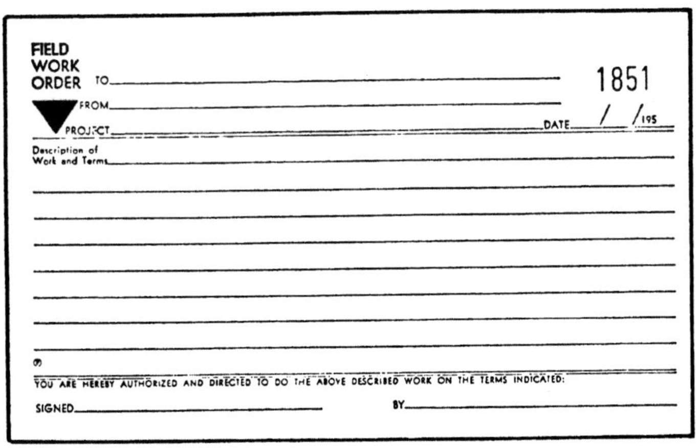

FIELD WORK ORDER TO ______ 1851

FROM ______

PROJECT ______ DATE ___/___/195__

Description of Work and Terms ______

YOU ARE HEREBY AUTHORIZED AND DIRECTED TO DO THE ABOVE DESCRIBED WORK ON THE TERMS INDICATED:

SIGNED ______ BY ______

Figure 2-7 Field Work Order. This form was designed to make it easy to reduce to writing agreements between the prime contractor and subcontractors, and thus prevent the disputes that arise from oral instructions and unrecorded "deals."

Special services to subcontractors

Lending labor to subcontractors or others is a practice that deserves special mention because of the accompanying hazards. It usually happens in one of two ways. A subcontractor's employees may be carried on the prime contractor's payroll, or employees on the prime contractor's payroll may be assigned to work for the subcontractor.

In either situation an industrial injury may subject the employer to liabilities which he never intended to accept, including liability beyond the coverage of workers' compensation insurance. An even greater risk for the prime contractor is involved because he is responsible for the work of all employees on his payroll. Thus the prime contractor may be relieving the subcontractor of responsibilities that the subcontractor should rightfully assume.

Often a prime contractor may find it necessary or desirable to provide subcontractors with such services as hoisting, mess hall, employee transportation, or housing. If some such service is to be rendered, it should be mentioned in the subcontract. Arrangements made either to offset the cost against progress payments, or to collect it separately, should be clearly spelled out in the subcontract unless, of course, such services or facilities are to be provided to the subcontractor without charge by the prime contractor.

Often when camp, mess hall, and commissary services are involved, collection can be made directly from the subcontractors' employees. Otherwise accounting procedures must allow for keeping records of the charges to be made. Thus if hoisting services are provided for a building, the hoist operator should keep a record of each lift. Usually a simple mimeographed sheet can be devised for such a record.

Second- and third-tier subcontracts

In accounting for the costs of second- and third-tier subcontracts, a subcontractor has the same problems and meets them in the same way as a prime contractor. However, as the contracts get smaller there is an increasing tendency to treat them more like small installations or service jobs and less like subcontracts.

From the prime contractor's viewpoint, no special accounting recognition need be given to second- and third-tier subcontracts. As a practical matter the prime contractor is responsible to the owner for their work, their conduct, and—what is even more serious—any liens that result from their activities, even though such liens may be guaranteed by the intervening subcontractor. If trouble brews and the subcontractor is unable or unwilling to cope with it, it may be up to the prime contractor's controller or office manager to take whatever steps are necessary to protect the owner and the prime contractor.

Security precautions

When a contractor has a job on some United States government installation where security restrictions are in effect, it is usually up to the contractor's jobsite office force to see that all men on the job are cleared. As a rule, if the security officer is cooperative, no serious problems are involved. It is necessary to see that the contractor's office personnel are fully briefed on the regulations and that the subcontractors are similarly informed. One common error is to fail to anticipate the arrival of new men who must be given security processing both at the government office and by the contractor before they can start to work. For example, a subcontractor, on whose performance the work of others depends, can hold up an entire job for half a day by sending out a crew that has not been cleared, for by the time the guard at the gate has called the security office to clear the newcomers an hour or so will have been lost and by the time a crew of from six to twelve workers has been processed it will usually be too late to start work until after lunch.

Since it is usually necessary to coordinate each subcontractor's schedule with those of the prime contractor and other subcontractors, it is necessary to anticipate the arrival of subcontractors' crews. On jobs covered by security regulations the subcontractors will undoubtedly try to send to the job only employees who have been cleared. If new employees are to be sent, it is often possible to obtain their clearance a day in advance.

When the size of the job warrants it, a special security officer for the contractor can be a worthwhile addition to the field crew, but it is usual to see the security work handled by the job accountant. In any event security is usually under the jurisdiction of the job office manager.

In accounting for the cost of subcontracts to the prime contractor, there is a tendency to overlook the overhead occasioned by subcontract administration. How much of this kind of overhead there is will vary with the size, kind, location, and complexity of the jobs. The presence of a security problem may, in itself, increase the subcontract administrative cost to such a point that it is worth making an overhead cost allocation to subcontract costs. If such an allocation is already being made, the security cost may well be considered in establishing the rate or formula.

Labor relations

The cost of supervising labor relations may well affect the overhead charge allocated to subcontract costs. Large jobs usually have a combination personnel manager and labor relations director whose job it is to handle administrative-level labor relations matters. If jobs are small but numerous, such an employee may serve the entire company instead of just one job. When labor relations problems of the subcontractor affect the prime contractor, the prime contractor's labor relations man usually handles the first negotiations.

It is fairly common practice for a labor union to put pressure on a subcontractor by picketing the entire job of a prime contractor. If this policy is followed, the prime contractor may find himself in the position of either backing the union's demands against the subcontractor or taking part in negotiations on behalf of the subcontractor. In either event, costs are involved, and when they are substantial they should be distributed as a part of the job overhead properly applicable to subcontract costs.

Change Order Routines

Most construction contracts make provision for changes in plans, specifications, amount of work to be done, completion time, and total amount to be paid. Legally (barring provisions to the contrary in the contract) there is no reason why such changes may not be made on oral instructions to the contractor. From a practical viewpoint, oral change orders are an open invitation to disputes and often to litigation. Ordinarily, an experienced contractor will refuse either to give or accept such oral instructions.

Change orders create both problems and opportunities for the contractor. When there are too many of them, they slow down the job and increase the unit costs. Numerous change orders will usually delay completion of the job, but, more significantly, they can affect labor productivity on other and later portions of the work; this is usually called the "ripple effect." On the other hand, they offer opportunities to negotiate changes in the contract price which can prove quite profitable. The best way to take advantage of the opportunities and avoid as many of the problems as possible is to develop a routine for handling changes.

Changes in the prime contract

When the owner's representative notifies the prime contractor that a change is to be made, the prime contractor should stop as much as possible of the work that would have to be torn out or changed if the order were actually issued. He should also notify his subcontractors to take the same steps. The procedure may be reduced to routine by a "notice of pending change order" (Figure 2-8).

Under ideal circumstances, all the necessary plans and drawings would be available for immediate preparation of a change estimate

ECKS & WYE No. 123
General Contractors
NOTICE OF PENDING CHANGE ORDER

Contract No.. . . . . .Date . . . .
Job description . . . .
. . . .
We have been notified by the owner's representative on the above-described job that the following change(s) are to be ordered:

Location of change(s)	Nature of change(s)

To the extent reasonably possible, you are requested to avoid performing work that will be affected by such change(s).

Signed. . . . .

Figure 2-8 Notice of Pending Change Order.

and negotiation of the change order. It is often necessary, however, to delay the entire change until the necessary drawings and specifications are ready or to go ahead with the work under a "notice to proceed" and leave negotiations of the formal change until later.

When the notice to proceed is used, it is better, in any event, to go ahead with the normal routine of preparing a change order (Figure 2-9) signed by both parties. However, if work on the change starts before the formal change order is issued, a work order (Figure 2-10) should be prepared.

As soon as notice of a pending change order is received, a change estimate number should be assigned and a change estimate folder (Figure 2-11) should be set up. In this folder are placed all working papers and other data supporting the change estimate (Figure 2-12). Before negotiation of the change order is completed, the estimate may be revised one or more times.

When the change order affecting the contract price is issued, the job estimate should be adjusted accordingly. As a practical matter, this is rarely done more than once each week or, more commonly, once or twice a month. There are two ways in which changes affecting the

CHANGE ORDER NO. ________

Date ....

To...............................
....
....

Sirs:

Please make the following changes in the contract between

....
(owner's name)

and yourselves dated...and
more particularly described as...............................
....

Description	Contract Increase (Decrease)
	$
Total this change order	
Previous contract amount	
Revised contract amount	$

Signed: ....Title ....

We hereby accept the foregoing changes and ratify all parts of the subject contract as amended thereby.

Signed: ....Title ....

Figure 2-9 **Change Order.** This form is prepared in quadruplicate on $8\frac{1}{2} \times 11$ paper. The original is accepted and returned to the owner by the contractor. The first copy is retained by the contractor. The second copy is sent to the job office. The third copy is sent to the architect.

price may be reflected. First, they may be treated separately, as if each were a separate work item. Second, they may be reflected by a direct adjustment of the work items. When the changes are few, it makes little difference which method is used. When they are numerous, it may be necessary either to incorporate them into the related work items or to lump them together into "additive changes," "deductive changes," and "changes not affecting initial work items."

Of course when the contract provides, as some do, for separate billing of each change order, then the overhead included in the price should be sufficient to cover the extra cost of separate billing. Change order overhead is a very touchy subject, because most owners and

WO Date.................... WO No....................

WORK ORDER

Job No............Description..............................

....

Location...

Description of work — Give full details

Charge this work order for all labor, materials, equipment rentals, and other direct costs incurred in performing the work described below:

Work ordered by..........................Title...........

Work to be done for.......................................

Accepted by...........................for.................

Figure 2-10 Work Order. This form is prepared in quadruplicate on $8^1/_2 \times 11$ paper. The distribution is: original to the person or firm that is to do the work; first copy to the person ordering the work; second copy to the job office; and third copy to the architect. This same form may be used to cover extra work requested by the prime contractor from the subcontractor or vice versa, as well as to compile costs on a pending change order.

their representatives are likely to feel that any overhead charge on a change order is excessive. From the contractor's viewpoint, overhead is a very real cost in change orders, and its addition should be vigorously defended.

When a change order is issued, there are several ways in which it can affect job accounting:

1. If the amount budgeted for each work item is adjusted to reflect change orders, then the total contract price and percentage of completion, by work items and in the total, will be changed. This change must be carefully noted.
2. If, in computing the amounts of progress payments, the contract requires that change orders must be accounted for as if they were separate work items, there is a problem of segregating costs. For example, if a change order merely increased the number of cubic yards of earth to be removed from a particular excavation, or increased the number of times a certain concrete form could be

CHANGE ESTIMATE FOLDER

Change estimate No. . . . . Job name. . . . . Contract No. . . . .

Description of changes. . . . .

. . . .

Estimate made by. . . . . Date. . . . . Requested by. . . . .

Date of request. . . . . Date quoted. . . . . Amount quoted. . . . .

Date change order issued. . . . . Amount of change order. . . . .

Date job advised. . . . . Date purchasing department advised. . . . .

Record of subcontract and purchase order changes made necessary by this change:

Vendor or subcontractor	Amount		P. O.		Subcontract		Work order
	Add	Deduct	No.	Date	No.	Date	

Instructions:

When a change estimate is requested, all supporting data will be compiled in this folder until the estimate is submitted. At that time the file will be sent to the purchasing department. When written instructions to proceed are received the necessary changes in the purchase orders and subcontracts will be made and the file sent to accounting.

Figure 2-11 **Change Estimate Folder.** This $8^1/_2 \times 11$ tagboard folder is used to accumulate all the documents supporting a change estimate. If the job is large enough to require a job engineer, the estimate is usually prepared on the job. Otherwise, it is prepared in the contractor's office. The use of such a folder will generally not be justified on very small jobs, but in this event some record must be kept to ensure that subcontractors are notified of any changes in their work.

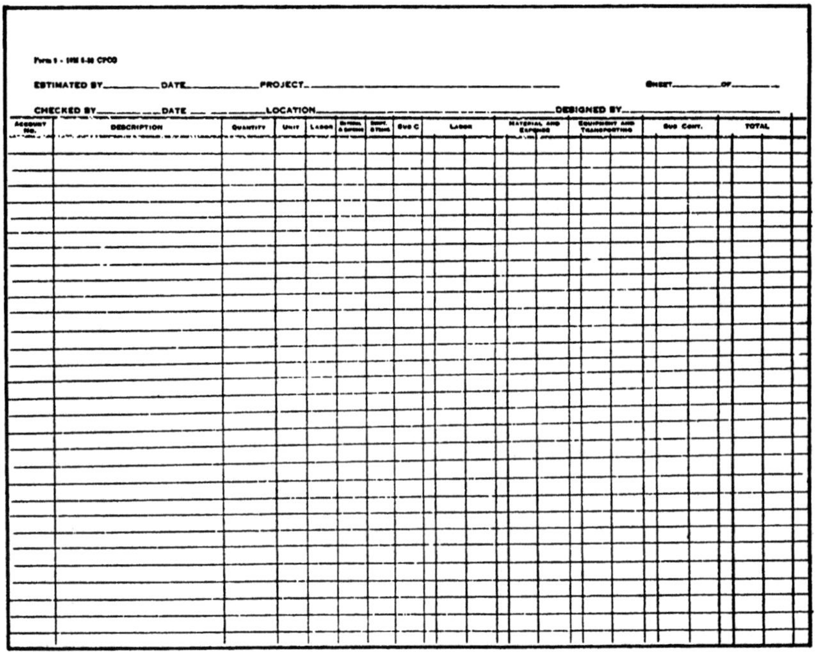

ESTIMATED BY_____ DATE_____ PROJECT_____ SHEET_____ OF_____

CHECKED BY_____ DATE_____ LOCATION_____ DESIGNED BY_____

Account No.	Description	Quantity	Unit	Labor	[illegible]	[illegible]	Sub C	Labor	Material and Expense	Equipment and Transporting	Sub Cont.	Total

Figure 2-12 Change Estimate. Change order estimates are often required when the cost of the change is an important factor in deciding whether the change is to be made.

used, it would be necessary for the job accountant to segregate the costs applicable to the principal work item from those applicable to the change order.

3. Some change orders would not affect the contract price and others would not affect any existing work item. The former would have no effect on accounting, and the latter would add another page to the cost ledger, regardless of whether the procedure in use calls for the adjustment of the budgeted amount for existing work items.

Changes in subcontracts and purchase orders

Prime contractors, and subcontractors who have sublet a portion of their subcontracts, are constantly facing the problem of determining to what extent, if any, a change in their work will affect their subcontractors. Usually this determination is made when the original change estimate is prepared, but there still remains the job of following through and issuing change orders to the subcontractors. The rulings on the face of the change estimate folder (Figure 2-11) are designed to give a measure of control.

The same follow-up needs to be carried out for materials purchase orders as well as for subcontracts. It is a good idea, whenever practical, to put copies of the subcontract change orders and the changes on purchase orders in the change estimate folder. When it is not practical, it is necessary to have some cross reference to the change order and purchase order files.

Once a proposed change order has been either abandoned or definitely ordered, a "disposition of pending change order" form (Figure 2-13) should be given to the subcontractors. A similar form adapted to fit purchase orders should be used to make changes in purchase orders.

When a change is actually ordered, a change order form (Figure 2-9) may be employed. If there is any question at all regarding the subcontractor's willingness to accept the change, it would be desirable to send notice by registered or certified mail. The change order forms, when prepared and submitted, should be accompanied by a transmittal letter, stating exactly how and by whom the change order is to be

ECKS & WYE
General Contractors
DISPOSITION OF PENDING CHANGE ORDER

Contract No.......................... Date.................
Job description...
....
Please refer to our "Notice of Pending Change Order No..........,"
dated...................., issued in connection with the above-described job.

The proposed change has been:

☐ Abandoned
☐ Ordered in the form originally described
☐ Ordered in an amended form, see attached.

Will you please proceed as follows:

☐ Continue under present plans and specifications
☐ Proceed at once with change. Formal change order is being prepared
☐ Do not proceed with change pending issuance of formal change order.

Signed...

Figure 2-13 **Disposition of Pending Change Order.**

signed and the date it is to be returned. A follow-up file should be kept, to make certain that all change orders are signed and returned.

When the contracts and the bond forms are correctly drawn, there should be no need for notice to the subcontractor's bonding company of any ordinary changes. Whether such notice is actually necessary is a decision that should be made by the contractor's attorney, and his written opinion on the point should be on file. If notice is necessary, it should be given by registered mail in a form approved by legal counsel. Failure to give notice when required could operate to discharge the surety.

Payments on change orders

Ordinarily, payment for work done under a change order will not be made until the change order is signed and returned by the contractor. It is usually the most businesslike procedure for the prime contractor to deal in the same way with subcontractors. When plans are available for prompt submission of change estimates and issuance of change orders, this procedure will create no undue hardship. However, unreasonable delay in issuing change orders coupled with the withholding of payment can cause trouble in several ways.

First, the job could be delayed because a contractor or subcontractor may justifiably refuse to proceed with work until he knows how much he is to be paid for doing it. Second, friction and dissension could (and usually do) arise on the job. Third, a liability for damages might result. Fourth, if a subcontractor is in financial difficulties it might make him insolvent, thus creating the problems and expenses that result from a subcontractor default.

If the issuance of change orders is being unduly delayed, it is a matter of concern to prime contractor and subcontractor alike, and there should be complete cooperation in locating and curing the causes of such delays. If they result from deliberate attempts to delay making payment, the party whose payments are being withheld should review his contract to see if he may not properly decline to start work on the change until change orders are issued.

If the contractor is required to start work before the changes are formally ordered, as frequently occurs with United States government contracts, the contractor should price his estimates so as to compensate him for the waiting time. Whether prime contractor or subcontractor, the more vigorously he demands compensation for waiting, the more contracting authorities will try to eliminate delay.

Often, the contracting authority (and here again the government is among the principal offenders) will deliberately refuse to issue a change order until the changed work is completed, and then will seek to negotiate the price so as to pay only for direct labor and direct material, and insist that no additional overhead has been incurred. If contracting authority follows such tactics as a matter of policy, the best practice is to try to provide a cushion for it in the bid. The next best procedure is to try to negotiate an acceptable formula for pricing change orders at the start of the job. With nongovernmental clients, it is usually possible and desirable to write the formula into the contract.

If the price of a change is to be negotiated after completion of the work, the only safe way to account for the costs is to set up a work order (Figure 2-10) and to charge it with all costs incurred as if it were a separate work item. When this procedure is followed, extraordinary care must be taken to see that the work order is charged with all the costs. If arbitrary segregations cannot be avoided, it is better to err on the side of overcharging the work order than it is to undercharge it.

In selecting a base to which an overhead rate is to be applied, the best choice is usually "total direct cost." Many cost auditors and owner's representatives, mindful of the questions they will be called upon to answer, are more concerned with the size of the overhead percentage rate than with the base to which it is applied. They are usually quite willing to see the contractor adequately reimbursed for his overhead costs, but they need a percentage rate that seems small so that it can be more easily justified. Total direct cost includes materials, equipment rentals, and subcontracts as well as direct labor; and from a strictly technical point of view it is probably the most accurate base against which to apply an overall overhead rate. In addition, it has the advantage, when accounting for change order costs, of being the broadest base.

When there are numerous change orders, there is often a need to summarize them in one place and on a single schedule. A job change order control sheet (Figure 2-14) is useful for this purpose. The manner in which the sheet is used is apparent from the column headings. Usually such a sheet is kept by the job engineer, or by the job office engineer on larger jobs.

In negotiating price adjustments for delays resulting from an excessive number of change orders, it is sometimes useful to have a set of plans marked with each point in the work that is affected by a change order.

One final word of caution in dealing with difficult third parties: Where there is a change in the scope or nature of the work, be sure to keep complete and accurate records on all aspects of the transaction. Many a claim or renegotiation has been lost primarily because of inadequate record keeping.

CHANGE ORDER RECORD

P. O NO.	P. O. DATE	C. O. NO.	VENDOR	REASON	AMOUNT OF C. O.	DATE C. O.	DATE MAILED	DATE REC'D	REMARKS

Figure 2-14 **Job Change Order Control Sheet.**

Chapter

3

Accounting for Labor*

Basic hiring forms and procedures

Before an employee is hired, someone has to decide that his services are needed. Just how much paperwork is required to start the new employee working depends on such things as the size of the contractor's organization, the nature of the construction project, the nature of the particular job the employee is to do, the requirements of the tax and labor laws, and the extent to which employment is subject to labor union rules and union contracts.

In a small organization there would be few formalities, and the contractor himself would probably do the hiring. In a very large operation the superintendent would issue a manpower requisition (Figure 3-1) which would be initialed by the project manager and sent to the personnel department to authorize hiring the number and classification of workers needed.

Regardless of the size of the organization, most of the people hired will come from substantially the same sources and will be recruited by one of the following methods:

1. Employing foremen who bring their own crews
2. Calling people who have worked for the contractor before on other jobs or are otherwise known to the contractor
3. Calling applicants from applications in the files
4. Employing applicants who happen to come in at the time of hiring

*Rapid changes are taking place in payroll and labor cost accounting through the use of computers. A more complete discussion of computer applications to payroll and labor cost accounting appears in Chapter 12. The forms illustrated in this chapter can easily be converted for use in a computerized system.

ECKS & WYE CO.
General Contractors No.........

Manpower Requisition Date...............

To: Personnel Dept.
Please provide the following personnel for........................
(Name of Job)

Number Needed	Classification	Date Needed

Signed.....................................
Title.....................................

OK to hire.................................... Project Mgr.

Figure 3-1 **Manpower Requisition.** The original is sent from the project manager to the personnel department. The duplicate remains in the superintendent's book.

5. Calling employment agencies
6. Calling the union hiring hall for unionized classifications
7. Advertising in newspapers and trade journals

The information to be obtained from a new employee varies with the type of employee, the tax laws, and the needs of the employer. Schedule 3-1 presents a list of forms used in the hiring process. Some of them are required; some of them may or may not be required; and some of them are entirely optional.

Most small contractors limit their hiring procedures to entering an employee's name, address, social security number, and dependency credits on a W-4 form and to the documentation required by the U.S. Immigration and Naturalization Service. But as the size and complexity of the contractor's organization increase, so do the formalities involved in hiring. Sometimes the kind of contract determines the amount of paperwork. For example, on United States government cost-plus-fixed-fee contracts, a manpower requisition would normally be required, together with applications (or signed data cards) for each employee, and, as a rule, evidence of each employee's citizenship. An order signed by someone in authority would have to be issued instruct-

SCHEDULE 3-1 CHECKLIST OF HIRING FORMS

Always required

Form W-4—employee's withholding allowance certificate (Figure 3-2). Note that the forms show name, home address, and social security number. Form W-4 is a U.S. Treasury Department form, but the data serve most state and local payroll tax purposes as well. Also required is the U.S. Immigration and Naturalization Service Form 1-9 plus supporting documentation (Figure 3-2c).

May be required

Union "clearance slip" may or may not be required, depending on the union agreement. If it is required, the unions will usually have their own forms.

State withholding tax certificate may or may not be required, depending on the state. California requires such forms (Figures 3-3 and 3-4).

Optional

Separate employee data card (Figure 3-5)

Combined employee data card and employee earnings record (Figure 3-6)

Application blank, used almost exclusively for executive and office employees (Figure 3-7)

Equipment issue receipts (Figure 3-8) are used when the contractor provides such items as hard hats, safety shoes, raincoats, boots, blankets, mattresses, pillows, sheets, tools of the trade, and so on. Such a receipt in the employee's file may be used to support a deduction upon his separation.

Camp and mess hall clearance slip (Figure 3-9) serves as a receipt when blankets, mattresses, bedding, and so on are turned in on termination, and when any mess hall charges are paid. If a camp and mess hall are maintained, this slip would have to be presented before the final check is issued.

Assignment slip for new employees (Figure 3-10)

Payroll change notice (Figure 3-11)

---

ing the paymaster to place the employee on the payroll. On some jobs, photographing, fingerprinting, and security clearances are also required. Of course, any property issues would be covered by receipts.

The test for hiring procedures is the same as for any other. Unless a procedure contributes more than its cost to the final profit, it is not justified. Usually the formalities observed at the hiring stage serve to comply with the law or with contractual requirements, and to prevent losses. If, under any given circumstances, a procedure does not serve either function, its justification may be seriously questioned.

A number of states have fair employment laws which prohibit questions on an application which disclose sex, marital status, dependents, pregnancy, birth control, names or addresses of relatives of adult applicants, or whether or not the applicant lives with parents.

Timekeeping procedures

Timekeeping serves two functions. First, it determines the total amount of time for which the employees are to be paid, and if the work is done at more than one rate, the amount of time to be paid at

No. 98 FORM W-4 131:1315

Form W-4 (1994)

Want More Money In Your Paycheck?

If you expect to be able to take the earned income credit for 1994, you can have part of it added to your take-home pay. For details, get Form W-5 from your employer.

Purpose. Complete Form W-4 so that your employer can withhold the correct amount of Federal income tax from your pay.

Exemption From Withholding. Read line 7 of the certificate below to see if you can claim exempt status. *If exempt, complete line 7; but do not complete lines 5 and 6.* No Federal income tax will be withheld from your pay. Your exemption is good for 1 year only. It expires February 15, 1995.

Note: *You cannot claim exemption from withholding if (1) your income exceeds $600 and includes unearned income (e.g., interest and dividends), and (2) another person can claim you as a dependent on their tax return.*

Basic Instructions. Employees who are not exempt should complete the Personal Allowances Worksheet. Additional worksheets are provided on page 2 for employees to adjust their withholding allowances based on itemized deductions, adjustments to income, or two-earner/two-job situations. Complete all worksheets that apply to your situation. The worksheets will help you figure the number of withholding allowances you are entitled to claim. However, you may claim fewer allowances than this.

Head of Household. Generally, you may claim head of household filing status on your tax return only if you are unmarried and pay more than 50% of the costs of keeping up a home for yourself and your dependent(s) or other qualifying individuals.

Nonwage Income. If you have a large amount of nonwage income, such as interest or dividends, you should consider making estimated tax payments using Form 1040-ES. Otherwise, you may find that you owe additional tax at the end of the year.

Two Earners/Two Jobs. If you have a working spouse or more than one job, figure the total number of allowances you are entitled to claim on all jobs using worksheets from only one Form W-4. This total should be divided among all jobs. Your withholding will usually be most accurate when all allowances are claimed on the W-4 filed for the highest paying job and zero allowances are claimed for the others.

Check Your Withholding. After your W-4 takes effect, you can use **Pub. 919, Is My Withholding Correct for 1994?**, to see how the dollar amount you are having withheld compares to your estimated total annual tax. We recommend you get Pub. 919 especially if you used the Two Earner/Two Job Worksheet and your earnings exceed $150,000 (Single) or $200,000 (Married). Call 1-800-829-3676 to order Pub. 919. Check your telephone directory for the IRS assistance number for further help.

Personal Allowances Worksheet

A Enter "1" for **yourself** if no one else can claim you as a dependent . . . . A ______

B Enter "1" if:
- You are single and have only one job; or
- You are married, have only one job, and your spouse does not work; or
- Your wages from a second job or your spouse's wages (or the total of both) are $1,000 or less.

. . . B ______

C Enter "1" for your **spouse**. But, you may choose to enter -0- if you are married and have either a working spouse or more than one job (this may help you avoid having too little tax withheld) . . . . C ______

D Enter number of **dependents** (other than your spouse or yourself) whom you will claim on your tax return . . . . . . D ______

E Enter "1" if you will file as **head of household** on your tax return (see conditions under **Head of Household** above) . E ______

F Enter "1" if you have at least $1,500 of **child or dependent care expenses** for which you plan to claim a credit . . F ______

G Add lines A through F and enter total here. **Note:** This amount may be different from the number of exemptions you claim on your return ▶ G ______

For accuracy, do all worksheets that apply.
- If you plan to **itemize or claim adjustments to income** and want to reduce your withholding, see the Deductions and Adjustments Worksheet on page 2.
- If you are **single** and have **more than one job** and your combined earnings from all jobs exceed $30,000 OR if you are **married** and have a **working spouse or more than one job**, and the combined earnings from all jobs exceed $50,000, see the Two-Earner/Two-Job Worksheet on page 2 if you want to avoid having too little tax withheld.
- If **neither** of the above situations applies, **stop here** and enter the number from line G on line 5 of Form W-4 below.

.... **Cut here and give the certificate to your employer. Keep the top portion for your records.** ....

Form **W-4** — Department of the Treasury, Internal Revenue Service

Employee's Withholding Allowance Certificate

▶ **For Privacy Act and Paperwork Reduction Act Notice, see reverse.**

OMB No. 1545-0010 — **1994**

1 Type or print your first name and middle initial | Last name | 2 Your social security number

Home address (number and street or rural route) | 3 ☐ Single ☐ Married ☐ Married, but withhold at higher Single rate.
Note: If married, but legally separated, or spouse is a nonresident alien, check the Single box.

City or town, state, and ZIP code | 4 If your last name differs from that on your social security card, check here and call 1-800-772-1213 for more information . . . . ▶ ☐

5 Total number of allowances you are claiming (from line G above or from the worksheets on page 2 if they apply) . 5 ______

6 Additional amount, if any, you want withheld from each paycheck . . . . 6 $ ______

7 I claim exemption from withholding for 1994 and I certify that I meet **BOTH** of the following conditions for exemption:
- Last year I had a right to a refund of **ALL** Federal income tax withheld because I had **NO** tax liability; **AND**
- This year I expect a refund of **ALL** Federal income tax withheld because I expect to have **NO** tax liability.

If you meet both conditions, enter "EXEMPT" here . . . . ▶ 7 ______

Under penalties of perjury, I certify that I am entitled to the number of withholding allowances claimed on this certificate or entitled to claim exempt status.

Employee's signature ▶ **Date** ▶ , 19

8 Employer's name and address (Employer: Complete 8 and 10 only if sending to the IRS) | 9 Office code (optional) | 10 Employer identification number

Cat. No. 10220Q

12-93 Published by The Bureau of National Affairs, Inc. 131

Figure 3-2a Employee's Withholding Allowance Certificate—Face.

Form W-4 (1994) Page **2**

Deductions and Adjustments Worksheet

Note: *Use this worksheet only if you plan to itemize deductions or claim adjustments to income on your 1994 tax return.*

1	Enter an estimate of your 1994 itemized deductions. These include: qualifying home mortgage interest, charitable contributions, state and local taxes (but not sales taxes), medical expenses in excess of 7.5% of your income, and miscellaneous deductions. (For 1994, you may have to reduce your itemized deductions if your income is over $111,800 ($55,900 if married filing separately). Get Pub. 919 for details.)	1	$
2	Enter: { $6,350 if married filing jointly or qualifying widow(er); $5,600 if head of household; $3,800 if single; $3,175 if married filing separately }	2	$
3	**Subtract** line 2 from line 1. If line 2 is greater than line 1, enter -0-	3	$
4	Enter an estimate of your 1994 adjustments to income. These include alimony paid and deductible IRA contributions	4	$
5	**Add** lines 3 and 4 and enter the total	5	$
6	Enter an estimate of your 1994 nonwage income (such as dividends or interest)	6	$
7	**Subtract** line 6 from line 5. Enter the result, but not less than -0-	7	$
8	**Divide** the amount on line 7 by $2,500 and enter the result here. Drop any fraction	8	
9	Enter the number from Personal Allowances Worksheet, line G, on page 1	9	
10	**Add** lines 8 and 9 and enter the total here. If you plan to use the Two-Earner/Two-Job Worksheet, also enter this total on line 1, below. Otherwise, **stop here and** enter this total on Form W-4, line 5, on page 1.	10	

Two-Earner/Two-Job Worksheet

Note: *Use this worksheet only if the instructions for line G on page 1 direct you here.*

1	Enter the number from line G on page 1 (or from line 10 above if you used the Deductions and Adjustments Worksheet)	1	
2	Find the number in **Table 1** below that applies to the **LOWEST** paying job and enter it here	2	
3	If line 1 is **GREATER THAN OR EQUAL TO** line 2, subtract line 2 from line 1. Enter the result here (if zero, enter -0-) and on Form W-4, line 5, on page 1. **DO NOT** use the rest of this worksheet	3	

Note: *If line 1 is **LESS THAN** line 2, enter -0- on Form W-4, line 5, on page 1. Complete lines 4–9 to calculate the additional withholding amount necessary to avoid a year-end tax bill.*

4	Enter the number from line 2 of this worksheet . . . . 4 ______		
5	Enter the number from line 1 of this worksheet . . . . 5 ______		
6	**Subtract** line 5 from line 4	6	
7	Find the amount in **Table 2** below that applies to the **HIGHEST** paying job and enter it here	7	$
8	**Multiply** line 7 by line 6 and enter the result here. This is the additional annual withholding amount needed	8	$
9	Divide line 8 by the number of pay periods remaining in 1994. (For example, divide by 26 if you are paid every other week and you complete this form in December 1993.) Enter the result here and on Form W-4, line 6, page 1. This is the additional amount to be withheld from each paycheck	9	$

Table 1: Two-Earner/Two-Job Worksheet

Married Filing Jointly				All Others	
If wages from LOWEST paying job are—	Enter on line 2 above	If wages from LOWEST paying job are—	Enter on line 2 above	If wages from LOWEST paying job are—	Enter on line 2 above
0 - $3,000	0	39,001 - 50,000	9	0 - $4,000	0
3,001 - 6,000	1	50,001 - 55,000	10	4,001 - 10,000	1
6,001 - 11,000	2	55,001 - 60,000	11	10,001 - 14,000	2
11,001 - 16,000	3	60,001 - 70,000	12	14,001 - 19,000	3
16,001 - 21,000	4	70,001 - 80,000	13	19,001 - 23,000	4
21,001 - 27,000	5	80,001 - 90,000	14	23,001 - 45,000	5
27,001 - 31,000	6	90,001 and over	15	45,001 - 60,000	6
31,001 - 34,000	7			60,001 - 70,000	7
34,001 - 39,000	8			70,001 and over	8

Table 2: Two-Earner/Two-Job Worksheet

Married Filing Jointly		All Others	
If wages from HIGHEST paying job are—	Enter on line 7 above	If wages from HIGHEST paying job are—	Enter on line 7 above
0 - $ 50,000	$370	0 - $ 30,000	$370
50,001 - 100,000	690	30,001 - 60,000	690
100,001 - 130,000	780	60,001 - 110,000	780
130,001 - 220,000	880	110,001 - 220,000	880
220,001 and over	970	220,001 and over	970

Privacy Act and Paperwork Reduction Act Notice.—We ask for the information on this form to carry out the Internal Revenue laws of the United States. The Internal Revenue Code requires this information under sections 3402(f)(2)(A) and 6109 and their regulations. Failure to provide a completed form will result in your being treated as a single person who claims no withholding allowances. Routine uses of this information include giving it to the Department of Justice for civil and criminal litigation and to cities, states, and the District of Columbia for use in administering their tax laws.

The time needed to complete this form will vary depending on individual circumstances. The estimated average time is: **Recordkeeping** 46 min., **Learning about the law or the form** 10 min., **Preparing the form** 69 min. If you have comments concerning the accuracy of these time estimates or suggestions for making this form more simple, we would be happy to hear from you. You can write to both the **Internal Revenue Service**, Attention: Reports Clearance Officer, PC:FP, Washington, DC 20224; and the **Office of Management and Budget**, Paperwork Reduction Project (1545-0010), Washington, DC 20503. **DO NOT** send the tax form to either of these offices. Instead, give it to your employer.

Figure 3-2b Employee's Withholding Allowance Certificate—Reverse.

U.S. Department of Justice
Immigration and Naturalization Service

OMB No. 1115-0136
Employment Eligibility Verification

Please read instructions carefully before completing this form. The instructions must be available during completion of this form. ANTI-DISCRIMINATION NOTICE. It is illegal to discriminate against work eligible individuals. Employers CANNOT specify which document(s) they will accept from an employee. The refusal to hire an individual because of a future expiration date may also constitute illegal discrimination.

Section 1. Employee Information and Verification. To be completed and signed by employee at the time employment begins

Print Name: Last | First | Middle Initial | Maiden Name

Address *(Street Name and Number)* | Apt. # | Date of Birth *(month/day/year)*

City | State | Zip Code | Social Security #

I am aware that federal law provides for imprisonment and/or fines for false statements or use of false documents in connection with the completion of this form.

I attest, under penalty of perjury, that I am (check one of the following):
- ☐ A citizen or national of the United States
- ☐ A Lawful Permanent Resident (Alien # A ________
- ☐ An alien authorized to work until ___/___/___
  (Alien # or Admission # ________

Employee's Signature | Date *(month/day/year)*

Preparer and/or Translator Certification. *(To be completed and signed if Section 1 is prepared by a person other than the employee.) I attest, under penalty of perjury, that I have assisted in the completion of this form and that to the best of my knowledge the information is true and correct.*

Preparer's/Translator's Signature | Print Name

Address *(Street Name and Number, City, State, Zip Code)* | Date *(month/day/year)*

Section 2. Employer Review and Verification. To be completed and signed by employer. **Examine one document from List A OR examine one document from List B and one from List C** as listed on the reverse of this form and record the title, number and expiration date, if any, of the document(s)

	List A	OR	List B	AND	List C
Document title:	________		________		________
Issuing authority:	________		________		________
Document #:	________		________		________
Expiration Date *(if any)*:	___/___/___		___/___/___		___/___/___
Document #:	________				
Expiration Date *(if any)*:	___/___/___				

CERTIFICATION - I attest, under penalty of perjury, that I have examined the document(s) presented by the above-named employee, that the above-listed document(s) appear to be genuine and to relate to the employee named, that the employee began employment on *(month/day/year)* ___/___/___ **and that to the best of my knowledge the employee is eligible to work in the United States. (State employment agencies may omit the date the employee began employment).**

Signature of Employer or Authorized Representative | Print Name | Title

Business or Organization Name | Address *(Street Name and Number, City, State, Zip Code)* | Date *(month/day/year)*

Section 3. Updating and Reverification. To be completed and signed by employer

A. New Name *(if applicable)* | B. Date of rehire *(month/day/year) (if applicable)*

C. If employee's previous grant of work authorization has expired, provide the information below for the document that establishes current employment eligibility.

Document Title: ________ Document #: ________ Expiration Date (if any): ___/___/___

I attest, under penalty of perjury, that to the best of my knowledge, this employee is eligible to work in the United States, and if the employee presented document(s), the document(s) I have examined appear to be genuine and to relate to the individual.

Signature of Employer or Authorized Representative | Date *(month/day/year)*

Form I-9 (Rev. 11-21-91) N

Figure 3-2c **Immigration and Naturalization Service Form I-9.**

CALENDAR YEAR

19____

Certificate of California Residence

CALIFORNIA FORM

590

▶ **For Privacy Act Notice, see Form FTB 1131.** | **To provide certification of residence to payer for this calendar year only.**

Type or print full name of payee | Social Security Number

Home Address

City | California, Zip Code

☐ Married ☐ Single | Number of exemptions claimed on Federal Form W-4:

PAYER — Keep a copy of this certificate with your records. Forward the original on or before the engagement to **Withholding At Source, P.O. Box 1673, Sacramento, CA 95808-1673. Telephone (916) 369-4900.** Payees must complete a new Form 590 certificate annually. If the payee becomes a nonresident at any time, tax must be withheld at source and Form 591 must be completed.

To ________________
Employer

I hereby declare that I am a resident of the State of California, and that I reside at the address shown above. Should I become a nonresident at any time, I will promptly inform you of that fact.

Signature ____________ Date ________

Form 590 1986 Side 1

The term "resident" includes every individual who is in California for other than a temporary or transitory purpose. Generally, an individual who comes to California for a purpose which will require a long or indefinite period will be considered a resident. However, an individual who comes to California to perform a particular contract of a short duration generally will be considered a nonresident. For assistance in determining residence status, call the toll-free telephone number for your area.

TELEPHONE ASSISTANCE

Southern California	Northern California	Sacramento Metropolitan Area
(Area Codes 213, 619, 714, 805, 818)	(Area Codes 209, 408, 415, 707, 916)	and Out of State
(800) 852-5711	**(800) 852-7050**	***(916) 369-0500**

For hearing impaired with TDD **(800) 822-6268**

*If this is a toll call from your Sacramento location, call **(800) 852-7050.**

Side 2 Form 590 1987

Figure 3-3 **California Residence Tax Form.** This form and that shown in Figure 3-4 are withholding tax forms used by the state of California. Other states, and even some cities, have personal income taxes, which may be collected by withholding.

each. Second, it determines which work items (cost accounts) are to be charged for the work done. The smaller and simpler the operation, the less chance that any employee will be absent for any substantial period of time without that fact coming to the contractor's attention.

The simplest time record is the foreman's time book. These books usually have about 50 pages (shown later in Figure 3-12) with a heavy tagboard cover so that they can be carried in the foreman's pocket without getting too badly crumpled. The form is self-explanatory. However, when the only time record kept is the one in the foreman's book, paychecks must be prepared directly from it, and that means that the foreman must either prepare the checks himself or be without his book for a while. If a time book is lost, the only record of the crew's time is also lost. This loss might be costly at the time of a compensation insurance or payroll tax audit.

0 592-B Personal CALIFORNIA 11,495

TAXABLE YEAR CALIFORNIA FORM

19____ Report of Tax Withheld at Source on Nonresidents 592-B

Attach to Form 592 for each recipient. See the separate instructions for Form 592, 592-A and 592-B.

Copy A
FOR FRANCHISE TAX BOARD

Part I Recipient	**Part II Withholding agent (Payer/Partnership)**
Recipient's name, address (number and street), city, province or state, postal code, and country	Withholding agent's (payer's/partnership's) name, address (number and street), city, state, ZIP code and telephone number
Social security no., California corporation no. or federal employer identification no. (F.E.I.N.)	Social security no., California corporation no. or federal employer identification no. (F.E.I.N.)

Part III Preparer (If other than withholding agent)
Preparer's name, address (number and street), city, state, ZIP code and telephone number

Social security no., California corporation no. or federal employer identification no. (F.E.I.N.)

Part IV Type of income subject to withholding. Check the applicable box(es).

☐ Payment to Independent Contractor ☐ Rents or Royalties ☐ Estate or Trust Distributions ☐ Payment to Foreign Partner ☐ Payment to Domestic (Nonforeign) Nonresident Partner ☐ Other ________ (describe)

Part V Tax Withheld

1 Total amount subject to withholding 1

2 Total tax withheld 2

Form 592-B (REV. 1993) **Side 1**

Figure 3-4 **California Nonresident Tax Form.**

For these reasons it is quite common for the information in the foreman's time book to be copied on time sheets or time cards. Attempts to make a carbon copy of the foreman's time book have not been very successful, because a foreman usually carries the book in his pocket. The copies resulting from such attempts are often illegible and badly smeared. One alternative is to copy the data from the time book on a weekly time sheet (Figure 3-13). Sometimes this work is done by the foreman, sometimes by a timekeeper, payroll clerk, or bookkeeper.

If weekly time sheets are used, they are usually prepared by the foreman. This procedure, however, permits collusion between the foreman and one or more of his workers. If the crew is small enough so that the contractor would normally find out about unauthorized absences, no further check is needed. However, when the crew is too large, or when, for some other reason, the contractor would not know of absences that did not appear on the time records, then occasional surprise visits by a timekeeper should be made, and the time sheets should be inspected when they are turned in, for any absences noted on the surprise visit.

Sometimes weekly time sheets are prepared by the timekeeper, who enters the time daily for each employee and summarizes it for the weekly payroll. Cost distributions are made on the same sheet.

EMPLOYEE DATA CARD

Date employed			Job classification	Rate	Per	Rate OK'd by	Tax status		Separation		OK to rehire
Mo.	Day	Year					Date	No. exemp.	Date	Reason	

Married ☐ Single ☐ Separated ☐ Work is: Permanent ☐ Temporary ☐
Male ☐ Female ☐ Date of birth . . . .
Place of birth . . . . U. S. Citizen: Yes ☐ No ☐
In case of accident notify . . . . Withholding exemptions . . . .

Address . . . . Phone . . . .
Name . . . . Badge No. . . . . Social Sec. No. . . . .

Figure 3-5 **Employee's Data Card (Separate).**

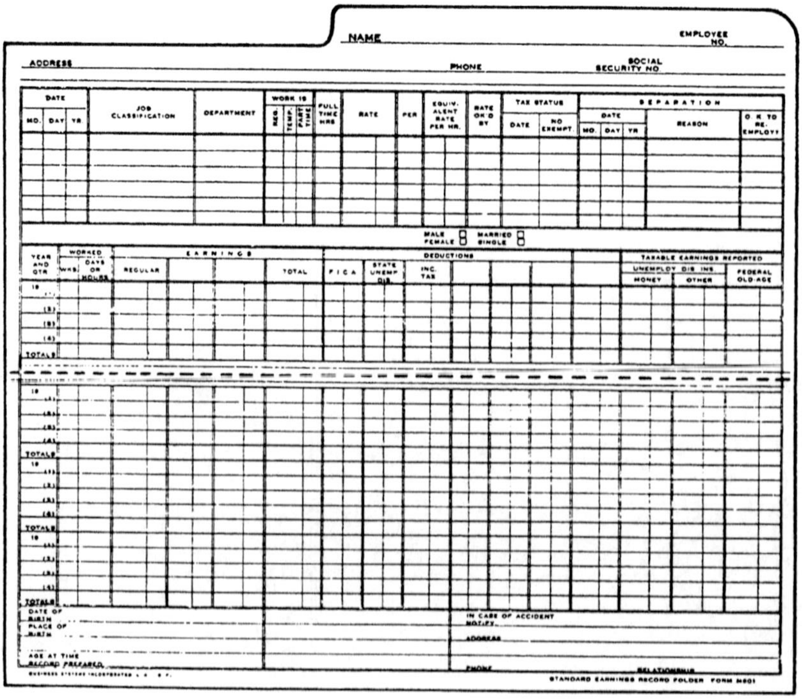

Figure 3-6 **Combined Employee's Data Card and Earnings Record.** This form is used as a file folder for copies of the payroll checks as well as a data card and earnings record.

The principal disadvantages of weekly time sheets are:

1. Labor cost figures are not available daily.
2. Errors may be as much as eight or nine days old before they are noticed; and usually the older they are the harder they are to correct.
3. If a foreman carries a time sheet around for a week it gets messy and illegible, and is a source of errors in pay records.
4. On a construction job, there may be a considerable turnover of personnel, and a weekly time sheet is not very flexible.
5. If a dishonest foreman wants to insert fraudulent names on the payroll or to carry a worker laid off in the middle of the week for an extra day or two, the chances of detection are somewhat poorer with a weekly time sheet than with other forms.

To overcome the disadvantages of the weekly time report (Figure 3-13), daily time sheets (Figure 3-14) or daily time cards (Figures 3-15 to 3-17) are frequently used.

Application for Employment

GENERAL INFORMATION

Print name in full____________________ Soc. Sec. No.__________

Address ____________________ Telephone __________

Date of birth________________ Are you a U. S. citizen__________

Name and address of parent or guardian (if applicant is a minor) __________

____

The company does not permit employment of relatives or spouses in the same department __________

Position applied for____________________ Monthly salary expected__________

Are you employed at present?__________ How soon can you report if engaged__________

EDUCATION

Name and location of school	Dates From — to	Did you Graduate	Nature of course taken, or degree	Studying now?
Grade school				
High school				
College				
Other				

What studies did you like best?____________________

EXPERIENCE

Give below, *starting with present or most recent position, complete record of your business experience.*

Firm	Period	Describe your duties and position	Name immediate superior and his title	Why did you leave?
Name	From			
Address	To			
Business	Salary			
Name	From			
Address	To			
Business	Salary			
Name	From			
Address	To			
Business	Salary			

— over —

Figure 3-7a **Application Blank—Face.**

The use of either time sheets or time cards is based on the assumption that someone (usually the foreman or the timekeeper) is going to be responsible for seeing to it that the employees are docked for the time lost if they are late to work in the morning or if they leave the job before quitting time at night. In a factory, this function is performed by time clocks. However, on construction jobs, time clocks have never proved to be very satisfactory, and they are rarely used. When a check-in and check-out system is required, the so-called "brass" system is probably the most common.

Why do you desire to change your position? ______________________________

Were you ever discharged from any position? ________ Explain ____________________

REFERENCES

GIVE THREE REFERENCES OTHER THAN RELATIVES OR PAST EMPLOYERS

Name	Business or profession	Position	Address	Telephone	How long known
1					
2					
3					

MISCELLANEOUS

Will you be engaged in any other income producing activity, if employed? ____________

Explain ______________________________

Have you ever been bonded? ________ Has bond ever been refused you? Explain ____________

The answers to the foregoing questions are true and correct and I have not knowingly withheld any fact or circumstance which would, if disclosed, affect my application unfavorably.

Applicant's signature ____________________

Date of application ____________________

Figure 3-7b **Application Blank—Reverse.**

In concept and operation the brass system is simple. Each worker is assigned a payroll number or badge number. Sometimes these numbers are grouped by crafts, and sometimes they are simply issued in the order of employment. On a board at the timekeeper's window are a number of metallic (brass) disks, each bearing a number corresponding to the employee's badge number. As employees report for work they show their badge and are handed their corresponding brass

BEDDING RECEIPT

Badge No. . . . .

Wool blankets	. . . .	Cotton blankets	. . . .
Pillows	. . . .	Pillow slips	. . . .
Sheets	. . . .	Mattress	. . . .

I received the above items on: . . . .

Payroll deductions for meals and lodging are hereby authorized. . . . .
(Signed)

Figure 3-8a **Equipment Issue Receipt—Face.**

Name Check-in date Badge No.

Home address

Contractor

Occupation

Barracks No. Room No. Bed No.

Check-out date

Figure 3-8b **Equipment Issue Receipt—Reverse.**

Ecks & Wye Construction Co.

CAMP AND MESS HALL CLEARANCE

Date. . . . . .Hour. . . . .

To the Paymaster:

This is to certify that. . . . .

. . . . .Badge No. . . . .

has turned in all items charged to him except: . . . .

. . . .

. . . .

He is to be charged/credited as follows: . . . .

. . . .

. . . .

. . . .

Signed: . . . . Camp Mgr.

Figure 3-9 **Camp and Mess Hall Clearance Slip.**

Assignment Slip for New Employees

(GIVE THIS TO YOUR FOREMAN)

EMPLOYEE'S P. R. No. ....

Name.. Ident. No..................

Occupation.. Rate.............. per..............

Report to Foreman..

(name)

At Div.................................. To Start Work.................. P.M. A.M.

WILL HAVE MEALS AND LODGING IN CAMP YES NO

Figure 3-10 **Assignment Slip for New Employees.**

PAYROLL CHANGE NOTICE

Please check —

☐ Enter on payroll
☐ Change rate
☐ Transfer to.............. Entered (Time Dept.).... Date.........
☐ Pay off and remove from payroll

Name.................................S.S.A/C No...............

Dept...................Shift...............Clock No.............

Date effective.............................Hour............. A.M. P.M.

Old rate..............Per.......New rate..............Per.......

Remarks:..

....

Record ☐ Discharged ☐ Left ☐ Laid off Would you re-employ Yes ☐ No ☐

	Excellent	Good	Fair	Poor
Ability				
Conduct				
Attendance				
Production				

Approved.......................... ....
Supt. *Foreman*

Figure 3-11 **Payroll Change Notice.** This form is used to hire a new employee, to change his assignment or rate after he is employed, or as a termination.

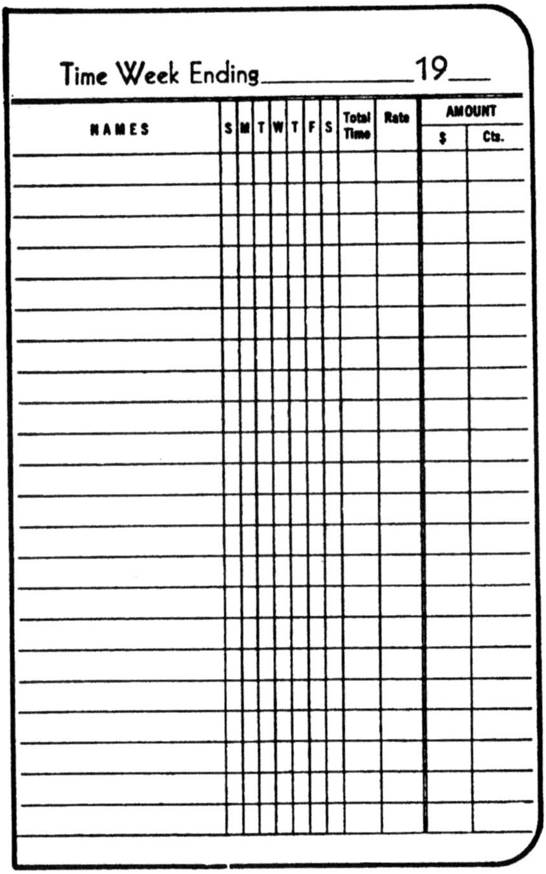
Time Week Ending__________19___

NAMES	S	M	T	W	T	F	S	Total Time	Rate	AMOUNT $	AMOUNT Cts.

Figure 3-12 **A Sheet from a Foreman's Time Book.**

tag. Brass tags still on the board when work starts indicate that the workers to whom the corresponding badges have been issued are either late or absent. As latecomers arrive, the brass is picked up and recorded on the time record for that day.

When the employees leave the job, their brass is turned in and put back on the board. If a person turns in his brass early, the time is noted and recorded on the time record. If he fails to turn in his brass at night, it is necessary to find out why. He may have just forgotten; he may have gone to lunch and not returned; or he may have left the job before quitting time. During the day it is not uncommon to use the brass as a tool check. When brass is used for this purpose, the worker gives his brass to the tool room upon checking out his tools. At the end of the shift he turns in his tools and retrieves his brass. Failure to turn in his tools means that he is unable to "brass out" upon leav-

Job No........ **Ecks & Wye Construction Co.**

TIME REPORT FOR THE WEEK ENDING

Badge No.	Name	S O	Sat	Sun	Mon	Tue	Wed	Thu	Fri	Total
		S								
		O								
		S								
		O								
		S								
		O								
		S								
		O								
		S								
		O								
		S								
		O								
		S								
		O								
		S								
		O								
		S								
		O								
		S								
		O								
Signed		S								
.... Foreman		O								

Figure 3-13 **Weekly Time Report.** This form is designed for a payroll week ending Friday evening. An employee who works on more than one job during the week gets a separate paycheck for each.

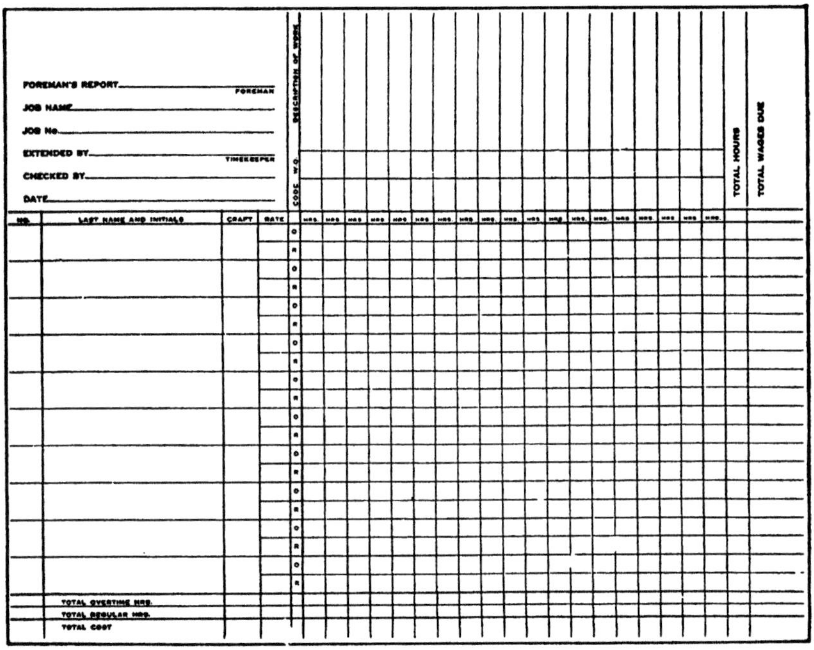

Figure 3-14 **Daily Time Report.** This form is prepared by the foreman and is then completed in the payroll department.

ing the site and will call attention to the fact that he has not returned his tools.

One of the most important disadvantages of the brass system is the fact that it automatically introduces "portal-to-portal" pay. If the work area is a long way from the time office, this system can cause the loss of a good many worker-hours each day, and thousands of dollars on a large job. It is common to find this system in use on United States government cost-plus-fixed-fee jobs for which paper accountability is more important than actual economy.

In some of the smallest construction firms there is no timekeeping procedure as such. At the end of the week the employees turn in weekly cards or sheets showing the hours worked and the names or addresses of the jobs on which they worked during the week; the paychecks are made from the time records turned in by each individual. On some very large jobs where the work area is close enough so that the portal-to-portal problem is not serious, it is quite possible to use time clocks satisfactorily. However, because time clocks traditionally have not been used much in the construction industry, attempts to use them are sometimes met by strong union opposition. Even when they are acceptable to the employees time clocks do not, as a rule,

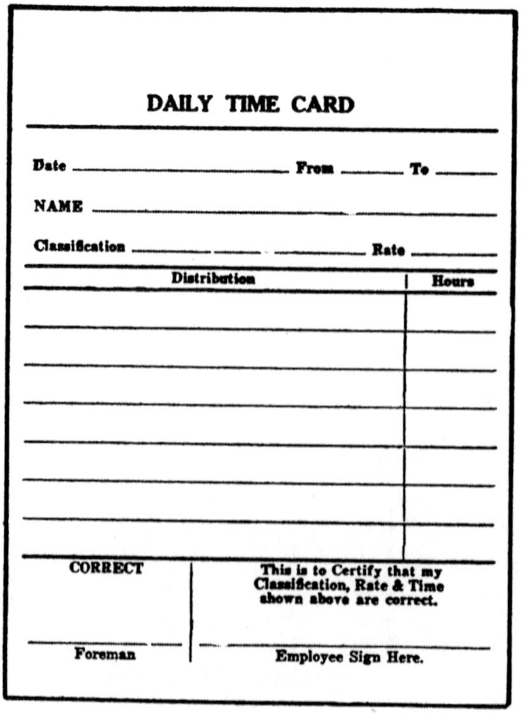

DAILY TIME CARD

Date ____________ From ______ To ______

NAME ____________________

Classification ____________ Rate ______

Distribution	Hours

CORRECT

This is to Certify that my Classification, Rate & Time shown above are correct.

____________ Foreman

____________ Employee Sign Here.

Figure 3-15 **Daily Time Card.** This simple card is suitable for almost any kind of job except one that is so large that mechanical sorting would be required because of the number of employees involved.

prove very practical in making the cost accounting segregations required for construction jobs.

Special timekeeping problems

If work is scattered over a wide area, it is often necessary to set up "area" time shacks, where an area timekeeper or the area foreman or superintendent makes his headquarters. If time on remote crews is kept by the foreman or an area timekeeper, surprise face-checks should be made at frequent but irregular intervals, and occasionally someone from the paymaster's department should distribute the paychecks.

The same problems arise when work is in an isolated area. Lack of direct supervision is common under such conditions, and sometimes employees believe that no one will know if they are absent. Such absenteeism can become an important item of increased cost. If the superintendent will arrange to make frequent visits at irregular

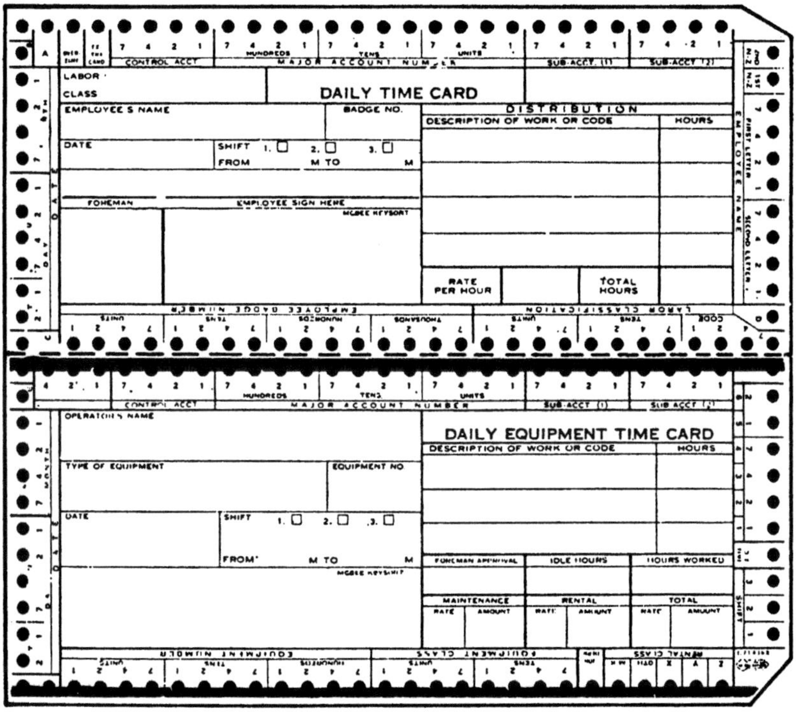
CONTROL ACCT — MAJOR ACCOUNT NUMBER (HUNDREDS, TENS, UNITS) — SUB-ACCT. (1) — SUB-ACCT. (2)

LABOR CLASS

DAILY TIME CARD

EMPLOYEE'S NAME — BADGE NO.

DISTRIBUTION: DESCRIPTION OF WORK OR CODE — HOURS

DATE — SHIFT 1. ☐ 2. ☐ 3. ☐ — FROM M TO M

FOREMAN — EMPLOYEE SIGN HERE

McBEE KEYSORT

RATE PER HOUR — TOTAL HOURS

CONTROL ACCT — MAJOR ACCOUNT NUMBER (HUNDREDS, TENS, UNITS) — SUB-ACCT (1) — SUB-ACCT (2)

OPERATOR'S NAME

DAILY EQUIPMENT TIME CARD

DESCRIPTION OF WORK OR CODE — HOURS

TYPE OF EQUIPMENT — EQUIPMENT NO.

DATE — SHIFT 1. ☐ 2. ☐ 3. ☐ — FROM M TO M

FOREMAN APPROVAL — IDLE HOURS — HOURS WORKED

MAINTENANCE (RATE, AMOUNT) — RENTAL (RATE, AMOUNT) — TOTAL (RATE, AMOUNT)

Figure 3-16 **Keysort Labor and Equipment Time Card.** This card is used in large operations where mechanical sorting is required. The daily time card may be used independently.

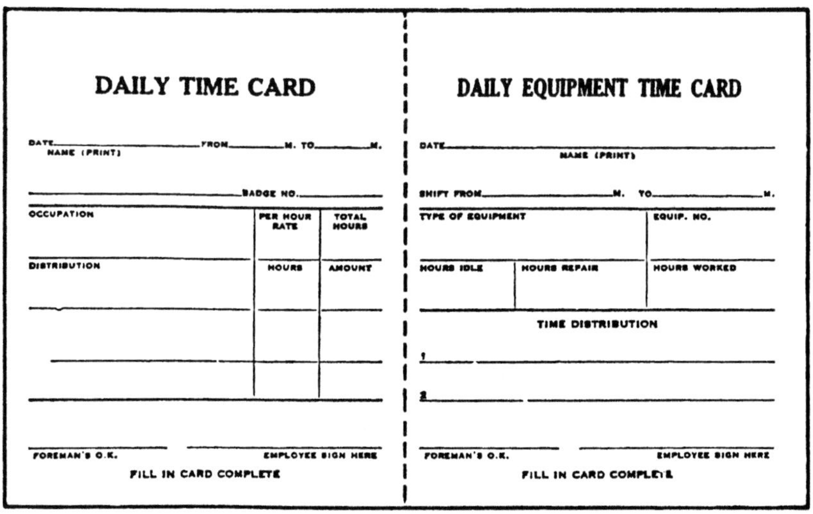
DAILY TIME CARD

DATE ____ FROM ____ M. TO ____ M.

NAME (PRINT)

____ BADGE NO. ____

OCCUPATION	PER HOUR RATE	TOTAL HOURS
DISTRIBUTION	HOURS	AMOUNT

FOREMAN'S O.K. — EMPLOYEE SIGN HERE

FILL IN CARD COMPLETE

DAILY EQUIPMENT TIME CARD

DATE ____

NAME (PRINT)

SHIFT FROM ____ M. TO ____ M.

TYPE OF EQUIPMENT		EQUIP. NO.
HOURS IDLE	HOURS REPAIR	HOURS WORKED

TIME DISTRIBUTION

1 ____

2 ____

FOREMAN'S O.K. — EMPLOYEE SIGN HERE

FILL IN CARD COMPLETE

Figure 3-17 **Daily Time Card.** This form is used by equipment operators and has an equipment time card attached.

intervals to check work progress at a remote jobsite or work area, such visits will also discourage unauthorized absences. If unit costs for labor are running too high it usually indicates a need for more and better supervision, and this observation is just as true when the overrun is caused by misconduct as it is when the overrun is caused by inefficiency and faulty planning.

Another special timekeeping problem is created when people are employed at shifting rates. There are three different situations under which such problems may arise:

1. A whole crew does work which varies in classification from hour to hour within the day. It is customary for the crew to be paid for the whole day at the rate for the highest classification in which the employees worked that day.
2. One or more crew members will be employed on work requiring a higher rate of pay than usual. For example, a laborer may use a jackhammer for all or part of the day and become entitled to an "air tool operator" rate for that day.
3. One or more individuals in a higher classification may work with a labor crew for one day and thus increase the rate which the crew foreman is entitled to receive for that day. For example, union rules in some areas require that a crew foreman must receive a wage differential above the highest paid worker on his crew. If, under such rules, a labor foreman during all or part of one day had a pipe fitter on his crew, his rate for that day would go up to the point where it exceeded the pipe fitter's rate by the required differential.

When shifting rates are possible, there are several things to be watched for. First, there is the possibility of collusion between a superintendent and a foreman to increase the pay of the foreman. Second, particularly when foremen report the time, there may be similar collusion between the foreman and individual workers. Of course an excess of work outside of classification may be due to faulty planning; no amount of timekeeping effort can remedy such a defect. However, when certain crews seem to be doing a disproportionate amount of work outside of their regular classification, a change in supervisory personnel may be indicated. Sometimes a mere request for an explanation will stop the irregularities. The same comments apply to excessive transfers of workers between crews.

If higher rates to certain individuals for the use of special tools appear on the payroll too often, it may be worthwhile to find out if the tool shack records show that such tools were actually charged out to those individuals on the days in question. It might also be in order to find out the kind of work the crew was doing to see if any work with special tools was required. For example, grounds for investigation would exist if a jackhammer operator showed up on a crew that was doing hand excavation in an area of soft or filled earth.

Any possibility of shifting rates calls for extreme care in timekeeping, and all persons responsible for timekeeping and payroll should know the rules thoroughly. Naturally, if things are not as they should be, it is necessary to find and correct the trouble. However, regardless of other circumstances, it is of paramount importance that the payroll checks be correct. There are very few timekeeping or payroll situations so vulnerable to misinterpretation as a failure to compute time at the correct rate. Anything more than an occasional error of this kind is damaging to job morale and is an open invitation to professional troublemakers.

Another special timekeeping problem is that created by rapid turnover of help. On very small jobs, and in very small firms, the payroll is so small that there is usually no problem. On larger jobs it is important that, when a new employee is placed on the payroll, the person responsible for the timekeeping be notified of that fact immediately. If the foreman keeps the time, this procedure is automatic. If timekeepers are used, there should be some prompt and completely dependable way for the timekeeper to learn of new additions to the payroll so that time can be correctly recorded. One method is to give the timekeeper a copy of all assignment slips (Figure 3-10). Another is to have all new employees report to the timekeeper to receive their assignment slips.

It is equally important that, whenever a person is laid off, his time up to the moment of termination be in the hands of the paymaster. For this purpose, time must be recorded daily through the last shift of the preceding day. Then when the individual is laid off, there need only be added the time for the current day.

There are several ways of summarizing time data coming in from the job. One of the easiest and simplest is to record it on a payroll form. However, a much more efficient way is to record it on a combination payroll ledger and earnings record (Figure 3-18). This sheet serves as the basis for making up the paychecks and also as the employee's earnings record.

A construction company's work volume varies widely. A person who works as a foreman or superintendent when there is plenty of work may work with his tools when there is little work. To provide for such a situation, it is important that the individual's personnel file show authorizations for the changes, and the dates when they are effective. Records permitting exact determination of service time are important, also, in applying the rules in the Employee Retirement Income Security Act of 1974. Various forms (Figure 3-11) are useful for this

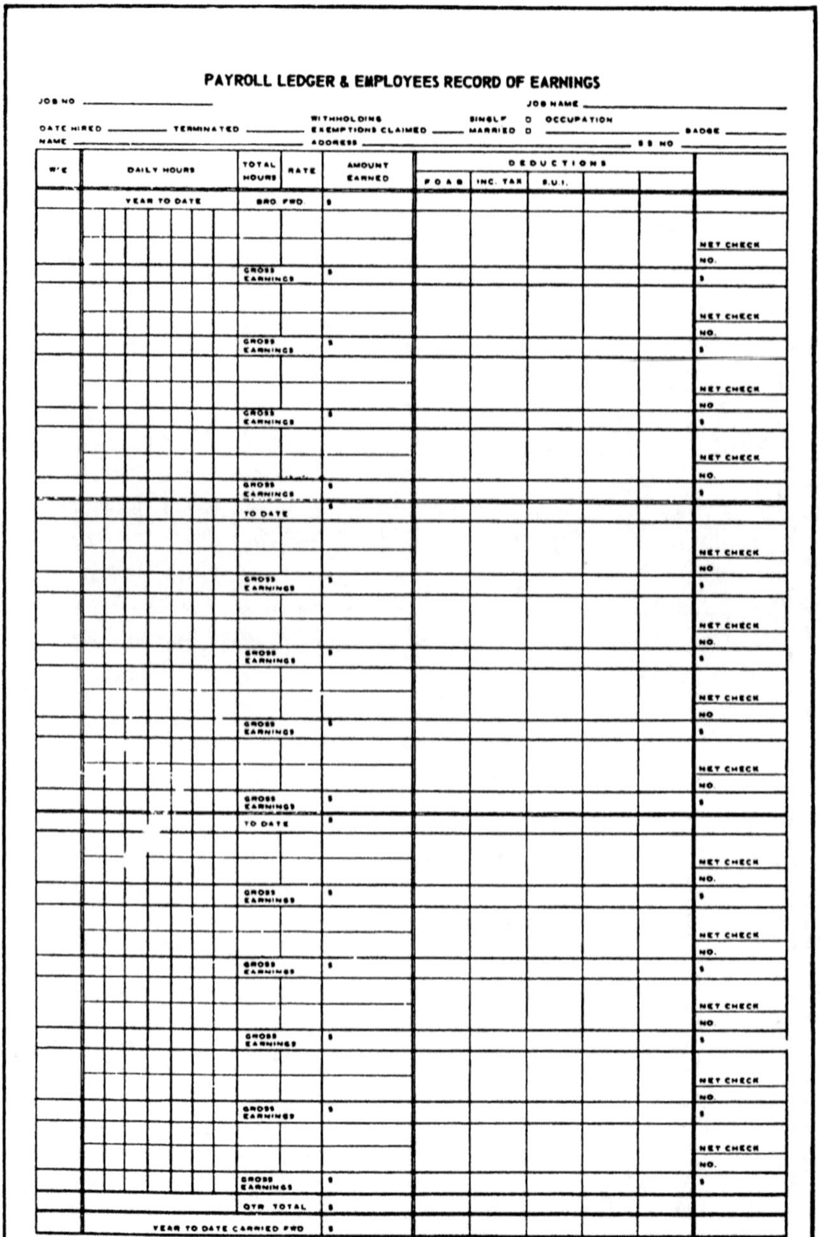

PAYROLL LEDGER & EMPLOYEES RECORD OF EARNINGS

JOB NO ______ JOB NAME ______

DATE HIRED ______ TERMINATED ______ WITHHOLDING EXEMPTIONS CLAIMED ______ SINGLE ☐ MARRIED ☐ OCCUPATION ______ BADGE ______

NAME ______ ADDRESS ______ S S NO ______

W/E	DAILY HOURS	TOTAL HOURS	RATE	AMOUNT EARNED	DEDUCTIONS F O A B	INC. TAX	S.U.I.			
	YEAR TO DATE	BRO. FWD.		$						
										NET CHECK NO.
		GROSS EARNINGS		$						$
										NET CHECK NO.
		GROSS EARNINGS		$						$
										NET CHECK NO
		GROSS EARNINGS		$						$
										NET CHECK NO.
		GROSS EARNINGS		$						$
		TO DATE		$						
										NET CHECK NO
		GROSS EARNINGS		$						$
										NET CHECK NO.
		GROSS EARNINGS		$						$
										NET CHECK NO
		GROSS EARNINGS		$						$
										NET CHECK NO.
		GROSS EARNINGS		$						$
		TO DATE		$						
										NET CHECK NO.
		GROSS EARNINGS		$						$
										NET CHECK NO.
		GROSS EARNINGS		$						$
										NET CHECK NO
		GROSS EARNINGS		$						$
										NET CHECK NO.
		GROSS EARNINGS		$						$
										NET CHECK NO.
		GROSS EARNINGS		$						$
		QTR TOTAL		$						
	YEAR TO DATE CARRIED FWD			$						

Figure 3-18 **Combination Payroll Ledger and Employee's Earnings Record.** In a computerized system, this record would normally be a by-product of check preparation.

purpose, regardless of the size of the contractor's operation. As a matter of fact, a small contractor may have more need for this type of record than a large operator whose volume of work tends to remain steady.

One of the standard timekeeping techniques is known as "face-checking." Theoretically, a timekeeper should know "by sight" all the workers on the crews for which he keeps time, and he should go out on the job and actually see them at work at least once, and preferably two or three times each day. Actually, this procedure is rarely followed in detail. On small jobs it is unnecessary. On large jobs, when the turnover is rapid, it may be impossible. One variation, similar to the brass system already mentioned, is the "badge number check," in which the timekeeper goes out on the job and checks off the badge numbers of the men he can account for on the job. On many cost-plus-fixed-fee jobs, when the time control is very tight, a time control record (Figure 3-19) is used. Spaces are provided for noting the man's presence in the morning, the afternoon, and at quitting time. The badge number at the left is circled for any man who failed to "brass in" in the morning.

Actually, such a form is well adapted to making spot checks on crews in the field when the actual timekeeping is done by the foremen. Unassigned badge numbers would be struck out by the payroll department on the basis of payrolls. Finding those badge numbers in use in the field, or failure to locate a worker with a listed badge number, would call for investigation.

Timekeeping routine

Basically the routine work of timekeeping involves these steps:

1. An up-to-date list of employees is maintained in the payroll department. In a very small business, payroll might be only one of a number of functions performed by the company bookkeeper. In a very large business, or even on a large job, payroll might be the responsibility of a number of payroll clerks working under a paymaster. Regardless of size, it is a payroll function to maintain the list of individuals currently employed and, where necessary, classify them according to the job they do and the supervisor or foreman under whom they work.
2. Either the foreman or a timekeeper makes a record of the number of hours worked each day by each employee. The record may, in very small operations, be turned in once each week. In larger operations it is essential that it be turned in daily, so that layoffs can

ECKS & WYE CO.
Time Control Record

Date. . . . .

01				51				101				151				201				251			
02				52				102				152				202				252			
03				53				103				153				203				253			
04				54				104				154				204				254			
05				55				105				155				205				255			
06				56				106				156				206				256			
07				57				107				157				207				257			
08				58				108				158				208				258			
09				59				109				159				209				259			
10				60				110				160				210				260			
11				61				111				161				211				261			
12				62				112				162				212				262			
13				63				113				163				213				263			
14				64				114				164				214				264			
15				65				115				165				215				265			
33				83				133				183				233				283			
34				84				134				184				234				284			
35				85				135				185				235				285			
36				86				136				186				236				286			
37				87				137				187				237				287			
38				88				138				188				238				288			
39				89				139				189				239				289			
40				90				140				190				240				290			
41				91				141				191				241				291			
42				92				142				192				242				292			
43				93				143				193				243				293			
44				94				144				194				244				294			
45				95				145				195				245				295			
46				96				146				196				246				296			
47				97				147				197				247				297			
48				98				148				198				248				298			
49				99				149				199				249				299			
50				100				150				200				250				300			

Figure 3-19 **Time Control Record.** This form is used by field timekeepers checking badge numbers on the job.

be handled promptly, and so that all differences can be resolved every day as they arise, instead of accumulating for a week and delaying the preparation of the payroll.

3. The daily (or weekly) time record is turned in by the person who keeps it (usually either the foreman or a timekeeper) to the person who makes the weekly payroll (either the paymaster or the bookkeeper).
4. The paymaster (or bookkeeper) examines the time record for errors. On large jobs any discrepancies are noted on a "trouble sheet" and are sent to the field by a timekeeper to be reconciled. Typical of the kinds of errors which occur regularly are the following:
   a. The time for one or more crews is missing.
   b. A worker or a crew shows working time less than that usually turned in.
   c. A person does not appear on the payroll who does appear on time record, or vice versa.
   d. A worker or a crew shows working overtime under circumstances where no overtime should be worked, or overtime as recorded appears excessive.
   e. No time, or an obviously wrong time, is reported for a person known to be working.
   f. Time of an individual or a crew is reported as chargeable to a work item in which no work is being done, or which is being done by another crew.
   g. Time is shifted from work items on which costs are running over the estimate to work items on which there is a gain.

Alert checking of time data coming in from the field can catch and correct errors which might otherwise be incorporated into the cost records. Of course, if good cost records are kept, substantial errors in timekeeping can distort unit costs, and in this way may be discovered by the accounting department. Nevertheless, errors on the payroll, if they are discovered at all, are often not discovered until the employee complains.

It is necessary that all time data be reviewed, immediately upon being brought in from the field, by someone who knows, at least in general, the parts of the job where work is being done, which crews should be doing it, what operations must coincide with what others, and approximately how much time should be going into each operation. On some jobs the project manager or superintendent will insist on personally coding the time cards. On larger jobs the coding may be done by the project engineer, the office engineer, or, if such an employee is available, a cost engineer.

Timekeeping cost data

The most variable of all construction costs is labor. Hence control of labor cost is one of the most important functions of job management, and labor costs are the most closely watched of all construction costs. It is not uncommon among the more progressive construction firms to find project managers requiring a daily labor cost report (Figure 3-20). This form requires the use of the so-called "effective production" or "equivalent units" methods of calculating the various units completed in any one day. As a practical matter, at the end of any day but the last, there will be work units in various stages of completion. The work units in various stages of completion are stated in terms of a comparable number of completed units.

When such a report is required, it usually must be in the hands of the project manager or superintendent shortly after work starts early the next day.

Following is a typical routine set up for a job with about 400 employees. A timekeeper familiar with the crews and the work done started checking and pricing the preceding day's time reports about 5 A.M. By 5:30 A.M. there were enough reports priced and checked to permit someone in the paymaster's department to start checking the wage rates and making the extensions. By 6 A.M. the cost engineer started: coding the cards to show the work items to be charged; sum-

DAILY LABOR COST REPORT

Job:.............................Date:......................

Work Item	Unit	YESTERDAY				JOB TO DATE			
		Units Compl.	Man Hrs.	Total Cost	Unit Cost	Units Compl.	Man Hrs.	Total Cost	Unit Cost

Time checked & priced by......	Pricing checked, extensions made by...............	Cost coded, unit priced by..................

Figure 3-20 **Daily Labor Cost Report.**

marizing the costs by work items; and scanning the reports for extension errors. As soon as the cost engineer finished with the sheets for any individual crew, he turned them back to the payroll department where they were totaled by crews; and a grand total of labor cost for the day was taken. When the cost engineer had finished his work sheet, which summarized the costs for the day, the totals by crews and the grand total had to balance (within a reasonable amount) with the total labor costs distributed.

Experience with this routine has indicated that on payrolls up to 700 or 800 men, a daily labor cost report can be available by 10 A.M. of the following day. This schedule can be accelerated if budgets and actual data are kept on a man-hour basis, thereby eliminating the task of pricing and extending the payroll for cost control purposes. Where computer equipment is available, these functions can be, and usually are, performed in whole or in part by the computer, but at this stage of recording and reporting labor costs knowledgeable participation by people is a necessity.

Only the larger jobs have a timekeeping department separate from the payroll department, and even some of these do not have a separate cost engineer. Therefore, on medium-sized and small jobs the functions described above must be performed by fewer people. Even if the job is relatively small, and the entire operation is performed by one person, there is often a need for daily labor costs. Daily labor cost figures are no longer worth what it costs to compile them only when the job is small enough for the contractor himself, or the superintendent or foreman who is running the job, to determine and control labor costs by direct personal observation on the job without the aid of statistical reports and the computation of unit costs. Even then, a weekly cost statement can be extremely useful in helping the person in charge to check his own observations.

It seems trite to say that the usefulness of cost figures is limited by their accuracy. It is equally trite to observe that if costs are largely based (as they admittedly are) on human judgment, they can be no better than the on-the-spot judgment of a thoroughly qualified construction engineer. On the other hand, even an experienced construction superintendent can be in only one place at a time, and, if work is going on in a number of places, the previous day's labor figures can tell him at a glance the places where his knowledge and experience can be used to the best advantage on that particular day. If he has an opportunity to see all the work every day, as in smaller operations, a weekly labor cost report helps to check his thinking and to guard against human error.

To summarize, the steps in preparing the daily labor cost report are as follows:

1. The time reports are assembled and checked.
2. They are priced and extended.
3. The extensions are coded to the appropriate work items.
4. The total costs thus distributed are compared with the total earnings of all employees for that day to provide an overall check on the cost distribution. Fractional cents and even minor distribution errors usually prevent an exact balance, and as a practical matter these small differences are usually absorbed in an overhead item without any distortion.

There remain several important practical questions. First, should the labor figures on the daily cost report include payroll taxes, insurance premiums, and the cost of various "fringe benefits" based on payrolls? Second, if such costs are included in the labor figures how should they be distributed? Third, will the labor figures so compiled be entered in the records or are they memoranda only? Fourth, if they are to enter the records, what records do they reach, how do they get there, and when? Fifth, how are adjustments and corrections handled?

1. Whether payroll taxes, insurance premiums, and fringe benefits should be included in the cost of labor or treated separately depends on the policy of the company. If these items were included with labor at the time the job was bid, it is better that they be treated in the same way in the job cost report, so that actual unit costs may be comparable with the estimate. If they were not included directly in labor costs, they should be included in an overhead account which is distributed either on the basis of labor cost or labor hours. From a practical standpoint, this treatment is usually acceptable when all overhead is distributed on the basis of total direct cost.
2. The problem of how to distribute costs such as those just discussed, when they are to be included in the labor costs, is solved by spreading them equally over the total labor figure, so that each dollar of labor bears the same burden. To attempt an exact segregation would not only entail unnecessary accounting costs, but would result in figures that are not comparable.
3. The question of treating the labor cost figures as memoranda or as record entries is ordinarily answered by having only one analysis of these figures by work items. If the daily labor cost reports are prepared correctly, it is a simple matter to combine them for the week and to let the totals for the week serve as the analysis for the weekly payroll.

4. There is usually no need for a separate entry to be made in the records for each individual paycheck other than to show the check number on the payroll. One generally accepted method of getting the cost figures into the books is through the voucher prepared to reimburse the payroll bank account. Usually a payroll voucher (Figure 3-21) is prepared to support the check issued to transfer the funds to cover the paychecks from the general account to the payroll account. This method brings the labor cost into the general and cost ledgers in a reasonable way—through the voucher register. In very small concerns with no separate payroll bank account, few jobs, and cost segregated only by jobs, the entry would be made directly through the check register (Figure 3-22).
5. Adjustments of the daily reports can usually be made on the report of the following day without serious distortion. On those unusual occasions when there is distortion, an explanation accompanying the adjustment would be required.

Some of the more progressive contractors, whose cost records are increasingly being used in the preparation of bid estimates, are experimenting with keeping labor costs in terms of man-hours as well as money, so that variations due to rate changes may be eliminated.

Timekeeping and paychecks

The importance of job morale, and the contribution that prompt, accurate paychecks can make to it, has already been noted. The point cannot be overemphasized. However, there are even more immediate and compelling considerations. Often state laws, union rules, or both, assess penalties against employers for late payments, dishonored checks, or underpayments. Union hiring halls are slow to supply workers to contractors whose reputation is bad, and even among the workers themselves the news travels fast. Therefore, it is most important that the paychecks are prompt and correct.

If the timekeeping data have been brought in from the field and checked, extended, and incorporated into the daily cost report, any major errors will usually have been discovered and corrected. Nevertheless, before the time cards are entered in the daily time accumulation, someone should be able to answer "yes" to all of these questions:

1. Have we accounted for every person on the payroll?
2. Do we know that the base rates are correct?
3. Is anyone working at a special rate?
4. If so, is it correctly shown on the time card?

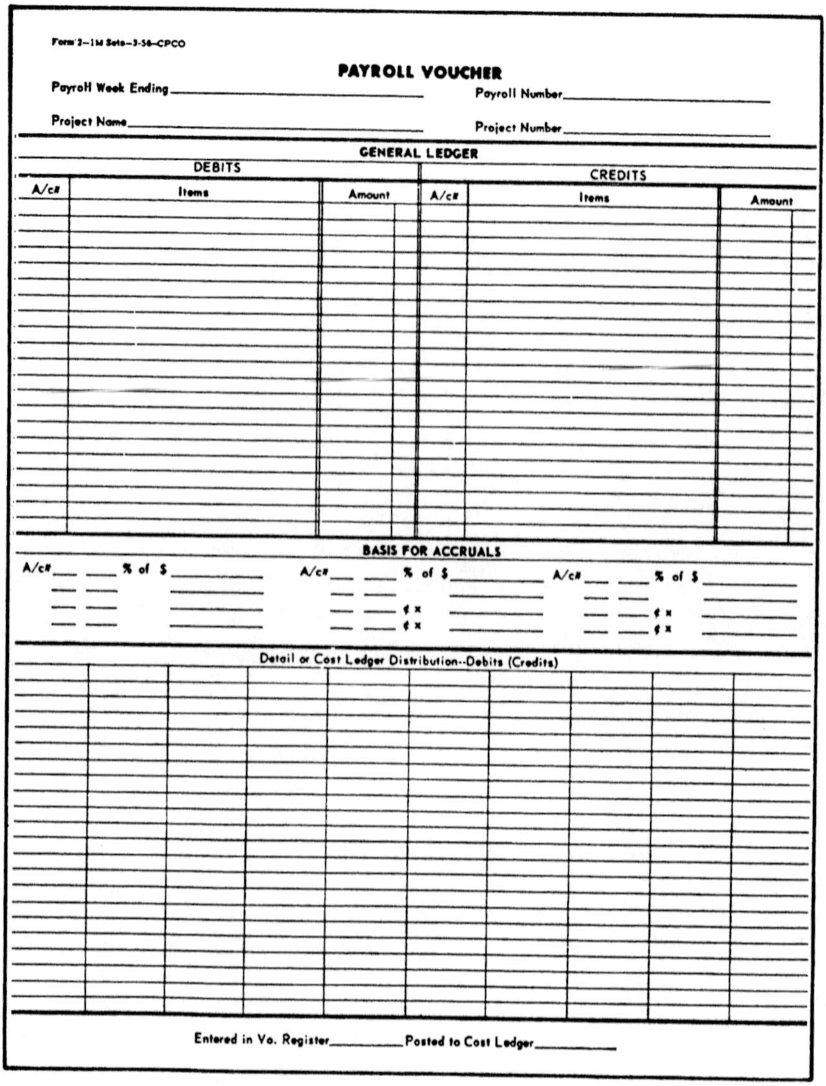

Form 2–1M Sets–3-56–CPCO

PAYROLL VOUCHER

Payroll Week Ending________ Payroll Number________

Project Name________ Project Number________

GENERAL LEDGER

DEBITS			CREDITS		
A/c#	Items	Amount	A/c#	Items	Amount

BASIS FOR ACCRUALS

A/c# __ __ % of $ ____ A/c# __ __ % of $ ____ A/c# __ __ % of $ ____

¢ x ¢ x ¢ x ¢ x

Detail or Cost Ledger Distribution--Debits (Credits)

Entered in Vo. Register______ Posted to Cost Ledger______

Figure 3-21 **Payroll Voucher.** To avoid unnecessary writing or typing on regular vouchers, many firms use specially printed vouchers to record the check that is issued to reimburse the payroll bank account. In computerized systems the payroll voucher is usually developed as a by-product of the payroll summary.

5. Have we checked all cards to see that apparent absences, time off, and overtime are explained?
6. Are any unexplained differences being investigated?

Again, if a daily cost report is made, there will be an overall check figure in dollars and possibly also in man-hours. If a daily cost report

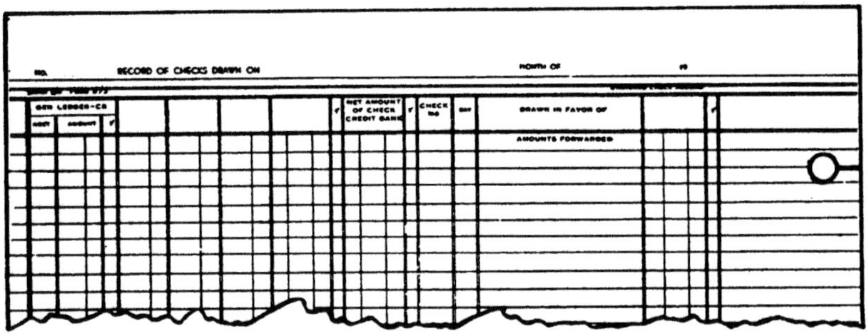

Figure 3-22 **Small Contractor's Check Register.** This form is suitable for payroll entries. If a separate bank account is used for payrolls, "payroll account" would appear in the "drawn in favor of" column.

is not made, it is sometimes worthwhile to compute the gross earnings of each crew in dollars to obtain a check figure. It takes comparatively little time and often catches serious and embarrassing errors. For example, if time sheets are being used and there is a crew of 10 working 8 hours at $6.50 per hour with one foreman at $7.25 per hour, the sheet would total 88 hours and $578.00 (80 hours at $6.50 equals $520.00, plus 8 hours at $7.25 or $58.00). After a few days the paymaster can look at the sheet and tell at a glance the proper amount to be extended, while the variations from the normal catch his eye and call for closer scrutiny. At the same time the total for each crew can be recorded on an adding machine tape to provide a quick checklist on which trouble can be quickly localized, by crews, if the total does not equal the gross pay on the payroll.

It is often said that tests like the one just described are all right for the big payrolls but cost more than they are worth on the small ones. This criticism is simply not justified. Any experienced payroll clerk can make the test in a matter of minutes and the smaller the payroll the less time it takes. On the other hand, frequent errors on a contractor's payroll can be, and often are, misinterpreted as evidence of dishonesty or lack of funds. Small contractors in small communities sometimes find that gossip travels fast. More than one contractor has found that a chance remark by a disgruntled employee has caused curtailment of credit—sometimes at a time when he could ill afford it.

As soon as the daily time has been proved, it is entered on some sort of accumulation work sheet (Figure 3-18). As often as not, particularly if the number of employees is small, the work sheet will be the final payroll form (Figure 3-23) itself. The former has the advantage of being much more flexible than a simple payroll form and can serve as an employee's earnings record. On the other hand, if the time is entered on a payroll form there may be no need to copy it over.

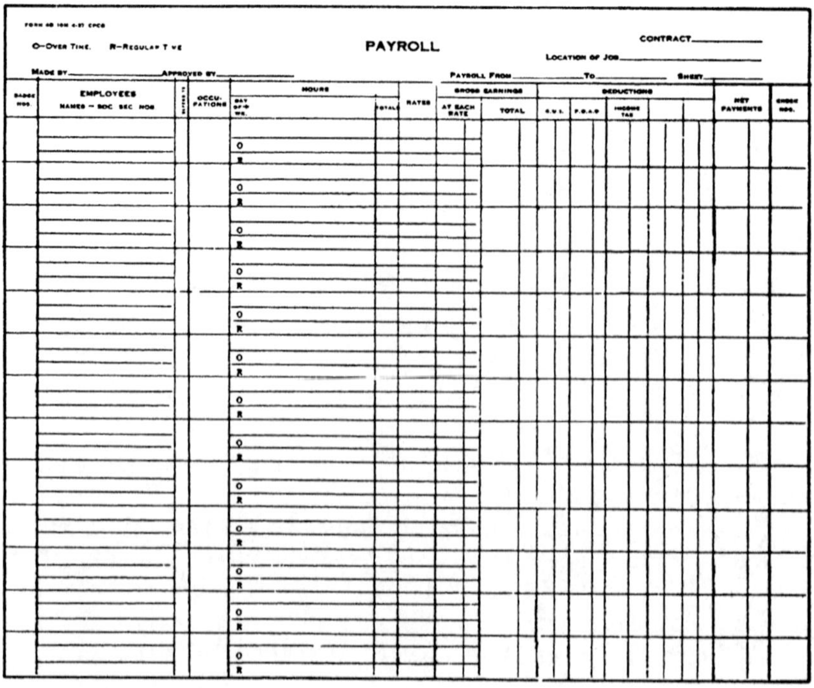

Figure 3-23 Payroll Form. This is one of the many forms on which payrolls are recorded. This form is more than usually complete, since it shows the hours worked each day. This kind of payroll is usually required on United States government jobs because of the need to comply with the various laws governing payment to employees on government work.

Whatever form is used, the time should be entered daily if it is submitted daily. Rate changes should be noted on the days they are effective so that, whenever a worker is laid off, the paymaster needs only the individual's time for that day to compute his termination pay.

Once the time for the week is in and entered on the work sheets, each employee's gross pay for the week is computed and a check tape is run on the adding machine to see if the gross pay, as extended, equals the total gross pay for each of the several days of the week. If they equal each other within the few cents that must be allowed for rounding off fractional-cent variations, the gross pay has probably been correctly computed. Here again it will be noted that an experienced payroll clerk will soon have many of the extensions memorized. The normal situation is for a person to work the regular work week at his regular rate. Thus, if a job is working six 9-hour days with time and a half for overtime, and the hourly rate is \$15.00 per hour, a payroll clerk will know, without computing it, that 54 hours at straight time is \$810.00. In addition, 14 of the hours were overtime, at a bonus rate of $7.50 or $105.00. The total of the full 54 hours at straight time plus the bonus for 14 hours overtime gives a grand total of $915.00 which is called the gross pay. Again, it is only when an exception to the normal situation occurs that a separate computation is necessary.

Some companies provide their payroll clerks and paymasters with charts of extensions at various rates and hours. Sometimes these are used but more often they are quickly discarded. The more common errors are not in the calculation of the extensions but rather in the pure human errors of failing to put the figures down or to add them correctly. Thus the bulk of the errors will come from the usual transpositions, slides, and misread figures rather than from errors in multiplication.

To catch this type of error and localize it quickly, it is a good idea to set up the payroll by trades, and by crews within the trades. If the overall total of gross pay shown on the payroll equals the total of all the time sheets, there is still the possibility of an offsetting error. However, if the adding machine tapes that were run each day to check the totals of each crew's time sheet against the total payroll for that day have been saved, the daily totals provide some assurance against such errors. Of course if the daily totals do not agree, errors can be quickly localized by crews so that a minimum of checking is required.

Obviously, the procedure just outlined is only one way of accomplishing the desired result. There are many others equally sound. Many oldtime paymasters are like artists who obtain results by highly individual methods. Often, too, they are just as temperamental. Thus, it is not hard to visualize such a person reading the example given above and snorting: "Huh! Anybody but a goofball from the front office would know that you figure six 9-hour days as 61 hours at straight time." But the same paymaster might spend several hours excluding the overtime bonus from a month's payroll to arrive at the correct figure for his compensation insurance report, which normally must be based on the full number of hours worked, extended at straight-time rates.

Deductions

Once the gross pay is computed and balanced, the next step is to compute the deductions. These will include, as a minimum, federal (and sometimes state) income tax withheld, and federal (and sometimes state) employment tax deductions. They may include a great many others, such as union dues, advances, mess hall, camp or commissary purchases, contributions, insurance, and sometimes payroll assignments, garnishments, attachments, or deductions pursuant to an arrangement under Chapter XIII of the Bankruptcy Act.

On certain public works contracts care must be taken not to make deductions prohibited on United States government jobs by the Davis-Bacon Act or the Copeland Act, or, on state or local projects, by similar state or local statutes. Usually permission can be obtained to make any legitimate deduction, but often it must be obtained in advance.

On the other hand, failure to make deductions for taxes, for advances, or for assignments, garnishments, and the like, may subject the employer to liability for the amount not deducted.

If a regular payroll form (Figure 3-23) is used, the exemptions will be recorded beside the employee's name, usually copied from the last payroll or from data cards (Figures 3-2 to 3-5). Sometimes the federal withholding tax and the social security deduction are combined in a single deduction, and the combined deduction is computed in the same manner as the withholding tax. State income tax withholding, when required at all, is usually similar to federal withholding and generally offers no more of a problem. Sometimes a combined federal and state chart is available.

Maximum deductions. Federal and state payroll tax deductions are usually computed from a chart, but they are often applied to a different maximum. If an earnings record (Figures 3-6 and 3-18) is used, the cumulative gross earnings may be brought forward as a memo figure each week so that the time when such deductions should be stopped will automatically come to the attention of the payroll clerk. Such information will appear on almost any of the numerous forms of employees' earnings records. It is important to be sure that it is called to the attention of the person computing the payroll in time to prevent an overdeduction and the extra work involved in a special credit when the error is finally discovered.

If the payroll is small and the earnings steady there is no problem. The payroll clerk can usually remember whose deductions are due to be stopped. However, construction payrolls are notoriously irregular and turnover is high. Probably the simplest way to control this problem is to have the person who posts the employee's earnings record prepare a list, as he computes his cumulative totals, of any employees whose earnings are approaching the point at which deductions should be stopped. When the payroll is prepared and ready to be extended, the payroll clerk takes the last payroll and inserts a dash in the deduction square for all employees whose contributions have been completed, and a light red dot in the square when only a partial deduction is to be taken. In a state like California, it would be necessary to go through this routine twice for each employee—once at each level.

Union dues, bond deductions, contributions, and insurance may be deducted on each or every other payroll. If they are deducted each time, it is a simple matter to spot errors because the omission would attract attention. Otherwise it is necessary to put them on the paymaster's checklist along with the irregular deductions.

Irregular deductions. It is standard procedure for paymasters to maintain some sort of checklist to call attention to irregular payroll adjustments that might otherwise be overlooked. Any item that will be on some payrolls, but not on all of them, goes on the checklist, and no payroll is released until it is certain that all items on the checklist have been considered and either checked into the payroll or the reason for their omission determined.

Thus, items like union dues, insurance contributions, and bond deductions are sometimes deducted on one payroll each month or on alternate payrolls. Accordingly, these deductions will be on the checklist, and if they are not to be deducted on the current payroll, they will be cleared until the next one. Advances would be entered on the payroll from the file of requests, and the total amount deducted would be balanced with the total of the checks written for advances during the preceding week. The few advances that are to be deducted on more than one payroll will show up as exceptions, and can be covered by a memo in the file of requests for advances. Commissary purchases, mess hall charges, and equipment charges arise only on large jobs away from cities or towns. They are generally entered on the payroll from the camp manager's weekly report, which should, by payroll time, have been checked and approved by the accounting office. Payroll assignments, garnishments, and attachments must be kept in a current file and entered on the payroll only at the last minute. Even then, other such items may come in before the checks are distributed. Advice of counsel regarding these items should be obtained, but it is better to hold an individual's check than to release one in error. Deductions under a wage earner's arrangement (Chapter XIII of the Bankruptcy Act) are usually few enough so that if they are indicated by the checklist they will be properly handled.

Verification. Once all the deductions have been entered on the payroll, they are added across to find the total to be deducted from each paycheck. Then the total of the deductions is subtracted from the gross pay to find the amount of the paycheck (usually called net pay). When this has been done for each man, the deduction columns, the "total deductions" column, and the "net pay" column are all totaled. The totals of the individual deduction columns must equal the total of the "total deductions" column; and the total of the "gross pay" column less the total of the "total deductions" column must equal the total of the "net pay" column. To those familiar with payrolls, this is elementary routine. Anyone who is unfamiliar with payrolls may find it helpful to study a payroll form (Figure 3-23).

If a payroll has more than one page, it is better to total and balance each page and then carry the page totals to a summary page where a final balance is obtained. This not only makes it possible to localize errors more easily, but it also makes it possible, on large payrolls, for more than one person to work on balancing the payroll sheets.

Once the payroll is balanced the checks can be written. Sometimes, when more than one person is working on the payroll, the check writing starts as soon as one page is balanced so that the check writing and the balancing can proceed at the same time. If one person is doing the whole job, the balancing is usually completed first.

Payroll and paychecks

When time is being compiled and net pay computed there are comparatively few variations in forms and, as pointed out above, the work sheet itself may serve as a payroll. Often, however, it is necessary to provide copies of the payrolls to owners, architects, or government agencies. Sometimes, too, the payrolls so furnished must be in some special form. Also, the requirements of the federal and state payroll tax and withholding tax laws must be met. For example, each employee's earnings record must show:

1. Name, address, and social security account number
2. The total amount (including any deductions) earned during each pay period, each calendar quarter, and the calendar year
3. The amount subject to social security tax and other payroll taxes
4. The amount of taxes withheld or collected from each paycheck

It is standard practice, based on a requirement of the law, to give each employee a memorandum, with each check, of the total earned during the pay period and the amounts deducted.

Before the end of January each year (or within 30 days after termination if his employment ends before the end of the year) each employee must be given a form W-2 (Figure 3-24) showing the total compensation for the year and the income and social security (FICA) taxes withheld.

To satisfy all these requirements with a minimum of paperwork, a large number of manual "one write" systems have been devised. Some prepare the paycheck, the payroll, and employee's earnings record at one writing by means of carbon copies held in position by a pegboard (Figure 3-25). This system seems quite logical, but among construction paymasters pegboards have never gained much popularity. They are not well adapted to rapid turnover of employees, and the carbon copies have not proved very satisfactory under field conditions.

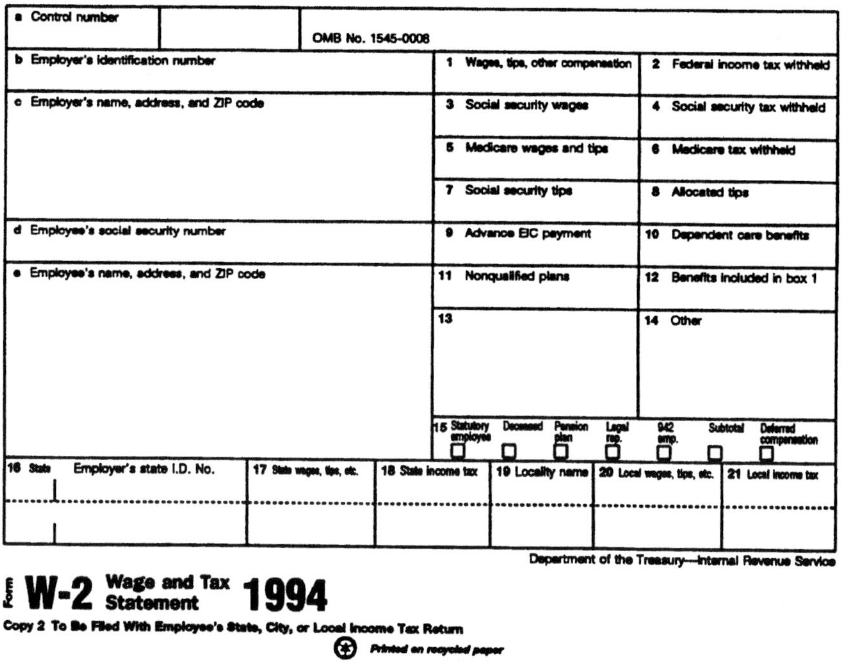

a Control number
OMB No. 1545-0008
b Employer's identification number
1 Wages, tips, other compensation
2 Federal income tax withheld
c Employer's name, address, and ZIP code
3 Social security wages
4 Social security tax withheld
5 Medicare wages and tips
6 Medicare tax withheld
7 Social security tips
8 Allocated tips
d Employee's social security number
9 Advance EIC payment
10 Dependent care benefits
e Employee's name, address, and ZIP code
11 Nonqualified plans
12 Benefits included in box 1
13
14 Other
15 Statutory employee / Deceased / Pension plan / Legal rep. / 942 emp. / Subtotal / Deferred compensation
16 State / Employer's state I.D. No.
17 State wages, tips, etc.
18 State income tax
19 Locality name
20 Local wages, tips, etc.
21 Local income tax
Department of the Treasury—Internal Revenue Service
Form W-2 Wage and Tax Statement 1994
Copy 2 To Be Filed With Employee's State, City, or Local Income Tax Return
Printed on recycled paper

Figure 3-24 Form W-2—Withholding Tax Statement. This form may be privately printed so as to provide a copy for the employer's records, unlike the forms furnished by the government. The form is available in snap-out style with single-use carbon paper inserted to save time in preparation.

Moreover, although theoretically all records in this system would always be in agreement, often an error is corrected on one copy and not on the others, with the result that there is often trouble at the end of the quarter when the social security and withholding tax returns are reconciled with the payrolls. Generally speaking, the pegboard systems work best on the smaller payrolls when there is very little turnover and only one copy of the payroll is required.

A similar principle is used by bookkeeping machine and computerized payroll systems. The problems are also similar. If a contractor has a large payroll, and numerous terminations, a special payroll can be devoted to the terminations, with the result that there are generally fewer errors caused by hurried workmanship. For the average small construction company, bookkeeping machine payroll systems (Figure 3-26) are not practical because of their cost, but in many areas, banks are applying data processing equipment effectively and economically to small construction payrolls.

When payrolls are kept in the field, a machine system, even on a fairly large job, exposes the contractor to the risk of serious delay if

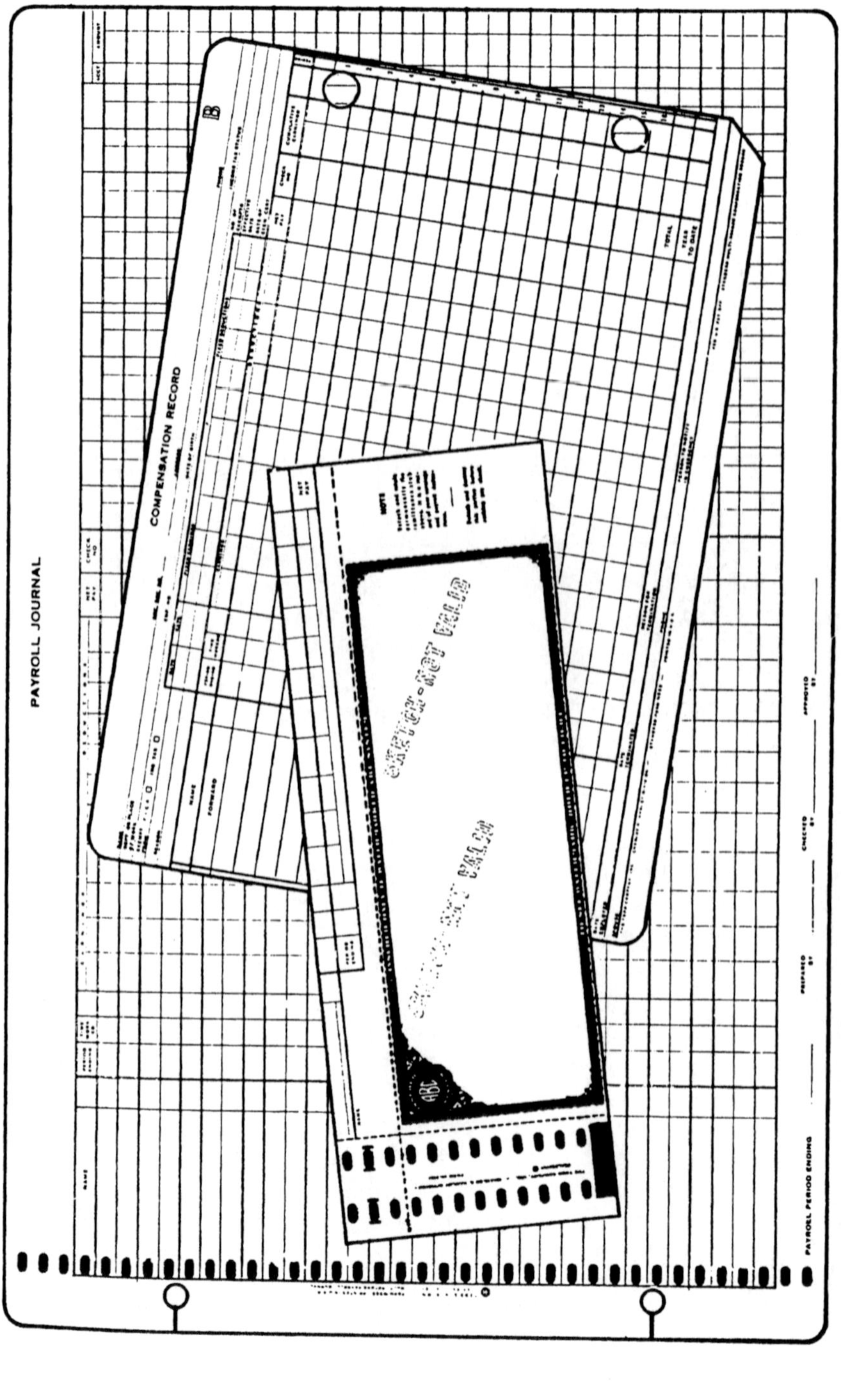

Figure 3-25 Timemaster Pegboard Payroll System.

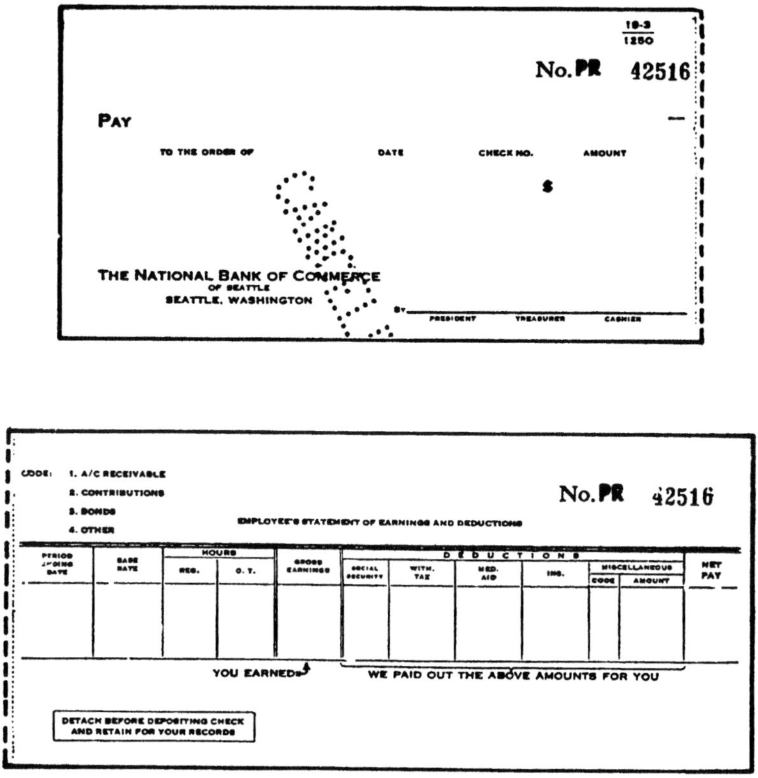

Figure 3-26 **Paycheck for Use with a Bookkeeping Machine Payroll System.** The headings on the check correspond to columns on the payroll records. The information is transferred by means of a strip carbon. Bookkeeping machines are no longer common in the construction industry.

the machine breaks down and specialist repairmen are not available on short notice.

One system which has proved quite successful in a variety of small construction operations calls for the use of a check with an original and three carbon copies printed, with one-use carbons, in sheets of four or five checks (Figure 3-27). If a combination payroll ledger—earnings record (Figure 3-18) is used, there is no need for a separate employee's earnings record, and if a separate payroll is not required, the third carbon copy of the checks, which remains in the sheets, can be used as a combination payroll and a payroll check register. When a separate payroll is required it must, of course, be prepared from the work sheets or from the checks themselves. Under such a system, the first carbon copy of the check, printed in the form of a withholding notice, accompanies the original to the employee and serves as the employee's record. The second carbon copy is filed in an employee's

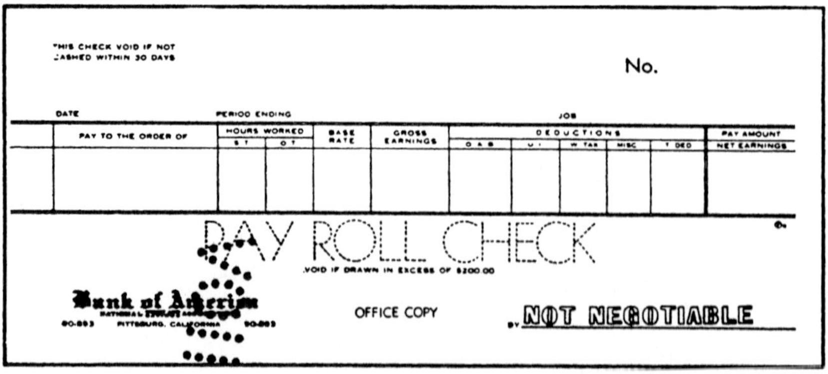

Figure 3-27 **Multiple-Copy Paychecks.** These checks are available with as many as five carbon copies and in sheets of as many as five checks.

folder (Figure 3-6). The third carbon copy is left in sheets and bound in numerical order.

Still another form of payroll check is the two-stub type (Figure 3-28). The second stub is merely the employee's record, and is to be torn off before the check is cashed or deposited. This type of check works quite well on very small payrolls and, since it can be obtained preprinted at most banks, is quite inexpensive. It would be rather costly to use on a large payroll because of the time required to prepare it. When this type of check is used, the payroll is usually a check register, and any addi-

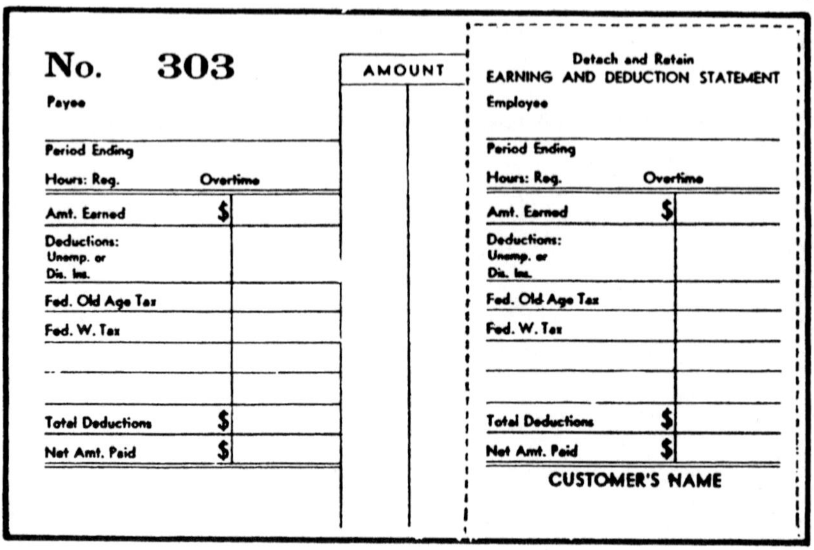

Figure 3-28 **Stub for Two-Stub Paycheck.** This type of stub is commonly used by contractors with fifteen to twenty employees. The check itself is not shown.

tional copies of the payroll that might be required must be separately prepared.

When the paychecks have been prepared, two adding machine tapes should be run. The first should be run on the amount of the checks in the figures that usually appear in the right-hand margin. The second should be run on the amount in words or, if a check writer is used, on the amount perforated into the check. Naturally the two tapes must agree.

In summary, it should be stated that the procedures and forms described are only illustrative, and not exhaustive. The variations are endless. The checkpoints and the procedures used in reaching them vary as widely as the people who use them and the conditions they are designed to meet. There are, however, five points which all sound payroll systems have in common:

1. The basic time data and rates must be reviewed by someone who knows what is being done, where it is being done, who is doing it, and the proper hourly rates applicable to each class of work.
2. Actual work done, and its cost as represented by the payrolls, should be compared regularly with the estimated progress as reflected in the bid estimates. The payrolls should be designed to facilitate preparation of comparative unit labor costs.
3. Each step in the computations must be subject to some form of internal check in terms of both arithmetical accuracy and reasonableness.
4. Checklists or tickler files need to be maintained to see that unusual and nonrecurring items are not overlooked.
5. Forms must be designed so that the unit cost of preparing each paycheck is the minimum practical for the contractor's individual circumstances.

No matter how carefully payrolls are made nor how many times the data are checked before they finally reach the completed paychecks, there will still be errors. Usually they will be minor ones, and often they will be traceable to facts that the person preparing the payroll did not know and possibly had no way of knowing. Still, a fair proportion of these errors can be eliminated by a critical reading of the payroll and comparison of the payroll with the actual checks. When such a final precaution is feasible, it is quite fruitful. Often it is done by the project manager or superintendent. For example, on one occasion it led to a conversation something like this:

"Hey! Where did Whitey and his gang work last week?"

"They were down clearing on the power line."

"Why aren't they getting paid then?"

Somebody had slipped. Whitey had failed to turn in his time daily because he and his crew were working several miles away. A separate page in a payroll that averaged over 20 pages had been reserved for that crew. Somehow it had been missed and, except for the project manager's final scrutiny, the error would not have been discovered until the crew members complained.

Distributing paychecks

One positive invitation to payroll fraud is to permit the foremen to distribute the paychecks. They may be distributed at a pay window as the man leaves the job, or by the paymaster, but if the foreman delivers the checks he is being tempted to commit fraud through collusion with some of his crew. When a paycheck is delivered by someone who does not know the payee personally, identification should be required. Usually a badge number or identification card will suffice. Sometimes it is advisable to require a payroll receipt and a thumb print. The method used should be reasonably calculated to prevent misdeliveries. Sometimes, through error, or through fraudulent identification, a misdelivery will occur. When it does, payment on the check should be stopped as soon as the misdelivery is discovered.

At frequent but irregular intervals the office manager, an auditor, or someone else outside the payroll department should distribute the checks. Checks that cannot be delivered should be listed immediately and the reason why the employee was not on the job should be determined. Usually it will be found that he was merely absent for some explainable reason. However, if nonworking employees are being carried, their presence on the payroll will be discovered, and if fraud is being contemplated it will be discouraged.

Sometimes employees die while they have wages due them. When that happens, the check may usually be paid to the deceased employee's estate (Estate of John Doe). It should be paid in no other way except on written instruction of the probate court or on advice of counsel.

Occasionally an employee is in jail or otherwise unable to call for his pay. On such occasions a great deal of care must be exercised. Often creditors, divorced wives, relatives of the employee, and sometimes complete strangers try to claim the check. If it is delivered to such persons without the employee's written authorization, the employer may find he is liable for misdelivery of the employee's property.

When a paycheck is reported lost or stolen, an immediate attempt should be made to stop payment on it at the bank. If the check has been stolen the police should also be notified. In most areas near a large construction job there are one or two bars or taverns where a great many of the employees collect in their off hours. Often a direct call to these places will be effective in heading off the cashing of lost or stolen paychecks. The same procedures are in order when blank paychecks are stolen. The bank, the police, and the local merchants should be notified. If possible, the numbers of the missing checks should be supplied and, if too many are missing, it may be necessary to change banks. If this is done, arrangements should be made with the bank in which the payroll account is carried to honor any legitimate checks that may be outstanding. Needless to say, all blank checks should be kept under lock and key.

Termination forms and procedures

Not too many years ago when an employer wanted to fire an employee, he fired him and that was all there was to it. Today, for better or worse, the wise employer keeps records of terminations as well as hires. He makes a record, among other things, of the employee's correct name, his badge or payroll number, his social security number, the exact date and reason for his termination, and wherever possible, his signature. This information is usually recorded on a termination slip (Figure 3-29) and is useful in many ways.

For example, a termination slip signed by the employee and the person who ordered his termination, and giving the reason for the termination, can be extremely important if an employer is later charged with unfair labor practices, illegal discrimination, or violation of a labor contract. The necessity of making out and sending to the head office a termination slip on all employees who are laid off, who quit, or who are discharged for cause often keeps a paymaster honest who might otherwise be tempted to carry the man on the payroll for a few days after he leaves. The reason for termination may, at times, serve as a guide if there is occasion to consider rehiring the employee—a very common occurrence in the construction business.

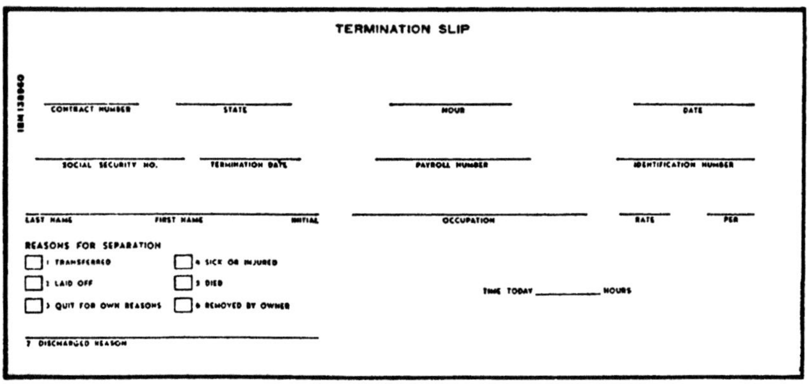
TERMINATION SLIP

CONTRACT NUMBER ______ STATE ______ HOUR ______ DATE ______

SOCIAL SECURITY NO. ______ TERMINATION DATE ______ PAYROLL NUMBER ______ IDENTIFICATION NUMBER ______

LAST NAME ______ FIRST NAME ______ INITIAL ______ OCCUPATION ______ RATE ______ PER ______

REASONS FOR SEPARATION

☐ 1 TRANSFERRED ☐ 4 SICK OR INJURED
☐ 2 LAID OFF ☐ 5 DIED
☐ 3 QUIT FOR OWN REASONS ☐ 6 REMOVED BY OWNER

TIME TODAY ______ HOURS

7 DISCHARGED REASON ______

Figure 3-29 Termination Slip.

One of the most important reasons for having an accurate record of terminations, and particularly the reasons for making them, is tax saving. All states now have unemployment insurance laws. Most of them have some provision for reducing the unemployment insurance tax of those employers who maintain steady employment. Usually this is done by maintaining an account with each employer and charging his account for all unemployment payments made to his former employees. In computing the amount in an employer's "reserve" account, the employer is not charged, and the employee is assessed a waiting period, if the employee quit voluntarily or was fired for misconduct. In a dispute over this type of item, good records, regularly kept, are very valuable. Moreover, unless the employer has a record of the reason for the termination, he may never know which unemployment insurance applications to protest. As a result he may protest none at all and have to pay a higher rate of payroll tax than some of his competitors.

Termination paychecks

When an employee is laid off he should be paid to the end of the half day. In other words, if he is laid off in the morning, he should be, and probably must be, paid for the full half day. If he is terminated in the afternoon it is usual to pay him for the full day. Some union contracts may provide for a full day's pay on the day of termination. In any event, a summary schedule should be kept by the paymaster setting forth company and union termination policies.

If time has been reported, checked, and posted daily, there is no problem in securing the time for the preceding days. There remains only the problem of determining whether or not there are any exceptions in hours or rate for the day of the termination. If time is reported weekly, it will be necessary for the foreman or superintendent to get the information to the paymaster in time and in enough detail to permit accurate preparation of the employee's final payroll check. If the termination checks are late there may be a waiting-time assessment. If they are short, there may well be claims made through the state labor commissioner. If they are over, there is virtually no chance to recover the excess.

The most effective procedure is, whenever possible, to plan the termination one day in advance and give the paymaster the names of the employees to be laid off. Then, on the last day, the superintendent can usually arrange the work so that the employees to be laid off do nothing to justify an exception in the time or rates. The paymaster can then prepare the checks on the basis of the last day being a full day at regular rates.

If the job is small and close to the head office, or if there is a decentralized payroll, the information necessary for the termination checks is readily available and there is no problem. If the payrolls are centralized and the job is large enough to justify such a measure, the project manager or superintendent can have a special bank account for issuing termination checks. However, when checks are written on the job without adequate control, the contractor must be prepared to accept the consequences. The results are often entirely satisfactory, but then again they also may include incorrect payments, payroll fraud, and excessive costs or losses. One variation, in which the project manager writes termination checks from an imprest bank account, is to let the superintendent telephone the information to the paymaster as early as possible on the day of the termination. Then the paymaster can compute the amounts due and telephone the figures back to the job.

In discussing preparation of the regular payroll checks, a checklist of deductions and other items was mentioned. This list is equally applicable, if not more so, to the preparation of termination checks. However, there are additional factors to be considered when drawing termination checks. Extra care must be taken to see that all property issued to the employee is returned, as well as all company identification. If any bond or other savings funds are being withheld the employee is entitled to receive any balance to his credit. Company policy or the union contract may entitle the employee to pay in lieu of vacation. Notices may be required to the union, creditors, group insurance carriers, or others. In short, if there are a substantial number of special considerations to be taken into account, a special termination checklist is in order. This list is even more necessary in very small companies where the routine is inclined to be hurried and the person in charge of payrolls has many other duties.

Once the paymaster has collected all the data, the preparation of the check itself sometimes creates certain problems. If there is much turnover, it is often advisable to prepare termination checks from a separate number series and to record them on a separate termination payroll. In very large operations it is even a good idea to have termination checks of a different size, shape, or color, so that they will show up quickly in bank reconciliations.

Most termination checks are handed to the employee directly. Sometimes the terminated employee will receive his W-2 form and his copy of the termination slip with the check. When this procedure is followed, the employee may be asked to sign off on the termination payroll. Some companies mark either the check or the voucher with such words as "termination" or "final check." Usually, however, the check is merely handed to the employee, and when he cashes it, the cancelled check serves as a receipt for the payment. This procedure is generally adequate.

When the employee is not present to accept the check in person there are other special considerations. If the check is sent through the mail, it is well to send it by registered or certified mail to minimize the risk of forgery or delivery to the wrong party. If the check is sent through the employee's labor union, certified mail will provide evidence of the fact that it was received by the union, and the identity of the person receiving it. Often retroactive paychecks are sent to the union, and it is well to have someone in authority at the union hall sign a receipt for the whole retroactive payroll.

Special payroll problems

Wage regulation. It is not the purpose of this volume to supply detailed coverage of special wage regulation. However, it is worthwhile mentioning the more important federal laws affecting construction payrolls. Foremost of these is the National Labor Relations Act, which controls all dealings between employers and any employees who wish to bargain, collectively or otherwise, regarding wages, working conditions, or any other pertinent matter.

Second in importance to the National Labor Relations Act is the Fair Labor Standards Act (the so-called "wage and hour law"). It has been held that, for some new construction, even when United States government financed, the wage and hour law is inapplicable because it deals wholly with intrastate commerce. However, this rule has been subjected to so many exceptions that it is unsafe to assume it applies without advice of legal counsel.

On United States government contracts the Davis-Bacon Act, the Copeland Act, and the Walsh-Healey Public Contracts Act all deal with measures to assure everyone concerned that all employees are paid at hourly rates equal to or greater than those set by the contract; that there are no "kickbacks" required; and that no unauthorized deductions are taken from the employees' pay. On all United States government work and much state and local work it is well to determine the effect of these acts on the contract, at or before the time the contractor's bid for the work is submitted. Certainly such a determination should be made no later than the time that the contract is signed. Sometimes rates set by the Secretary of Labor upset local differentials or otherwise create difficulties. Very often it is necessary to get special permission in advance to make such payroll deductions as charitable contributions or even camp or mess hall deductions.

Many state and local governments have similar laws governing payrolls on their contract jobs. It is wise to inquire at the time of bidding, and again at the time of signing public contracts, exactly what the rules are for that particular job.

Attachments. Every contractor deals with a certain number of creditor claims which, by law, must be deducted from employees' paychecks. Creditors and ex-wives are the source of most executions, garnishments, or attachments of wages and, if any of these documents are served on an employer, he renders himself personally liable if he pays the employee without satisfying the levy. The same is true of assignments of wages.

It is important to know the rules of the state where the payroll is being made when such claims arise. It is good business practice, from the employer's point of view, to use, where possible, every available means to protect employees from levies of this type, and to know every legal means available, without undue expense, to nullify or mitigate their effect. However, once an employee's creditors are levying on his paycheck regularly, it is better business ordinarily to lay him off and employ someone else. There are several reasons why this course should be followed.

First, an employee in financial or domestic difficulties, or both, is not likely to be a good employee. Second, if it finally becomes necessary to lay an employee off because the garnishments are taking all his paycheck, it is better for job morale to have taken strong steps for his protection at the outset. Third, a construction company, particularly a large one, can easily let itself become a collection agency for every credit store and finance company in the area if it cooperates too freely with its employees' creditors. Fourth, a lineup of collectors, process servers, book makers, and ex-spouses at the pay window can deter capable employees from working for a contractor who permits such a situation to develop.

Any attempted collusion between employees and paymasters, foremen, and others to defeat the creditors and save the harassed employees' jobs by extra-legal means can result only in additional trouble for the employer. Often such actions may result in actual financial liability for the employer. Such misdirected sympathy should be strictly forbidden, and when discovered should be grounds for discipline. All employees should have careful instructions to immediately report to the head office the service or attempted service of any legal document.

Advances. Payroll advances are a necessary evil. They should be discouraged as much as possible because, when used to excess, they can materially increase the cost of preparing payrolls. One way to discourage them is to require written authorization. Also, a company loan fund to provide wage and salary loans at reasonable interest rates might be provided.

Payday. If the job is working six days a week, payday should be Saturday afternoon. If the job is working five days a week, payday should be Friday. Hangover inefficiency and absenteeism may be held to a minimum by this means.

Fraud. Payroll fraud has occurred in one form or another any time people are paid for time not spent or work not done. If labor productivity falls below a reasonable standard, one of the possible explanations is payroll fraud, which must therefore be considered when the unit costs of labor go up. Often, however, the amounts involved are either not large enough, or cost keeping on labor is not accurate enough to cause fraud to show up in this way.

The next best substitute is alert auditing. Surprise face-checks for absenteeism; surprise distributions of paychecks by someone outside the payroll department; transferring accountants, timekeepers, and foremen so that collusion will tend to be broken up before it starts; regular reconciliation of payroll bank accounts; regular comparison of endorsements on checks with known signatures—all these and many similar devices are effective in creating the greatest safeguard of all, the knowledge by all concerned that the payrolls are being checked and that fraud will probably be discovered if it is occurring.

One possible source of control that is often overlooked is the employee's own records. Some companies now send out with the employee's withholding receipt (Form W-2) an information card (Figure 3-30). When these forms reach the employee they occasionally turn up payroll padding that has gone undetected for months, or even years. A single disclosure of this kind can prove a more potent deterrent to payroll fraud than hundreds of hours of auditing.

IMPORTANT

Enclosed is a Withholding Tax Statement (Form W-2), showing your earnings on the contract indicated. If your record of earnings is different from the enclosed Form W-2, please fill out this card and return.

Name ______________________________

Social Security Number ______________________

Correct Earnings ______________ **W-2 Earnings** ________

Termination Date ______________________

Location of Work ______________________

Figure 3-30 **Information Card.** If this notice is enclosed with the employee's copy of the Form W-2, payroll errors and frauds may be discovered.

Chapter

4

Accounting for Materials and Supplies*

The object of accounting for materials and supplies is to help management recover their cash value. Management cannot ordinarily collect for materials or supplies that are stolen, lost, wasted, or spoiled by exposure, nor can it collect for materials or supplies not included, in one way or another, in the amounts billed to customers. This means that physical as well as financial controls are needed, and one without the other cannot protect the contractor from losses. The fact that materials and supplies are usually charged directly to cost does not mean that they have been fully expended at the time they are sent to the job. They represent a very real, and usually a very substantial, value and unless that value is protected it can easily be lost.

Physical Control

Physical control of materials and supplies involves four steps:

1. Control of purchasing
2. Control of receiving
3. Protection against misappropriation or theft
4. Protection against loss, shortage, spoilage, or exposure

Purchasing has been discussed in Chapter 2. It is enough here to point out that theft, loss, waste, or spoilage cannot cost the contractor any money unless it occurs after he has received the goods. Buying

*Practical procedures used in manually accounting for materials, supplies, and small tools have been difficult to devise because of the cost of performing the great mass of detailed work involved. The increasing application of computers in developing such procedures is more fully discussed in Chapters 12 and 14.

and scheduling deliveries only as materials and supplies are needed can avoid a great many of the problems of physical control before they arise. When material shortages, rising price levels, or other special conditions call for buying in advance of requirements, the costs of storage, handling, and protection must be considered in deciding what, when, and how much to buy.

The problems of receiving and protecting materials and supplies vary with specific conditions. Therefore, it is usually an oversimplification to say that any particular contractor has a good or bad system for handling materials and supplies. Any system may be good, bad, or indifferent depending upon the circumstances. That is why problems of materials and supplies accounting are frequently troublesome in the construction industry.

This difficulty does not mean that a construction company cannot adopt standard forms and procedures for its materials and supplies accounting. On the contrary, such action is absolutely essential in maintaining the required controls. It does mean that the establishment of a system is only the first step. Continually changing conditions call for corresponding variations in the application and use of the standard forms and procedures, for supplementing them, suspending some control routines, extending others, and so on.

Unlike manufacturing systems which, once established, can be policed and operated satisfactorily for indefinite periods, the construction industry must devise its accounting systems for flexibility and apply them in accordance with current job requirements. For that reason, this chapter confines itself to those problems, principles, procedures, and, most of all, to those objectives that a contractor or his accountant must consider in controlling and accounting for materials and supplies.

Physical control at the jobsite

In approaching the problems of physical control over materials and supplies delivered to the site of a construction job, these questions must be considered:

1. What can be done to make sure we receive everything we sign for?
2. Is it easy to steal?
3. Is it worth stealing?
4. Can the weather hurt it?
5. Does the risk of loss justify the cost of protection?
6. At the time it is physically used on the job, will the resulting paperwork cause the proper cost account to be charged with the correct amount of cost?

A few examples of how these questions are analyzed and answered under a variety of job conditions will serve to suggest the way similar problems arising on specific jobs or under specific circumstances may be treated.

Stone. The least vulnerable of all construction materials is stone or riprap rock. Its weight is determined at the quarry; it is not worth stealing; and it is completely impervious to weather. It is almost invariably charged directly to the job at the time of purchase and is extremely difficult to divert to other jobs. If it is hauled by rail to the jobsite, care must be exercised to see that it is unloaded as quickly as possible to save demurrage on the railroad cars, although sometimes it is cheaper to pay demurrage than it is to handle the rock twice.

Steel. Prefabricated steel beams, which are usually placed when delivered, rarely cause any protection problems. They would be difficult and costly to steal and weather cannot harm them. If the risk is substantial it will be necessary to evaluate the potential delay to the job as well as the cost of replacing the beams in deciding how much could or should be spent for protection.

Cement and concrete. Ready-mix (and usually dry-batch) concrete is placed immediately upon arrival at the jobsite. Here the contractor faces either the receiving problems inherent in pouring the concrete or excessive waiting-time charges for failure to unload it in the time allowed, or both. Insofar as the receiving problems are concerned, it is usually sufficient to compare the cubic content of the forms filled in each day's pour with the yardage shown on the delivery tickets for that day. It is necessary to scale the forms anyway in order to determine the amount to order. Sometimes the requisite physical control of concrete can be secured at the batch plant; sometimes weighing is required. Either way, it must be remembered that it always takes more than a cubic yard of material to make a cubic yard of concrete. Usually, however, scaling the forms is adequate. Excessive waiting time is essentially an operational problem, but since it increases the cost of materials, it becomes a matter of concern to anyone responsible for materials and supplies accounting.

Concrete may be mixed at the jobsite: if the volume required is large enough to make it profitable; if the forms cannot be reached with ready-mix trucks; if the haul is too long; or if, for some reason, ready-mixed concrete is not economical. Jobsite mixing calls for stockpiles of sand, gravel, and sometimes crushed rock. On large jobs, if a local deposit can be found, sand and aggregate are often produced at the site and materials of this kind rarely create any storage or protection problems.

Even if such materials are hauled in by truck and stockpiled, the value per cubic yard is, as a rule, not large enough to justify the cost of any substantial security measures. Usually it is enough to scale the stockpile occasionally and compare it with the amount that should be on hand after allowing for the amount that had to go into the volume of concrete already poured. Sand used for gunite operations must, of course, be kept dry, and it is sometimes necessary either to store it indoors or to cover it with tarpaulins. Bulk cement is usually delivered into silos where its quantity can easily be checked and where it is protected from the weather. It is a simple matter to padlock the loading gates on the chutes to prevent theft.

Materials requiring medium protection. Rough lumber, reinforcing wire or steel, pipe (whether concrete, clay, steel, or cast iron), conduit, steel scaffolding, steel concrete forms, form ties and clamps, drums of form oil, bricks, concrete blocks, clay tile, and metal lath are typical of a large group of items which may or may not be stored out of doors depending upon local weather conditions. Usually when the weather is good enough to permit construction to proceed, it is good enough to permit such items to be stored outside. However, many of these items are vulnerable to theft and if the job location is such that it can be readily reached by truck, it frequently pays to store such items inside a fenced enclosure protected by a watchman, floodlights, a burglar alarm, or all of these. The receiving problems generally encountered in dealing with the types of items just discussed, and those encountered in dealing with items in the group to follow, are enough alike so that the two groups may be discussed together insofar as receiving procedures are concerned.

Materials requiring much protection. Finish lumber, sash, doors, nails, finish hardware, plumbing and electrical fixtures, sacked cement, plaster and lime, finish tile, wire, switchboxes, acoustical tile, floor coverings, heating and air conditioning units, hand tools, surveying instruments, pneumatic and power tools, hose, hard hats, raincoats, rubber boots, extension cords, small motors, small pumps, bench grinders, and light welding equipment and supplies—all are typical of the kinds of materials and supplies which require maximum protection from the weather and are extremely vulnerable to theft. Another group of items frequently classed with this group is gasoline, lubricating oil, repair parts, and tires. Since gasoline and oil are usually stored in drums or in underground tanks, they do not require as much protection against the weather but their vulnerability to fire and theft is very high.

When stored at the jobsite, items in these groups should be stored under cover and under lock and key. Materials to be incorporated into the work should be issued only at the request of the foreman whose crew is installing them and usually in quantities not exceeding one day's requirements. Tools should be checked out either on tool check, brass (Chapter 3), or charge-out slip. Supplies should be signed for when taken. If problems of time, distance, or volume make it impractical to charge out tools individually, they may be checked out to the foreman, who is then made responsible for their return.

It is foolish for the contractor to think that the job is too small or that he knows his men too well for this precaution. A small operation where tools and supplies in this class are loosely handled can lose as much (or more), proportionately, as a large one. It is true that many smaller construction companies issue such items from a stockroom, but if this practice is not feasible, even a small job cannot afford to be without a locked tool shack and a padlock on the spigot of the gasoline drum. Where, as frequently happens, lumber, bricks, concrete blocks, or similar materials are stockpiled without physical protection, it is usually wise to employ a local watchman service to cover the area in its regular rounds. Sometimes, also, the area may be floodlighted. In the absence of such precautions, a contractor may find himself providing the entire neighborhood with materials for outhouses and chicken coops.

Receiving materials and supplies

A great deal of money may be saved by taking proper precautions when receiving jobsite deliveries of materials and supplies of the kinds mentioned in the last two groups. To be certain the items delivered correspond with the delivery receipt, someone must count or measure, inspect, and sign for each incoming delivery. On large jobs where one or more warehousemen are employed, incoming deliveries can be, and usually are, checked and counted. The more difficult problems arise on smaller jobs where volume does not justify the employment of a receiving clerk.

Ordinarily, the key to adequate receiving procedures is knowing when deliveries are to be made and having someone there to receive them. In other words, the problem is primarily one of timing and scheduling so that responsibility can be narrowed down to one or two people. If, as so often happens, there is no fixed responsibility for receiving, material shortages will inevitably develop.

It is standard commercial practice for the person making the delivery to secure the signature of someone acknowledging receipt of the shipment. It is entirely practical to require the person signing to take whatever steps may be reasonable in the circumstances to check any deliveries for which he signs. This procedure is basic. If such a check is not made, there can be little or no control of receiving materials or supplies.

If the operation is too small for a job warehouse but large enough for a field office, the field clerk or timekeeper will usually be the logical person to check in materials and supplies. If a repair or maintenance engineer is on the job, he can be made responsible for receiving spare parts, tires, gasoline, and oil. On very small jobs, the foreman may be made responsible for receiving, although, generally speaking, a foreman's time can be more profitably spent in planning and supervision. Deliveries of materials to jobsite locations may have to be checked and signed for by the foreman for lack of anyone else to whom the work can be delegated.

If traveling timekeepers or job accountants are used to cover more than one job, it is often possible to schedule deliveries and the timekeeper's visits so that the two coincide at least most of the time. The foreman can then handle such current deliveries as those of ready-mix concrete. It is not uncommon for a subcontractor to find that his part of the work on a particular job is not large enough to justify having his receiving done by his own personnel. In such cases, it is often possible to arrange with the prime contractor or one of the larger subcontractors to accept his deliveries. Occasionally a foreman will find that he has a truck driver or a laborer on the job to whom he can safely assign receiving duties. In this event, or if the foreman handles receiving personally, even the smallest job can have the protection to be obtained from checking the materials and supplies actually received against the items shown on the delivery receipts. The main thing is to recognize receiving as a major source of potential savings; to see that safeguards that are practical under the circumstances are embodied in a set of procedures to meet the problems peculiar to the job; and then to see that those procedures are enforced.

In this connection it should be stressed that no procedure, by itself, will operate successfully. It must be policed. If the contractor is on the job personally, he should check occasionally to see that the receiving routines are being followed. If the company is large enough to use a general superintendent or a field auditor, then that person should make jobsite receiving procedures one of the points on his regular checklist.

Jobsite storage

In discussing physical control of jobsite deliveries, six questions were posed as tests to determine what controls and protective measures were justified on any particular job. The same tests are applicable where materials, supplies, and small tools are kept in a stockroom, a warehouse, or a yard. Basically, any successful system for physical control follows these four rules:

1. If you are charged for something, be sure you receive it.
2. Once you have it, store it where you can find it.
3. Be sure it is safe from theft and the weather.
4. When you release it, be sure you know who took it, what he got it for, and whether he has a right to have it.

Even the smallest of construction companies has a back room or a shed where some materials, supplies, and small tools are kept. The larger companies, of course, have yards and warehouses in which may be found a large variety of materials, supplies, tools, and equipment. Between the two extremes are all sizes and kinds of off-job storage facilities. Some types of contractors (plumbers and electricians, for example) may keep a considerable inventory of new materials and supplies on hand for delivery, as needed, to the jobs. Others use their stockrooms, warehouses, and yards principally to store used materials, equipment, and supplies brought in off the jobs as surplus or salvage. In most construction companies, off-job storage facilities are used for both purposes.

If practical and economical, there is no better way to control materials and supplies at the jobsite than to send them out only as needed, which may well require a warehouse stock. An excellent example of this type of operation is found among plumbing and painting contractors working on small jobs. Such a procedure will be satisfactory for some, but not all, types of materials and jobs. For example, a plumber's repair truck will usually be a small stockroom on wheels. However, if a particular job calls for setting a water heater, that item will come from general stock, as will the soil pipe or galvanized pipe required for the work scheduled for a particular day. With such repair trucks, it is not unusual for the operator to turn in a daily or weekly stockroom order for the items used, so that a standard stock can always be maintained on the truck. Or, the operator may turn in, for each job, a report (Figure 4-1) from which a stockroom requisition may be prepared in the office.

There are other examples of jobsite storage in which the problems and procedures are so similar to those of the off-job stockroom or warehouse that they may be considered together. The supply room (or supply barge) on a hydraulic dredge, the truck-mounted rolling warehouse used by road and pipeline contractors, and, of course, the jobsite warehouse often maintained on large building jobs may well all be operated in the same way as a stockroom at the site of the contractor's general office.

One fact cannot be overemphasized. If it is worthwhile to maintain a stock of materials, supplies, and equipment, it is worthwhile to main-

PLUMBER'S REPORT

DATE__________

NAME__________

ADDRESS __________

Work Order No.__________ __________ Workman

Figure 4-1 **Plumber's Report.** This is commonly used as a source of materials and time control, and of billing data, from repair truck operations.

tain adequate control over it. A stockroom or warehouse to which everyone has access is an open invitation to theft, misappropriation, and waste. In such a stockroom or warehouse the contents will tend to be continuously disarranged so that, even when desired articles are there and in usable condition, it takes so much time and effort to find them that it is easier (and sometimes cheaper) to buy more than to try to locate those on hand. Even when the desired materials or supplies can be found in such a warehouse, the chances are good that when they reach the job they will be found to be in unusable condition, with the result that the job is delayed and more cost is incurred than if the warehouse stock had been ignored in the first place.

Usually, unless a contractor's operation is very small indeed, it pays to have one employee stationed in the stockroom or warehouse to receive incoming shipments, clean up and repair the equipment, check outgoing shipments, maintain whatever stock and equipment records may be necessary, and keep whoever is doing the buying advised of the items that are on hand so that they will not be duplicated.

In very small operations, these functions are usually performed by the contractor himself or by a combination bookkeeper, purchasing agent, and stock clerk, in much the same way that a field clerk might operate a small stockroom on a jobsite. However it may be done, unless someone is on duty in the warehouse or stockroom full time, only one, or at most, two people should have access to the key.

One common mistake, when operating an unattended warehouse or stockroom, is to give a job superintendent a key and tell him to send an employee down to pick up whatever equipment or supplies he may need. It would be an unusual superintendent who did not take more than he needed and an unusual truck driver who did not see numerous articles for which he could find a good use or a ready market. Eventually there would be an unknown number of duplicate keys made and the contractor's inventory would be there for the taking by any dishonest employee who wanted to buy a key. Also, if such a policy had been in effect for any substantial period of time, the stock would usually be in such confusion that stolen items would never be missed. In correcting such a situation, the first step is to change the locks.

Thus far, little or nothing has been said about protection from the elements of materials and supplies stored off the job. Stockrooms and warehouses present no problem so long as the roof and windows are tight. Sometimes, in taking over a different warehouse, the roof and windows will be checked for leaks, but no attention will be paid to the possibility of the floor being wet. It is a point worth considering, particularly if cement, nails, or any other materials or supplies that would be damaged by moisture are to be stored.

A contractor's yard sometimes includes a warehouse or shed. The outdoor portion would then be used to store the things which cannot be harmed by the weather. The yard should be protected by a wire mesh fence, at least seven feet high, and, depending on circumstances, the fence may be supplemented by a watchman service, floodlights, a burglar alarm, or all three.

Control forms and procedures

The preceding pages have dealt with the purely physical aspects of controlling inventories of materials and supplies. But specific forms and procedures—paperwork—are also important. The exact procedures described illustrate methods that have been used successfully. Naturally, they do not cover all possible situations. They do indicate how the problem may be approached, the tests for determining the effectiveness of any particular form or procedure, and how it may be modified to fit a wide variety of circumstances.

The materials budget. One of the best overall checks on materials is the materials budget. Many contractors prepare a standard bill of materials (Figure 4-2), a materials takeoff, or some other detailed list of materials required for a particular project in order to "buy the job" at its inception. When such a list is drawn up, it is a simple matter to require that any further materials purchased or requisitioned for that project be specially authorized. Requests for further purchases or additional requisitions on the warehouse are, of course, investigated and usually reveal weaknesses at some point in the procedure. Even if the jobs are not "bought" at their inception, the principle can still sometimes be applied if requirements are determined in advance of use.

There are two principal objections to the use of the materials budget. First, it is argued that it causes extra work when there are change orders. But if a control is useful it is worth a certain amount of additional work. Also the materials required by an additive change order, or for which credit must be given on an order, must be determined in negotiating the price, and hence any additional work would be negligible.

Second, it is argued that the cost of keeping a current record of materials charges is excessive in relation to the benefits received. This argument is obviously not sound since the materials list, once made, constitutes the basic control. If a separate sheet of the general type (Figure 4-2) is used for each kind of material to be controlled, it takes only seconds to adjust the balance on the sheet by the amounts of incoming and outgoing deliveries.

Users of the materials budget point to some dramatic positive results. One west coast tract builder, for example, has claimed that his materials budget led to elimination of losses equal to 9 percent of his total material costs.

Purchasing forms and procedures. Purchasing policies, as noted above, may contribute heavily to the effectiveness of accounting for materials and supplies, by so scheduling deliveries as to minimize the need for physical storage and control at the jobsite. However, attempts to tie materials and supplies accounting to purchasing routines, as distinguished from purchasing policies, generally result in costly and time-consuming paperwork, the results of which do not justify the cost.

There are three important exceptions to this general rule. The first exception is the practice of pricing purchase orders (Chapter 2). This routine provides valuable internal auditing data and is the most practical check on the pricing of invoices.

Form 107A

LUMBER, MILLWORK, AND OTHER CARPENTRY MATERIALS (7A)

JOB NO.

COST ESTIMATE

DESCRIPTION	AMOUNT	UNIT PRICE	TOTAL ESTIMATED COST
A. BUILDING LAYOUT			
TOTAL			
B. FORMS AND SCREEDS			
TOTAL			
C. WOOD FLOORS AND FRAMING			
TOTAL			
D. EXTERIOR WALLS			
TOTAL			
E. ROOF FRAMING AND SHEATHING			
TOTAL			
F. EXTERIOR TRIM			
TOTAL			
SUB-TOTAL -- ESTIMATED COST OF LUMBER, MILLWORK, AND OTHER CARPENTRY MATERIALS (TRANSFER TO FORM 107B)			

Figure 4-2 **Materials List for Lumber.** This form is often used as a takeoff sheet. If the material takeoff has been prepared on working paper, this kind of sheet is used to summarize the materials required so that purchase orders may be written as needed. If a materials budget is used, one sheet of this type may be used for each kind of lumber included in the budget.

The second exception is the preparation of the job copy of purchase orders and purchase contracts. By sending a copy of every purchase order and purchase contract to the job superintendent as soon as the order is placed or the contract negotiated, it is possible to give the jobs notice of what has been bought and when and how it is to be delivered.

If there is a delay of more than 24 hours between placing an order or negotiating a purchase contract and the preparation of the order form or contract document, a weakness exists and should be corrected.

The objection has been made that, in small operations where jobs are run by foremen, the purchasing information should not go to the foremen for two reasons: If deals with suppliers are known to foremen, confidential information may leak out to competitors or may be misused. Also, the foremen do not use the information anyway so it is a waste of time.

These objections will not stand up under analysis. The first can readily be overcome by omitting price information from the copies of the purchase orders and contracts sent to the job. The second, if true, constitutes an organizational weakness that should be corrected without delay.

The third exception is the printing of provisions on the purchase order. Every purchase order should prominently display a legend to this effect: "No back orders will be permitted. If any item on this order cannot be delivered delete it and notify us at once." Without such a provision in his purchase orders, a contractor may find himself forced to accept, and pay for, materials especially ordered by the supplier for some job, long after that job has been completed with materials purchased from some other supplier.

Receiving forms and procedures. Receiving procedures must vary with what is being received, where it is being received, and for what purpose. It is a matter of universal practice for the person making the delivery to obtain a signed delivery receipt. Often this signature is perfunctorily given without any examination of the shipment, thus constituting an open invitation for short deliveries as well as substituted, and sometimes unsatisfactory, products. A minimum receiving procedure practiced by soundly managed firms, however small, is to inspect every shipment before signing the delivery receipt. If possible, the quantity, quality, and condition of the items should be checked. If a receiving copy of the purchase order is available, it should be checked for items ordered but not delivered. Any attempts at back ordering should immediately be called to the attention of the person or department responsible for purchasing.

It is common, in modern shipping and billing systems, for the delivery receipt to be either a copy of the invoice or the invoice itself. If the person receiving the shipment has to sign the document, then it is a delivery receipt, regardless of the facts that it may be headed with the word "invoice" and that no other invoice may be given by the supplier. A single piece of paper may do double duty as both invoice and delivery receipt. Needless to say, if the contractor requires additional copies of the invoice they should be clearly marked "copy."

Many contractors, particularly those who do business in highly competitive trade areas, find it worthwhile to require the person receiving shipments to prepare his own count on a receiving report which later can be checked against the purchase order. If competition among suppliers is stiff, it is not unknown for them to bid for orders at cost or less and make up the loss of overhead and profit by shortening the loads or delivering inferior goods. Receiving documents are used to control payments in such a way as to combat any such practices.

A typical minimum receiving procedure would involve these steps:

1. Unloading the materials or supplies at the job or warehouse
2. Counting them to see how many are there (or how much)
3. Inspecting them for condition and noting any damaged items
4. Checking the delivery receipt to see that the items signed for are the same as those received
5. Verification that the items received meet the type, specification, or characteristics set forth in the purchase order

Other refinements such as receiving copies of purchase orders or separately prepared receiving memoranda may be required. Basically, however, the steps will parallel those just given.

Receiving procedures for similar items will normally be the same whether delivery is at the jobsite or elsewhere. The required procedure should never be relaxed because deliveries are made to a job.

Warehouse or stockroom. Once materials or supplies are received at the warehouse or yard, the accounting records kept and the forms and procedures used (though they may vary widely to fit a wide variety of actual situations) should be such that the following questions can be answered at a moment's notice: (1) Have we got it? (2) How much (or many) do we have? (3) Where is it? (4) What condition is it in?

In a small warehouse, stockroom, or yard under the control of a full-time employee, or even a part-time employee, the warehouseman can usually answer these questions from memory. This method is adequate if the employee is available, or if the stock is small enough and well enough arranged so that anyone familiar with construction materials and supplies can answer the questions by going to the warehouse and inspecting the contents. Many small stockrooms and warehouses get along quite well with nothing more. As the number and variety of items increase, to the point where memory no longer serves the purpose, records become necessary. Often, when a construction company finds its warehouse becoming increasingly cluttered, and finds that it is purchasing articles which are already on hand in adequate supply, it is a sign the company has outgrown the memory stage and is in need of records.

For power or pneumatic hand tools and similar small, serially numbered equipment one of the most practical controls is a visible card record (Figure 4-3). The Kardex System is one of several simple, practical methods.

Frequently it is useful to keep a central control file of this type for all small power and pneumatic tools, using one or more drawers for the tools on each job and separate drawers for the tools on hand in the warehouse. A similar card record (Chapter 5) is useful for equipment of all kinds.

The first step in keeping track of warehouse stock is to establish a permanent arrangement with marked bins or racks so that similar articles are always found in the same place. In this way anyone familiar with construction materials and supplies can tell at a glance whether a particular item is in stock and the approximate quantity on hand.

If the warehouse is kept in good order it is frequently unnecessary to carry physical accounting further. When the person or department which does the purchasing receives a requisition to purchase certain materials or supplies normally in warehouse stock, the warehouseman can usually tell which items are on hand. In very small operations, the buyer will simply go to the warehouse or stockroom and see for himself, if he does not already know.

EQUIPMENT LOCATION CARD

DATE	TRANSFER NO.	FROM	TO	CONDITION

DATE PURCHASED______ FROM______

COST______ NEW ☐ USED ☐

MONTHLY DEPRECIATION RATE______ MONTHLY RENTAL RATE______

DESCRIPTION | SERIAL NO. | EQUIP'T. NO.

Figure 4-3 **Equipment Location Card.** This card is used to control power or pneumatic hand tools and similar small equipment. One or two drawers in a cabinet usually suffice for the cards listing equipment used on any one job, and the cards on any additional items in the warehouse are kept in drawers reserved for warehouse stock. Thus it is possible, at a moment's notice, to tell what items are immediately available at the end of each job.

If jobs are large enough to have purchasing agents at the jobsite, some companies supply them with monthly lists of materials, supplies, and equipment on hand in the central warehouse, in order to minimize delays in ordering and long-distance telephone calls from the jobs to the warehouse to learn what items are in stock.

As a device for physical control, hand-recorded perpetual inventories of stock in the warehouse (even those kept in terms of units without reference to cost) are uncommon. Usually they represent an expenditure of time that could be more usefully spent elsewhere. If they are kept, the most practical method is to have an "in and out" type of card (Figure 4-4) fastened to the front of the bin. Items removed are recorded in the "withdrawn" column and all items received are recorded in the "received" column. Each movement results in a new total carried forward. Other types of bin cards (Figure 4-5) are available. Construction contractors whose operations are large enough to justify the cost frequently find that computerized records are practical in controlling warehouse inventories.

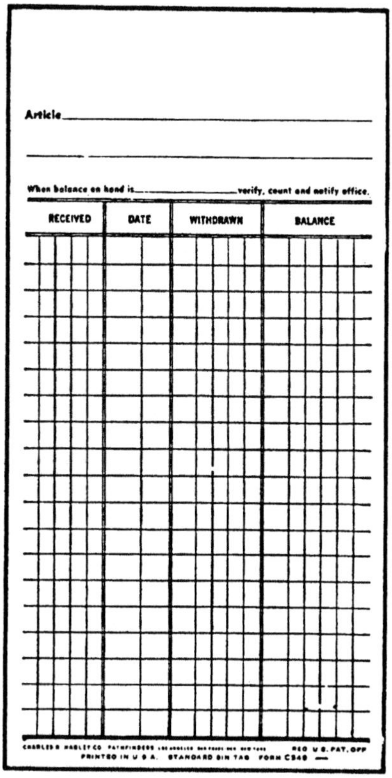

Article ______

When balance on hand is ______ verify, count and notify office.

RECEIVED	DATE	WITHDRAWN	BALANCE

Figure 4-4 "In and Out" Bin Card.

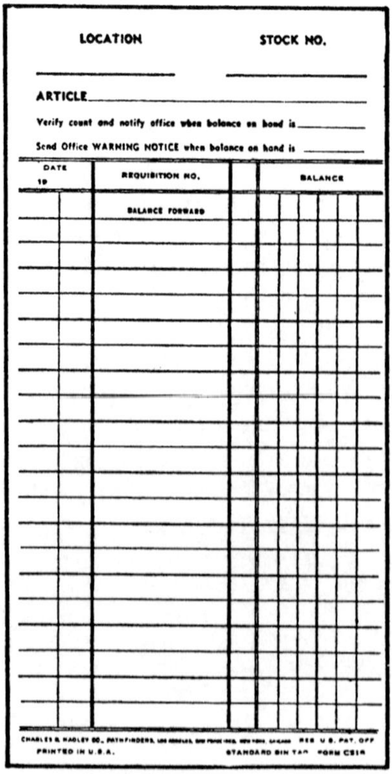

Figure 4-5 **Inventory Bin Card.** Additions or deductions can be made in the balance column, or the adjusted balance can merely be carried forward.

Variations. The physical control thus far discussed is intensely practical. It is elusive in the sense that no single practical, uniform system to secure and maintain it on construction jobs ever has been or ever will be devised. The materials, the job conditions, the kind and amount of equipment, the personnel, and the nature and location of the jobs are but a few of the variables that make physical control of materials and supplies a problem to be solved on the basis of experienced judgment applied to individual facts, rather than by any preconceived system of procedures. That a system has worked well on one job is no guarantee that it will work well on the next job. In fact, one of the strongest features in a program of physical control over materials and supplies is flexibility and variation. When the routine becomes too well established, someone will find a way to beat it.

It must be repeated for emphasis, however, that the extreme variability in the forms and procedures required for adequate controls on materials and supplies cannot, under any circumstances, justify failure to establish standard procedure. A company may, within the scope of its normal operation, establish those forms and procedures best calculated to meet its individual problems. However, once such forms and procedures are established it cannot stop there. The system itself must be flexible enough to permit wide variations and it must be subjected to constant review to gauge its effectiveness under prevailing conditions. Failure to follow this process of constant reassessment has caused more breakdowns in materials and supplies accounting than any other single factor.

On the other hand, flexible as the detailed procedures for physical control of materials and supplies may be, the objectives never change. These may be summarized as follows:

1. When you receive a shipment, make sure you are getting what you ordered, and that it is all there and in good condition.
2. After you have received it, be sure you know where you put it and that it is as well protected physically from loss, damage, or theft as is reasonably possible in the circumstances.
3. Whether it is delivered directly to the job for immediate use, or is stored in a yard, warehouse, or stockroom, be sure you have a written record of who is getting it and for what, before you release it. You signed for it when you got it. The person to whom you deliver it should do the same. He should also tell in writing what he expects to use it for, and if the question is applicable, the source of his authority to request its delivery to him.

These steps sound simple and they are. The only complications are those that arise when no one takes the trouble to see that the steps are taken.

Accounting for direct charges

In its most rudimentary form, accounting for materials and supplies follows these steps:

1. The foreman on a small job (who is often the contractor) orders whatever he needs for the job from his suppliers. He may use a materials takeoff and from it write a purchase order, but often he will simply take the plans in to the materials dealer and make out his order with the help of a salesman.
2. When the materials arrive on the job, the foreman, or someone designated by him, shows the driver where to unload. He may or may not save the consignee's copy of the delivery receipt.

3. When the material dealer's statement comes in at the end of the month, the contractor pays it and charges the amount to material cost on the books.

So long as everyone is honest and accurate and no one wants to know the cost of any particular job, there is nothing wrong with such a system. Unfortunately, life is not that simple. Some people are dishonest; everyone makes mistakes occasionally; and in the construction business it is usually necessary, for one reason or another, to know the cost of each job. Be that as it may, there are hundreds, possibly thousands, of small contractors who account for their materials and supplies in about the way the three steps above indicate.

The better (and more successful) contractors—even the very smallest—usually add several more steps to the procedure. They prepare their own material list and put it on one or more purchase orders to the supplier. They keep copies of their purchase orders and drop them into a folder labeled with the job name and number. They then see that the loads of materials are checked in on the job, and that each load is separately billed showing the address of the job to which it is delivered. They save the delivery receipts and place them, together with the purchase orders and the invoices, in the job folder.

Every ten days, or before the tenth of each month, depending on how the discount dates fall, the invoices, the delivery receipts, and the purchase orders are matched up. Items shown on the invoice or the purchase order, but not on the delivery receipt, are deleted, and a separate check is written to cover the items purchased and actually received on each job. If the disbursements record is properly arranged, only one check need be written to each supplier even though his deliveries covered more than one job. Thus, almost automatically, the jobs are charged with the materials and supplies that are delivered to them, and unless management wants separate costs for the several work items into which the jobs are usually broken down on the specifications, the accounting for materials and supplies is completed.

In the operations of smaller construction companies, if a job is carefully bought there will be very little salvage and what there is can usually be returned for credit or transferred at salvage price (or more often at no price at all) to another job. Only a few power hand tools are kept in stock and these the contractor knows from memory. However, as the size of the operation increases, so do the number and variety of items and at some point accounting controls become necessary.

The system just described is used with great success by many contractors—even the smallest. Basically, the same system, expanded in some particulars because of larger volume, is used by most of the largest contractors regardless of whether materials and supplies are bought directly for the job or are passed through a warehouse, yard, or stockroom inventory account. Many of the larger jobs have separate sets of books (particularly if a job is being done by a joint venture) and some have extensive cost breakdowns. However, there are numerous multimillion-dollar jobs on which only a single account is kept for materials and supplies. The final test in determining the system to be used is the end use of the figures. There is no merit in paying for records that will not be used.

Those contractors who use cost data from one job as a guide in bidding other jobs frequently break their jobs down according to work items, which constitute a basis for compiling unit costs (Chapter 1). Thus a building contractor might have a cost account for "wall forms" and compute a cost per square foot. A tract housing builder, on the other hand, might have a cost account called "cupboards and cabinets" and compute a unit cost for each house of a particular design. In such cases the "materials" account in the general ledger becomes a controlling account for the "materials" column in the cost ledger (Chapter 9). The only difference in accounting treatment would be in coding the charge to the cost account as well as to the general ledger.

The essence of any system of cost accounting for materials and supplies is coding—the accounting term used to describe the job of placing on each invoice, purchase order, requisition, or other original document, the number of the general ledger and the cost ledger accounts to be affected by the transaction. Thus, in the voucher system flowchart (Figure 4-6) it will be noted that the voucher is coded so that account 145 in the general ledger is debited and account 205 is credited. In the cost ledger, account Z-2-27 is charged, which would normally mean that the "materials" column in the cost ledger sheets for Job Z is designated as column 2 and that the materials were used on work item number 27—this work item having its own separate sheet in the cost ledger. Of course, in a system without cost accounts, the charge would simply be applied to Job Z.

The best method of coding materials and supplies that are to be charged directly to a job is to code the purchase order; no one knows better than the person placing the order the purpose for which the material is being bought. If coding the purchase orders is not feasible, it is usually necessary to code the invoices. Ordinarily invoice coding has to be done by a foreman, superintendent, or cost engineer who is familiar with the job and who knows where the invoiced materials were used.

The most rudimentary accounting system maintains a simple unpaid bills file which the owner checks at regular intervals to determine what bills are to be paid. Under such a system, materials and supplies are accounted for only when the bills are paid and no formal

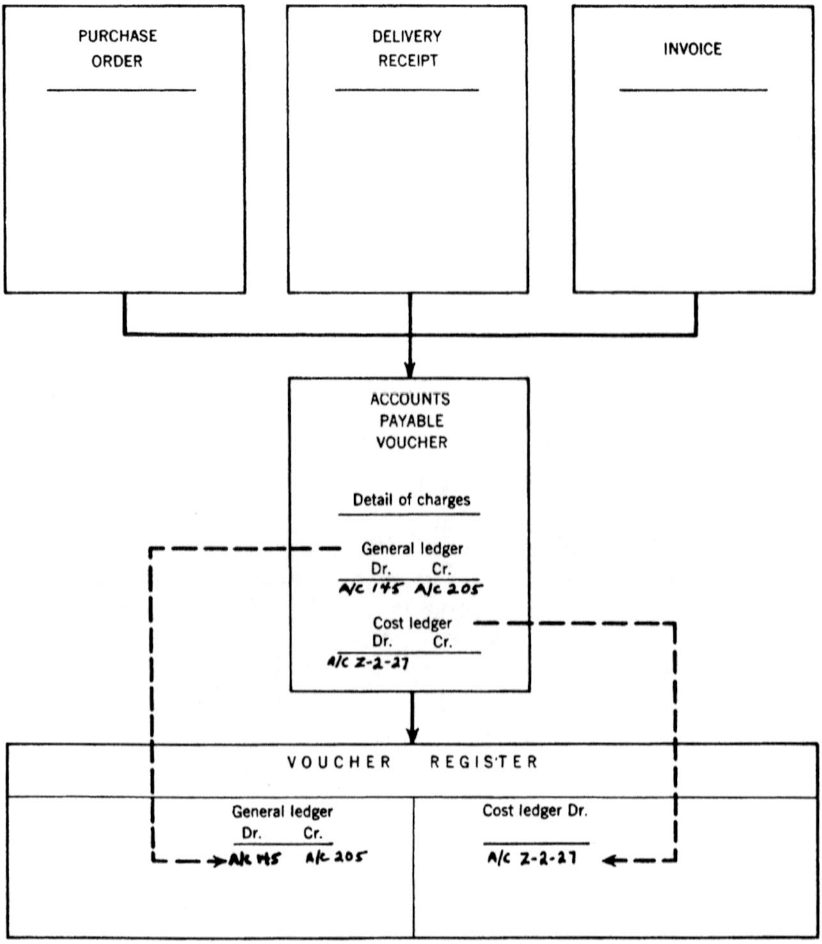

Figure 4-6 Voucher System Flowchart. Three basic documents—purchase order, delivery receipt, and invoice—support a voucher. The information in these documents is summarized on the voucher, entered in the voucher register, and posted to the cost ledger (daily). Monthly voucher register totals are posted to the general ledger.

account is taken of unpaid items. This method may be converted to an accrual method at the end of an accounting period by entering all unpaid items through the general journal and reversing the entry after the new period begins. This method is used on many construction jobs (even some very large ones) where separate books are kept on the job, and where interim financial statements are not important.

A better procedure for construction companies, either large or small, is the voucher system, and its use will be assumed in further discussions and illustrations. Some contractors still keep an accounts payable ledger, either because they use it for reciprocity information or because a voucher system becomes unwieldy when bills are not being paid currently. Nevertheless, there are many contractors, particularly small ones, who distribute costs on a cash basis and only record liabilities at the end of an accounting period. The flowchart illustrated (Figure 4-6) shows a typical voucher and, in illustrating the voucher system, it follows a single transaction from the purchase order to entry in the records. It also demonstrates one method (probably the most common) of getting cost data from the original documents into the general and cost records. This procedure is discussed in greater detail later in this chapter.

Various procedures may be used under the voucher system. First, the material may be charged directly to the job and not to a warehouse account. Second, costs may be recorded net of discount. The first is standard procedure and should be followed whenever the facts permit. The second is a matter of accounting judgment. If the amount of discount earned is important to management, then discounts should not be netted. If management has no need for the discount figures, then the handling should be as illustrated. In reaching a decision on this point, the extent, if any, to which material cost figures are used in bidding may be significant.

When distribution is made only at time of payment, a cash disbursement record is ordinarily used instead of a voucher register.

If a blanket purchase order is used, or if more than one delivery is made on one order, the order is marked to show the items received and the voucher number on which they appear and is kept in the file under the supplier's name for the remainder of deliveries. If these are delayed too long back orders must be suspected.

One important variation of the procedure illustrated in the flowchart arises when all, or substantially all, of a particular type of material (framing lumber, for example) is bought at one time, as in the construction of tract housing. If no cost ledger is kept, the charge would, of course, be simply to materials. If, on the other hand, costs are kept on each unit, the lump sum purchase must be prorated to the several cost accounts on the basis of the original materials takeoff. If more of the same type of materials is required toward the end of the job, that cost, too, must be prorated over all units even though some may be finished. Such additional purchases usually mean errors in takeoff or buying or that there has been loss or waste of materials.

Accounting for items out of inventories

Basically, there is no difference between accounting for materials and supplies taken from stock, and accounting for those that are bought for the job. It has been pointed out that nothing should leave the warehouse or stockroom without written authorization, whether in the form of a materials list (Figure 4-2), a material order (Figure 4-7), a storeroom requisition (Figure 4-9), or a copy of a direct billing made on a counter billing machine (Figure 4-11). A material credit form is illustrated in Figure 4-8, a storeroom credit form in Figure 4-10. All of these forms are, in effect, requisitions, and are referred to as such for convenience.

The requisition is the basic document for stockroom or warehouse accounting. The first copy, which should be signed by the foreman, or someone designated by him, takes the place of the purchase order. The copy signed by the person receiving the order is comparable to

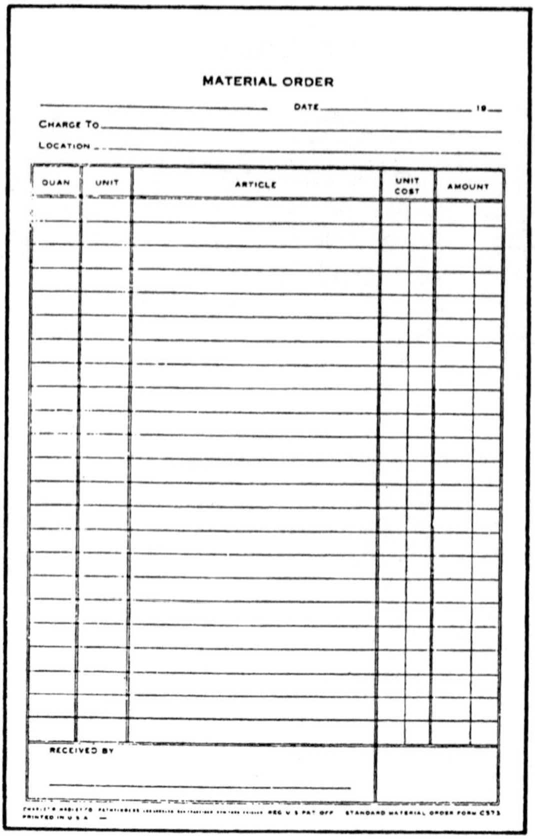

MATERIAL ORDER

DATE ______ 19__

CHARGE TO ______

LOCATION ______

QUAN	UNIT	ARTICLE	UNIT COST	AMOUNT

RECEIVED BY ______

Figure 4-7 **Material Order.** This form and the one shown in Figure 4-8 are frequently used to authorize deliveries from, and returns to, the stockroom or warehouse. They may be made in as many copies as the accounting needs of the business dictate.

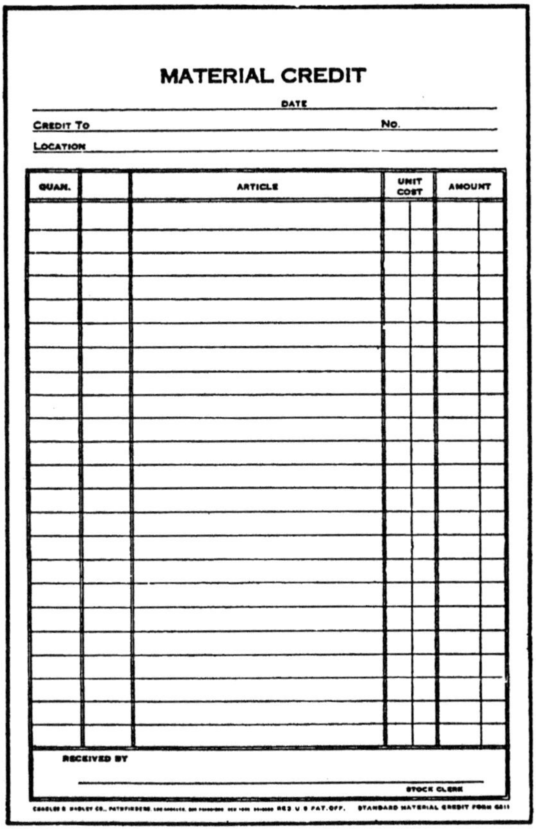

MATERIAL CREDIT

DATE

CREDIT TO   No.

LOCATION

QUAN.		ARTICLE	UNIT COST	AMOUNT

RECEIVED BY

STOCK CLERK

Figure 4-8 **Material Credit.**

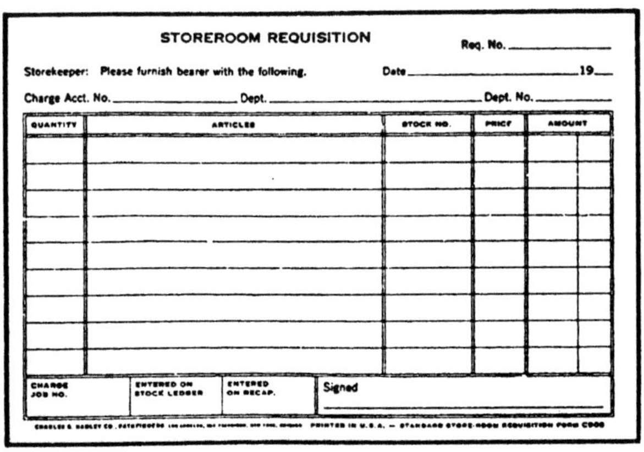

STOREROOM REQUISITION   Req. No.

Storekeeper: Please furnish bearer with the following.   Date 19

Charge Acct. No.   Dept.   Dept. No.

QUANTITY	ARTICLES	STOCK NO.	PRICE	AMOUNT

CHARGE JOB NO.	ENTERED ON STOCK LEDGER	ENTERED ON RECAP.	Signed

Figure 4-9 **Storeroom Requisition.** This form and the one shown in Figure 4-10 are similar in use to the material order and credit.

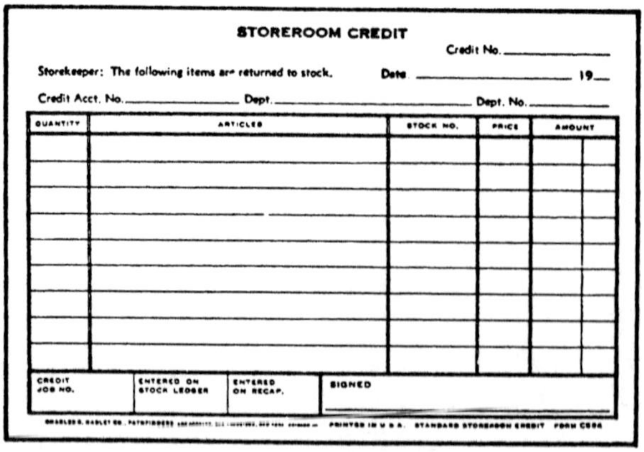

STOREROOM CREDIT

Credit No. ______

Storekeeper: The following items are returned to stock. Date ______ 19__

Credit Acct. No. ______ Dept. ______ Dept. No. ______

QUANTITY	ARTICLES	STOCK NO.	PRICE	AMOUNT	

CREDIT JOB NO.	ENTERED ON STOCK LEDGER	ENTERED ON RECAP.	SIGNED

Figure 4-10 **Storeroom Credit.**

the delivery receipt. The second copy, which goes directly to the bookkeeping department and is priced, extended, and often costed by the bookkeeper, takes the place of the vendor's invoice. Instead of supporting an accounts payable voucher (Figure 4-6), these copies may be summarized in a warehouse disbursement record from which postings are made to a general ledger and the cost ledger or may be summarized on work paper, and the work paper used to support a journal voucher. A journal voucher form for general journal entries is preferable to the usual general journal form. The "credit" forms illustrated in Figures 4-8 and 4-10, of course, operate as reductions of the requisition totals.

When materials or supplies are sold from the warehouse, the procedure for charging out is much the same as when they are used on a job, if actual cost is known or estimated, or if warehouse stocks are priced out at market.

To illustrate, if cost is known, and is to be accounted for currently, the entries in the sales record will be as follows:

Dr. Cash (or Accounts receivable) for selling price

  Cr. Warehouse sales for selling price

Dr. Cost of warehouse sales for actual cost

  Cr. Warehouse inventories for actual cost

If the cost is not known, but is estimated, the procedure is the same, except that at the end of the fiscal year an inventory will be taken and the overage or shortage will be adjusted by charging or crediting a "Warehouse inventory adjustment" account (or "Materials cost" if no detailed job costs are kept), as follows:

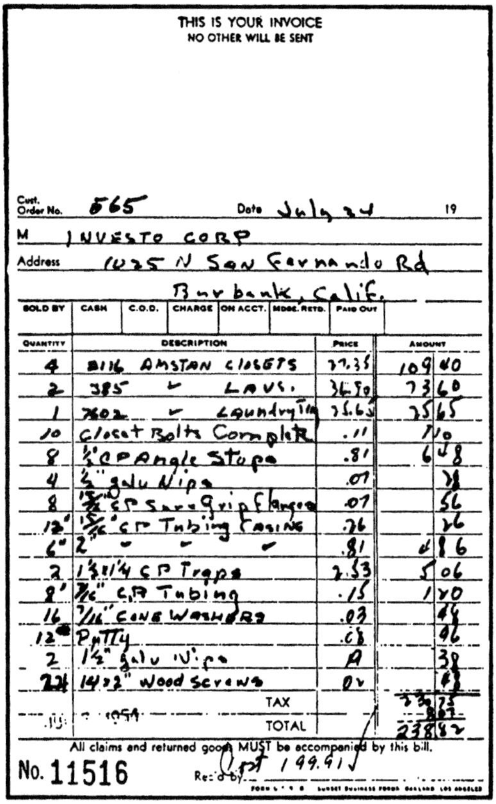
THIS IS YOUR INVOICE
NO OTHER WILL BE SENT

Cust. Order No. 565 Date July 24 19

M INVESTO CORP

Address 1025 N San Fernando Rd
Burbank, Calif.

SOLD BY	CASH	C.O.D.	CHARGE	ON ACCT.	MDSE. RETD.	PAID OUT

QUANTITY	DESCRIPTION	PRICE	AMOUNT
4	2116 AMSTAN CLOSETS	27.35	109 40
2	385 " LAVS.	36.80	73 60
1	7602 " Laundry Tray	35.65	35 65
10	Closet Bolts Complete	.11	1 10
8	3/8 CP Angle Stops	.81	6 48
4	3/8" galv Nips	.07	28
8	1/2 CP Sure Grip Flanges	.07	56
12'	1/2 CP Tubing Casing	.26	26
6"	2" " "	.81	4 86
2	1 1/2 1/4 CP Traps	2.53	5 06
8'	3/16" CP Tubing	.15	1 20
16	3/16" Cone Washers	.03	48
12	Putty	.08	96
2	1 1/2" galv Nips	[illegible]	38
24	14 x 2" Wood Screws	[illegible]	[illegible]
	TAX		[illegible]
	TOTAL		238 82

All claims and returned goods MUST be accompanied by this bill.

No. 11516 Rec'd by Cost 199.91

Figure 4-11 **Billing Form.** This is the second (accounting department) copy of a form used by a small local plumbing contractor. When materials are taken from stock, they are recorded on a hand-operated billing machine on the stockroom counter. Four copies are made. The first is the customer's copy of the invoice. The second is sent to the bookkeeper and serves as the office copy of the customer's billing. The third copy is the cost copy. The fourth copy remains in the machine as a chronological record.

Dr. Inventory adjustment account (or Materials cost) for amount of shortage

Cr. Inventory for amount of shortage

Many contractors who carry substantial inventories (plumbers, electricians, and so on) cost their warehouse sales and warehouse requisitions from a current catalog. This is substantially the same as estimating cost and the procedure followed is the same.

If, as is often the case, the warehouse is operated as if it were a separate business, the cost of sales would be found by taking inventory once each year and using the formula (beginning inventory plus purchases less ending inventory) to determine cost of warehouse sales. Cost of sales on interim statements would, in a system of this kind, be estimated as a percentage of sales or estimated from the supplier's catalog as described above.

It will be noted that a separate account called "Warehouse sales" is used. Under normal circumstances any additional cost incurred in keeping such an account is far outweighed by its usefulness when tax returns are to be prepared.

There are a number of special problems which arise in working with inventories of materials and supplies in yard, warehouse, or stockroom. The most universal of these are pricing and costing.

Theoretically, there are three general methods of costing out materials and supplies taken from inventory: (1) First-in-first-out (FIFO), (2) Last-in-first-out (LIFO), (3) Average cost.

Practically, the use of these methods in the construction industry is rare because they cost too much to operate and, particularly with used items, are too inaccurate. Many warehouses and stockrooms are operated as if they were separate businesses. The warehouse is charged with the cost of materials and supplies purchased for stock or brought into stock as salvage, and, in turn, "sells" them to the jobs. The so-called "profit" from the warehouse offsets the overhead. This method distributes the cost of operating the stockroom or warehouse among the jobs on a roughly proportionate basis. The prices charged to the jobs closely approximate the going market for the items supplied.

This method has an additional advantage when a contractor takes some jobs on a lump sum bid, others on a unit price basis, and still others on a time-and-materials or cost-plus basis. So long as the price charged to the job for material is not greater than the price the contractor would pay some outsider for the same items, no one has grounds for complaint. The contractor, on the other hand, has a system which works the same way for all jobs and effectively distributes the cost of operating the warehouse or stockroom to the jobs. The system costs less to operate because current market value is quickly and simply determined, whereas any kind of pricing based on cost means additional work and expense.

For income tax purposes, it is usually necessary to take some sort of physical inventory regardless of the method of handling the warehouse or stockroom. New stock is customarily priced for inventory purposes, according to the most recent invoices, in keeping with traditional first-in-first-out accounting. Used materials or supplies are usually priced either at market, less an estimated cost to recondition, or at an estimated 40 to 50 percent of current market. Consistently followed, such a method is rarely questioned when used for income tax reporting purposes because, when properly applied, it results in a realistic valuation of warehouse stock and a substantially correct determination of income.

Similar problems arise with minor equipment. When not bought directly for the job, hand tools, parts, and supplies are normally "sold" to the jobs from the warehouse, as described above. Power concrete buggies, table saws, power hand saws and drills, pneumatic tools, and similar equipment may also be "sold" to the job and then "repurchased" at the end of the job at going rates. When these items are "sold," the prices are usually at or near the market price for the "sale" and normally range from 20 to 50 percent of market for the repurchase, to allow for handling and reconditioning. Market, here, is what the contractor would have to pay if he bought the item from someone else. When this procedure is followed, the repurchase at a reduced price is often used instead of a formal depreciation charge on minor equipment. Even when the "sale" system is used, these items should be charged to the job and accounted for. The visible index files are usually adequate for this purpose and constitute, in effect, a perpetual inventory of minor equipment. When such items are rented to a cost-plus job, it is usually necessary to stop the rentals when an amount equal to 80 percent of the going market price has been charged on any piece of equipment, or—as for form ties, whaler braces, shores, and so on—when the rentals reach about 80 percent of the cost of that class of equipment on the job.

If power and pneumatic hand tools are rented to the jobs, it is common practice to keep the control and central depreciation records on the cards, although these records are sometimes kept in an equipment ledger. In most cases the card file alone is less costly and just as effective as control cards used in conjunction with an equipment ledger. The determining factors in selecting the system to be used are usually the location of the warehouse and the personnel used to run it.

One factor strongly favoring the "selling" of small tools and supplies and the "renting" of power and pneumatic tools to the job is the psychological effect on foremen and superintendents. Such people are almost invariably cost-minded and will return unused equipment or supplies for credit rather than allow them to remain idle at the jobsite and increase the costs charged to their jobs.

Perpetual inventory, as an accounting device, is frequently too costly to be practical when applied to construction materials or supplies. When such perpetual inventories are required on certain types of cost-plus-fixed-fee jobs, as they sometimes are, the forms and procedures are almost always prescribed beforehand. On ordinary commercial work the test is usually one of volume. The bin type of perpetual inventory card (Figures 4-4 and 4-5) is practical for smaller warehouses. Also shown is a more complicated card (Figure 4-12) used by one of the largest construction firms in the country. Notice that while data on costs, including freight and handling, are provided for on this card, the inventory itself is kept in units only.

If tabulation equipment is available, and the volume of transactions is very large, perpetual inventories in terms of quantity and

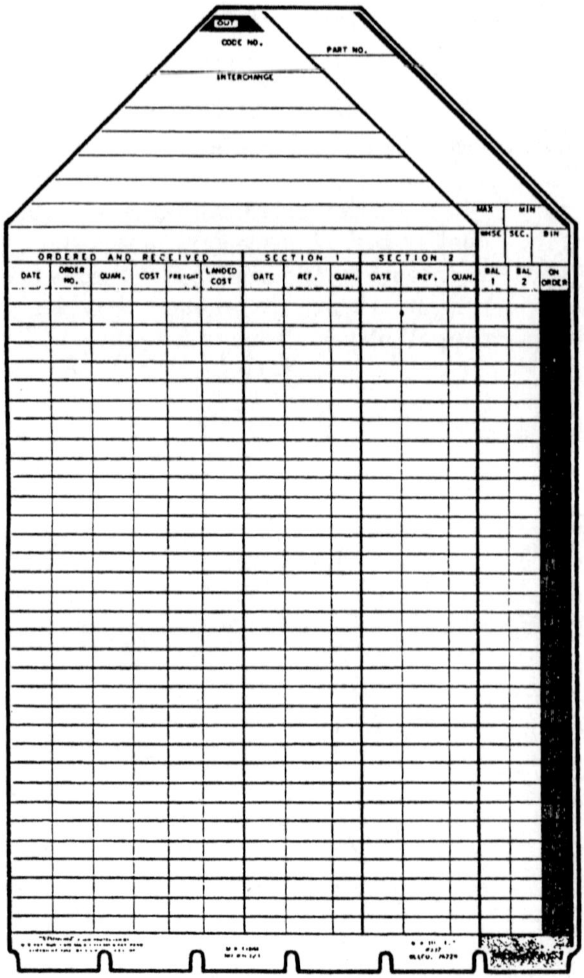

Figure 4-12 **Perpetual Inventory Card.** This card is used by one of the largest construction companies in the United States. It is reported to be very accurate and a great timesaver if volume is heavy. (See Chapter 12.)

value can be used successfully. However, they are usually kept in terms of units only and the value is either ignored entirely or is limited to unit price figures.

If a detailed quantity control is required and the volume of transactions is so large that it makes sorting and grouping of data a problem, a modified form of punched card accounting may be useful. In this system the only equipment needed is a hand punch and a sorting needle.

If a large volume of materials and supplies is disbursed from a warehouse or stockroom, the requisition can be prepared on cards designed to be used with one of these modified punched card methods (Figure 4-13).

If perpetual inventories are kept, whether in terms of money or units, or both, the problem of physical inventories can be greatly reduced by taking them for a portion of the warehouse or stockroom each month and adjusting the cards to show the actual quantities on hand, with corresponding general ledger adjustments of the total inventory value. An even more common method is to count any particular bin whenever the quantity is low.

Regardless of the type of accounting controls used, a physical inventory of materials and supplies, including spare parts, hand tools, pneumatic tools, and power tools, should be taken at least once a year in the general warehouse or stockroom if the amount on hand is substantial. If jobs are large enough to have their own warehouses, repair shops, or stockrooms, it may or may not be practical to take a physical inventory, depending on the length of the job, the quantities on hand, and the degree of physical control exercised.

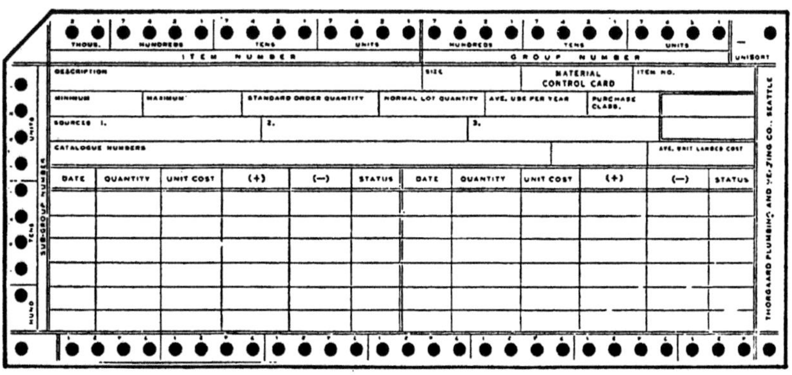

Figure 4-13 Unisort Punch Card. If classification of data is a problem and the volume to be processed does not justify the cost of machine sorting equipment, this type of punched card can often be used to good advantage. A hand punch and a sorting needle are the only equipment required. Electrically driven punches are available. Any clerk can be trained to use this type of card in a very short time.

Physical inventories should be taken on forms prepared and provided by the general office so that terminology will be uniform and proper designations and descriptions used. Because most contractors have both new and used materials, supplies, and equipment on hand, it is important that the forms provide space to indicate the condition (new or used, and serviceability) of each item.

One fairly common method of taking physical inventory is to have the warehouseman or (if none is employed) the person making the count tag the bins and piles and to have some other person list the tags and make test counts. When this method is used, the tags should be prenumbered so that a positive check may be made to be sure that no tags have been missed. If withdrawals are necessary during the count, or if items of the same kind are stored in more than one place, this information can be marked on the tags and taken into account when the tags are summarized.

Many contractors will object strenuously to a physical inventory of warehouse and storeroom, because most of the contents have been charged to the cost of jobs, and the setting up of an inventory on the books would require payment of income tax on an amount equal to the cash value of the inventory. If the practice has been followed for a number of years, and if there is no substantial change in warehouse stock from year to year, there would seem to be no serious distortion of income and under present income tax laws no harm is likely to be done. However, if inspection indicates a trend toward substantial increase, it may mean overbuying and tying up working capital to such an extent that it would be cheaper to pay the tax. It may also mean the contractor is paying for the storage and handling of worthless junk which should be scrapped.

One final admonition about physical inventories: If the item is worth keeping it is worth listing even though it may be assigned no value. If it is not worth listing, it probably should be sold for scrap.

Prevention of theft

A system has yet to be devised which will completely ensure against theft of materials, supplies, and small tools from construction jobs. Jobsite conditions usually make complete physical protection so difficult and sometimes so costly that certain risks are justified. These problems have already been discussed, as have the normal accounting procedures designed to develop records of accountability from the time an item is ordered until it is used or consumed on the job. Inevitably, such accountability procedures must be designed with the possibilities of fraud and theft in mind. We turn now to the consideration of methods used by dishonest employees and the extent, if any, to which accounting records may be expected to help control losses of this kind as well as the ways in which they may be used for that purpose.

At the outset, it should be understood that no accounting system can prevent a dishonest warehouseman from stealing nails, screws, fittings, small tools, and similar small items. It may eventually disclose the shortage, and even point to the culprit, but the system alone cannot prevent him from stealing in the first place. Moreover, the economic limit to accounting controls is the point at which their cost exceeds the amount which may be saved by maintaining them.

Bonding is an effective psychological preventive against major thefts and all employees responsible for materials and supplies should be bonded. Bonding, however, is not strictly an accounting device.

One of the most effective overall accounting controls is the materials budget, which has been discussed previously. This control forces the job superintendent or foreman to justify any requests for purchases in excess of the budget and will cause him to be alert to the possibilities of jobsite losses. However, the method also leads to attempts on the part of superintendents to disguise the purchase of items on which the budget is exhausted by having suppliers bill them as other items, for which an adequate budget has been established, or for which there is no budget and hence no basis for control.

When purchasing and payment are handled from the general office, it is usually possible to detect this practice at the time the receiving tickets are matched up with the invoices and purchase orders. There is comparatively less incentive for a supplier to conspire with the superintendent when purchases are made from the general office, and almost none at all when irregularities might delay payment. If, however, purchasing and payment are handled at a job office, the only effective check on misrepresentation of purchases by the superintendent is an effective field audit procedure that includes thorough physical checks on receiving routines made at irregular intervals, and is followed by tracing back individual shipments to the purchases and then tracing the same shipments forward into the accounts.

When purchases are delivered directly to the jobsite, and charged directly to costs, it is common for dishonest employees to have some of the incoming loads of materials diverted to a site where personal construction projects are in progress or to other contractors' jobs where the goods have been sold at bargain prices. In the absence of collusion between the supplier, or his truck driver, and the job superintendent, it is fairly simple to discover this type of theft by instructing the voucher clerk to notice the point of delivery on the delivery ticket and to exclude from the payment any invoices covering shipments not delivered to the jobsite. The ensuing demand by the supplier for payment will usually bring to light any improper diversions of shipments. If there is collusion, the detailed check-in at irregular intervals, followed by detailed tracing of individual items, will usually provide the best means of discovering misdeliveries as well as most other types of dishonesty at the receiving point. Even forged delivery receipts may be caught by an alert auditor.

A materials budget will disclose diversions of deliveries, but rarely soon enough to be very helpful. If the superintendent is gone, and the structure built from stolen materials has been sold to a bona fide purchaser, the only recourse is to the employee, or the bonding company if he is bonded.

Although surprise audits of incoming shipments will discourage fraud or theft at the delivery point, they will have little effect on violations of the procedures for charging out materials and supplies from warehouse stock. The simplest method of stealing stores from a warehouse or stockroom, and the one least likely to be detected, is to take them out without making any record. Therefore, regardless of the charge-out method in effect, there must be some document, originating at the time of withdrawal, which shows what is withdrawn by whom. Careful inspection of these documents alone can be an important deterrent to theft, particularly when coupled with an accurate knowledge of the warehouse stock.

If there is no one in charge, or if a number of people have access to the stockroom or warehouse, there is no way to allot responsibility for materials or supplies removed without a record. If one person is in charge, any theft that might occur must be with his acquiescence or through his negligence. In any event, if the goods leave the warehouse without a charge, the shortage is thrown into the inventory where it will not usually show up until the next physical count.

One of the best ways of avoiding a failure (either negligent or intentional) to prepare a charge slip when withdrawing stock from warehouse or stockroom is to have the requisition signed and coded by the person who wants the item withdrawn, and to have a copy of the requisition go directly to the accounting department where it is ultimately matched with the slip on which the person receiving the goods had signed. Absence of this control is the most serious weakness of the system illustrated in Figure 4-11 of this chapter. Figures 4-9 and 4-10 illustrate forms designed for this type of use and the forms in Figures 4-7 and 4-8 could easily be redesigned to permit it. The comparable control for purchases that are to be charged directly would be a purchase requisition which can, as a rule, be a duplicated copy of the material takeoff sheets for the job, or in standardized structures like tract houses, the standard materials list. A similar result can be obtained by having the purchase orders reviewed and signed by one of the principals or by some person in a position to know what materials and supplies are needed but with no direct contact with the suppliers or their truck drivers.

Such procedures do not prevent someone, totally unauthorized, from backing his truck up to the loading dock and, with the cooperation or acquiescence of the person in charge of the warehouse, loading it up and driving off with company property. Nor does it prevent the taking, by authorized persons, of items not included on the requisition. Surprise auditing may disclose such activity if the auditor arrives about the time a loaded truck is ready to leave and then checks the load against the documents. Another way to discover shortages is to spotcheck certain items. The auditor (public or internal) selects a single item (such as claw hammers, picks, steel tapes, aluminum or copper tubing) that might be particularly subject to misappropriation; checks all movements in and out to arrive at a theoretical inventory in terms of units; and compares the result with a physical count. This procedure, of course, is useful only when one person is responsible for the warehouse or stockroom.

The most effective deterrent to theft from a stockroom or warehouse is not an ironclad accounting system but rather a belief on the part of employees that if they steal they will be caught. Some of the more successful methods of inducing such a belief include: (1) a resourceful internal auditor making surprise checks at irregular intervals, (2) bonding of employees with access to the warehouse, (3) insistence on careful record keeping, (4) some practical system of checking persons and vehicles leaving the warehouse or the jobsite.

The last is very difficult because, if injudiciously done, searching vehicles or individuals can lead to lawsuits, union intervention, and sometimes even criminal charges against the contractor. Although it has nothing to do with accounting, it might be pointed out that if the employees' parking area is as far as possible from the warehouse, the chance of an employee carrying off any substantial amount of materials, supplies, or equipment without detection is minimized.

Employees often walk off the job with small tools issued to them. If a "brass" system (Chapter 3) is being used, the brass can be used as a tool check. However, it is often equally satisfactory to make the foreman responsible for tools issued to his crew.

On government jobs where there is some form of security patrol, it is often possible to obtain the cooperation of the security officer for physical inspections of outgoing automobiles and personnel.

Many old-time construction men ridicule accounting methods as a means of controlling theft of materials and supplies from the jobs or from a central warehouse. Usually such ridicule is accompanied by tall tales of brazen stealing. However, when these stories are traced to the source it will almost always be found that one indispensable ingredient was missing from the system—alert management. An accounting system, like a shovel, never works by itself. Someone who knows labor, materials, and construction must be constantly alert for "on the spot" conditions that can open the door to laxity, error, or dishonesty. Given that kind of supervision, almost any accounting system, reasonably tailored to the circumstances, will operate satisfactorily to control and account for materials and supplies.

Vouchering and Disbursement Control[1]

Disbursement control is the paperwork a company does to make sure it does not buy anything it does not need and does not pay for anything it has not received. A good many of the relevant procedures have been described in earlier chapters. On a companywide basis disbursement control starts with a cash forecast or a budget or at the job level with a "job estimate." In small companies it may begin with an estimated balance sheet as of some future date with an estimated income statement for the intervening period. Whatever form it may take, disbursement control starts with an estimate of how much money the company will have to spend, where it will come from, and what it will be spent for.

When applied to purchasing, disbursement control, starts with the decision that some particular article is needed. In a large firm, there would normally be a written requisition followed by a purchase order bearing properly authorized signatures. In a very small firm the contractor might originate and place the order himself. At either extreme, management is doing what it needs to do to prevent unnecessary buying, and to get the best possible quality, prices, and terms when it does buy. If these objectives are being accomplished then disbursement control is effective for purchasing.

When applied to receiving, disbursement control means a report to the person who pays the bills on each incoming shipment. That report should show exactly what was received, when it arrived, and what condition it was in. It should also be possible to tell the number of the purchase order that authorized the supplier to ship the goods. As with most other accounting procedures, the routine will vary widely according to the size of the business and the nature of the work.

In the very small, one-person firm, the contractor himself does the purchasing, receiving, and paying, in addition to actually using the materials. His memory will provide most of the information. As the

[1]Particularly in those branches of the construction industry where nonlabor costs are of major importance, the speed and accuracy with which such costs are transferred to job cost records are assuming increasing importance. In many construction companies computers (see Chapter 12) are being called upon to speed up and refine such distribution.

number of jobs and the number of transactions increase, he can rely on memory less and less. As soon as more than one person is needed to do the work, it becomes necessary to substitute documents for personal knowledge. So, at the other extreme, the receiving routine of a very large contractor, with numerous employees and jobs, would include a receiving report on which the warehouseman reports his independent check on merchandise received. If the shipment was not exactly as ordered, the warehouseman may also prepare an "over, short, and damage" report.

As applied to financing, disbursement control means knowing where the money to pay the bills is coming from before the bills are incurred. It involves cash forecasting, and also following up on the forecast so that the money is actually in the bank when the checks paying the bills are written.

As applied to job management, disbursement control means not only checking the invoices, purchase orders, and receiving records carefully, but also staying within the job cost estimate. Any substantial difference between the job estimate and the actual cost should be accounted for in detail, since there may be some improper or unnecessary disbursements.

As applied to supplier and subcontractor relationships, disbursement control means two things. First, it means the usual follow-up on shipments and invoicing, and making sure that no incorrect bills are paid. Second, and equally important, is the prompt processing and payment of bills that are correct. This step involves not only the taking of all allowable discounts, but obtaining the price advantages that come with a reputation for prompt, accurate payments. Suppliers and subcontractors pay for their money like everyone else. If they know that their invoices will be promptly paid, they can afford to give better prices, terms, or service than if they have reason to believe their collections will be slow.

As applied to credit standing (closely related to the foregoing), disbursement control means the procedures established to make certain that all obligations are paid on or before the due date and strictly in accordance with their terms. These procedures include the cash forecasting, collection follow-up, and the financial planning necessary to cover payment, as well as processing invoices to assure prompt and accurate payment. Disbursement control also requires procedures for preventing errors and misunderstandings and for correcting or adjusting those that do occur.

One of the least-well-recognized aspects of disbursement control is its effect on job morale. Few experienced construction men will question the value of a spirit of cooperation and enthusiasm among the men on the job, and the feeling that they are a part of a first-rate organization. A large part of that feeling is the pride that comes from the knowledge that the firm is important and respected in the community. One of the quickest ways to destroy that pride is to let the company get the reputation of being "slow pay." Slow or inaccurate payrolls can have the same result. This is not to say that the company needs to be "easy." A reputation for sloppiness can be just as bad as one for being slow. Prompt, accurate payments are a factor in developing the confidence essential to a reputation for strength and reliability. Such a reputation can contribute a great deal to job morale.

Support for disbursements

No matter where disbursement control starts for any particular company, it almost always utilizes some variation of the voucher system (Figure 4-6). This is another way of saying that eventually all contractors learn that it pays to have the correct supporting documents before disbursing money. It also means that in most construction businesses an accounts payable ledger is unnecessary unless the company falls behind in paying its bills. A great many special forms (Figures 4-14 to 4-17) are available for voucher system control. The mechanics of vouchering are affected by such things as the type of contract, accounting methods used by the company, and special job conditions, but the requisites of a good voucher are the same under any system. In one way or another the documents should show:

1. Evidence that the goods or services were ordered on proper authority or that the cost was properly incurred
2. Evidence that the goods or services ordered were actually received
3. Evidence that the prices and discounts are correct
4. Evidence of authority to make the disbursement

In its simplest form a voucher (used here to mean the documents supporting a cash disbursement) might consist only of two pieces of paper—a combination delivery ticket and invoice (Figure 4-18), and a check. The invoice indicates a delivery of materials to the job. It bears the signature of the employee at the jobsite who received it. It shows the items delivered, the quantities, the prices, and the extended totals. In a very small operation the contractor himself would check and approve the invoices. He would know the job and recognize the type and quantity of materials as being proper for the job. He might even remember the delivery, and there would be few enough deliveries so that any duplications would be noticed at once. Since the contractor himself probably made the purchase, he would know the prices quoted and be able to check them from memory. The extensions could be checked by recomputations or by simple inspection. When

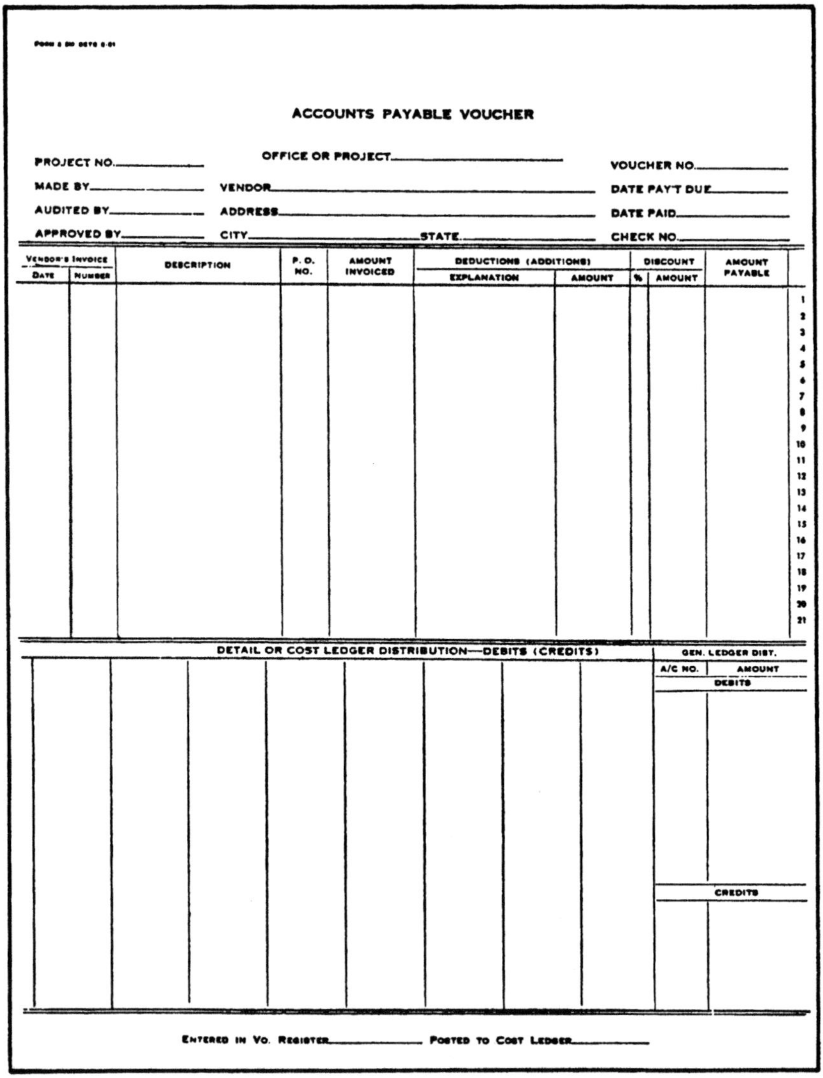

ACCOUNTS PAYABLE VOUCHER

PROJECT NO.______ OFFICE OR PROJECT______ VOUCHER NO.______

MADE BY______ VENDOR______ DATE PAY'T DUE______

AUDITED BY______ ADDRESS______ DATE PAID______

APPROVED BY______ CITY______ STATE______ CHECK NO.______

VENDOR'S INVOICE		DESCRIPTION	P. O. NO.	AMOUNT INVOICED	DEDUCTIONS (ADDITIONS)		DISCOUNT		AMOUNT PAYABLE	
DATE	NUMBER				EXPLANATION	AMOUNT	%	AMOUNT		
										1
										2
										3
										4
										5
										6
										7
										8
										9
										10
										11
										12
										13
										14
										15
										16
										17
										18
										19
										20
										21

DETAIL OR COST LEDGER DISTRIBUTION—DEBITS (CREDITS)

GEN. LEDGER DIST.

A/C NO.	AMOUNT
DEBITS	
CREDITS	

ENTERED IN VO. REGISTER______ POSTED TO COST LEDGER______

Figure 4-14 Accounts Payable Voucher. This is a simple voucher form suitable for a small contractor. Supporting documents are stapled to this form after it has been completed.

the contractor writes and signs the check, and puts his initial and the check number on the invoice, the documentation is complete and it remains only for the paid bill to be recorded and filed with the others for the month.

As the volume of transactions grows, and the work is divided among more people, the personal knowledge of the contractor is

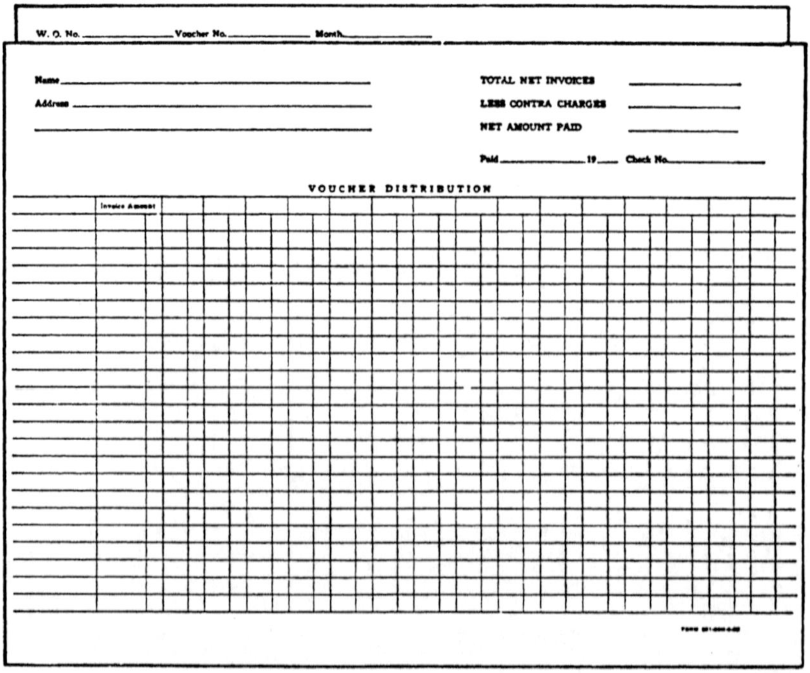
W. O. No.________ Voucher No.________ Month________

Name________________

Address________________

TOTAL NET INVOICES ________

LESS CONTRA CHARGES ________

NET AMOUNT PAID ________

Paid________19___ Check No.________

VOUCHER DISTRIBUTION

Invoice Amount

Figure 4-15 **Voucher Envelope.** One of the functions of vouchers is the assembling in one place of the various papers and documents supporting a transaction. A convenient way of accomplishing this end is to use a manila envelope for a voucher.

replaced by documentary evidence. As the size and need for accountability increases, so does the need for documents, until (as on United States government cost-plus-fixed-fee contracts) every transaction, however small, must be documented in the most minute detail. Even so, the basic requirements are those stated above. The only variations are of degree.

In the discussions that follow, typical forms and procedures, although they have been tried and found satisfactory, are used as illustrations only.

Data for vouchering

The information for accounts payable vouchers comes from four principal sources:

1. Operating departments indicate the goods and services that are to be provided. These may or may not be documented by requisitions signed by authorized individuals, depending upon who does the purchasing. If purchasing is done by a separate person or depart-

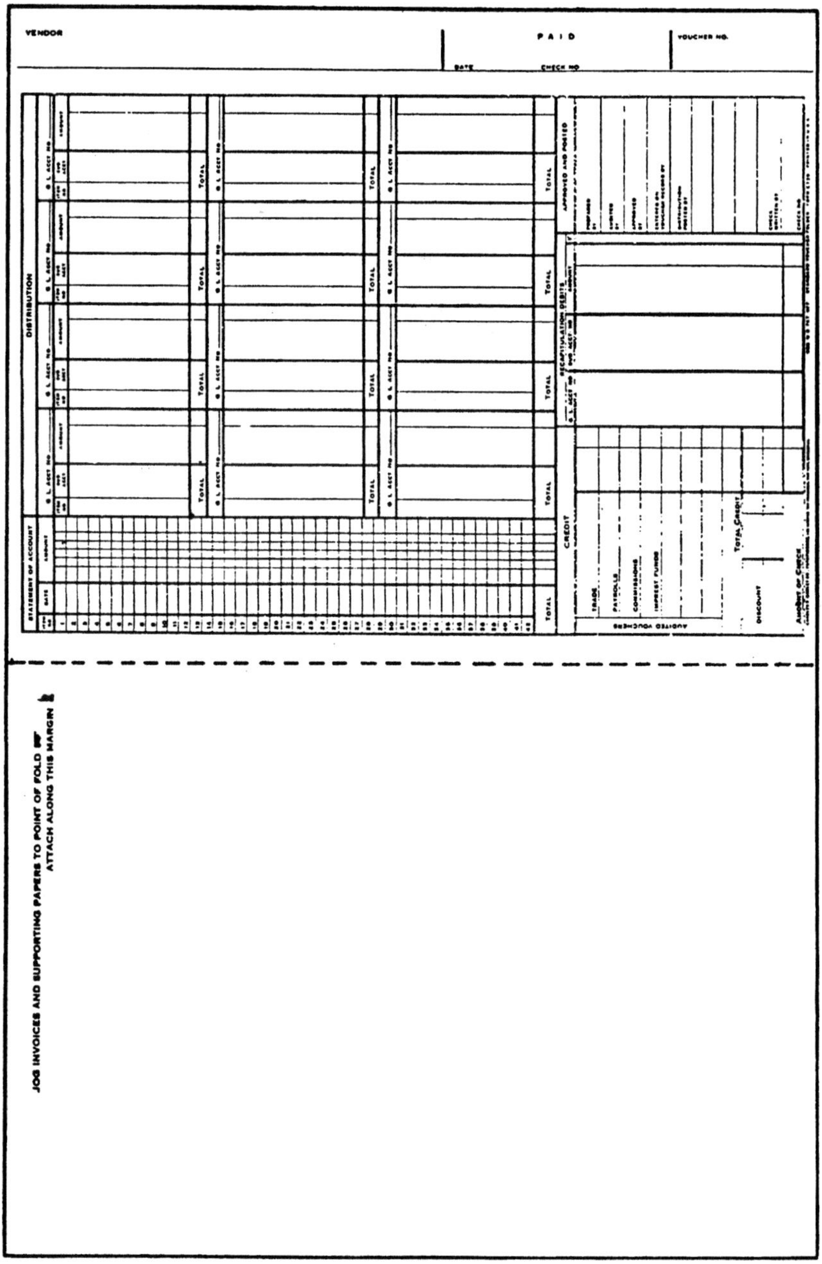

Figure 4-16 **Accounts Payable Voucher Folder.** This voucher jacket is a file folder imprinted with spaces for analyzing the transactions with a particular vendor for the month. Supporting papers are stapled or fastened inside.

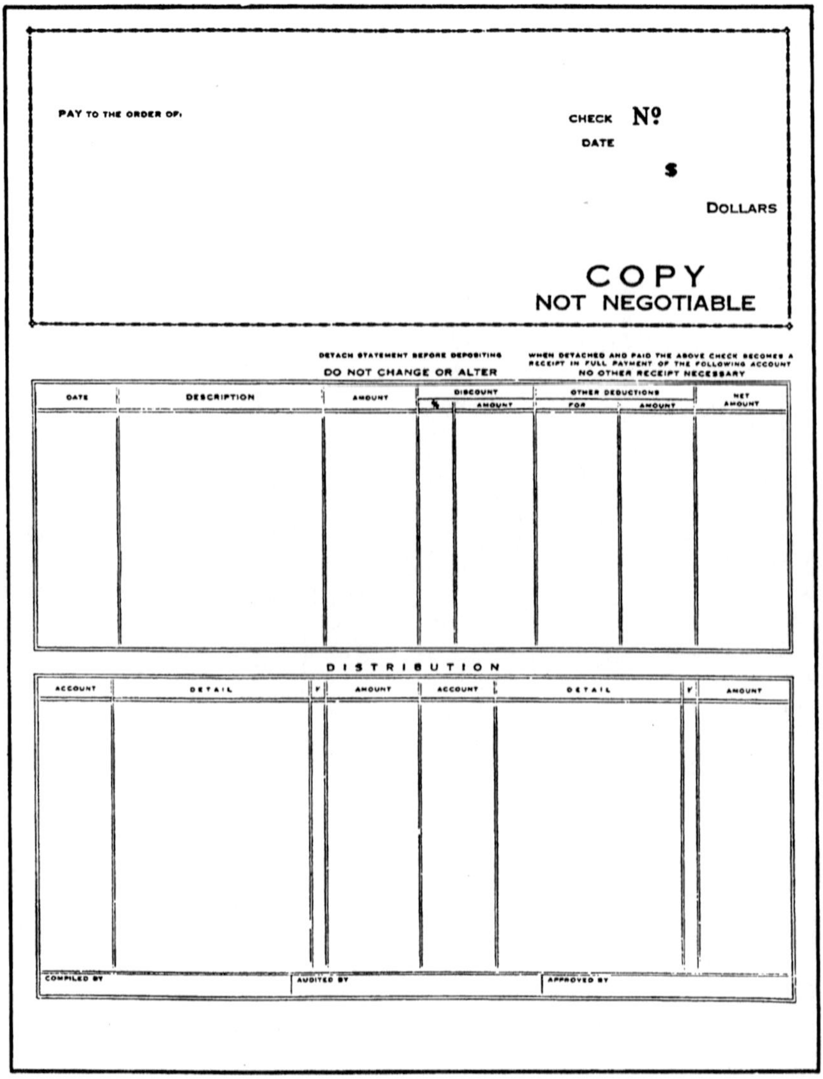

PAY TO THE ORDER OF:

CHECK Nº

DATE

$

DOLLARS

COPY

NOT NEGOTIABLE

DETACH STATEMENT BEFORE DEPOSITING

DO NOT CHANGE OR ALTER

WHEN DETACHED AND PAID THE ABOVE CHECK BECOMES A RECEIPT IN FULL PAYMENT OF THE FOLLOWING ACCOUNT

NO OTHER RECEIPT NECESSARY

DATE	DESCRIPTION	AMOUNT	DISCOUNT		OTHER DEDUCTIONS		NET AMOUNT
			%	AMOUNT	FOR	AMOUNT	

DISTRIBUTION

ACCOUNT	DETAIL	✓	AMOUNT	ACCOUNT	DETAIL	✓	AMOUNT

COMPILED BY | AUDITED BY | APPROVED BY

Figure 4-17 **Voucher Check.** The analysis of the payment is on the face of the check and the stub is used as a simple voucher. If a carbon copy is made, the copy can serve as a voucher and the stub as a remittance advice.

ment, the buyer needs evidence of his authority to make the purchase, and a signed requisition (Figure 4-19) is the simplest and surest way to provide that evidence. A copy of the requisition is sent to the voucher clerk in the accounting department and provides the needed documentation without having to go to the originator's files.

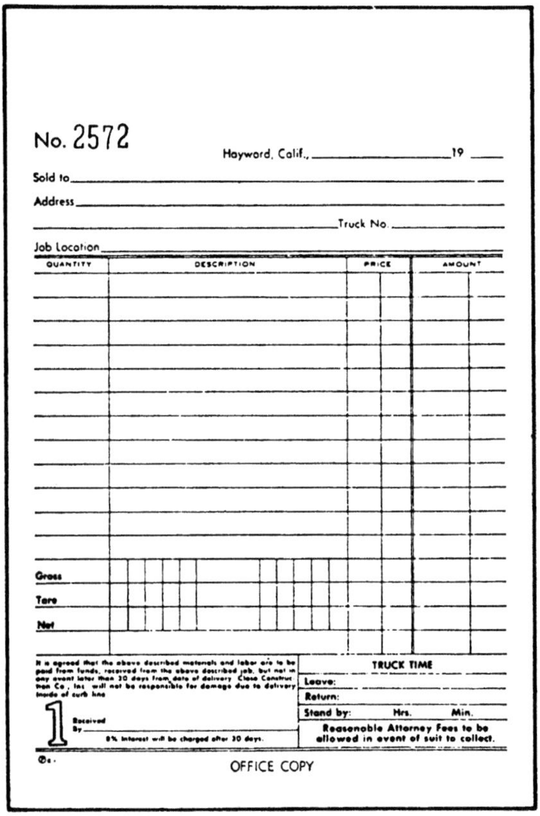

No. 2572 Hayward, Calif., ______ 19 ___

Sold to ______

Address ______

______ Truck No. ______

Job Location ______

QUANTITY	DESCRIPTION	PRICE	AMOUNT
Gross			
Tare			
Net			

It is agreed that the above described materials and labor are to be paid from funds, received from the above described job, but not in any event later than 30 days from date of delivery. Close Construction Co., Inc. will not be responsible for damage due to delivery inside of curb line.

1 Received By ______

8% Interest will be charged after 30 days.

TRUCK TIME

Leave:

Return:

Stand by: Hrs. Min.

Reasonable Attorney Fees to be allowed in event of suit to collect.

OFFICE COPY

Figure 4-18 **Combination Delivery Ticket and Invoice.** This snap-out form published by Moore Business Forms, Inc., has single-use carbon paper. The original is the office copy, the first copy is the invoice, and the second and third are delivery memoranda.

2. The second basic document is the purchase order. Chapter 2 discusses its preparation in some detail and illustrates typical forms. In its most useful form, from the standpoint of documenting cash disbursements, the purchase order is priced and is sent to the vendor before the invoice arrives. In theory (and often in practice), it contains all the terms of the purchase including any special arrangements and understandings. It should be serially numbered and, among other things, should include all the instructions the seller will need to render his invoices in the required form. It should state clearly that all invoices must show the number of the purchase order authorizing the shipment. A copy of the purchase order should always go to the voucher clerk in the accounting

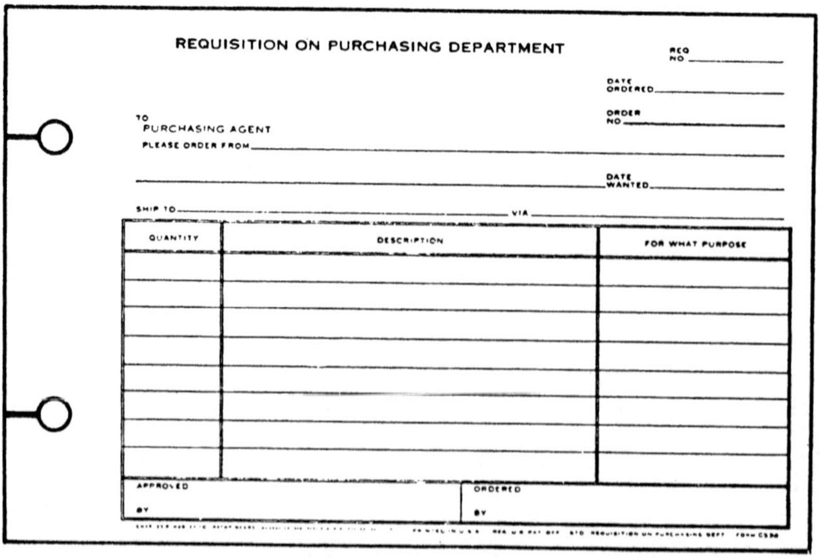

REQUISITION ON PURCHASING DEPARTMENT

REQ NO. ______

DATE ORDERED ______

ORDER NO. ______

TO PURCHASING AGENT

PLEASE ORDER FROM ______

DATE WANTED ______

SHIP TO ______ VIA ______

QUANTITY	DESCRIPTION	FOR WHAT PURPOSE

APPROVED BY ______ ORDERED BY ______

Figure 4-19 **Purchase Requisition.**

department for subsequent comparison with the invoice when it is received. In many systems, a copy of the purchase order is sent to the warehouse to serve as a receiving memo. This latter procedure is sometimes criticized on the ground that it creates a temptation for the warehouseman not to check the incoming shipments physically but merely to compare the delivery memo with the receiving copy of the purchase order. This can be avoided by designing the forms so that the receiving copy does not indicate quantities.

3. The project manager or the project engineer (or other person given the responsibility of checking them) receives the vendors' invoices. They come in by mail and are sent to the operating department to be checked and coded for accounting distribution before being paid. Sometimes this step is bypassed, and they go directly to the voucher clerk for comparison of items, price, and accounting distribution with the purchase order, receiving report, and notation of exceptions. When this method is followed, either the purchase order or the requisition must show the accounting distribution ("coding") of the charges. As indicated elsewhere (Chapter 2), back orders are noted at once so that appropriate steps can be taken to avoid duplicate shipments.

4. From the warehouse, or from the point of delivery at the actual site of the work, come the documents evidencing receipt of the goods or services. These may take the form of a receiving report, a

receiving copy of the purchase order, or sometimes just a copy of a delivery ticket signed by the person receiving the shipment. Whatever documents come in must show not only what was received, but what condition it was in. If there were overages, shortages, or damage to the goods received, an "over, short, and damage" report (Figure 4-20) should accompany the receiving memo so that the item can be reordered, if necessary, and so that proper insurance claims can be filed. When vendors' invoices first come in they should be stamped or otherwise identified so that only the original or the payment copy can form the basis for payment. Discrepancies between the purchase order, the receiving memorandum, and the invoice must be reconciled before an invoice is placed in line for payment. On subcontractor invoices the subcontract takes the place of the purchase order, and the engineer's computations (Figure 2-3) of the subcontract payment take the place of the receiving memo.

OVER, SHORT AND DAMAGE REPORT

No.........

Req. No..................

Vendor.......................... P.O. No..................

Address.......................... Date received....................

Shipper.......................... Delivery ticket No...............

Car No................ Collect ☐

Name of carrier.................. Prepaid ☐ Charges (if collect)....

Quantity		Unit	Description of Material			
Received	Charged		Items	Units over	Units short	Nature of damage

Report prepared by............Checked and received by...............

....

Signature

Figure 4-20 Typical "Over, Short, and Damage" Report.

Preparation of vouchers

When all the documents are in and all discrepancies reconciled, the invoices to be paid and the discounts to be claimed are summarized on vouchers. It is at this point that an experienced voucher clerk can earn many times his cost, for it is he who must notice differences and exceptions and know which ones are important. For example, he must know whether charges for demurrage or waiting time on ready-mix concrete trucks are proper and authorized. He must know how to recognize partial shipments and how to catch and call attention to improper freight charges. He must know the difference between a freight bill and a bill of lading, and he must know when to deduct sales tax on the discounts he claims. He must watch the cash discount invoices and see that they are paid within the discount period and, most important of all, he must see that bills are paid promptly, correctly, and only once. The title "accounts payable clerk" is not very impressive, but it is no measure of the importance of the job in maintaining a smooth-running, economical operation.

Several different kinds of vouchers have been illustrated and some of them have been described in some detail. The actual preparation of most accounts payable vouchers starts with a vendors' alphabetical file in which the purchase orders, invoices, delivery memos or receiving reports, and correspondence affecting the vouchers are accumulated.

When an invoice, delivery memo, and purchase order are found in agreement, they are stapled together and placed in another alphabetical file with other invoices ready for payment. If partial deliveries are made against a single purchase order or purchase contract, that fact is noted on the face of the invoice and the purchase order is marked with the invoice number, date, and items shipped, and returned to the accumulation file. The invoice and delivery memo are checked, and after any errors have been corrected they are sent to the file of invoices to be paid. Blanket purchase orders or purchase orders issued against a purchase contract are treated like partial shipments in that the voucher clerk's copy is held in the file of unfilled orders until the invoice covering the last part of the order comes through for payment.

Some companies process invoices carrying cash discounts ahead of all others, and such an invoice is filed for final processing the day before payment is due. Other companies merely go through their "ready for payment" file once a week and pick out the discount invoices for immediate payment. Under this system the nondiscount items are allowed to accumulate until the end of the month, when they are processed for payment by the due date, which is usually the tenth of the month following delivery.

Either at that time, or better yet, as the invoices are made ready for payment, they are entered on a voucher. Sometimes the voucher is a separate sheet of paper. With modern machine methods, the entry is often made on the carbonized remittance advice section (of which the voucher is the copy) of a voucher check. Whatever the mechanics of the system, as soon as the invoices to be paid are summarized on the voucher, and the supporting documents attached, the voucher is complete and ready for final checking and payment.

As a final precaution before preparing the check, it is a good idea to check the vouchers to see that all invoices are authorized for payment, and that all supporting documents are attached. At the same time, the amounts on the invoices and the amounts on the vouchers are compared for possible errors. The totals can be inspected for obvious errors. The vouchers themselves should be serially numbered and the number, if it cannot be the same as the check number, should appear somewhere on the check.

As a matter of good office practice each company should have its own vouchering procedures reduced to writing. These procedures must take into account that such items as utility and telephone bills come in only one copy with no delivery receipt. Subcontractor invoices must contain the engineer's computation in lieu of the delivery receipt. Depending on the system in use, invoices for equipment rentals may or may not have a copy of the equipment time report for the items included. In short, there will be a number of normal variations from the typical accounts payable pattern, and these variations and exceptions need to be covered by specific instructions. Special instructions are also needed for such items as freight bills, insurance reports, tax returns, rent on office or warehouse space, and similar items which often are not invoiced. The same would be true of payroll account and petty cash reimbursement vouchers. It almost goes without saying that where there is a need for petty cash expenditures, an "imprest" or revolving fund should be set up for a fixed amount, and reimbursed whenever it becomes low. The receipts for expenditures form the support for the petty cash voucher, which in turn supports the reimbursement check. Whenever possible, the petty cash fund should be kept in the office and not in the superintendent's pocket.

None of these special types of payment vouchers offers any unusual or complicated problems so long as they are recognized for what they are, and procedures in keeping with the needs of the company are set up to pay them when they fall due.

Voucher checks

When the voucher is complete, all the data necessary for the preparation of the check, including the serial number of the voucher, appear on the voucher itself, and writing the check is reduced to mere copying. Three rules governing check writing, so obvious that they would hardly seem to need mentioning, are these:

1. No check should ever be drawn payable to cash.
2. No check should ever be prepared and sent through for signature without the name of the payee shown on the face.
3. No check should ever be presigned.

These rules are stated only because of the frequency with which they are broken. When they are not observed, an open invitation to theft exists, and an inference of collusion on the part of those who allow them to be ignored may arise.

One type of check (Figure 4-17) has a voucher attached. When this kind of check is used it is common to use the same number for the voucher and the check. When the two numbers are different, both should appear on the voucher. The function of the voucher itself is to summarize the individual items or invoices paid and to show the accounting distribution. Usually, both a general ledger and a cost ledger distribution are shown (Figure 4-6).

Unless cash balances are adequate to pay all bills currently, it is usually not practical to have the voucher number and the check number the same, because there will be vouchers unpaid at the end of the accounting period, and this will tend to upset the numerical control of the checks. It is, of course, essential to maintain a strict numerical control on both the vouchers and the checks. If only the payment copy or original of each invoice, supported by original or payment copy documents, is included in the original voucher, and payment is made only on numerically controlled vouchers, all bills will be presented for payment, but it would be difficult for a bill to be paid more than once without attracting attention.

The practice of prenumbering checks serially and then accounting for all numbers is vital to any ordinary system of bank reconciliation which, in turn, is virtually indispensable to control of cash. Checks should always be printed with serial numbers, and all voided checks should be retained and accounted for. This principle seems so elementary that it hardly deserves mention. However, like the three rules stated above, it is violated often enough to justify stating it. If voided checks are thrown away or destroyed, there is very little to prevent a dishonest employee from writing a check to himself, cashing it, and withdrawing the canceled check from those returned with the bank statement. This would leave a shortage which could be concealed by omitting a check of like amount from the list of outstanding checks. This omission would not be too difficult if numerous checks are being written for the same amount as, for example, when a contractor is renting dump trucks from independent sources.

The simplest way to maintain the numerical control in either series is to file them numerically and account for all missing numbers immediately.

After the vouchers and voucher checks are prepared, completed, and ready for signature, they should be given a final scanning for proper documentary support, mathematical accuracy, and general reasonableness. If a two-signature check is used (and this is recommended), the first signer is usually the controller, office manager, auditor, or chief accountant, and his signature means the check and the supporting voucher have been preaudited and are technically correct. The countersignature by the general manager or one of his assistants (preferably someone in operations) means that the general propriety of the transaction has been reviewed and approved.

In smaller operations where only a single signature is required, or in very large operations where checking and verification are a separate function and the single signature is usually applied with a machine, the same two types of preauditing are required. In the small company it is often done by one person. In a very large operation it may be done by a whole department or, to some extent at least, it may be done by machines. Either way, if errors in payments and in the release of checks are to be avoided, the accuracy and the overall reasonableness and propriety must be checked by someone in a position to know.

The most nearly foolproof and tamperproof system for distributing voucher checks is to mail them. This procedure is most effective when a multicopy voucher check is used. It is practical to type the name and address of the payee on the face of the check and to mail it in a window envelope. Preferably, the check should be mailed immediately after the last signature has been obtained, by a person independent of all cash or vouchering functions. However, even when this separation of work is not possible, the system of mailing the checks works well. The check copies are difficult to tamper with and manual delivery of checks (when the system calls for mailing) is peculiar enough to attract attention.

Voucher register and cost ledger

If transactions are few it is possible to post them directly from the vouchers to the general ledger and the cost ledger. The usual procedure, however, is to enter them in a columnar voucher register (Figure 4-21) and post the totals monthly. Figure 4-6 charted the flow

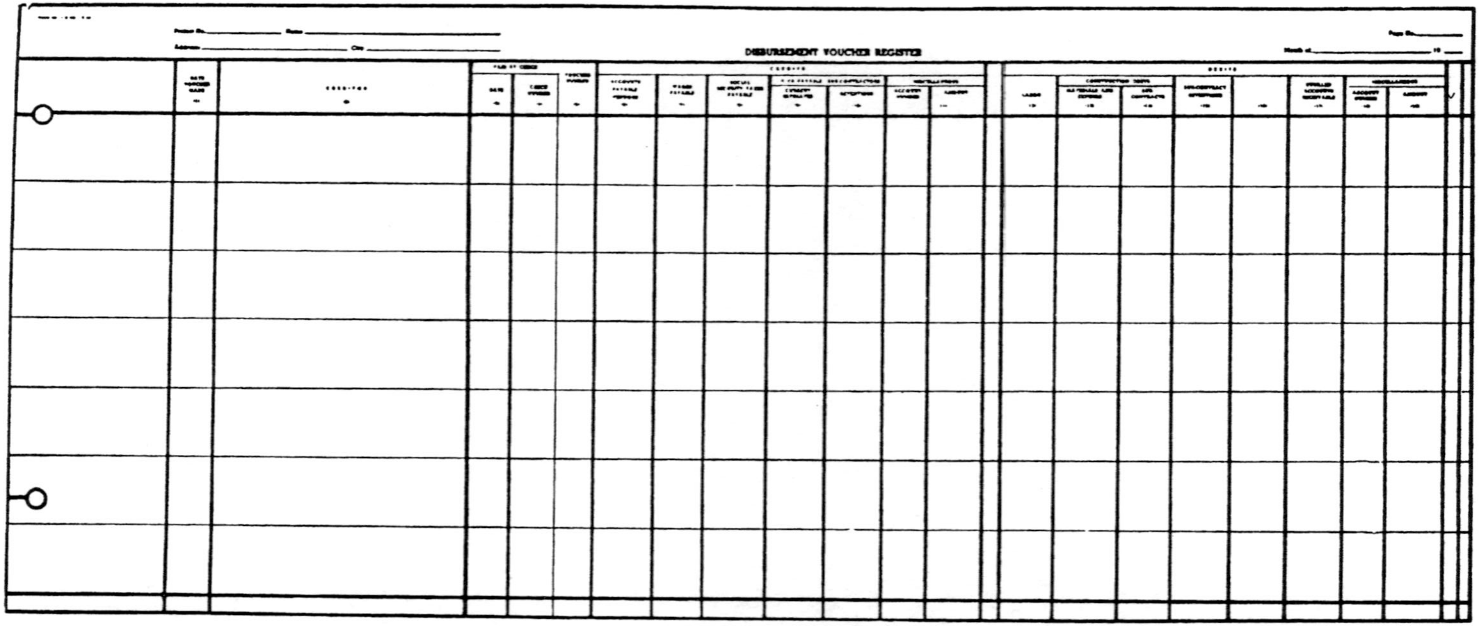

Figure 4-21 **Disbursement Voucher Register.** This is one of the many types of columnar distribution voucher registers. It is made in duplicate to permit a copy to be sent to the general office in lieu of a job report.

of data from the basic documents through the voucher into a columnar voucher register.

Before entering the voucher, the accounting distribution should be checked. Some firms post the vouchers directly to the cost ledger in order to keep the record completely current, and when day-to-day cost reports are made from the cost ledger there is merit to this procedure. Still, most construction jobs move too fast even for this method. If the books can be closed and statements completed by the tenth of the month, it is usually better to use memo day-to-day costs on the larger variables like labor and equipment, and post the vouchers to the cost ledger through the voucher register in the same way that the general ledger is posted.

If the general ledger closing is delayed for some reason or if the general ledger is kept in the general office and the cost records are kept on the job, the direct posting of vouchers is often the only practical way to get accurate costs at the earliest possible time. It might be added that proper accounting machine or data processing applications can often make the posting of the cost ledger a by-product of writing up the voucher register. There are numerous forms of machine-kept cost ledgers but most of them follow the principle of columnar analysis.

Many smaller contractors do not keep a cost ledger as such. Instead, they merely use an analysis type ledger sheet (Figure 4-22) for their general ledger accounts "Jobs in progress" and "Cost of completed jobs."

Unpaid vouchers

Theoretically, as soon as a shipment of goods is received it should be entirely possible to pay for it at once. As a practical matter, there are many factors that can and do prevent the immediate payment of

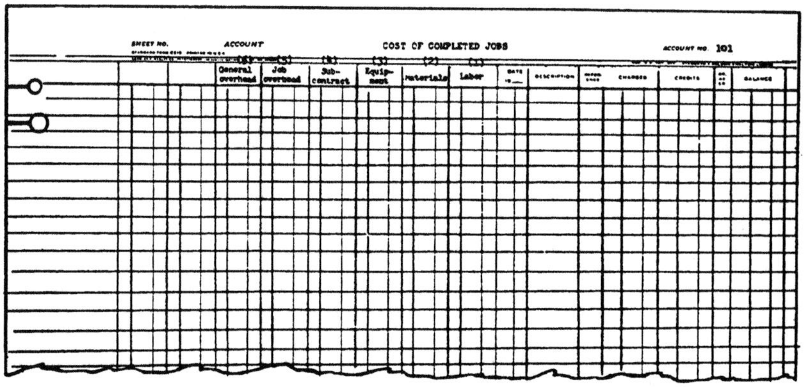

Figure 4-22 **Analysis-Ledger Sheet for Cost of Completed Jobs.**

invoices. Not the least of these is lack of money, but there are many others such as errors in shipment, errors in billing, back orders, disputed prices, and inadequate personnel. The result is that at the end of any accounting period there are always some unrecorded liabilities arising from unpaid invoices. If these are taken into account only when paid (on the so-called "cash basis") the books will reflect a more favorable financial condition than actually exists.

If such unrecorded liabilities are small in comparison with the whole operation, they have little significance. If money is short, or if for some other reason these liabilities are proportionately large, cash-basis accounting may mislead management into thinking a job is profitable when, in fact, it is operating at a loss.

There are several ways of taking unpaid vouchers into account at the end of an accounting period. The obvious way, of course, is to summarize the unpaid items on a work sheet which then becomes the basis for a journal entry. Such an entry is posted as one of the closing entries, and reversed after the close of the period. In the following period, the vouchers then are entered in the voucher register in the usual manner. This method, although quite standard, is subject to the criticism that it makes for more accounting work, particularly if the number of vouchers is large.

Another method commonly used to reflect unpaid vouchers in the financial statements at the end of the accounting period is to hold the books open until they are paid and then record them on a cash basis. If the time involved is not too great, this method has the merit of being simple and easy. Usually the time tends to get longer each period until the delay in obtaining financial statements more than offsets the ease and simplicity of the system.

Among the more effective and practical methods is to enter all vouchers in the voucher register as if they had been paid. Then in separate columns, enter (1) the amount of each voucher for which a check is issued, (2) the amount of the check, (3) the amount of any other credits such as discounts taken. The total amount of all vouchers is credited to "Vouchers payable." The total of vouchers actually paid is debited to "Vouchers payable." In the voucher register for the following month the checks paying the unpaid vouchers of the prior month are entered in the columns provided for the checks, but there is no corresponding voucher distribution since one was made in the preceding period.

Under one variation of the method just described, only the number of the check paying the voucher is entered. At the end of the month the unpaid items are totaled to determine how much of the total credit amount would go to "Vouchers payable" and how much to other general ledger accounts such as "Cash" and "Discounts earned."

Whatever method is used, one precaution is applicable. When unpaid vouchers are entered, the debits to the various cost, expense, and asset accounts are made at that time, and care must be taken to see that when the voucher is paid the debit goes to "Vouchers payable" instead of being distributed to the various accounts a second time. Any system which does not have automatic safeguards against such an error is deficient.

One such safeguard is to use a prefix to the voucher number that shows the month in which the voucher is first entered. Another is to make no distributions through the cash disbursements record, if such a record is maintained. Still another is to reserve a separate sheet for payments of prior months' vouchers in the current month's cash disbursement record. Safest and easiest of all is a system that combines all three of these safeguards.

Commitment control

A great deal of emphasis has been placed in this book on financial planning. To make such planning work, a good system of disbursement control is essential, but it is not, by itself, enough. There must also be a control on commitments. One of the most effective tools for controlling commitments is a system of pricing purchase orders and sending one copy to the voucher clerk immediately upon placing the order. The objections to this system have already been discussed in Chapter 2. It was also pointed out in Chapter 2 that, in most situations, the objections are without substantial merit and boil down, essentially, to a dislike on the part of some purchasing agents for the routine work required by such pricing. The situations in which pricing will not work are generally those involving a "black" or "grey" market. In such cases the requisitions can be used with estimated prices, and actual prices paid can be filled later. An alternative is the "field purchase order" prepared as a matter of record at (not after) the time of purchase and showing the actual price paid.

If a file of voucher documents is maintained, the first document should be the purchase order. Therefore, a summary of unfilled purchase orders in such a file would serve quite adequately as the starting point for a commitment control. The function of a commitment control is to make sure management knows how much money, not yet included in either cash disbursements or accounts payable, has been committed for payment.

The standard treatment for blanket purchase orders, and orders against which partial deliveries have been made, is to note the amounts received and the balance still to come, and to leave them in the file in order to reflect the commitment for deliveries still to be received. There is some question of the extent to which subcontracts should be taken into account in computing commitments, particularly when there are numerous change orders pending. It is usually better practice to include the unpaid subcontract balances, because any forecast of the cost to complete a job would have to take them into account, and any unbalancing in subcontract payment (Chapter 2) would be discovered by including them.

If a budget is used, commitments are deducted from available funds as soon as the money is spent, regardless of when it is paid. When the budget is tied directly into the accounting, an adjustment has to be made when the amount of cash paid out differs from the obligation as it appears on the purchase order. For the majority of contractors it is unnecessary to integrate the budget with the books. Usually, a weekly report showing how much has been spent on each item and how much remains uncommitted is enough. In separate accounts for a job, such a statement would usually treat each work item as a budget item. In a general office budget, the budget items are usually the same as the balance sheet and profit-and-loss accounts.

Commitment control in cash forecasting simply means that the forecaster makes provision for paying all future bills. If the contractor has a schedule of amounts which have been spent and which will have to be paid as soon as delivery is received, it takes just one more element of guesswork out of the forecasting.

Sometimes cash forecasting on jobs takes the form of an estimate of cost to complete. Such an estimate usually starts with the cost ledger, adds known liabilities and commitments by distributing them to individual work items, and then estimates the additional money that will be required for each work item in order to complete the job. One of the most common errors in making estimates of cost to complete is failure to take into account the obligations reflected by open purchase orders and subcontracts. It can be avoided by using the unpaid voucher file as a source of commitment data.

Chapter

5

Accounting for Equipment Costs*

Accounting for equipment costs starts when the equipment is purchased, but the problems peculiar to the construction industry begin when an item is sent to a job or is put into service at the head office. On small jobs equipment costs are often negligible, and there is no formal accounting for them. However, when such costs are substantial, some kind of equipment cost accounting is essential. The first step in controlling equipment costs is to make a complete and accurate record of each piece of equipment at the time it is purchased.

Physical control

Some firms use various special forms (Figures 5-1 to 5-4) to achieve physical control. Other firms use an ordinary warehouse receiving report. In addition, some contractors use an equipment location card for controlling their own equipment at the jobsite. Sometimes, too, a similar record is necessary for equipment rented from others when there is much of it.

A standard way to keep track of equipment is to keep an equipment ledger with a separate sheet or card for each unit. It is to this record that the cost of each unit is posted and it is in the equipment ledger that control is kept of shifting combinations. For example, bulldozer blades, power control units, and similar attachments may be shifted back and forth between tractors. Motors and bodies may be exchanged between trucks. Many other shifts and combinations may be made as the job requires. In a well-kept system these changes will

*Control of equipment costs has been one of the major problem areas in construction accounting. Applications of computers to equipment accounting are now becoming increasingly practical. For a more complete discussion, see Chapter 12.

THE PRIME COMPANY
Equipment Receiving Report

No. . . . .

Job No. . . . . Location . . . . Date . . . .

Equipment No.							
Job No.	Owner's No.	Motor or serial No.	Owner	Description	Condition — Specify any condition not satisfactory in detail	Insurance value	

Received by . . . . Condition checked by . . . .

Figure 5-1 Equipment Receiving Report. This form is prepared in quadruplicate, in different colors. The white copy is sent to the job file; the yellow to the general office; the pink to the equipment timekeeper; and the blue to the accounting department.

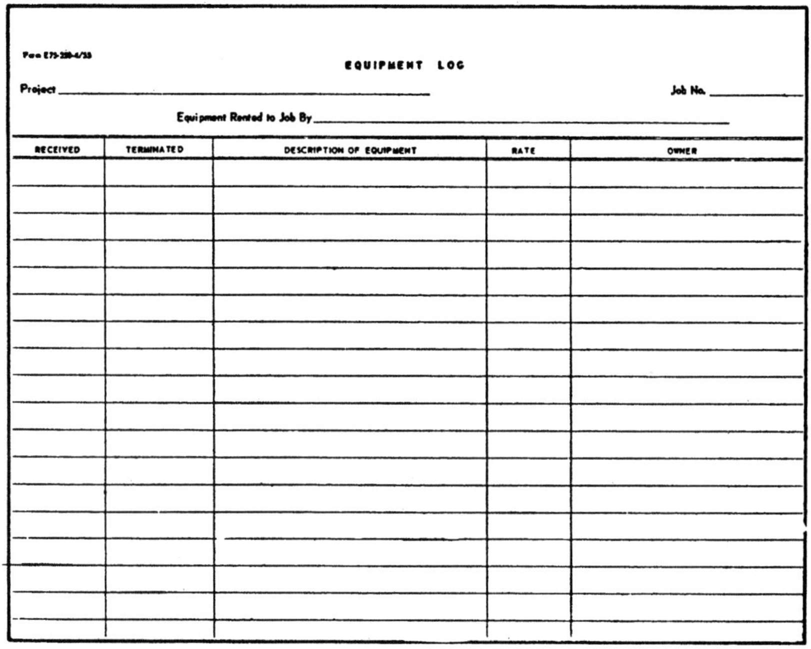

EQUIPMENT LOG

Project ____________ Job No. ________

Equipment Rented to Job By ____________

RECEIVED	TERMINATED	DESCRIPTION OF EQUIPMENT	RATE	OWNER

Figure 5-2 **Equipment Log.**

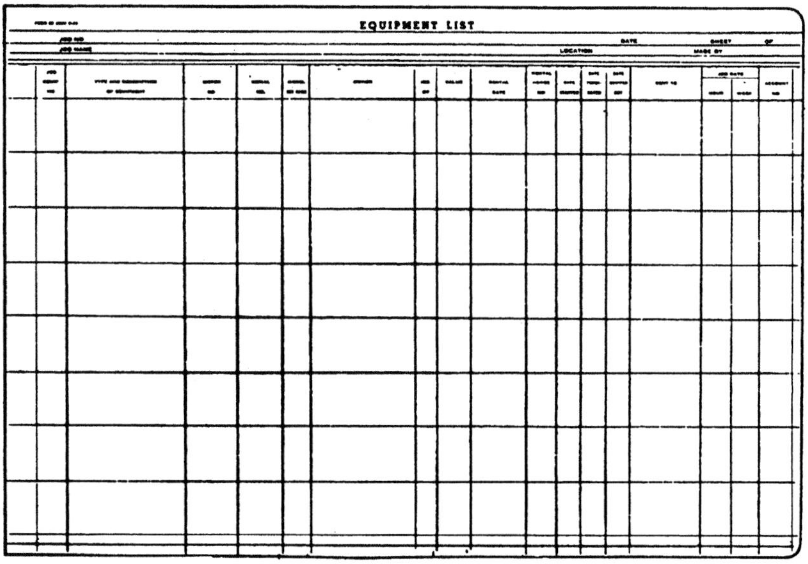

EQUIPMENT LIST

Figure 5-3 **Equipment List.** This form illustrates one of the methods used to keep a current control of equipment on hand.

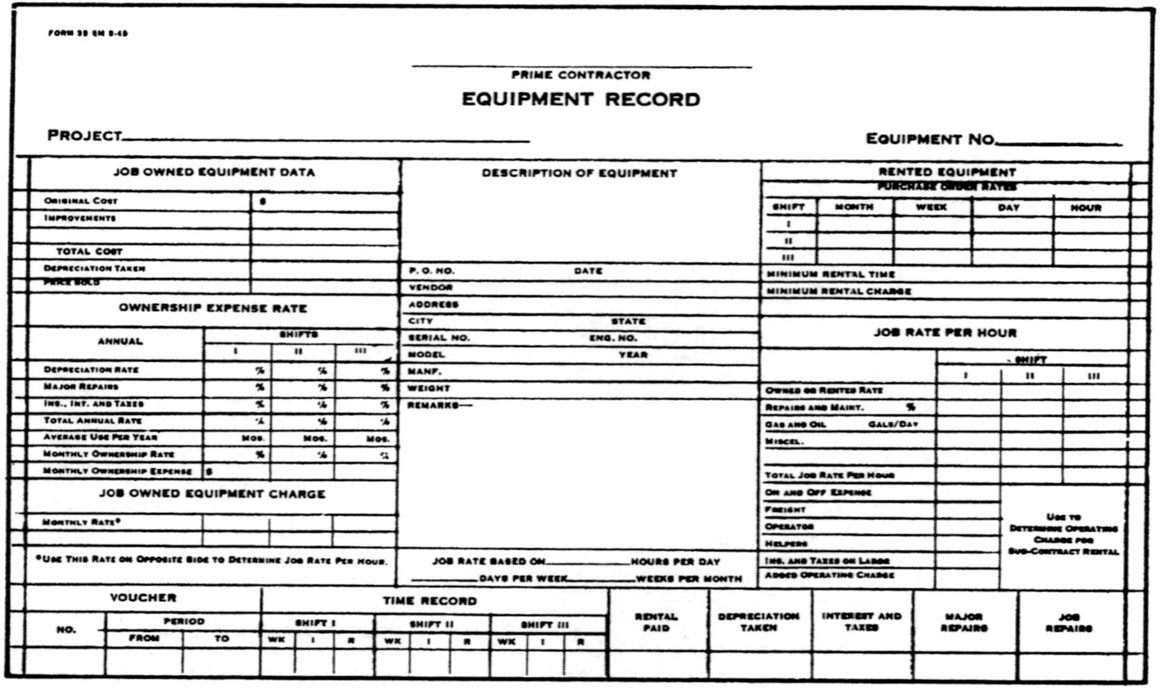

Figure 5-4 Equipment Record. Another method of controlling equipment is to keep a ring binder, with a sheet like this one for each piece of equipment.

show on work orders, and when the cost of the work order is posted, the necessary changes will also be made in the equipment ledger. At least once each year the physical equipment should be checked against the equipment ledger to see that the record is complete. Income tax considerations alone more than justify the cost of an inventory of equipment.

One of the most important steps in gaining the necessary physical control over equipment on a job is inspection on arrival at the jobsite. It is fairly standard practice for those who rent equipment, and sometimes even for departments or jobs of the same firm, to make only the barest minimum of repairs before releasing equipment. Thus inspection of condition on arrival is vital, and the equipment receiving report should provide for it. In addition, the equipment record should show make, model, accessories or extra equipment, serial numbers if any, and any other information which the company may find useful (Figure 5-4).

Often, the shipping records show that equipment was shipped with attachments and accessories that are missing when the equipment arrives on the job. Such discrepancies require either an insurance claim or correction of the shipping records.

As soon as equipment reaches the job it should be given a job number, or the owner's equipment number should be recorded and used for identification.

Basic accounting

Keeping equipment costs is essentially simple. Whether the equipment is owned or rented, all relevant costs are charged to an account called "Cost of equipment operation" or something similar. As the equipment is used on the job, its use is charged to the cost accounts on an hourly basis (sometimes referred to as "rentals") with the credit going either to "Cost of equipment operation" or to a contra account with a title such as "Equipment costs charged to operations." Under normal circumstances there will tend to be more money charged to the work items than to the "Cost of equipment operation," and the resulting credit balance is used to absorb the cost of overhauls and other extraordinary repairs.

Example: Perhaps a somewhat oversimplified example will illustrate the principles involved:

> Assume that a piece of equipment is assigned to a job and the job cost accounts are charged $20 per hour. On the job books the entry for a month of 22 eight-hour days would be:

Dr. Job Costs		$3520	
Cr. Equipment costs charged to operation			$3520

To record equipment charges as follows:

One 365-cfm compressor for the month

Charge work items:

Cost account No. 3	100 hours	$2000
Cost account No. 5	76 hours	1520
		$3520

The entry clearing actual costs might look something like this:

Dr. Cost of equipment operation	$3000	
Cr. Depreciation		$1800
Repairs and maintenance		500
Provision for taxes and insurance		700

To transfer costs for the month to cost of equipment operation

Note that at this point there has been $520 more charged to operations than the actual costs incurred. Suppose that roughly this rate of accumulation continues for 12 months and the excess builds up to $7,000. Then assume a major overhaul costing $7,500. The excess credit would be wiped out and a debit balance of $500 would be created which would be wiped out by the next month's operations. Meanwhile, however, the cost accounts properly chargeable with the cost of the overhaul have already been charged with their proper share.

Demonstrating with a single piece of equipment tends to oversimplify the example and also the illustrative journal entries, but the principle would be the same regardless of the amount of equipment or the number of cost accounts involved. It would also be the same whether the job accounts were centralized or decentralized.

The method described, like all others, has a great number of variations. Some companies prefer to provide their reserves for major repairs through larger depreciation provisions, thus building up an excess credit balance in the depreciation reserve. Other companies, instead of compiling costs by individual units, compile them according to types of equipment. Thus, compactors, pumps, and gunite equipment might form three classes of equipment for cost accounting purposes. This method has the advantage of reducing the paperwork and at the same time producing averages that tend to smooth out the differences between individual pieces of equipment. If the averages are used in estimating, this feature could be important.

Keeping equipment time

Whatever the procedures used in accounting for equipment costs, it is necessary to keep time on equipment in much the same way as it is kept on labor. For the many small firms whose operations require little or no equipment aside from a pickup truck and a few power-operated hand tools, the need to have any special accounting for equipment costs is negligible. When some portion of a job requires equipment the company does not have, they either subcontract that portion or rent the needed equipment and return it as soon as their need has been satisfied. For contractors in this category, it would be a waste of money to keep time on equipment.

For any contractor whose equipment costs are a substantial part of the cost of his work, keeping time on equipment is essential. Equipment time records are the starting point for equipment cost control because most of the costs are incurred on the basis of time. If equipment is rented, the rentals are generally based on some time period. If the equipment is owned, such ownership charges as depreciation, taxes, and insurance are clearly based on time. Even such items as maintenance, repairs, and provision for major overhauls are provided for in the accounts on the basis of operating hours.

One of the simplest forms for equipment timekeeping is a daily time sheet (Figure 5-5) showing each piece of equipment on the job and the hours spent on each of the several kinds of work or on each of the separate work items in the cost code (Chapter 1). It is submitted daily with the labor time sheets and is checked against the operators' time, then it is sent to the person who maintains the equipment cost records and who possibly prepares an equipment payroll (Figure 5-6).

Another form for keeping time on equipment is the daily equipment time card (Figure 5-7). Another form (Figures 3-16 and 3-17) has the operator's time card attached. After these cards have come in for the day they may be checked for accuracy and then coded and punched for sorting. Ultimately they will be summarized on an equipment payroll. This system can be modified so that the equipment costs automatically follow the labor cost distribution the operator has shown on his time card.

There are many other forms for reporting equipment time but those described illustrate the main principles. On jobs where there are many pieces of equipment, and costs are compiled according to classes of equipment rather than by individual units, the time is also compiled by classes of equipment. With less equipment, however (50 pieces or less), it is simpler to compile costs by individual units.

One form of equipment record sheet (Figure 5-4) controls both individual pieces of equipment and the costs of their operation. Whether all of the data called for by this sheet would be compiled for any one

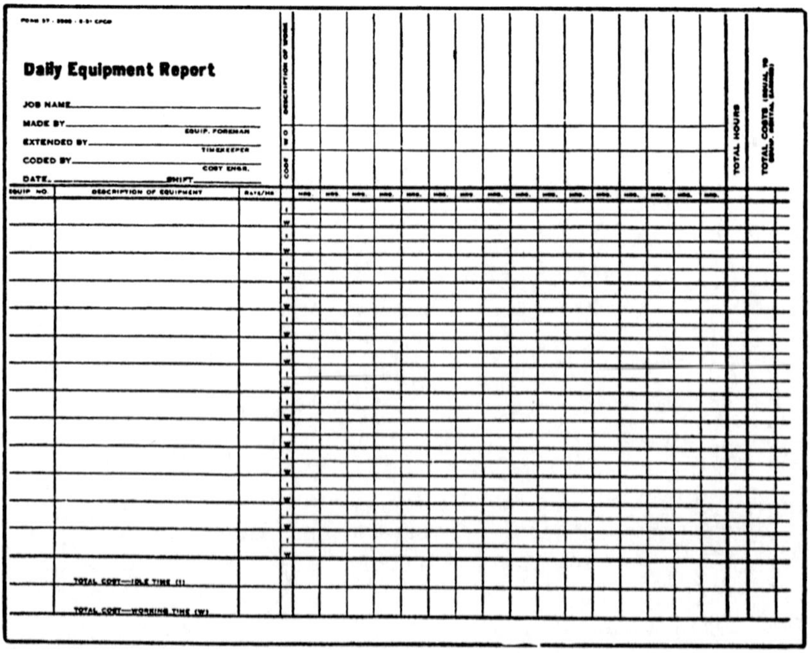

Figure 5-5 **Daily Equipment Report.** An equipment payroll can be prepared from this timesheet.

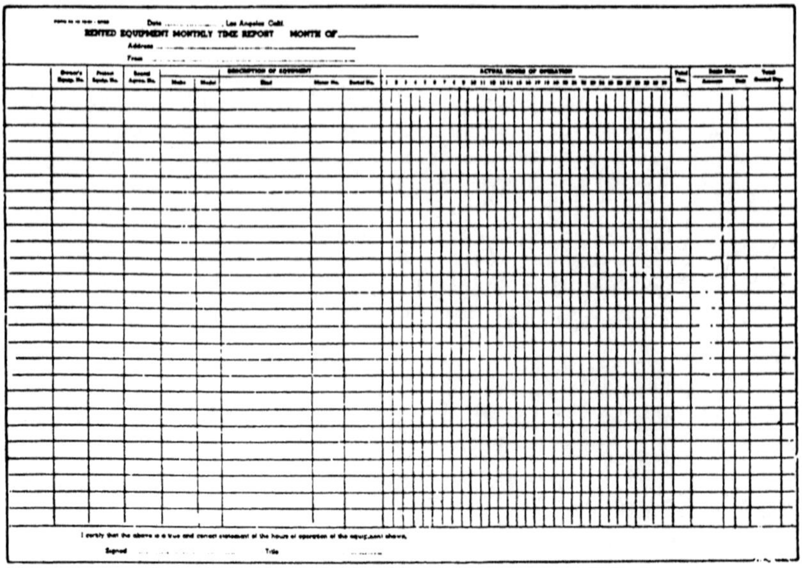

Figure 5-6 **Equipment Payroll.** This form provides all the data needed when equipment is rented by the month. It is somewhat more complete than those used by many contractors.

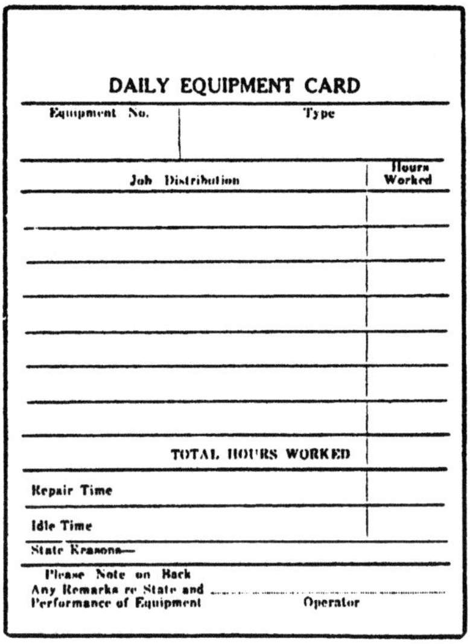
DAILY EQUIPMENT CARD

Equipment No.	Type

Job Distribution	Hours Worked
TOTAL HOURS WORKED	
Repair Time	
Idle Time	

State Reasons—

Please Note on Back Any Remarks re State and Performance of Equipment ______ Operator

Figure 5-7 Equipment Time Card. This card does not have an operator's time card attached. A combination equipment time card and operator's time card is shown in Figures 3-16 and 3-17.

piece of equipment or not, the form is a good checklist of the things a contractor might want to know about the equipment on any particular job. This form would not, as a rule, replace the equipment cost ledger (Figure 5-8). A simplified form (Figure 5-9) shows the accumulation of equipment time on a daily basis.

Repair and maintenance costs

If a contractor has very little equipment it may pay him to have repair and maintenance work done by an outside agency, which will, of course, submit repair orders to cover all work done. If a contractor uses any substantial amount of equipment, he normally repairs it in his own shops. One of the easiest ways to keep track of repair costs in the contractor's own shops is to use a shop order or repair order (Figure 5-10). Each day the time of the shop crew is balanced against the labor time shown on the repair orders and the differences accounted for. When this procedure is followed, care must be taken to see that jobs are not loaded with idle time.

The entire cost of operating the shop is charged to an account called "Equipment repairs and maintenance" or a similar title, and this account is credited with amounts charged to "Cost of equipment operation." From this account, as previously indicated, the costs are charged out to work items in the cost ledger.

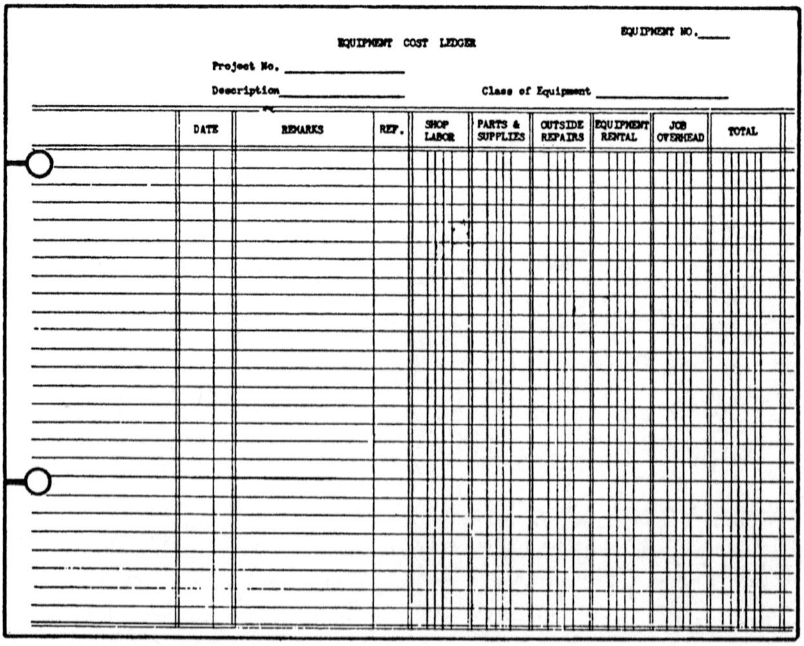

EQUIPMENT NO.____

EQUIPMENT COST LEDGER

Project No. ____________

Description____________ Class of Equipment ____________

DATE	REMARKS	REF.	SHOP LABOR	PARTS & SUPPLIES	OUTSIDE REPAIRS	EQUIPMENT RENTAL	JOB OVERHEAD	TOTAL

Figure 5-8 **Equipment Cost Ledger.** This form provides a convenient way of summarizing information if costs are kept by individual pieces of equipment or by groups of similar equipment.

If repair and maintenance costs are charged to individual pieces of equipment in the equipment cost ledger, it becomes a relatively simple matter to learn when costs of maintenance and repair on any individual piece of equipment exceed its earning power. When such a unit is found it must be disposed of before its costs eat up the profit made by a good machine.

Depreciation

In this discussion no consideration is given to such special income tax provisions as the former Class Life Asset Depreciation Range System (ADR), the Accelerated Cost Recovery System (ACRS) adopted in 1981 as a part of the Economic Recovery Tax Act (ERTA), and the depreciation changes in the Tax Reform Act of 1986 (TRA 86). These are discussed in Chapter 13.

There are five principal methods of depreciation commonly in use in the construction industry:

1. Straight line
2. Production hours

FORM 84 5M 5-48

PRIME CONTRACTOR

EQUIPMENT EARNINGS RECORD

Owner__________
Street__________
City__________ State__________
Date Started . . __________
Date Terminated __________

P. O. No.__________ Equip. No.__________
Description__________
Earnings From__________ To__________

P. O. Rate	Day	Week	Month
1st Shift			
2nd Shift			
3rd Shift			

DAILY TIME RECORD

DATE	1ST SHIFT			2ND SHIFT			3RD SHIFT			DATE	1ST SHIFT			2ND SHIFT			3RD SHIFT		
	WK	I	R	WK	I	R	WK	I	R		WK	I	R	WK	I	R	WK	I	R

Amount Earned:

Remarks and Repair Charges:

Figure 5-9 Equipment Earnings Record. Essentially a time report on which space is provided to compute total earnings and offset repair and maintenance costs, this form is best adapted to systems in which equipment costs and time are kept on a memo basis.

3. Production units
4. Fixed rate on declining balance
5. Sum-of-the-years'-digits

To compare their operation, assume a contractor has a power shovel which cost $21,000; that it has a salvage value, when worn out, of

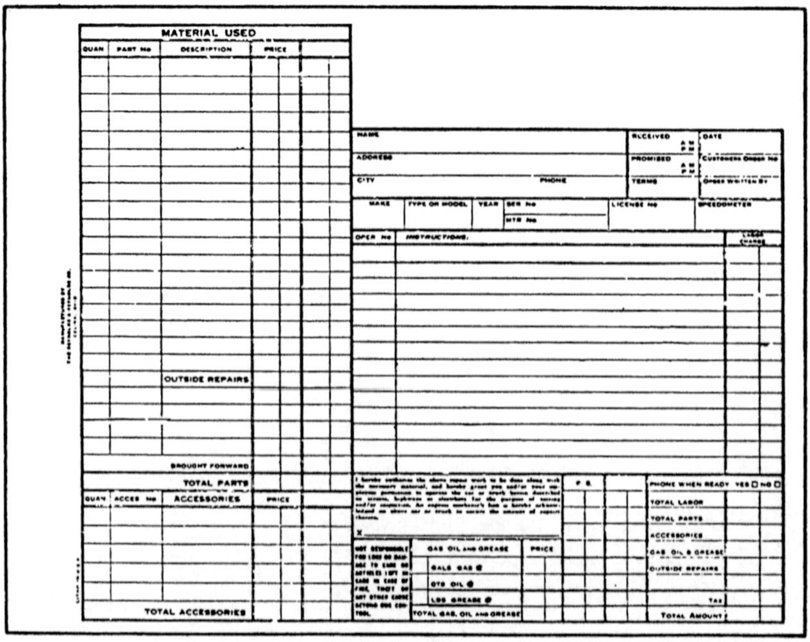
MATERIAL USED
QUAN | PART No | DESCRIPTION | PRICE
OUTSIDE REPAIRS
BROUGHT FORWARD
TOTAL PARTS
QUAN | ACCES No | ACCESSORIES | PRICE
TOTAL ACCESSORIES
NAME
ADDRESS
CITY
PHONE
RECEIVED
DATE
PROMISED
TERMS
MAKE | TYPE OR MODEL | YEAR | SER No | LICENSE No | SPEEDOMETER
MTR No
OPER No | INSTRUCTIONS
GAS OIL AND GREASE | PRICE
PHONE WHEN READY YES ☐ NO ☐
TOTAL LABOR
TOTAL PARTS
ACCESSORIES
GAS, OIL & GREASE
OUTSIDE REPAIRS
TAX
TOTAL AMOUNT

Figure 5-10 **Equipment Repair Order.** This form is manufactured principally for use in repair shops maintained by automobile dealers. However, it is an excellent form for recording the cost of construction equipment repairs and maintenance.

$1,000; and that it has an estimated useful life of five years, or 8,000 operation-hours, or 250,000 yards of material moved.[1]

By the straight-line method, one-fifth of $20,000 (cost less salvage) depreciable value, or $4,000, would be written off during each of its five years of useful life.

By the production-hour method, the $20,000 depreciable value would be apportioned over 8,000 hours of expected operation, and so each hour could be charged with $2.50 to cover depreciation.

By the production-unit method, the $20,000 depreciable value would be apportioned over 250,000 yards of material which the shovel should move during its expected life, and the charge for each yard would be 8 cents.

The declining-balance method is illustrated in Schedule 5-1. This method first came into popular use after the 1954 Internal Revenue Code approved it at 200 percent of straight-line rates for new equipment and 150 percent of the straight-line rates for used equipment.

---

[1]The figures used in this example are for illustrative purposes only and are not intended to be typical.

SCHEDULE 5-1 DOUBLE DECLINING BALANCE WRITE-OFF

Depreciable cost	$21,000.00
Less first year's depreciation (40% of $21,000)	8,400.00
Book value after one year	$12,600.00
Less second year's depreciation (40% of $12,600)	5,040.00
Book value after two years	$ 7,560.00
Less third year's depreciation (40% of $7,560)	3,024.00
Book value after three years	$ 4,536.00
Less fourth year's depreciation (40% of $4,536)	1,814.00
Book value after four years	$ 2,722.00
Less fifth year's depreciation (40% of $2,722)	1,089.00
Book value after five years	$ 1,633.00

Under this method, salvage value is ignored on the theory that the mechanics of the computation will always leave a residual balance. The method is sometimes known by the term "double declining balance" which, of course, applies only to new assets.

By the sum-of-the-years'-digits method, the $20,000 depreciable value would be written off as shown in Schedule 5-2. This method was also virtually unknown until authorized by the 1954 Internal Revenue Code for new assets only.

A sixth method, sometimes used for large jobs and joint ventures, is to charge the entire cost of the equipment to the job and to sell or appraise the equipment when the job is over, crediting job costs with the sales price or appraisal value at the end of the job. This method, however, can hardly be called depreciation in the strict sense of the term. This method is found frequently in use on large "cost-plus" contracts.

SCHEDULE 5-2 SUM-OF-THE-YEARS'-DIGITS WRITE-OFF

Year No.	1	5/15 of $20,000	$ 6,666.00
	2	4/15 of $20,000	5,334.00
	3	3/15 of $20,000	4,000.00
	4	2/15 of $20,000	2,666.00
	5	1/15 of $20,000	1,334.00
TOTAL	15		$20,000.00

What is the "correct" rate of depreciation to charge on equipment of a given type? This question does not have a categorical answer because conditions vary too widely on construction jobs. For example, dump trucks, which might last several years on some jobs, may be completely worn out in a few months on others. The method of depreciation and estimated useful life should be selected on the basis of which method and estimate of useful life most closely approximates the rate of wear, tear, and obsolescence for each piece of equipment.

There is a tendency, particularly on the part of accountants and other nonoperating personnel, to set too high an estimate of the useful life of equipment. As a rule of thumb, some of the larger engineering contractors work on the principle that equipment loses half of its value on each job, and in the long run, if jobs last from one to two years, this rule often proves surprisingly accurate.

Some firms, instead of establishing a rate for each piece of equipment, will establish a composite rate for each class of equipment. Composite lives have the advantage of convenience in accounting, but if much money is involved there are disadvantages that are not to be overlooked. The greatest of these is the difficulty of establishing reasonable hourly rates for charging equipment costs to the job. If the amount of equipment is small, and equipment costs are not kept, then the single composite rate is justified by its simplicity. Otherwise there seems to be little to recommend it.

Actual depreciation charges, even on similar pieces of equipment, will tend to vary, and this variation can cause some puzzling differences in unit costs. One of the ways to overcome this problem is to use standard hourly rates to distribute equipment costs to work in process. In fact, many companies use the Allied Equipment Dealer (AED) Greenbook to arrive at standard rates. This book is a compilation of average hourly, daily, monthly, and yearly rates charged for equipment rentals in the preceding year. Another helpful book is the Associated General Contractors of America's "Computing Owner Equipment Cost." The effect on unit costs is to eliminate substantially such variations, particularly if overhauls are charged to the reserve for depreciation or to some special credit balance account.

The use of group rates for charging equipment costs to operations has already been mentioned. The practice of using group depreciation rates for all pieces of equipment of the same kind or of the same life is well established and is quite useful in a large spread. If this method is followed, each year's acquisitions must, of course, be segregated by groups in maintaining fixed asset and depreciation records.

If there is very little equipment, and no equipment costs are kept, depreciation is not important enough to justify any substantial amount of accounting. The greater the investment in equipment, the greater the effect of depreciation policies on cost keeping.

Many construction executives take the attitude that "you can't pay bills with tired iron." In other words, until the equipment has been depreciated so that its cost is charged off to operating costs, the books will show an inflated figure for the value of such equipment, thus distorting the profit picture and perhaps misleading the contractor into thinking that he has more cash for working capital than is actually available. This viewpoint has a great deal to recommend it. However, it also has these weaknesses. If equipment is charged off there is a tendency to buy more when there is no real need to do so. Also, if costs have been burdened by unjustifiably heavy depreciation, unit costs will tend to be high, and in a competitive bidding market the company may lose bids it should win. Some apt rejoinders to the "tired iron" quip might include:

1. "When you're bidding for work, they don't pay off on second place."
2. "If you write it all off against this job there won't be anything to write off against the next one."
3. "Just because your equipment is all written off there's no reason why you should leave all your profits on the table."

These observations are just another way of saying that if your estimators are going to get any help from your past experience, they must have accurate unit costs to work with. In the final analysis, the most profitable depreciation rates are those that are most realistic in terms of charging the cost of equipment to the operations that benefited from its use.

Tax aspects. Discussion of the tax aspects of depreciation usually centers on income taxes. In some states the tax on personal property is a factor to be considered, and where "book value" forms the basis of assessment, depreciation is involved. Still, income taxes (both federal and state) are the most important of all. Many property tax assessors ignore book value and use fair market value or, alternatively, estimate the future value of used equipment to the business. This is sometimes referred to as "percent good."

To the extent that it is currently deductible in determining taxable income, depreciation increases costs and correspondingly reduces profits; even in years when there are no profits to deduct it from, however, the depreciation must still be claimed. Sometimes the depreciation cost of loss years can be recovered through net operating loss carrybacks and carryovers, but this method involves a great many uncertainties; and under some state tax regulations, operating losses may not be carried to other years.

Except for the special statutory cost recovery methods mentioned at the start of this chapter, most advantageous methods for tax purposes are in the so-called accelerated methods—double declining balance and sum-of-the-years'-digits. Such methods are advantageous because they accelerate tax deduction, thereby deferring taxes to later periods. It is acceptable to use one method of depreciation for taxes (an accelerated method) and another for financial and cost accounting purposes (straight line). For tax purposes, double declining balance depreciation may be applied only to completely new equipment and may be elected on a piece-by-piece basis. The Tax Reform Act of 1986 made substantial changes in the income tax rules on depreciation, and more changes are likely when the final regulations are written. For the present it seems practical to continue the present method after making the changes in allowable lives required by existing statutes.

It is no longer acceptable for tax purposes to overdepreciate equipment, sell it, and then pay tax on the gain at capital gains rates.

One advantage of the currently accepted depreciation methods for tax purposes is that a contractor who is short on cash may, to a large extent, finance new equipment purchases with "depreciation money." This is not to say that depreciation reserves are, or should be, set aside in cash, but merely that where a large percentage of the cash revenue goes for taxes, there is often too little to meet current payments from current revenue.

Equipment cost reports

Equipment costs on most construction jobs usually arise from three general types of equipment operations. First is the operation of production plants, such as aggregate plants, concrete batch plants, and asphalt plants. The actual costs of operating these plants are charged to an account bearing a title such as "Plant operations" and the charges to the job cost accounts are based on an amount per yard of material produced. The credits contra to these charges would be either to "Plant operations" or to a special account set up to take the credits.

Second is the operation of heavy or major equipment. This includes power shovels, tractors, carry-alls, trucks, cranes, compressors, motor graders, compactors, and similar equipment. Accounting for these costs has been described under the heading "Basic accounting" early in this chapter. Essentially it is similar to that just described for plant costs except that the basis for charging the job cost accounts is usually hours worked.

Third is the operation of a very large number of items which are too small to be classed as heavy or major equipment, but too large to be treated as tools and supplies. Some examples are power-operated concrete buggies, air tools, vibrators, chain saws, office equipment, engineering equipment, carpenter shop and machine shop equipment, warehouse equipment, and camp and mess hall equipment.

The easiest way to account for the cost of minor equipment is to give it a nominal salvage value of perhaps $10 to $15 per unit, and write the balance off at the rate of 40 to 50 percent per year. If the jobs are large, the entire value may be written off over the life of the job and any minor equipment remaining can be appraised and credited back to the job. On smaller jobs, where centralized control is kept in the general office, such equipment is usually charged out to the job at a daily or monthly rate and a single cost account kept to receive the costs of its ownership and operation. Under Internal Revenue Code Sec. 179, a taxpayer may elect to charge off small items to expense. The amount allowable has been increased by the Tax Reform Act of 1986 from $5,000 to $10,000, but the application is somewhat more strict.

Various forms (Figures 5-4 to 5-7) for keeping equipment time have been illustrated. The purpose of each form is to show the time charged to each work item or cost account. As the time is compiled and the accounts are charged for it at the standard rates, job management often needs current comparisons between equipment costs incurred and those estimated for each work item. The weekly cost statement is a form designed to provide this comparison for both labor and equipment costs. The fact that one form can serve for the reporting of both classes of costs points up the similarity in the procedures for recording and reporting them.

It must be emphasized that forms and procedures for reporting vary as widely as the reporting requirements. The important point is that equipment cost, like labor cost, is an important item in the total cost of many construction jobs. It is referred to by accountants as a variable cost because, like labor cost, it tends to vary directly with the volume of work. Also, like labor, it tends to respond quickly to the efforts of management. The cost statement (Figure 5-11) may be needed daily rather than weekly. On the other hand, if labor control is good, it may not be needed at all, because the control on the operators' labor serves a double purpose. Like all reports, equipment cost reports should be made only when and if they are useful to job management in determining where and how to direct its efforts.

Often, instead of an overall report on all job equipment costs, a special report for those operations where heavy concentrations of equipment are used will serve job management's purposes better. Thus a batch plant may report the daily yardage run, or a spreader operation, the yardage of material placed. The grade foreman's report (Figure 5-12) illustrates such a form for an earthmoving job. Note that if work is being done at scattered locations a superintendent can

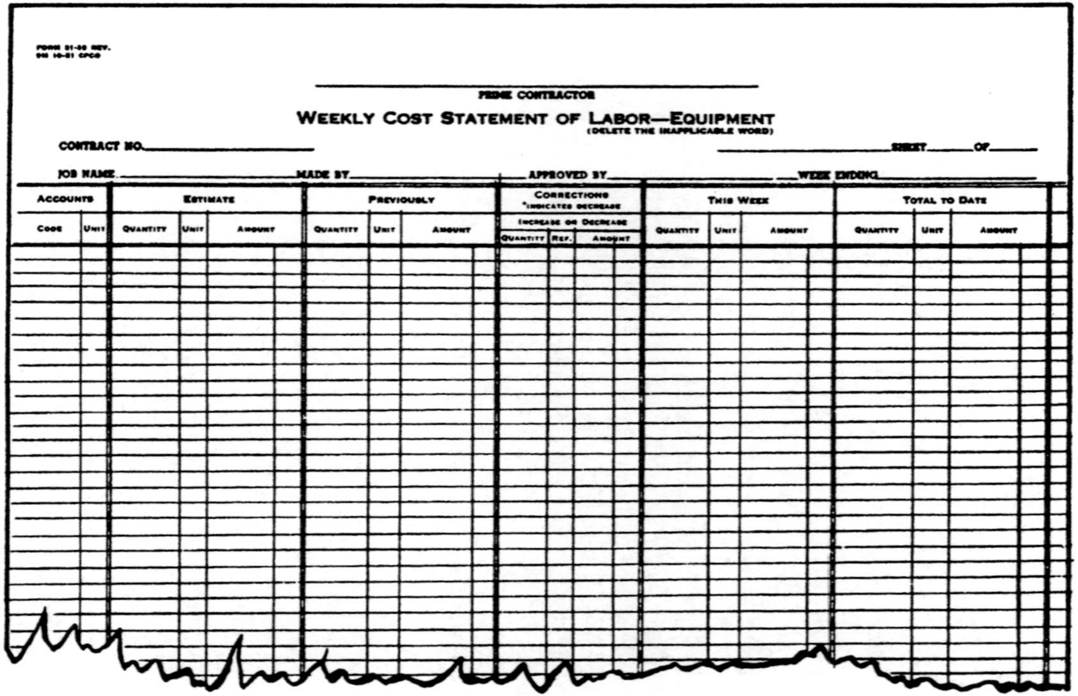

Figure 5-11 Weekly Cost Statement. The function of this weekly report is to compare the costs actually incurred with those provided for in the estimate. It can be used for either equipment or labor costs.

GRADE FOREMAN REPORT

Shift: ______ to ______ Date: ______

Work Acct. No. ______ Work Acct. No. ______

Location: ______ Location: ______

Type Equip.	Equip. No.	Total Loads	Size Load	Total Cu. Yds.	Type Equip.	Equip. No.	Total Loads	Size Load	Total Cu. Yds.

Total ______ Total ______

Remarks: ______

______ Foreman: ______

Form 1—Desert Trail—M 5-52

Figure 5-12 **Grade Foreman's Report.** An example of the specialized forms that each company may devise for its own particular needs, this form is designed to provide job management with current information on an earthmoving job.

tell at a glance from this form: (1) what equipment is at the location, (2) the specific units in use, (3) whether they are working or idle, (4) whether they are working to capacity, and (5) the total yardage moved for the day.

When a substantial amount of equipment is being rented, one of the ever-present problems is to remove it from the job and stop the rental as soon as the equipment is no longer needed. On the other hand, if equipment is taken off the job too soon it is very expensive to bring it back temporarily, and it is often even more expensive to try to do the work with inadequate tools. Basically, the answer to this problem lies in effective planning for equipment use, but if a superintendent is failing to plan adequately, one way of detecting his error is through the daily equipment cost report (Figure 5-13). Such a daily report is the basis for the monthly "payroll" (Figure 5-6) for rented equipment.

Moving costs

One of the pitfalls of construction cost keeping is the cost of moving (mobilization). All too often it is overlooked in bidding because of a hope that someone else will pay for it. However that may be, the costs of equipment mobilization and demobilization are very large and very important items in any job where a substantial amount of equipment is used. In fact, on a $3 million earthmoving job, it is not unusual to see a budget of from $100,000 to $150,000 for move-in and move-out expenses.

DAILY EQUIPMENT COST REPORT

Job:............................Date:......................

Work Item	Unit	YESTERDAY				JOB TO DATE			
		Units Compl.	Equip. Hrs.	Total Cost	Unit Cost	Units Compl.	Equip. Hrs.	Total Cost	Unit Cost

Time checked and priced by	Pricing checked, extensions made by	Cost coded, units priced ....

Figure 5-13 **Daily Equipment Cost Report.** This form is useful on jobs for which equipment costs are large in relation to total cost.

If accurate unit costs are to be kept, the moving costs must be written off to job costs in proportion to the amount of equipment used on each job. Actual mobilization cost will normally be charged to an account of that name or to a special subdivision of the "Equipment operating cost" account. To the standard equipment rate per hour, or per unit, a certain amount or percentage is added to cover the moving costs. When the mobilization costs have been fully absorbed, a credit balance develops in the account and it is against this credit balance that costs are charged.

There are, of course, other bases for writing off the move-in, move-out costs. Sometimes these costs are added to overhead and absorbed on the basis of total direct costs. At other times, direct labor hours or direct labor cost provides a better or more accurate measure. Generally, however, adding a percentage to the equipment operating rates provides the most satisfactory basis for distribution. If there are loading and unloading costs or assembly and dismantling costs, they are usually handled in the same way as move-in, move-out costs and often as a part of them.

Equipment rentals

The problems that accompany the rental of equipment fall into the following general categories:

1. Rates
2. Move-in and move-out
3. Loading and unloading
4. Assembling and dismantling
5. Rental period
6. Condition and repairs
7. Insurance and valuation
8. Idle time

Of course, in any well-drawn equipment rental contract these points must be carefully covered.

Rates. Ordinarily equipment is rented on a monthly basis. Rentals for fractional months are usually stated in days, and fractional days are usually stated in either shifts or hours. It is not unusual, if equipment is to be worked for more than one shift, to find the rental rate stated in terms of a daily rate on a one-shift basis. For two shifts the rate would be one and one-half times the rate for one shift. If the equipment worked three shifts there would be no extra charge for the third shift. Usually, however, rates are stated at a flat amount per month and it is up to the contractor to determine the usage and to keep the equipment running. If special attachments are normally used with the types of equipment being rented it is important to know what attachments the contract rate includes.

Moving costs. The cost of moving the equipment from the owner's yard to the user's job and back again is borne by the user. However, problems arise when the equipment is not being moved from (or to) the owner's yard but instead is sent from one job to another. If the hauls from job to job are greater than the distance from the owner's yard to the user's job, a question arises as to who pays the difference. One common provision found in equipment rental contracts provides that the user will pay the cost of transportation to the job and return but that this charge is not to exceed the cost of transporting the equipment from the owner's yard to the job and back again.

Even with this type of provision, an alert owner can sometimes collect twice for some part of the transportation charges. Users sometimes guard against this by providing that the owner must certify that return transportation is not being paid for by the next user. It is fairly common to limit the amount to be paid for transportation.

Loading. Unloading and reloading at the jobsite are usually performed at the user's expense. If the equipment comes from or returns to the owner's yard, the owner may have some loading or unloading expense to pay, but if it moves from one job directly to another, the owner is saved this cost. Obviously an owner will consider this fact when deciding whether to remove a piece of equipment from a job immediately upon notice from the user that he is through with it.

Assembly. If production plant or heavy equipment must be assembled after it arrives on the job and disassembled before it leaves, the costs are often quite heavy. It is standard operating procedure for the user to pay these costs except when a seller sends his own crew to assemble a new plant and see that it is properly placed in operation.

Rental period. For light equipment, rented for a comparatively short time, the rental period customarily starts when the equipment leaves the owner's yard and ends when it is returned. On heavier equipment that is rented for longer periods, it is customary to find the rental period starting when the equipment is loaded for shipment and ending a stated time (say three days) after notice of termination and availability (Figure 5-14) is given by the user. Often, to avoid the cost of unloading and reloading at the owner's yard, an owner will leave equipment stored at the user's jobsite until it is to be shipped elsewhere. When this happens, the temptation is great to use the equipment without paying rent for it. On the other hand, the user must be careful to see that no rent is paid for idle equipment which is merely awaiting shipping orders.

Request for Equipment Termination

To:.................................(Equipment timekeeper)
From...
Please terminate the following piece of equipment effective:
....
Equipment No........................Serial No..............
Owner..
Description.......................................
....
Condition...

Requested by......................................
Approved by.......................................
Date..........................

Figure 5-14 **Request for Equipment Termination.** When equipment is being rented, it must be kept busy or its cost will be lost to the job. Most rental contracts require notice of termination of rented equipment. This form is satisfactory for that purpose.

A special problem, which usually is the subject of a special provision in the equipment rental contract, is the user's liability for rent during time lost through strikes, storms, or other reasons beyond the user's control. Some contracts terminate the rental entirely, others provide for a special "standby" rate which is often about half of the regular rate. Still others make no provision at all for such contingencies. The alert user will usually demand some protection against this kind of situation.

Condition. Equipment that arrives at the jobsite in such poor condition that it cannot reasonably be operated does not usually incur rental charges until it is repaired, and the cost of such repair is customarily borne by the owner. Because, on the other hand, the user is normally responsible for ordinary repairs during the rental period and for returning the equipment in as good an operating condition as when it arrived, most equipment rental contracts make provision for inspection on arrival and departure. It is usual for such contracts to provide that if one party fails to have an inspector on the site, that party agrees to accept the findings of the other party's inspector.

Insurance. Most equipment rental contracts require the user to insure the equipment against accidental loss or damage and, because most insurance policies of that type are based on the value of the equipment, the value is usually agreed upon in the contract.

Idle time. Whether equipment is owned or rented, it costs money whenever it is idle. To charge idle equipment time to cost accounts or work items is to distort costs and conceal a source of waste. Hence most equipment time records call for an accounting for idle time. The easiest way to report on idle time is to account for it separately in the cost ledger as if it were a separate item of job overhead cost (which in fact it is). The account set up is charged with the out-of-pocket costs of ownership (taxes, insurance, depreciation, and sometimes storage) for all hours except the "normal" idle time, which is figured into the hourly rate.

Owning or renting

One of the greatest risks any business can incur is that of having a large portion of its capital tied up in nonproductive assets. Construction equipment sitting idle in the contractor's yard or warehouse is more than just nonproductive. It demands outlay for insurance and for maintenance, and depreciation continues just about as fast as if the equipment were working. Worst of all, a contractor who has all his profit tied up in equipment at the end of a job finds himself in the unhappy position of having income taxes to pay and no money to pay them with.

Therefore many contractors (particularly those in lines where heavy equipment is needed) prefer either to subcontract those portions of the work that require equipment or to rent whatever equipment is needed to do the job on hand. By so doing, these firms avoid not only the heavy capital commitments required by equipment ownership but also the temptation to "make do" with inadequate or unsuitable equipment simply because it is available.

One alternative to this policy is sometimes used on very large jobs, particularly those run by joint ventures. That is to buy new equipment at the start of the job and sell it at the end of the job.

Another alternative commonly followed is for a construction company to own a separate equipment company. The combined group buys at dealer prices, makes the maximum profit, and, by renting to others, manages to cut idle time to a minimum.

On the other hand, when the work volume is reasonably steady and the variety of equipment required is not too great, many contractors find it both convenient and profitable to own equipment. Also, many prospective clients for construction work require that all bidders submit a brochure showing, among other things, a list of equipment owned by the contractor. Rightly or wrongly, there is a tendency on the part of some prospective clients and their architects, as well as some public bodies, to question the stability of a contractor who owns no equipment.

There are many compromises between the extremes of owning all the necessary equipment or none of it. Many excavation contractors own their power shovels and bulldozers but rent their trucks from independent truck owners who drive and maintain their own vehicles. The drivers are placed on the contractor's payroll and therefore are employees, but they take care of their own licensing and maintenance. This practice sometimes avoids transportation taxes, but payroll tax liabilities are then incurred.

Another compromise is to own the minor equipment and either rent or subcontract to secure the major items. Still another is to own the major equipment and to accept only subcontract jobs where that type of equipment is required.

Very little equipment is rented on an "operated and maintained" basis. Usually, if an operator is to be furnished, it is safer, from the standpoint of compensation insurance if nothing else, for the contractor to put the operators on his own payroll and rent the equipment "bare."

It has been pointed out that the construction contractor is in a peculiar position relative to ownership of his work in process. Most of this work consists of improving the real property of others, and as soon as the work is done, legal title to it vests in the owner of the realty. To protect the contractor, the federal government and all states (but not all foreign governments) have lien laws or bond acts to provide, as security, the power to seize the property improved, or in lieu of such seizure, a claim against the contractor's bond. The lien laws provide for a statutory procedure to place a lien on the real property improved or (as in government contracts) to provide a solvent bondsman to whom the laborer or material supplier may turn in the event he cannot collect from the person with whom he has contracted.

Most of these lien laws and bond acts refer to "labor" and "material" as the items secured. Nevertheless, the Federal Bond Act of 1935 (Miller Act) and most, if not all, of the state laws also include equipment rentals among the items covered so long as the equipment is used directly at the jobsite or is otherwise connected with the work in some fashion direct enough to come within the language of the particular statute. However, equipment rented to a material supplier who uses it to load or process materials would not, under most statutes, be covered.

Equipment problems in joint ventures

A joint venture, like any other organization which has contracted to do a construction job, may either buy its equipment or rent it. As long as it rents from some third party there are no problems peculiar to joint venture equipment. As soon as it rents equipment from one of the members of the venture, questions of rental rates and repair and maintenance policy arise. As soon as it purchases equipment the question of ultimate disposal is raised, and if the purchase is made from a member of the venture the additional questions of price and condition arise. It is not uncommon to hear a contractor remark that he "got the joint venture" to put his equipment "in good shape."

Needless to say, it is most important that joint venture agreements deal in detail with the questions just raised. It is outside the scope of this book to discuss in detail the contents of joint venture agreements, but it is appropriate to point out the trouble spots from an accounting viewpoint. Accordingly, Schedule 5-3 is included as a checklist of the accounting aspects of equipment provisions in a joint venture agreement.

If these ten questions are discussed and answered in advance there will be few arguments about joint venture equipment that cannot be settled by reference to the agreement. On the other hand, no agreement, however carefully discussed and drafted, can take the place of fair dealing between the parties. This is especially true of agreements on joint venture equipment.

SCHEDULE 5-3 CHECKLIST OF EQUIPMENT PROVISIONS OF A JOINT VENTURE AGREEMENT

1. Is there a schedule of rental rates to be charged by members to the venture for equipment rented to it?
2. Does the schedule, if present, provide for additional rentals for special attachments or devices?
3. Does the agreement define ordinary repairs and maintenance for which the venture must pay, and extraordinary repairs which may be charged back to the owner? A system of repair orders (Figure 5-10) is virtually indispensable if the job is to handle its own repairs.
4. Does the agreement include special rates for use of the equipment on more than one shift?
5. Does the agreement specify who is to pay for transportation to and from the job and what happens when the equipment is to be brought from or shipped to a point unusually distant from the jobsite?
6. Does the agreement specify when rentals are to begin, when they are to end, and whether they are to continue during periods when the equipment is in the shop or during delays due to weather, strikes, and so on?
7. May the job refuse to accept from a venturer equipment that is in poor condition on arrival? How is condition to be determined?
8. How is joint-venture-owned equipment to be handled on the venture books, and how is it to be valued at the end of the venture? Particularly, how fast is it to be written off?
9. Will members of the venture have the first chance to buy it? If so, at what price, and who has the preference if more than one member of the venture wants to buy the same equipment?
10. If equipment is to be transferred to another venture between the same parties, at what value is the transfer to be made?

Chapter

6

Accounting for Overhead and Distributable Costs*

Direct, Indirect, and Overhead Costs

The word "direct," when applied to construction cost, means costs which can be specifically identified with a construction job and/or with a unit of production within a job. An "indirect" cost is a cost which can be identified with jobs but not with a specific job or unit of production. "Overhead costs" are those which cannot be identified with or charged to jobs or units of production unless some more or less arbitrary allocation basis is used.[1]

How you apply these cost terms, however, depends on whether you are looking at these costs from a home office or jobsite point of view (the main difference in perspective centers on indirect costs).

From the superintendent's point of view at the jobsite, all costs which cannot be specifically identified with a bid or work item of the construction project are considered indirect, or "distributable," costs. This concept would also include those costs and those situations in which it is impractical or too costly to attempt to identify the specific cost items with a portion of the work. A few typical examples of such costs may include equipment supplies (including fuel and oil), compressed air, concrete hauling labor, drill steel and bits, explosives,

---

*The larger and more complicated the jobs become, the more accurate must be the segregations between direct, indirect, and overhead costs, and the more sophisticated must be the methods of distribution. Larger firms are now turning more and more to computers (see Chapters 12 and 14) to improve the quality of overhead distribution.

The Tax Reform Act of 1986 makes no important changes in the manner of determining distributable costs. It does make substantial changes in how such costs are to be allocated among the taxable years covered by each "long-term contract" as that term is defined in the new law. These changes are discussed further in Chapter 13. It should be noted, however, that smaller contractors are not affected.

[1]As to federal jobs, see pronouncements of the Cost Accounting Standards Board.

welding rods, telephone and utilities, warehousing costs, small tools, and jobsite overhead cost items such as move-in and move-out, temporary structures, accounting, purchasing, personnel and labor relations, and security guards. When viewed from the home office, however, these would be considered direct costs since they have been specifically identified with a job.

There are other costs incurred in the home office which the home office would consider direct costs, yet which the superintendent would consider an indirect overhead cost. For example, an accountant in the home office may charge each specific job for the time he or she spends in preparing progress billings for it. So far as the home office is concerned, that labor cost incurred by the accountant is a direct job cost. Yet when it is distributed to the job accounts, it is usually classed as job overhead, because rarely can the accountant's time be clearly identified with some individual bid item or work item on the project.

The distinction made between home office overhead and job overhead is a distinguishing feature of construction accounting. In manufacturing, a similar distinction is recognized by segregating the general, administrative, and selling expenses from the overhead incurred at the manufacturing facility. Among construction companies there is little uniformity in classification and treatment of these costs.

Besides the home office versus jobsite perspective problem, there is another reason for the lack of sharp distinction between direct, indirect, and overhead costs. It is that certain items of cost may be either a direct or an indirect cost on the same job. For example, the first set of aerial photographs made for a particular job might be a direct charge to a specific bid item such as a site survey. However, on the same job, the next set of aerial photographs might be for job progress pictures (an indirect or distributable job cost item), or to support claims for default by a subcontractor (another indirect job cost item), or to show to bankers or surety companies in connection with the general credit application (a home office overhead item).

These distinctions are not academic, even for the smallest contractor. How indirect costs are accounted for on a project and in the home office may have a very important bearing on contract pricing. This is especially true where the total bid is based on numerous individual bid items within one project with differing unit prices for each bid item. It will be a major factor once the project has begun in terms of "tracking" (i.e., allocating) actual cost against the original bid for the purpose of projecting final project outcome. It will be an important factor in the pricing of changes in the scope of the work, change orders, and special work orders. It will become a critical factor should disputes over individual bid items within a job arise between the contractor and the owner, or between the contractor and subcontractors. Literally thousands of claims and lawsuits annually center on assertions of delays, changed conditions, or changes in scope. Even when it has been agreed that there is a basis for a claim covering delays, changed conditions, or changes in scope, a dispute may continue over how much the contractor, owner, or subcontractor should be paid. These cost or price disputes will quite often center on the allocation of indirect and overhead items among bid items within the job and charges from the home office which have been allocated among bid items in the job.

Another important factor to consider in accounting for indirect and overhead costs is that, in allocating such items between years, there may be a major effect on the contractor's income tax. For example, subject to changes pursuant to the Tax Reform Act of 1986, a contractor who elects the completed-contract method of accounting will often charge into job costs all overhead that is directly applicable to jobs, and such overhead costs will be a part of the costs and therefore a deduction for tax purposes in the year in which the job is closed. Therefore, the contractor will have failed to take advantage of deductions in the year that the overhead was incurred; hence the manner of accounting for overhead can have a significant tax and financial impact.

The method of handling overhead must be weighed in terms of two perspectives—tax and financial reporting. If overhead is charged to expense when incurred rather than charging such costs to a job, the tendency in terms of financial reporting will be higher profits in high-volume years and lower profits in low-volume years.

As a practical matter, the completed-contract method of accounting, even under the Tax Reform Act of 1986, is used by a great many small contractors and is generally accepted without question for tax purposes because it reflects income more accurately than either the percentage-of-completion, "straight" accrual, or cash method.

Distributing Indirect and Overhead Costs

For the very small contractor, the simplest and most satisfactory segregation of direct, indirect, and overhead costs is to treat as distributable items all components of cost except:

1. Labor performed at the jobsite, including a pro rata share of payroll taxes and related employee benefits, such as workers' compensation, group insurance, holiday and sick leave, and similar "fringe" benefits
2. Materials and supplies actually consumed on a job, including related sales and use taxes
3. Rental of equipment actually used on the job
4. Subcontractor costs

These items constitute direct costs, and for most small contractors the sum of direct costs is usually a very satisfactory basis to use in distributing indirect and overhead costs.

For example, a small building contractor who has four jobs in progress during a given month might have incurred direct costs during the month as shown in Schedule 6-1. His indirect expenses are set forth in Schedule 6-2.

The total indirect costs for the month ($1,018) are 20 percent of the total direct costs incurred during the month. To allocate the indirect costs among the small contractor's jobs, a journal entry is prepared charging a general ledger account called "Job costs" and crediting another general ledger account called "Indirect cost distribution." This entry spreads the indirect costs to the job on the ratio of 20 cents for every dollar of direct cost. The entry to accomplish this is set forth in Schedule 6-3.

It should be noted that because this is a very small operation, there is no segregation of indirect costs between home office and jobsite or between overhead and indirect. Note also that for cost accounting purposes, material costs include sales and use taxes and similar

SCHEDULE 6-1 SMALL CONTRACTOR'S DIRECT COSTS

Job No.	*Labor*	*Materials*	*Equipment rentals*	*Total*
1	$ 375	$ 425	$ 60	$ 860
2	440	305	45	790
3	1,230	915	125	2,270
4	665	475	30	1,170
	$2,710	$2,120	$260	$5,090

SCHEDULE 6-2 SMALL CONTRACTOR'S INDIRECT EXPENSES

Office salaries	$300
Rent	75
Telephone	30
Stationery and office supplies	45
Postage	20
Repairs and maintenance	92
Small tools and supplies	107
Insurance	120
Taxes	66
Depreciation	150
Other expenses	13
TOTAL	$1,018
Overhead distributed to jobs	1,018
BALANCE	$-0-

SCHEDULE 6-3 JOURNAL ENTRY DISTRIBUTING OVERHEAD

Job costs	$1,018	
Overhead charges to jobs		$1,018
To distribute overhead for the month of ______.		

Cost ledger detail:

Job	*Direct costs*	*Overhead*
1	$ 860	$ 172
2	790	158
3	2,270	454
4	1,170	234
	$5,090	$1,018

related material costs. In the same way, payroll taxes and related employee benefits have already been included in direct labor by job. This is accomplished by establishing a so-called "provisional" fringe benefit rate. This provisional rate is generally developed by computing the percentage that the prior year's actual fringe costs bear to total labor, and adjusting it for known changes in fringe costs for the current year (for example, social security—i.e., FICA—percentage increase, or negotiated changes in union fringe benefits). Fringe costs related to payrolls usually include employer-incurred payroll taxes such as FICA, federal and state unemployment insurance, and other payroll-related costs such as workers' compensation insurance, group health and life insurance, profit sharing or retirement plan contributions, and holiday and sick leave. Other costs which may be treated in a manner similar to fringe benefits are such items as engineering supplies, telephone, and housing allowances.

Essentially the same principles of distribution should be followed regardless of the size of the operation, at least insofar as indirect cost is concerned. However, as operations increase in size and complexity, so also do the problems of indirect cost distribution. For example, the larger contractor will reallocate the indirect costs which the home office has allocated to a job, to individual bid items within the job or major components of the job. On top of the allocated cost from the home office or overhead accounts will be other indirect costs which must be allocated on some basis to bid items or components within the job. This intra-job allocation might also include components of direct cost which are impractical to record initially by bid item or component of the job. An example of a large contractor's cost ledger is set forth in Schedule 6-4.

In this example, the construction company had contracted to build the powerhouse for a large dam project. Because of the multiple number of bid items—in excess of 100—the contractor elected to "pool" many direct cost items which either could not be identified with specific bid items or would involve prohibitive cost and complexity to do so.

SCHEDULE 6-4 LARGE CONTRACTOR'S JOB COST LEDGER

Job 1638	Mt. Tamelpais Dam Cost Ledger			October 31, 1975		
	Bid item	*Labor*	*Materials*	*Equipment*	*Pool*	*Total*
100	Powerhouse excavation:					
	101 Draft tube	500,000	10,000	10,000	110,000	630,000
	102 Penstocks	600,000	15,000	30,000	150,000	795,000
	103 Powerhouse	1,100,000	150,000	290,000	400,000	1,940,000
	104 Powerhouse tunnel	300,000	20,000	15,000	80,000	415,000
	105 Exit tunnel	200,000	5,000	5,000	50,000	260,000
	106 Downh					
	107 Misc					
		300,000	200,000	50,000	125,000	675,000
		200,000	50,000	10,000	65,000	325,000
		100,000	10,000	10,000	30,000	150,000
	Total direct	18,000,000	3,500,000	2,500,000	7,500,000	31,500,000
	Indirect "pools"					
	Equipment, fuel, and supplies				900,000	
	Drill bits				140,000	
	Compressed air				60,000	
	Utilities				300,000	
	Explosives				100,000	
	Yarding and hauling				200,000	
	Concrete hauling				300,000	
	Batch plant				1,000,000	
	General administrative				4,500,000	
	Total indirects				7,500,000	
	Allocate to bid items				(7,500,000)	
	TOTAL COST	18,000,000	3,500,000	2,500,000	7,500,000	31,500,000

NOTE: In addition to the above columns, there will generally be additional columns for statistics such as bid item unit price, budgeted bid item unit cost, budgeted quantities, actual quantities purchased to date, actual cost per unit to date, and projected quantities and cost to go.

For example, it is usually impractical for a contractor to have his laborers keep track of bid items when hauling concrete from the batch plant to the pour. In this particular case also, the construction superintendent found it easier to control cost as the project progressed by keeping distributable items or pools by natural classification rather than by bid item. On the specific job used here for illustration, it became necessary to allocate the pools and indirects (see Schedule 6-5) only when it became clear that there was a basis for a major claim against the contract owner. In this case, the contractor elected to use a rather complex allocation basis because it was expected that the dispute with the owner would result in litigation. Because of the amount of undistributed pools and indirects, it was felt that the less arbitrary the allocation, the more defensible the financial aspects of the claim would be.

Selling and Bidding Expense

One of the most controversial of home office charges to jobs is that sometimes made for bidding or business development expense. It is

SCHEDULE 6-5 INDIRECT POOL ALLOCATION METHODS

Pool account	*Allocation basis*
Equipment, fuel, and supplies	Equipment hours by bid item
Drill bits	Cubic yards of excavation
Compressed air	Cubic yards of excavation
Utilities	Direct labor
Explosives	Cubic yards of excavation
Yarding and hauling	Ratio of cubic yards of excavation to cubic yards of concrete poured then allocated on basis of cubic yards in each category
Concrete hauling	Cubic yards of concrete poured
Batch plant	Cubic yards of concrete poured
General and administrative	
Procurement and purchasing	Direct materials dollars
Manpower service	Direct labor hours
Superintendence	Direct labor hours*
Accounting	Total direct cost
Estimating	Direct labor hours*
Engineering	Direct labor hours*
Facilities	Labor hours related to manpower etc. serviced
Miscellaneous income and expense:	
Interest income and expense	Total direct cost
Workmen's compensation premium return	Direct labor dollars
Purchase discounts	Direct materials

*Including subcontractor hours of effort.

commonly said that this cost is clearly incurred for the benefit of the job and that therefore the job should bear it. However, the same argument could be advanced on behalf of charging all bidding expenses to the jobs, since if there were no bidding there would be no jobs, and thus the successful bids must carry the cost of the unsuccessful ones.

The fallacy of this theory is demonstrated when a company submits a long series of unsuccessful bids and attempts to burden a single active job with the entire bidding cost. The result is a totally unrealistic figure for both job overhead and general overhead. The unit cost figures become useless and the expense incurred in keeping them is wasted. Actually, underabsorbed general overhead in construction accounting is roughly comparable to idle plant capacity, and from a management viewpoint both call for about the same treatment.

Closely related to the problems of allocating bidding expense is that of allocating such items as promotional expense and institutional advertising (magazine ads, etc.). These, along with expense accounts of top executives, are essentially general overhead costs and should be treated as such. That is, they should not be allocated to jobs and treated as a direct cost.

Home Office Charges

Another controversial issue arises in some companies (particularly those in which job accounts are decentralized) when the home office charges the jobs a flat percentage of their gross billings to cover general office overhead, plus interest at a fixed rate on the funds advanced to finance the project. This system has the merit of passing a reasonable proportion of the general office overhead to the jobs and thus keeping before management the fact that the income from jobs is not net income but must bear its share of general office expense. Skillfully handled, this method of accounting can provide a realistic basis for measuring the effectiveness of departmental management. However, in charging overhead to jobs, the items should be specifically identified and they should be allocated on the basis of some standard cost rate.

In a previous example, it was pointed out that in a decentralized system where the home office prepares the billings for job progress, the accountant may charge to the job the time spent in preparing the monthly billing from the job to the owner. When this procedure is followed, then when the payroll cost distribution is made at the end of the month, the accountant's time will be charged to the job as a direct job cost. In other words, if a service department can recover all or substantially all of its cost by charges to jobs, and still not charge the jobs more than would have to be paid to an outsider for the same work, then the service department is proving its worth by competing successfully with outside business. But if a service department is unable to meet or beat the prices by outside suppliers for the same service, and an arbitrary allocation system is used to charge all of its costs to jobs, the ability to measure the efficiency of that department is lost. When control can be established over service departments, and charge-outs to job costs are on a specific identity basis, then there is a great deal to be said for the practice of charging such costs to jobs.

It should be noted, however, that with a captive market and a paper record to make for himself, an unscrupulous department head can use the interdepartmental charges and charges to jobs to bury his own inefficiencies until the latter become so flagrant that they cause complaints from the jobs. Additionally, top management (unless it happens to be unusually accounting-minded) may confuse these internal charges with direct costs on the job and wonder why profits are so low. This is especially true when overhead is allocated on an arbitrary basis. For example, some small companies will allocate all home office overhead to jobs on the basis of direct labor incurred on the job. This leads to distortions among jobs due to the different mix of labor versus materials or subcontracts on any one job. Schedule 6-6 shows the distortion that can occur when jobs with equal gross profit before allocation of overhead (variable margin) are receiving home office charges on the basis of direct labor.

Of all those outside the construction industry on which construction companies regularly rely, probably the most important, to the construction companies at least, are the bankers and bonding companies. For this reason, it is unfortunate that too few bankers or bonding companies understand the accounting procedures of their contractor clients well enough to recognize transfers of head office overhead for what they are and to evaluate them accordingly. Too often they will view them as "trick" accounting devices designed to confuse and

SCHEDULE 6-6 DISTORTION CAUSED BY ARBITRARY OVERHEAD ALLOCATION METHOD

	Job A	*Job B*
Contract revenue	$2,000,000	$2,000,000
Direct cost:		
Labor	1,400,000	700,000
Materials	200,000	200,000
Subcontracts	50,000	750,000
Equipment	150,000	150,000
	1,800,000	1,800,000
Variable margin	200,000	200,000
Overhead at $0.25 per labor dollar	350,000	175,000
Operating profit (loss)	$(150,000)	$25,000

deceive. When this misunderstanding occurs, the internal charges and treatment of overhead will have done more harm than good. Because of this fact, many accountants nowadays advocate a direct cost system whereby there is no arbitrary allocation of overhead to jobs, and data on individual jobs are presented on a "variable margin" basis (i.e., contract revenue less direct contract cost).

No discussion of indirect and overhead costs would be complete without discussing the cost of temporary structures. Like the cost of mobilization and demobilization, the setting up of shops, batch plants, and similar service activities, the cost of temporary roads, buildings, bridges, etc. must be written off against the job. But equipment move-in and move-out are distributed by adding a factor to the hourly equipment rate. Shops charge off their setup and dismantling by adding a percentage to the labor charges. Batch plants cover their cost of setup and dismantling by adding a percentage to the cost of each cubic yard of material processed.

Temporary structures are usually set up at the beginning of the job and therefore become a cost which must be amortized over the life of the project. Usually, either total direct cost or total man-hours prove to be the most satisfactory basis for allocation. However, if the structure can be related to a particular activity or portion of the work, a suitable base should be selected for amortizing this cost. For example, a structure to house equipment spare parts might be amortized over the life of the job and allocated to bid items on the basis of construction equipment hours. For another example, a structure designed to house electrical parts and components might be amortized over the life of the job and allocated to various bid items on the basis of electrical craft labor hours. For maximum usefulness, a method selected to account for indirect and overhead costs should be consistent and comparable. Different reports for different purposes is the key. If the construction manager desires a cost ledger maintained on a direct cost basis whereby indirect costs are not allocated among bid items, then he should get that kind of report to properly manage the job. On the other hand, the home office may require submission of reports on a "full absorption" basis, whereby all indirect costs are apportioned to the various bid items. The real value of any cost record or system is that it generates reports that can be compared with some standard. Within the construction industry, the accountant should keep in mind that any accounting system or method of recording costs should be flexible so that reports can be generated to meet the diverse needs of home office management, on-site construction management, the owner, and in some cases, governmental and regulatory agencies.

Many construction accountants believe that standard rates can be developed for allocating indirect and overhead costs to bid items and other portions of the work. Whether such standard indirect and overhead rates are safe to use for competitive bidding purposes is debatable. Such rates generally represent an average of cost figures that are both higher and lower than the average. It is often argued that when they are too high they cause the company to lose jobs, and when they are too low "an awful lot of money is left on the table." Nevertheless, all bids are estimates and past experience is often the best possible guide. In any event, even when such standard rates are not used to price bid items directly, the standard indirect and overhead average rates may serve as an overall check on the estimator's current pricing.

Distribution Methods

Overhead is inherently indirect and must therefore be distributed to the bid or work items, or to the production cost accounts, on some more or less arbitrary basis. Unlike some distributable indirect and direct costs (batch plant, machine shop, equipment maintenance, and supplies), the basis is not always obvious. However, there are many diverse costs, such as mess hall and camp facilities, gasoline and supplies for management vehicles, premiums on performance and payment bonds, procurement, and other service group costs, which cannot be charged into direct cost components of the job by any reasonable method or for which there are obvious bases on which they can be allocated to the various bid or work items. If these costs are charged into job cost as they are incurred (and they generally are), job costs become distorted in two ways. First, the job appears to lose money in the early stages and then to recoup the losses later on. In addition to the fact that this causes difficulty in "tracking" actual costs against bid estimates, it also distorts the overall company's financial statements if the percentage-of-completion method of accounting is being used. If the "cost-to-cost method" (cost to date over forecast total costs) is the method utilized to determine percentage of completion, then obviously the job will show a higher percentage of completion in the early stages due to this "front-ending" of costs. Second, when such costs are distributed, they are usually distributed to bid and work items that are active at the time they are incurred, instead of to those that will actually be benefited. While all these costs must be charged to the job in the long run, it is not too hard to see that if such large items as surety bond premiums and temporary structures were charged directly to cost instead of being written off over the life of the job, the cost accounts in the early stages of the job would be seriously distorted, and the progress figures for bid or work items in the early stages of the job would be virtually meaningless except, possibly, as a means of estimating the job's cash requirements.

One way of distributing such front-end costs is to defer them in an account not unlike a work-in-process type inventory account and take them into job costs on the basis of percentage of completion. That percentage is usually determined each month by the job engineer when he prepares his progress estimates. For example, if the engineer estimates, at the end of the month, that the job is 47 percent complete, the amount of deferred indirect and overhead costs written off that month to job cost would be such as to make the total writeoff to date equal to 47 percent of the projected total indirect and overhead costs for the life of the job. Since percentages of completion will usually be computed on individual bid or work items, the bid or work items in progress would be burdened only with the share of total indirect or overhead costs that the related bid items bear (as a percentage) to the projected total cost of all bid items. Thus each bid or work item is charged with enough of the overhead to maintain a constant ratio of indirect and overhead cost to direct cost.

This method has the disadvantage of requiring constant adjustments and more complex accounting procedures. Often indirect costs are shifted between work items when there are no changes in the direct costs charged to them. Still, for the most part, the results are reasonably sound. Occasionally this method is criticized on the basis of mismatching revenue and cost. This criticism is often justified on the basis that the overhead incurred in any period is there to support the work performed in that period and therefore should not be deferred to a different period when different work is being supported by the overhead.

Total Direct Cost Allocation Method

Many construction companies allocate indirect and overhead costs on the basis of the total direct cost estimated for the job and adjusted for change orders.

There are some arguments against the theoretical soundness of using total direct cost as a measure of indirect and overhead cost distribution. It is claimed, for example, that subcontract costs should not be burdened with the same percentage of overhead expenditure as work performed by the contractor's own forces. It is also argued that by including direct materials in the base for indirect cost distributions, bid and work item costs can be distorted by the inclusion of large expensive units of material such as steam turbines, generators, or chemical processing installations. On the other hand, it could hardly be argued that subcontracts or costly direct materials should bear no indirect cost at all, although that would be the result if only direct labor cost or direct labor hours were used.

Theoretically, the correct answer to this question of indirect construction costs lies (as for manufacturing costs) in allocating various types of indirect costs on different bases and in selecting those bases according to the needs of the particular company schedule. For most construction companies, however, there are two serious practical objections to this answer. First, relatively few construction companies have developed the use of construction cost figures to such a point that the complexity and cost of using multiple bases for allocating indirect cost items could be justified. Second, the wider variations in individual jobs encountered by construction companies pose allocation problems rarely present in a manufacturing operation. For example, assume the company had decided what items of overhead were to be allocated to subcontract costs and then assume the company was awarded these four jobs:

1. A harbor development job for which the bulk of the cost is covered by a single subcontract for dredging
2. A large office building with some large and many small subcontracts
3. A steam-powered electric plant for which a large portion of the cost is distributed among two or three equipment installation contracts on which most of the cost is for equipment
4. A pipeline in which only a few small subcontracts are involved and the work will be performed primarily by the contractor's own work force

Clearly, any attempt to use the same indirect cost allocation basis on each of the four jobs would result in a serious distortion. Therefore, to be consistent, it would be necessary to vary the formula for each type of job. To do so is not impossible, but it is unlikely that the value of this procedure would justify its complexity for the small contractor because the next job might well have its own peculiarities that make all the other formulas inapplicable.

For this reason, a large part of the construction industry seems to prefer total direct cost as a standardized basis for indirect cost distribution. It is simple and easy to use and, for most jobs, it is accurate enough to accomplish all that is required of it.

As previously discussed, there is another approach to job costing widely accepted among construction accountants, certified public accountants, banks, and surety companies. This method requires no distribution of overhead at all and is commonly referred to as a "direct cost" system. The accountants who prefer this approach argue that, from the company's viewpoint, all costs which can be identified with the job are direct costs and should be cost estimated and accounted for as such. If there is no direct relationship between an item of cost and a particular job or an item of cost within a job and a particular bid or work item, then it is misleading to allocate such a cost item. Furthermore, proponents of the direct cost system point out that, if indirect job costs are distributed to bid or work items, their identity tends to be lost and the individual cost components forgotten. By the same reasoning, if general office overhead is distributed on bid or work items, it too becomes difficult to control as a cost center. Estimators claim it is better merely to allow a percentage of total direct cost to cover general office overhead. This procedure is certainly practical and possibly it is the one most likely to produce usable figures for the small contractor. In setting such a percentage, these estimators are in effect using "break-even analysis." They are saying that "in addition to recovering our direct cost, we need a certain percentage to provide for overhead recovery and a profit or we shouldn't take the job." The danger in this thinking is that when a contractor's volume is low, he probably should be bidding on any job where he will break even or better. First, every dollar in excess of his direct cost helps him to recover his overhead, which is a fixed cost. Second, when volume is low, he must keep his work force together in anticipation of higher volume. By taking a job at, or sometimes even below, his break-even point, he will maintain his work force as well as provide positive cash flow through unbalanced bid and front-end pricing procedures. The practice has its hazards but is so widely followed in the industry that its presence cannot be ignored.

In summary, when indirect costs are to be distributed to jobs or to bid or work items within a job, it is probable that the most satisfactory method is the use of the developed standard rates. Depending upon the way in which the company compiles its bids, the rates may be designed to apply to total direct cost, direct labor cost, direct labor hours, or to some other base to which the indirect cost component most closely relates. What base is selected is not as important as the consistency with which it is used. Standard rates may contain only items of on-site, indirect job cost, or they may contain an allocation of home office overhead as well. If the rates are consistently applied, they will generally reveal deviations from the standard, make it easier to price change orders and claims, and additionally will provide a basis for indirect cost and overhead cost control.

If the jobs last long enough, and the standard rates are used to control indirect cost, it is usually necessary to distribute the indirect costs from month to month in order to obtain work item costs comparable with those in the estimate. If the jobs are relatively short and indirect cost distribution is not currently used, it is common to see job indirect and overhead cost, and sometimes an allocation of home office overhead, applied in a lump sum figure in the cost ledger at the end of the job. When applied in this way, indirect cost distributions serve no purpose but to provide information for future bidding. If not used for this purpose, the figures have no function at all and no distributions should be made.

Control of Office Overhead

The control of overhead, like the control of any other cost, must inevitably be based on the contribution of each item of cost to the final profits. There is no other standard that can be justified. This test is easier to state than to apply. Where should the line be drawn between the extremes of short-sighted penny pinching and open-handed waste?

One guide is the potential for mechanization. Whenever mechanical equipment can save hours, it is usually good economy to buy the equipment. Another is employee turnover. It costs money to change help, and when turnover is high it means that working conditions, salaries, or the methods of personnel selection may be deficient. When high-priced people are spending their time on low-priced routine jobs, it means that there are "too many chiefs and not enough people to do the work."

One commonly used and completely ineffective method of cutting down overhead in many offices is to fire all the clerical help and leave the "key" personnel. Of course, the bulk of the overhead continues, less work gets done, and in a short time the clerical help has to be replaced. In the end this method costs more than it saves.

Another favorite (and expensive) technique for cutting down general office overhead when job offices are decentralized is either to charge out all the general office costs to the jobs or to require the job offices to do some of the work formerly done in the general office. Of course the result of this policy is that job overhead is raised or the work does not get done. Sometimes both results follow. This technique may make the office manager in the general office look good for a while but it rarely saves any money.

It is not possible to state any rule for the control of overhead that will fit all companies. Still, the following principles should be generally useful:

1. The more overhead functions that can be brought together in one place, the easier it is to break them down into routine functions

that can either be mechanized or performed by lower-level and lower-cost employees.

2. There needs to be someone who looks hard at every expenditure to see if it is justified in terms of its contribution to the final profit. This scrutiny includes a constant analysis of the jobs people are doing, the reports they are preparing, and the paperwork going over every desk in every office.
3. Telephones are a mixed blessing. The more valuable an employee's time is, the more he needs to be isolated from unnecessary telephone calls.
4. Mistakes cost money. The time spent in job analysis to make each person's work check that of someone else is well spent. So also is time spent in staff training—that is, teaching people how to do their jobs accurately without loss of speed. Most of all, interruptions result in more mistakes than any other cause; reduce interruptions and you reduce overhead.
5. Nobody knows a job as well as the person who is doing it. Recognition for time- and money-saving suggestions yields good results, but be sure the straw boss gets a pat on the back for the suggestions of people in his department. Otherwise, suggestion campaigns will die on the vine.
6. Many companies develop internal "yardsticks" for measuring indirect costs relative to construction volume. These yardsticks generally take the form of indirect cost dollars (accounting, legal, procurement, etc.) per hour of construction effort. This can be a good tool for top management to review and control indirect cost from one year to the next.

Controlling Jobsite Overhead

On very small jobs, jobsite overhead is negligible. As the jobs get larger, the ratio of indirect cost to total cost increases, and ultimately jobsite overhead becomes a problem. In general, the same methods suggested for control of general office overhead can also be used to hold job overhead to a minimum. However, the emphasis on preventing interruptions in the work of the project manager or superintendent, the job engineer, and the job accountant must be greater than ever. There will always be a host of details demanding attention, but the more they can be diverted from top personnel, the more time those individuals will have to concentrate on getting the job done quickly and with a minimum of delays and errors.

One way to prevent excessive interruptions of key personnel is to have adequate transportation on the job. It is poor economy for key jobsite personnel to have to leave the job to pick up spare parts or run errands in town because they are the only people on the job with a company vehicle.

Another equally prolific source of interruption is the steady stream of salesmen and miscellaneous visitors who ask to use the telephone in the job office. It is sometimes good economy to limit telephones within the job office and to have pay telephones outside. Still another source of interruption is the general office employee who regularly calls the jobs for trivial reasons.

There is no money to be saved by equipping job offices poorly. If anything, job offices should have better equipment than the general office, because the people who use it usually draw higher salaries than their clerical counterparts in the general office.

Accounting figures can be useful in the control of overhead if they are current enough to be significant and can be compared to some standard. However, the figures themselves do not control overhead. Control is a management function, and those who exercise it must be able to take the time to scrutinize the figures and to account for significant changes. Unless these steps are followed, the cost of compiling overhead data, either on the job or elsewhere, may be wasted.

Joint Venture Overhead

It has been pointed out elsewhere that the most efficient way to operate a joint venture is to have one of the members act as sponsor. When this method is followed the sponsor inevitably incurs overhead expenses in operating the joint venture. There are several ways of compensating for these expenses. One way is to allow, as a joint venture cost, a fixed allowance to cover overhead—sometimes stated as a flat monthly amount and sometimes as a percentage of contract volume. It is unusual for either of these methods to be entirely satisfactory. If charges are to be made, it is better that they be made on the basis of time reports by employees and specific charges that can be audited if need be.

There are two common alternatives to the use of specific charges. First, the sponsor may be allowed a somewhat larger percentage of interest. Second, if the members are more or less regularly engaged in joint ventures together, it is common to see sponsorship rotated among the members so that no one company assumes a disproportionate share of the overhead cost.

Many of the difficulties occasioned by joint venture overhead arise because of failure to preplan the venture carefully enough. However, some problems always arise, and it is usually a wise plan to designate one individual in each organization to be the liaison between that company and the other members of the venture. This group is not in any sense a management committee since it does not interfere with the sponsor's management. It does serve as an advisory group to which the sponsor can refer special problems. If such a group exists it is usually the first to hear any complaints regarding joint venture overhead.

Such complaints usually come from one of three sources. Either the job is complaining because the sponsor's charges are too high; the sponsor is complaining because the costs exceed the reimbursement; or one of the nonsponsoring members is questioning the amount of the charges.

When the sponsor is billing for actual costs, the questions concern specific items. When a percentage of contract volume is used as a measure, or when a flat amount is charged, it is up to the sponsor to support with specific costs any claims that the allowance is inadequate. If the sponsorship is rotated or if the sponsor is allowed a larger percentage of participation, there would seem to be little ground for discussion of regular overhead items. The only remaining questions would involve the cost of things furnished or services performed outside the scope of ordinary overhead and for which special reimbursement is being requested. For such situations an informal liaison committee can be extremely useful.

Special Overhead Problems

Special overhead problems are occasioned by such items as move-in, move-out, cleanup, temporary structures, camp and mess hall losses, job completion overhead, overhead costs in connection with the guaranty provisions of the contract, standby charges during shutdowns due to strikes or weather or other causes outside the contractor's control, and numerous other more or less special circumstances.

Some of these items have already been mentioned and a method of apportionment has been suggested. The others must be recognized and the job must be burdened with their costs. However, the wisdom of incorporating special or nonrecurring items in regular overhead distributions is questionable, to say the least. Generally, it is better to allow them to stand as special deductions in a separate section of the job operating statement, so that their true nature can be seen at a glance.

Chapter

7

Accounting for Contract Revenue and Cash Receipts*

A unique aspect of construction accounting centers on the problem of when profits are earned. For some construction contractors the problem is not significant since they have numerous short-term contracts and can account for revenue, cost, and income in much the same way as any other business. When the work is done or the goods are delivered, the customer is billed, the sale is recorded, and the revenue taken into account.

What makes the construction contractor's problem different from the accounting for revenues in other businesses is the fact that most construction contractors are engaged in contracts which stretch over more than one accounting period. There is no clear-cut definition of what constitutes a long-term contract, but by custom they are generally thought to exceed one year in length. However, the definition is somewhat relative since a small paving contractor whose jobs average three or four weeks in duration would consider a three- or four-month contract to be long-term. Conversely, a large design-build contractor whose contracts may average six to seven years in length would consider a one- or two-year contract to be short-term. Whatever the definition, the length of the job, timing of the progress billings, and cer-

*This chapter deals primarily with how a contractor should account for contract revenue (and therefore gross profit) and cash receipts for financial reporting purposes. Chapter 13 deals with the subject of accounting methods suitable for tax reporting purposes. The extent to which financial accounting methods should be governed by tax accounting requirements must be determined by each company on the basis of its own needs. Also, the importance of speed and accuracy in accounting for revenue and cash transactions has become so great in recent years that larger firms (and numerous small and medium-sized firms) are looking to computers (see Chapter 12) for help in solving their recording, billing, and reporting problems.

tain features of the contract, such as target penalties and incentive awards, change orders, etc., all tend to make what is simple in manufacturing industries a difficult task in construction.

The AICPA recognized this problem with the issuance of Accounting Principles Board Statement 4 which says:

> Revenue is generally recognized when both of the following conditions are met: (1) The earnings process is complete or virtually complete and (2) an exchange has taken place.
>
> Revenue is sometimes recognized on bases other than the realization rule. For example, on long-term construction contracts revenue may be recognized as construction progresses. This exception to the realization principle is based on the availability of evidence of the ultimate proceeds and the consensus that a better measure of periodic income results.
>
> The exception to the usual revenue realization rule for long-term construction-type contracts, for example, is justified in part because strict adherence to realization at the time of sale would produce results that are considered to be unreasonable. The judgment of the profession is that revenue should be recognized in this situation as construction progresses.

In manufacturing or marketing industries, the accounting problem created by the purchase or production in one year followed by a sale in the following year is solved by the use of inventories. In the construction industry, this is not technically possible because inventories are by definition property held for sale. The work in process of a contractor consists almost entirely of improvements made to someone else's land. When the improvement is completed, it becomes the property of the landowner. Moreover, there is substantial support for the proposition that profits on construction work are not really earned until the work of improvement is completed, accepted, and the retention has been released. So instead of a true inventory, the costs incurred by a contractor prior to the completion and acceptance of a job are more in the nature of an account receivable secured by a lien right or a public works bond.

On the other hand, many accountants feel that profits from long-term contracts should not be deferred until the end of the job, but rather that they should be recognized pro rata as the job progresses. This concept seems to denounce the long-accepted accounting principle that profits should not be anticipated. The answer to this, of course, is that profits are not being anticipated but are being recognized as the work progresses and the costs are incurred. The difference is that the profits to date are being recognized on the basis of how the job is doing in relation to its original estimate or revised forecast. The thinking here is that if the contractor is on budget he should recognize the pro rata share of profits that have been earned in relation to work performed. Should unanticipated events occur which knock the contractor off of his original budget, then the adjustment for these events, which were unknown in the previous accounting period, would be reflected in that period. This holds true for upward as well as downward revisions of profit forecasts.

Another unique aspect of the construction contractor's revenue accounting problem is the timing and amounts of progress billings in relation to work performed and actual costs incurred. Typically, contractors unbalance their payment schedule or their bid so that billings are well ahead of cost in the early stages of the job to help provide working capital to finance the job. Progress billings in the later phases of the job may be barely covering actual cost incurred or, in fact, may be less than the cost incurred depending on the ability of the contractor to sell his unbalanced bid or payment schedule. Figure 7-1 depicts this problem graphically.

It is obvious in Figure 7-1 that, if an income statement were prepared at the end of the year 1, and if billings were the basis for determining revenue, profits would be significantly overstated. Conversely, income would be significantly understated in year three.

Acceptable Accounting Policies

As a result of these timing problems, the construction industry and the accounting profession have developed two acceptable methods of accounting for revenue from long-term contracts for financial reporting purposes. These methods were formalized by the AICPA's issuance in 1955 of *Accounting Research Bulletin No. 45, Long-Term*

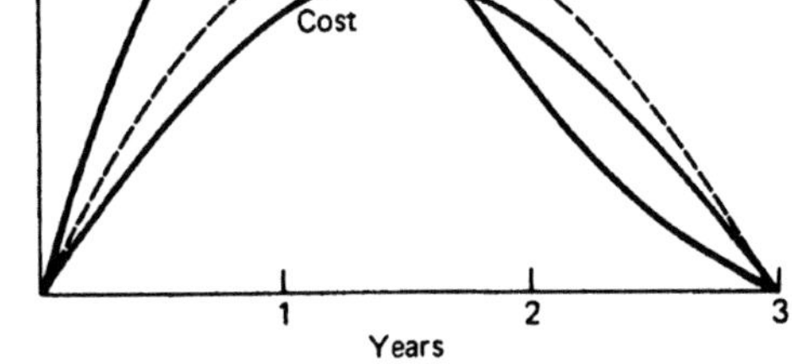

Figure 7-1 **Unbalancing the Bid and Payment Schedule.**

Construction-Type Contracts. This bulletin describes the two generally accepted methods of accounting for long-term construction-type contracts for financial reporting purposes as follows:

> *The Percentage of Completion Method* Recognizes income as work on a contract progresses; recognition of revenues and profit generally is related to costs incurred in providing the services required under the contract.
>
> *The Completed Contract Method* Recognizes income only when the contract is completed, or substantially so, and all costs and related revenues are reported as deferred items in the balance sheet until that time.

The Tax Equity and Fiscal Responsibility Act of 1982 and the Tax Reform Act of 1986, and the regulations proposed to be issued in connection with them, have made important changes in the completed-contract method.

Audits of Government Contractors, published by the AICPA, describes an alternative method which is referred to as the units of delivery and is a modification of the percentage-of-completion method of accounting for contracts.

> *The Units of Delivery Method* Recognizes as revenue the contract price of units of a basic production product delivered during a period and as the cost of earned revenue the cost allocable to the delivered units; costs allocated to undelivered units are reported in the balance sheet as inventory or work in progress. The method is used in circumstances in which an entity produces units of a basic product under construction-type contracts in a continuous or sequential production process to buyer's specification.

While the above quotation is specifically related to production-type contracts, it has some value in this discussion since it may be suitable for land developers who are developing individual pieces of property and will receive their revenue for these pieces as the improvements are completed and title is transferred. In general, the units-of-delivery method is not applicable to construction contractors.

The percentage-of-completion method of accounting for contracts is based on the theory that profits are, in fact, earned as the job progresses; it takes into account as job revenues a portion of the total contract price proportionate to the stage of completion of the job. In contrast, the completed-contract method of accounting for contracts is based on the theory that no profit has been earned until the job is completed.

Accounting Research Bulletin 45 gives guidance in determining which method is most suitable in given circumstances:

> The Committee believes that in general when estimates of cost to complete and extent of progress towards completion of long-term contracts

> are reasonably dependable, the percentage of completion method is preferable. When lack of dependable estimates or inherent hazards cause forecasts to be doubtful the completed-contract method is *preferable.*

In addition, then, to the theory that no profit has been earned until the job is completed, another reason for favoring the completed-contract method is "lack of dependable estimates or inherent hazards."

Inherent Hazards and Reasonable Estimates

When a company is determining what overall accounting policy is best suited for its business in terms of the two accounting methods set forth above (percentage-of-completion versus completed-contract), much of the thinking that goes into the decision concerns the meaning of the term "inherent hazards" and also the question of what constitutes "reasonable estimates." It is hard to fault the conservative thinking that until the job is complete, anything can happen.

In attempting to deal with the subject of inherent hazards the AICPA says the following in its Statement of Position (SOP) 81-1:

> The present business environment and the refinement of the estimating process have produced conditions under which most business entities engaged in contracting can deal adequately with the normal, recurring business risks in estimating the outcome of contracts. The division believes that inherent hazards that make otherwise reasonably dependable contract estimates doubtful involve events and conditions that would not be considered in the ordinary preparation of contract estimates and would not be expected to recur frequently, given the contractor's normal business environment. Such hazards are unrelated to, or only incidentally related to, the contractor's typical activities....
>
> Therefore, the division believes that there should be specific, persuasive evidence of such hazards to indicate that use of the percentage of completion method...is not preferable.

From a practical point of view, a contractor's ability to give reasonably dependable estimates are evidenced by the fact that the contractor is still in business. For most contractors dealing in the competitive bid segment of the construction industry, profitability is quite often related to the contractor's estimating ability in bidding on the job in the first place. For general contractors who subcontract most of their work, this is especially true. The AICPA publication SOP 81-1 discusses reasonably dependable estimates as follows:

> In practice, contract revenues and costs are estimated in a wide variety of ways ranging from rudimentary procedures to complex methods and systems. Regardless of the techniques used, a contractor's estimating procedures should provide reasonable assurance of a continuing ability

> to produce reasonably dependable estimates. Ability to estimate covers more than the estimating and documentation of contract revenues and costs; it covers a contractor's entire contract administration and management control system. The ability to produce reasonably dependable estimates depends on all the procedures and personnel that provide financial or production information on the status of contracts. It encompasses systems and personnel not only of the accounting department but of all areas of the company that participate in production control, cost control, administration control, or accountability for contracts. Previous reliability of a contractor's estimating process is usually an indication of continuing reliability, particularly if the present circumstances are similar to those that prevailed in the past. Estimating is an integral part of contractors' business activities, and there is a necessity to revise estimates on contracts continually as the work progresses. The fact that circumstances may necessitate frequent revision of estimates does not indicate that the estimates are unreliable for the purpose for which they are used. Although results may differ widely from original estimates because of the nature of the business, the contractor, in the conduct of his business, may still find the estimates reasonably dependable. Despite these widely recognized conditions, a contractor's estimates of total contract revenue and total contract costs should be regarded as reasonably dependable if the minimum total revenue and the maximum total cost can be estimated with a sufficient degree of confidence to justify the contractor's bids on contracts.

In summary, then, it would seem to be a rare situation in which a company's overall estimating ability cannot be reasonably depended upon or in which all contracts face inherent hazards to such a degree that the percentage-of-completion method would not be preferable as an overall policy.

Deviating from Overall Policy

Occasionally, a contractor will engage in a contract in which his estimates may not be dependable or inherent hazards may exist. When this is the case, the contractor may deviate from his overall policy of percentage-of-completion accounting to that of completed-contract accounting for the one contract. Obviously, if the contract is significant to the overall operating results of the company, then appending a footnote explanation on the financial statements would be in order.

The Break-Even Method as an Alternative

Many contractors, particularly those engaged in projects including heavy earth-moving procedures, use a variation of the percentage-of-completion method when the reliability of their estimates is questionable or when inherent hazards may exist. This variation is commonly referred to as the "break-even" or "zero profit" method. Under this variation, a zero profit outcome is assumed until the stage of completion when estimating can be considered to be reliable. At this point, results are revised and normal percentage-of-completion accounting ensues. This variation of percentage-of-completion as opposed to the completed-contract method of accounting has the advantage of running revenue and costs through the income statement rather than deferring them on the balance sheet and provides the reader of the financial statements with a better indication of the volume of a contractor's operations. Obviously, as with deviation to the completed-contract method, just discussed, a footnote explanation of the situation is warranted if the contract is significant. A typical footnote for a deviation to the completed-contract or the zero profit method from the normal percentage-of-completion accounting might read as follows:

> The company has entered into a fast-track contract, the terms of which have not been finalized. Final outcome of the contract, therefore, cannot be estimated at this time except to ensure that no loss will be incurred. Because of the inability to reasonably estimate the outcome of this contract, the company is using the completed-contract method of accounting for this contract. Revenue and cost are being deferred and included in unbilled work in progress (*or* a zero profit is being used) until such time as the company can adequately forecast the outcome on the contract. The effect of this deviation from normal company policy of percentage-of-completion accounting on financial position or results of operations for the current year cannot be estimated at this time. Included in revenue and cost on the income statement are $2,300,000 of revenue and cost relating to this contract.

Determining the Profit Center

Within the construction industry, the profit center, for purposes of applying the percentage-of-completion or completed-contract methods of accounting, is usually thought to be the contract itself. However, there are cases where, for purposes of determining revenue and profits to be recognized in an accounting period, the job or contract is not the profit center.

Combining contracts. There are cases where several contracts may be combined and treated as one profit center. For example, a contractor, knowing that once he has obtained a job there is a very good chance of obtaining additional contracts or of obtaining significant change orders, may bid a job at a lower profit margin. After the contractor has mobilized to the site and has many of his costs behind him, it becomes difficult for other contractors to compete on a competitive-bid basis since they will have to include such costs in their bid. In these situations many accountants would combine the first contract and subsequent contracts for purposes of profit recognition. The theory is that a bid with a profit considerably less than the contractor's normal rate of profit or, in fact, even a small planned loss, should be combined with the higher-than-normal profit margins which might be obtained with the resulting contracts or significant change orders. The AICPA discusses this subject in its SOP 81-1 as follows:

> A group of contracts may be so closely related that they are, in effect, parts of a single project with an overall profit margin, and accounting for the contracts individually may not be feasible or appropriate.
>
> For example, a group of construction-type contracts may be negotiated as a package with the objective of achieving an overall profit margin, although the profit margins on the individual contracts may vary. In those circumstances if the individual contracts are performed and reported in different periods and accounted for separately, the reported profit margins in those periods will differ from the profit margin contemplated in the negotiations for reasons other than differences in performance.

Obviously, the decisions concerning which contracts are to be combined for purposes of determining a profit center or whether or not to remain with the single job or contract as the profit center are not casual choices, but must meet certain criteria. The SOP 81-1 sets forth the criteria which must be met in order to combine contracts for accounting purposes:

> A group of contracts may be combined for accounting purposes if the contracts
>
> (a) Are negotiated as a package in the same economic environment with an overall profit margin objective. Contracts not executed at the same time may be considered to have been negotiated as a package in the same economic environment only if the time period between the commitments of the parties to the individual contracts is reasonably short. The longer the period between the commitments of the parties to the contracts, the more likely it is that the economic circumstances affecting the negotiations have changed.
>
> (b) Constitute in essence an agreement to do a single project. A project for this purpose consists of construction, or related service activity with different elements, phases, or units of output that are closely interrelated or interdependent in terms of their design, technology, and function or their ultimate purpose or use.
>
> (c) Require closely interrelated construction activities with substantial common costs that cannot be separately identified with, or reasonably allocated to, the elements, phases, or units of output.
>
> (d) Are performed concurrently or in a continuous sequence under the same project management at the same location or at different locations in the same general vicinity.

(e) Constitute in substance an agreement with a single customer. In assessing whether the contracts meet this criterion, the facts and circumstances relating to the other criteria should be considered. In some circumstances different divisions of the same entity would not constitute a single customer if, for example, the negotiations are conducted independently with the different divisions. On the other hand, two or more parties may constitute in substance a single customer if, for example, the negotiations are conducted jointly with the parties to do what in essence is a single project.

Segmenting contracts. Many contractors provide several different types of services or specialize in the construction of unique projects. In many cases these different services or projects carry different profit margins depending upon the degree of risk, the uniqueness of the service, or a monopoly of a particular type of specialty. For example, large design-build firms may sell to owners such services as engineering, procurement, financing, construction, or construction management. Such firms sell these services both separately and combined under one contract with one owner. Typically, engineering might carry a profit margin of 15 percent, procurement and financing services a five percent profit margin, and construction management and/or construction a four to five percent profit margin. Because these services are, in effect, different lines of business carrying different profits, many accountants and contractors believe that a contract that carries these combined services should be segmented as though there were a separate contract for each service. Another example might be a contractor who specializes in the construction of condominiums, swimming pools, and tennis courts. He might sell these jobs under separate contracts to various owners or he may sell all three products to one owner under one contract. The margin on the condominium jobs will tend to be much smaller than on swimming pools and tennis courts since there is a higher degree of competition in the building construction segment of the marketplace. In this case, it might be appropriate to segment one contract that includes all three products so that they are accounted for as separate components with their varying profit margins.

The AICPA in its SOP 81-1 sets forth certain steps which must be followed and criteria which must be met in order to segment the contract:

A project may be segmented if all the following steps were taken and are documented and verifiable:

(a) The contractor submitted bona fide proposals on the separate components of the project and on the entire project.

(b) The customer had the right to accept the proposals on either basis.

(c) The aggregate amount of the proposals on the separate components approximated the amount of the proposal on the entire project.

Paragraph (c) above recognizes the fact that there will be some economies of scale if a contractor negotiates successfully several phases of a contract with an owner. Since he will have common management on all the phases, more efficient purchasing, etc., he may pass on this savings to the owner through an overall discount on the aggregate profit margin in the contract. SOP 81-1 further states:

> A project that does not meet the criteria (quoted above) may be segmented only if it meets all the following criteria:
>
> (a) The terms and scope of the contract or project clearly call for separable phases or elements.
>
> (b) The separable phases or elements of the project are often bid or negotiated separately.
>
> (c) The market assigns different gross profit rates to the segments because of factors such as different levels of risk or differences in the relationship of the supply and demand for the services provided in different segments.
>
> (d) The contractor has a significant history of providing similar services to other customers under separate contracts for each significant segment to which a profit margin higher than the overall profit margin on the project is ascribed.
>
> (e) The significant history with customers who have contracted for services separately is one that is relatively stable in terms of pricing policy rather than one unduly weighted by erratic pricing decisions (responding, for example, to extraordinary economic circumstances or to unique customer-contractor relationships).
>
> (f) The excess of the sum of the prices of the separate elements over the price of the total project is clearly attributable to cost savings incident to combined performance of the contract obligations (for example, cost savings in supervision, overhead, or equipment mobilization). Unless this condition is met, segmenting a contract with a price substantially less than the sum of the prices of the separate phases or elements would be inappropriate even if the other conditions are met. Acceptable price variations should be allocated to the separate phases or elements in proportion to the prices ascribed to each. In all other situations a substantial difference in price (whether more or less) between the separate elements and the price of the total project is evidence that the contractor has accepted different profit margins. Accordingly, segmenting is not appropriate, and the contracts should be the profit centers.
>
> (g) The similarity of services and prices in the contract segments and services and the prices of such services to other customers contracted separately should be documented and verifiable.

Determining Percentage of Completion

The AICPA's *Accounting Research Bulletin 45* recommends that income to be recognized in a contractor's accounts be either:

> (1) That percentage of estimated total income that costs bear to estimated total costs, after giving effect to estimates of cost to complete based upon most recent information, or (2) that percentage of estimated total income that may be indicated by such other measure of progress toward completion as may be appropriate giving due regard to work performed.

The Cost-to-Cost Method

The cost-to-cost method is the most prevalent method of arriving at a percentage of completion for the purpose of recognizing profit on contracts in progress. It is fair to state that it is generally used by most medium-sized and small contractors since amounts are easy to determine from the contractor's normal accounting records, and the estimated total costs may not provide the necessary detail for other methods which will be discussed later in this chapter. In applying the cost-to-cost method, accountants recognize that, since work performed is a primary basis for income allocation, certain costs may—or should—be disregarded as a measure of performance in the early stages of a contract for the purposes of determining income allocation. *Accounting Research Bulletin 45* discusses this cost problem in general terms:

> Costs as here used might exclude, especially during the early stages of a contract, all or a portion of the cost of such items as materials and subcontracts if it appears that such an exclusion would result in a more meaningful periodic allocation of income.

In SOP 81-1, the AICPA further expands discussion of costs which should be excluded from the formula:

> Some of the costs incurred particularly in the early stages of the contract should be disregarded in applying this method because they do not relate to contract performance. These include the costs of items such as uninstalled materials not specifically produced or fabricated for the project or of subcontracts that have not been performed. For example, for construction projects, the cost of materials not unique to the project that have been purchased or accumulated at jobsites but that have not been physically installed do not relate to performance. The costs of such materials should be excluded from cost incurred for the purpose of measuring the extent of progress toward completion. Also the cost of equipment purchased for use in a contract should be allocated over the period of its expected use unless title to the equipment is transferred to the customer by terms of the contract.

Whether or not to include costs which have been incurred but have not directly benefited the project is a source of major differences between contractors and accountants. This is particularly true where the contractor's work is in remote areas of the world and the mobilization effort with regard to the job is almost of equal importance to performance of the work at the site. If material costs which have been incurred through purchase and delivery to the job are not included in the formula for the cost-to-cost method then the contractor will receive no profits to date for the major purchasing and expediting effort involved in getting the materials to the project site. In any event, costs to date must be carefully studied to determine whether they should be included in the cost formula for determining percentage of completion under the cost-to-cost method.

The formula for applying the cost-to-cost method for determining the percentage of completion is shown in the following equation:

$$\frac{\text{Cost to date}}{\text{Projected total cost}} \times \text{Projected total gross profit} + \text{Cost to date} = \text{Earned revenue}$$

or, alternatively, the formula may be:

$$\frac{\text{Cost to date}}{\text{Projected total cost}} \times \text{Contract price} = \text{Earned revenue}$$

In both equations, the cost to date is adjusted for costs incurred which have not benefited the progress of the job.

Efforts-Expended Method

Many contractors use some form of labor base for determining percentage of completion. These contractors believe that their profits are earned from their efforts expended rather than from the intrinsic value of materials, subcontracts, and other costs incurred on the project. This is particularly true in the case of some general contractors whose profit margins result more from their ability to manage subcontractors than from the value of the subcontracts themselves.

The formula most commonly used is direct labor hours of effort to date versus forecast total labor hours. In this context the term "direct labor" may include the time of home office employees whose efforts can be directly identified with the job (see Chapter 6). One advantage that this method has over the "cost-to-cost" method is that it does attach profit to the mobilization effort, since, for example, hours of effort expended by the purchasing department to purchase the materials and hours of effort expended by the personnel department and management to employ the labor are included in the numerator and denominator of the efforts-expended formula.

Other variations of the labor-based method of determining percentage of completion include total versus estimated (or "forecast") man-days, -weeks, or -months of effort. Some companies use labor dollars to date versus forecast total labor dollars. However, this latter method can cause some minor distortions when significant overtime and related overtime premiums are paid to hourly workers.

The formula for applying any of the labor-based efforts-expended methods for determining percentage of completion is shown in the following equation:

$$\frac{\text{Labor to date (hours, days, etc.)}}{\text{Project to labor (hours, days, etc.)}} \times \text{Projected total gross profit} + \text{Cost to date} = \text{Earned revenue}$$

The variation formula which was presented for the cost-to-cost method will not work when an efforts-expended (labor-based) method is used because it can result in significant distortions. In the efforts-expended method we are focusing more on the allocation of gross profit among periods based on labor rather than cost. Therefore, if the product of labor to date versus the forecast total labor is multiplied times total contract price, distorted results will occur when materials are purchased toward the end of the job and labor cost is incurred primarily in the early stages of the job. Note also that, under this method, a uniform profit percentage will not be achieved over the life of the job even if all costs are exactly as estimated when the job began. This again is due to the fact that profits will be a uniform percentage of labor over the life of the job but the timing of costs into the work stream will cause a variation in gross profit percentage.

Physical Observation Method

Perhaps the best method of computing percentage of completion is the physical observation method. For example, if the job is an excavation contract, then cubic yards of excavation to date versus forecast cubic yards of excavation would be a good measure of progress. Another example would be a highway job where square yards of concrete or asphalt laid versus projected total square yards would be the formula. However, on a complex job such as a nuclear electric power plant or a job where there are significant amounts of engineering and where components of the project are hard to measure, the physical observation method is difficult to employ.

It is important for the company accountant and the certified public accountant to independently verify the percentage of completion periodically. The physical observation method should be employed periodically as a check on the cost-to-cost method or labor-based method of determining physical percentage of completion. This is similar to what a typical accountant does in a regular manufacturing operation, where at month end he will use perpetual inventory records to record inventory for financial statement purposes. However, even where accounting records are used for statement purposes, a physical inventory is taken generally on an annual basis and adjustments are made where necessary to bring the perpetual inventory records into line with what is actually on hand.

Computing Earned Revenue

Once a method for determining the percentage of completion has been chosen, the next step in determining earned revenue is the application of the percentage. The AICPA's SOP 81-1 presents two alternative approaches to the mechanics of computing earned revenue:

Under Alternative A:

(a) *Earned Revenue* to date should be computed by multiplying total estimated contract revenue by the percentage of completion (as determined by one of the acceptable methods of measuring the extent of progress toward completion). The excess of the amount over the earned revenue reported in prior periods is the earned revenue that should be recognized in the income statement for the current period.

(b) *Cost of Earned Revenue* for the period should be computed in a similar manner. Cost of earned revenue to date should be computed by multiplying total estimated contract cost by the percentage of completion on the contract. The excess of that amount over the cost of earned revenue reported in prior periods is the cost of earned revenue that should be recognized in the income statement for the current period. The difference between total cost incurred to date and cost of earned revenue to date should be required on the balance sheet.

(c) *Gross Profit* on a contract for a period is the excess of earned revenue over the cost of earned revenue.

Under Alternative B:

(a) *Earned Revenue* is the amount of gross profit earned on a contract for a period plus the costs incurred on the contract during the period.

(b) *Cost of Earned Revenue* is the cost incurred during the period, excluding the cost of materials not unique to a contract that have not been used for the contract and costs incurred for subcontracted work that is still to be performed.

(c) *Gross Profit* earned on a contract should be computed by multiplying the total estimated gross profit on the contract by the percentage of completion (as determined by one of the acceptable methods of measuring extent of progress towards completion). The excess of that amount over the amount of gross profit reported in prior periods is the earned gross profit that should be recognized in the income statement for the current period.

The following illustrates more clearly the differences between alternatives A and B.

Alternative A:

Percent complete × Total contract price = Earned revenue

Percent complete × Total estimated cost = Cost of earned revenue

Earned revenue − Cost of earned revenue = Gross profit

Alternative B:

Percent complete × Estimated total gross profit + Cost to date =
Earned revenue − Cost to date = Gross profit

The gross profit in dollars will be identical but revenue and cost will generally differ. For example, if you assume the following:

Total price	$1,000,000
Total estimated cost	800,000
Gross profit	$ 200,000
Percent complete (labor hours method)	50%
Cost to date	$ 450,000

then the calculations will be as follows for alternative A and B:

Alternative A:

50% × $1,000,000	$500,000
50% × $800,000	400,000
Gross profit	$100,000

Alternative B:

50% × $200,000	$100,000
Add cost to date	450,000
Earned revenue	550,000
Less cost to date	450,000
Gross profit	$100,000

The presence of two alternatives has caused considerable confusion among contractors because few have seen or heard of Alternative A. Alternative A is really more appropriate for production-oriented contractors that are using the units-of-delivery method. The authors believe that Alternative B is more appropriate for construction contractors.

Adjusting to Earned Revenue

Using Schedule 7-1 as an example, if we assume the contractor has completed site clearance, excavation, and foundations on a billed basis, the contractor will have recorded $80,000 in billings, $45,000 in actual cost, and a profit of $35,000. Note that at this stage, on a billed basis, the contractor has reflected more profit than will be earned on the total contract.

The illustrations based on Schedules 7-1, 7-2, and 7-3 are obviously oversimplified for purposes of illustration. As a practical matter, at the end of any accounting period there would be some partially completed work items.

SCHEDULE 7-1 UNBALANCED BID AND PAYMENT SCHEDULE

Remlap office center	**Job 1098** *Payment schedule*		*Profit (loss)*
Site clearance	$ 20,000	$ 10,000	$10,000
Excavation	30,000	15,000	15,000
Foundations	30,000	20,000	10,000
Framing	40,000	35,000	5,000
Roofing	15,000	15,000	-0-
Plumbing	10,000	15,000	(5,000)
Electrical	8,000	12,000	(4,000)
Exterior finishing	20,000	25,000	(5,000)
Interior finishing	30,000	40,000	(10,000)
Cleanup and inspection	5,000	6,000	(1,000)
	$208,000	$193,000	$15,000

SCHEDULE 7-2 CONVERTING BILLED REVENUE TO EARNED REVENUE USING THE COST-TO-COST METHOD

	Job 1098		
Remlap office center	*Billed Revenue*	*Direct Cost*	*Billed Profit*
Site clearance	$ 20,000	$ 10,000	$10,000
Excavation	30,000	15,000	15,000
Foundations	30,000	20,000	10,000
	$ 80,000	$ 45,000	$35,000
Projected outcome	$208,000	$193,000	$15,000

$$\frac{45{,}000}{193{,}000} = .23 \times 208{,}000 = \$47{,}840$$

Earned revenue	$47,840
Cost incurred	45,000
Earned margin	$ 2,840

SCHEDULE 7-3 JOURNAL ENTRY TO ADJUST BILLED REVENUE TO EARNED REVENUE

	Dr.	*Cr.*
Contract revenue	$32,160	
Billings in excess of costs and earned profit (liability)		$32,160
To adjust revenue on Job 1908 to reflect earned revenue		

Schedule 7-2 shows the application of the cost-to-cost method to convert Schedule 7-1 from a "billed" basis to the percentage-of-completion method, using the cost-to-cost ratio to determine the percentage of completion.

Schedule 7-3 shows the journal entry required to adjust the contract from a "billed profit" basis to an "earned profit" basis.

Where multiple jobs are in progress, many contractors find that a simple worksheet (Schedule 7-4) will facilitate adjusting jobs from a "billed" to an "earned" profit basis to reflect income from contract operations. Many surety companies request that this schedule be included with the basic financial statements of the company and, in fact, many certified public accountants include such a schedule in their audit reports. Note that from this schedule the total of earned revenue will agree with total revenue in the income statement, the total of the cost-to-date column will agree with total contract cost in the income statement, and the totals of the two revenue adjustment columns will agree with the corresponding balance sheet accounts. Note also that, from this schedule, contract backlog and the profit contained in that back-

SCHEDULE 7-4 WORKSHEET FOR ADJUSTING JOBS FROM BILLED TO EARNED REVENUE

Job number and name	Contract price	Projected total cost	Margin	Cost to to date	Percent complete	Earned revenue	Billed to date	Dr. revenue	Cr. revenue
	(A)	(B)	(C)	(D)	(E)	(F)	(G)	(H)	(I)
1908 Remlap Office Ctr.	208,000	193,000	15,000	45,000	.23	47,840	80,000	32,160	-0-
1102 Coombs Tower	400,000	360,000	40,000	180,000	.50	200,000	250,000	50,000	-0-
1103 Redner Apartments	150,000	140,000	10,000	126,000	.90	135,000	130,000	-0-	5,000
1104 Bruzek Shopping Ctr.	600,000	540,000	60,000	135,000	.25	150,000	160,000	10,000	-0-
1105 Irene Ctr.	1,000,000	900,000	100,000	-0-	-0-	-0-	-0-	-0-	-0-
TOTALS*	$2,358,000	$2,133,000	$225,000	$486,000	N/A	$532,840	$620,000	$92,160	$5,000
	(J)			(K)		(L)			

Formulas	Journal entry		
E = D ÷ B	*Dr.* revenue (net)	$87,160	
F = E × A	*Dr.* earned revenue in excess of billed revenue (asset)		
H or I = F − G		5,000	
Contract backlog = J − L			
Profit in backlog = (J − L) − (B − D)	*Cr.* billings in excess of earned revenue liability		$92,160

*If a section were added showing completed jobs, the totals would tie in to the related figures on the financial statements. This type of schedule is commonly used for presentation to banks and bonding companies.

log can be readily determined. Profit in backlog is an important figure for use by surety companies in determining the amount of profit available to cover fixed overhead costs of the company.

Completed-Contract Method[1]

As previously mentioned, the generally accepted alternative method of accounting for revenue on long-term contracts is the completed-contract method of accounting. Under this method no profits or income are recorded on any contract until the job is substantially complete. During job construction, progress billings and costs are accumulated in balance sheet accounts and then are transferred to the income statement when the job is substantially complete. A contract is generally considered completed for accounting purposes if the remaining costs are minor in amount. For example, if physical construction is complete and the only remaining work is clean-up and demobilization, then the contract might be considered completed. It is important, however, that a contractor establish a consistent policy in determining when a contract is complete. The new regulations mandated by the Tax Equity and Fiscal Responsibility Act of 1982 spell out in greater detail when a job is complete for federal income tax purposes (see Chapter 13). As a general rule, many contractors do not consider a job complete until the job has been inspected and accepted by the owner or architect.

The principal advantage of the completed-contract method is that it is based on final results and it is clear what the actual revenue, cost, and profit are. It is also easier for accountants because they do not have to be concerned with the reliability of engineers' or architects' estimates of final costs. The main disadvantage of the completed-contract method is that it is difficult to evaluate the company as a whole when it has many jobs in progress, since all revenues and costs on jobs in progress are deferred. It also has the potential for significant distortion of events. For example, in the extreme, it is conceivable that a contractor could have no work during the year other than completion of a major long-term contract. If the contract is profitable his income statement will show a significant amount of revenue, cost, and income for the year, when in reality the profit had been earned in prior years.

[1]All comments here must be read in light of the effects on completed-contract accounting of the Tax Equity and Fiscal Responsibility Act of 1982, passed August 19, 1982, and the Tax Reform Act of 1986, passed October 22, 1986. Chapter 13 explains the effect on construction accounting of the Internal Revenue Code as amended by these two acts.

When using the completed-contract method, it is generally common practice not to accumulate, as deferred contract costs, general and administrative expenses or overhead (see Chapter 6). These expenses are usually treated as "period costs." For some contractors, period costs have to be allocated to contract costs under the 1982 tax law mentioned above. Not allocating such expenses to contract costs means, of course, that a contractor will have a disastrous-looking income statement if all of his work during the year is still in progress at year-end. There will be no revenue or profits from jobs and only overhead costs will appear on the income statement. Under the percentage-of-completion method, this difficulty does not occur because the revenue and profits earned on contracts in progress are reflected in the income statement for the period. The AICPA recognizes this problem and makes the following statement:

> When the completed-contract method is used, it may be appropriate to allocate general and administrative expenses to contract costs rather than periodic income. This may result in a better matching of costs and revenues than would result from treating such expenses as period costs, particularly in years when no contracts were completed.

This statement can be interpreted to mean that when a contractor has relatively few jobs in progress at any one time it may be preferable for him to allocate overhead to these contracts, thereby providing a better matching of income and expense when the completed-contract method is used. If he has numerous projects in progress at any one point in time, it is preferable to expense overhead as it is incurred (see Chapter 6).

Contract Losses

Conservatism is a basic byword of accounting practice. In other words, most accountants will "provide for all losses and anticipate no gains." The principle of conservatism is especially important in the construction industry, and it is generally an accepted accounting principle that when a loss on a job is forecast, the full loss must be reflected in the income statement in the period in which it is first learned that the job will result in a loss. This means that even though the percentage-of-completion method of accounting is being employed, the loss is still not prorated over the life of the job but is reflected in the earliest period when it is reasonably sure that the loss will occur. This is also true if the completed-contract method of accounting is being employed. In determining, for this purpose, whether a contract will result in a loss, it is general practice to take into account only direct contract costs and exclude general and administrative overhead from the calculation. However, as previously recommended, if a contractor has only one or two jobs in progress at any one time, and general administrative expenses are being allocated to jobs and therefore deferred until the contract is complete, these costs should then be considered together with the estimated direct contract costs.

Other Considerations

Where the percentage-of-completion method of accounting is being employed, the task of determining contract price can sometimes be difficult. This is especially true in contracts covering some of the larger, more sophisticated projects. When a job is completed ahead of schedule, there may be an amount added to the contract price based on the number of days of early completion. Where the project involves a producing asset, such as an electric generator, the increase may be based on the increased production. When the job is completed late, there may be a comparable reduction in price based on time or production. Where an estimated (or target) price is set, the contract price may be adjusted upward or downward, depending on whether the final cost is below or above the target price.

Other factors to consider are provisions for escalation, change orders, extras, and price redeterminations. Here again, the adequacy of the contractor's forecasting plays an important part, especially when there are target rewards or penalties for completion performance. Many accountants will argue that the rule of conservatism should apply in that the contract price should be adjusted down when penalties are forecasted but that rewards should be reflected only when such rewards have been received.

Progress Billing

Usually, most construction contracts permit the contractor to bill the owner for work performed up to a given date, percentage of completion, or stage of physical completion. Progress billings have been referred to previously as "billed" revenue as opposed to "earned" revenue. Because of the unbalanced bid or payment schedule procedure that a contractor will usually attempt to get into his contract to provide financing at early stages of the job, it is usually not correct to reflect income from the job on the basis of progress billings. Reflecting income based on progress billings ignores the fact that under many contracts, a progress billing is merely an advance against the amount which will be earned when and if the contract is completed. Strictly speaking, the excess of progress billings over cost to date, for financial statement purposes, is treated as a current liability representing the amount due the owner until the contract has been completed and accepted.

The percentage-of-completion method of accounting corrects and accounts for this advance payment concept for long-term construction projects.

On the other hand, complete nonrecognition of the profit element in progress payments can result in the equally anomalous position of showing, as a current liability, a credit balance which is in fact virtually all profit. Again, the percentage-of-completion method of accounting adjusts job results for this anomaly.

Another variation from the customary type of revenue recognition for fixed-price contracts occurs with progress payments on cost-plus contracts. Some accountants believe that all reimbursable costs other than the contractor's own costs should be excluded from the books of accounts. This means that in recording cost-plus progress billings, reimbursable costs such as subcontract payments, craft labor, and materials are excluded from revenue and contract cost in the contractor's accounting records. Other accountants believe that all reimbursable costs billed should be reflected in revenue and job cost. The answer to which alternative is correct probably lies in the contract itself or the intent of the parties.

It would seem that where the contractor is responsible for the specification, type, and characteristic of a reimbursable cost item or for the management of that cost item, or where the contractor's fee is based on the reimbursable cost item, those reimbursable costs should be included in the contractor's revenue and cost. The income statement of a contractor which excludes all reimbursable costs other than his own tends to overstate gross margin as a percent of revenue in his income statements; however, net income for his total operations will be the same whether such costs are included or excluded from revenue and direct cost.

An additional problem occurs with respect to client-furnished materials on cost-plus contracts. Some accountants believe that unless the contractor actually expends cash for materials or subcontractors performing work under his supervision, they should not be included in revenue and cost. Other accountants take the view that it doesn't matter whether the owner gives the money to the contractor who then buys the material or whether the owner in effect says "I can do a better job of buying, so I will give you the materials to install." Once again, the basic contract or intent of the parties should be the governing rule. That is, if the contractor is responsible for the specification, type, or characteristic of the materials being applied or if his fee is based or computed on a base including contractor-furnished materials or labor, then such costs should be included in the revenue and cost in the contractor's statements.

Retained Percentages

It is quite common in the construction industry for the owner to withhold a percentage on progress billings from the contractor until the contract is complete. This is commonly 10 percent, although it may be more or less. In any event, it is referred to as "retained percentage" or "contract retentions." Whether the collection of retained percentages presents any special accounting or billing problems depends upon how the accounting was done at the time the progress billings were made. Generally the contractor will segregate his receivables into contract receivables currently due and retained percentages receivable. When a special billing for the retained percentage is required by the owner, even though it has already been taken into account in the current progress billings, it is necessary to treat the special billing for the retention as a memo billing. Otherwise revenue and receivables will be duplicated. On the other hand, if the strictly legalistic concept of retained percentages is followed, and only the currently collectible portion of the progress billing is reflected in the contractor's accounts, then the special billing for the retained percentage must be entered like any other invoice. Most accountants support the former view, that progress billings should include retention and should be so recorded in the accounts.

Of course, if the accounts are kept on a completed-contract basis, the accounting treatment of the retained percentage is not a problem. No part of the billing becomes revenue until the job is completed and closed. Interim progress billings would have been credited to an account called "Progress billings of jobs in progress" and charged to a contra account such as "Progress payments receivable" and "Retained percentages receivable." When the job is closed the balance of the account called "Progress billings on jobs in progress" is transferred to an account generally called "Revenue from completed contracts."

Other Income

Aside from progress billings, construction companies may derive income (and ultimately cash receipts) from other sources such as those listed below. In later portions of this chapter, the accounting for these items of income is described in somewhat greater detail. In summary, however, sources of other income generally are:

1. *Work done for others at the jobsite:* The income from this source would usually enter the records through a field work order (Figure 2-7), which would be the source of the billing to the person for whom the work is done.
2. *Equipment rentals:* The income from this source would enter the records through the equipment rental record which forms the basis

for billing rentals. Sometimes the person renting the equipment is required to report the time the equipment is used, and then the reports form the basis for the billing.

3. *Sales of used equipment, scrap, and surplus materials and supplies:* Usually the charges derived from these sales can be most easily accounted for by a warehouse issue ticket (Chapter 4). However, when assets are being distributed at the conclusion of a joint venture, it is usually better merely to divide the surplus items and make only transfer records. Otherwise some states (such as California) will attempt to levy a sales tax on the transfer. In those cases in which final results on the job may be used as a tool for future bidding purposes, or for a claim against the owner on the existing contract, the credit arising from the sale of used equipment, scrap, and surplus materials should be applied to the applicable work item in the cost ledger in addition to the appropriate general ledger credit.
4. *Insurance refunds and proceeds of insurance claims:* The refunds are usually not taken into account until the check is received from the insurance company. The credit is usually applied to insurance expense or to miscellaneous income. However if, for example, workers' compensation has been part of an additive percentage applied to direct labor and therefore spread over numerous work items of the job, and if the amount of the refund is significant, the refund should be allocated back over the work items on the basis of direct labor. Insurance claims, on the other hand, are recorded by journal voucher as soon as proof of loss or the "over, short, and damage report" on the resulting claim is filed. This is a key internal control procedure to ensure that when the check is received it will be properly deposited and accounted for.
5. *Discounts earned:* Discounts almost always enter the records through the check register, where the liability account is charged for the invoice price and cash is credited only for the amount paid. The excess credits go in a miscellaneous income account called "Discounts earned." An alternative treatment, often followed where cash basis accounting is used, is to charge the items purchased, net of discount, to the appropriate job cost accounts. Generally speaking, this method is not too satisfactory, since it fails to reflect the cash value of having money on hand to pay bills.
6. *Camp and mess hall receipts:* Some of the larger contractors whose jobs are isolated may operate a camp, mess hall, commissary, company store, trailer camp, and sometimes all five. On large turnkey projects the authors have seen schools, hospitals, and other aspects of routine city life provided by the contractor.

Employees or other users may pay in cash or through payroll deductions for these services. The routines used in accounting for these items of income are described in greater length in this chapter and in Chapter 11.

7. *Repair service:* If a contractor sells repair services such as those provided by plumbers and electricians, in addition to regular contract work, how can these numerous small jobs be handled within the structure of a job order accounting system? The answer usually lies in lumping all small jobs into a single blanket work order which is closed out periodically. The customer receivables are entered from the job ticket, and both revenue and cost are charged to a single work order designated by a title such as "Repair jobs." Sometimes such contractors, for example plumbers or electricians, will operate a number of repair trucks. Then a separate blanket work order for each truck will show the comparative earning power of each of the repair trucks.

The Billing Process

The real starting point for most revenue accounting in the construction business is the preparation of invoices or progress billings to owners. There are a few exceptions, such as those arising in camp operations, but the bulk of revenue comes from transactions that are represented by billings to the owners. Once invoices are prepared, they must be summarized before being posted to the general ledger. One form of journal record (Figure 7-2) combines the cash receipts and contract billing records. This form is generally in use by the smaller contractors. There are many others. Sometimes cash receipts are recorded in one book called the "cash receipts book" and sales or progress billings are recorded in another book generally referred to as "revenue ledger." When progress billings and jobs are numerous, it is a good idea to post the details to a subsidiary "accounts receivable subledger" (Chapter 9).

In its simplest form, the request for progress payment is no more than an invoice. It will usually state the basis on which the billing is made and often it must be authorized by the owner or his representative. For example, if a contract for a single-family residence called for payment of one-third of the contract price when the rafters were in place, the billing might appear as in Schedule 7-5.

If the job is one in which payments are to be based on a percentage of completion of each work item, the request for progress estimate might appear as in Schedule 7-6.

Cost-plus contracts call for progress billings which merely list the reimbursable costs incurred or include the total of those costs along

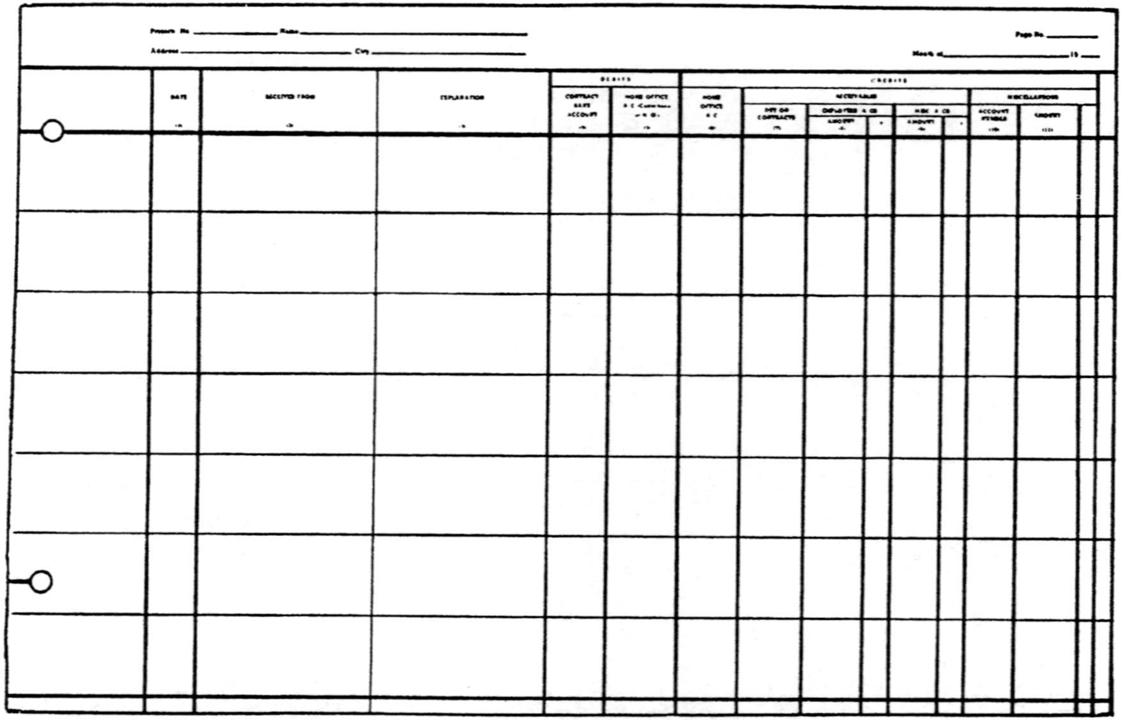

Figure 7-2 Cash Receipt Record. This form, prepared in duplicate, is designed for use in a system in which the copies of the books of original entry are sent to the general office for posting to the job general ledger.

SCHEDULE 7-5 SIMPLE PROGRESS BILLING

Per Article XI of Contract Dated March 4, 19_____

Contract price	$75,000	
One-third due when rafters are in place		$25,000
Less 10 percent retention		2,500
Due on this billing		$22,500

APPROVED:

________________________ *Architect*

SCHEDULE 7-6 PROGRESS BILLING ON PERCENTAGE-OF-COMPLETION BASIS

Work item No. Description	*Total amount*	*Percent complete*	*Amount earned*	*Previously billed*	*Earned this period*
1 Site preparation	$ 1,000	100	$1,000	$1,000	—
2 Excavation	2,500	100	2,500	1,500	$1,000
3 Forms	2,400	75	1,800	400	1,400
4 Reinforcing	1,500	60	900	100	800
5 Concrete	13,300	10	1,330	—	1,330
6 Roof	9,100	—	—	—	—
7 Plumbing	3,100	5	155	—	155
8 Electrical	4,300	5	215	—	215
9 Cement finishing	1,000	20	200	—	200
10 Sash and doors	3,900	—	—	—	—
11 Painting	2,500	—	—	—	—
12 Cleanup	500	—	—	—	—
	$45,100		$8,100	$3,000	$5,100
Less: 10 percent retention			810	300	510
TOTALS			$7,290	$2,700	$4,590

APPROVED:

________________________ *Architect*

with supporting documents. Usually the summary sheets, or "transmittals" as they are sometimes called, are accompanied by copies of checks, invoices, purchase orders, delivery receipts, and whatever other supporting data may be available. Generally, however, payroll checks are not provided with the progress billing but a photocopy of the payroll journal is included.

Some owners require that all supporting invoices on a cost-plus progress billing be marked "paid" and some owners, particularly government agencies, demand this procedure. However, a better procedure is to require copies of the checks, since as a practical matter, contractors generally require suppliers to receipt the invoices in advance so that they then can be submitted for payment. The supplier is then paid when the contractor receives the reimbursement check. This is another way in which some contractors in addition to unbalancing bids and payment schedules can acquire additional financing. If there is any question about the contractor's financial position, or ability to perform, it is quite in order to require a surety bond. This usually costs the owner less than the additional cost of the unnecessary paperwork involved in getting receipted bills. Many old-time contractors will dispute this assertion, but modern practice has proved that it is often true. Figure 7-3 illustrates a type of invoice

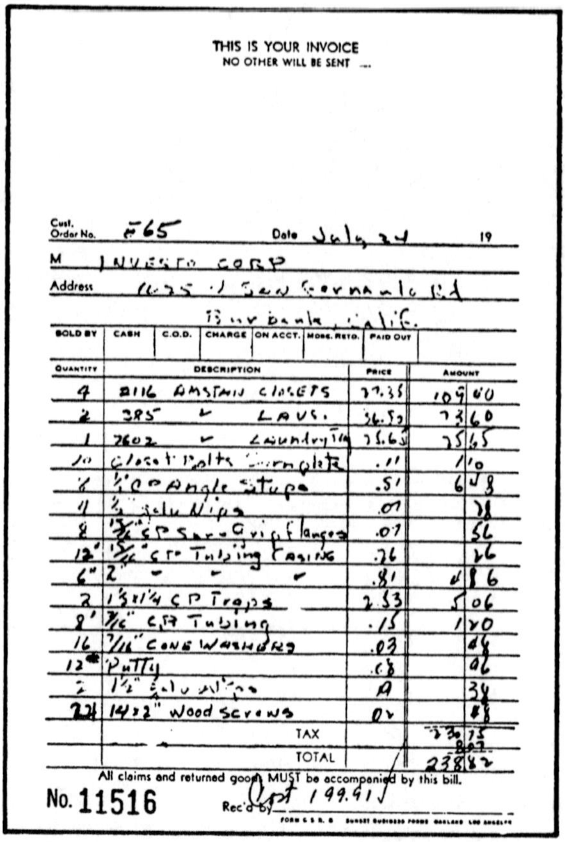

THIS IS YOUR INVOICE
NO OTHER WILL BE SENT

Cust. Order No. #65 Date July 24 19

M INVESTO CORP

Address

SOLD BY	CASH	C.O.D.	CHARGE	ON ACCT.	MDSE. RETD.	PAID OUT

QUANTITY	DESCRIPTION	PRICE	AMOUNT
4	2116 AMSTAN CLOSETS	27.35	109 40
2	385 " LAVS.	36.52	73 60
1	7602 " Laundry Tray	25.65	25 65
10	Closet Bolts Complete	.11	1 10
8	½ CP Angle Stops	.51	6 08
4	½ Galv Nips	.07	28
8	⅞ CP Screw Grip Flanges	.07	56
12'	1⅛ CP Tubing Casing	.26	26
6"	2 " "	.81	4 86
2	1½ × 1¼ CP Traps	2.53	5 06
8'	⅜ CP Tubing	.15	1 20
16	⅜ Cone Washers	.03	48
12#	Putty	.08	96
2	1½ Galv Nips	[illegible]	34
24	14 × 2" Wood Screws	.02	48
	TAX		230 75 / 8 07
	TOTAL		238 82

All claims and returned goods MUST be accompanied by this bill.

No. 11516 Rec'd by

Figure 7-3 **Combination Invoice and Receiving Report.** This form is prepared in triplicate. The original is signed by the person receiving it and placed in the job cost file. The duplicate is left with the client and serves both as a delivery memo and as an invoice. Illustrated is the triplicate, which is sent to the accounting department for inclusion in the client's account.

commonly used by small subcontractors (in this case a plumbing contractor) to charge for materials on a "time and material" contract. This type of invoice is also used to bill materials sold out of a warehouse or from a repair truck.

Back Charges

Back charges represent charges for materials, equipment, services, or other costs or expenses furnished to the contractor by the owner or to the subcontractor by the prime contractor. Using the field work order (Figure 2-7) is one way of authorizing back charges. Another method is to cover such charges in the basic contract itself. It is rarely satisfactory to accept back charges without written authorization made at the time of the specific transaction. If some charge appears to be proper and there is no authorization, the contractor (or in larger firms, some top executive) should personally authorize any back charge before it is accepted. The procedure for handling back charges, once they are authorized, is simple enough. The owner back-charging the contractor (or the contractor back-charging the subcontractor) merely prepares an invoice for the agreed price of the back charge and enters it as he would any other invoice—charging a receivable account specifically set up for back charges and crediting the applicable cost account or miscellaneous income account. In the next progress payment by the owner to the contractor (or by the contractor to his subcontractor), the owner will deduct the amount of back charge from the progress invoice amount. Entering the receipt of a progress payment, reduced by the amount of a back charge, the appropriate cost, expense, or asset account is debited for the amount of the back charge and the debit to cash is correspondingly decreased.

Equipment Rentals

In accounting for revenue from equipment rentals, the first step is to establish a file of rental agreements. These may not be in the form of written contracts, but some record must be kept to show what units are involved, the terms of the rental, the rental rates, and any special provisions with respect to downtime and maintenance and repairs.

The significance of such a file is that it can be used as a starting point for the preparation of equipment rental billings. Often it is a good idea to use some type of visible card record (Figure 7-4). Sometimes it is possible to design the equipment ledger itself so that rentals may be controlled by one of its sections. Whatever form the rental agreement takes, the best way to be sure that all equipment rentals are billed is to use the file to prepare billings on all equipment

EQUIPMENT RENTAL CONTROL						
Rented To	Time In Use		Total Time	Rate	Rental Earned	Date Billed
	From	To				
Item.			Equipment Number.			

Figure 7-4 **Index Card for Controlling Equipment Rental Billings.** If volume is heavy, Kardex or similar cards may be useful.

on rental during the month. When the amount of the rental charge depends upon an equipment time report, the billing can still be prepared so that the file of uncompleted billings will serve as a follow-up on the time reports still to come. The billings themselves are entered in a receipts record (Figure 7-2), and from there, the totals are posted to the general ledger accounts and the detailed customer ledger (Chapter 9). Some contractors, especially those who use decentralized accounting, distribute the cost of owning and maintaining equipment by "renting" to their own jobs. As previously indicated (Chapter 5), this method has virtually the same effect as a standard cost system for equipment operations. However, it is not uncommon to see the so-called "income" from "renting" equipment to the company's own jobs entered with the actual revenue derived from renting equipment to others. The two are not comparable and should always be segregated.

Requests for Final Payment

A request for final payment usually takes the form of an ordinary invoice or a progress billing, the latter being somewhat more specialized. The form is often determined by the provisions of the contract, and some owners and architects actually specify the form in detail.

Usually it must be accompanied by releases (Figure 2-4), and sometimes by affidavits and other evidence of having satisfied all possible lien claimants. (In some states, such as California, the form of the release is set by statute.) If all, or any part, of a final payment is to be held past the lien period to ensure performance of any guarantee provisions in the contract, then that amount should be transferred to an account such as "Guarantee deposits."

If the retained percentage, which usually makes up the final payment, has already been taken into income, the final request will form the basis for transferring the amount from "Retained percentages" to "Progress payments currently due." If the practice has been to record only the progress payment net of the retention, then the final payment request would be entered the same as any other billing.

On a cost-plus contract with a participation by the contractor in savings under a guaranteed maximum or target price, there might be one or two "final" billings. If the cost has been audited and the amount of participation has been determined at the time the final billing is made, or if costs exceed the guaranteed maximum, then only one final billing would be required. However, if costs are less than the guaranteed maximum and have not been audited, then a final billing for the previously nonreimbursed costs plus the unbilled portion of the agreed fee would be made, followed by a supplemental billing for participation in the saving.

Camp Revenue and Receipts

If a camp is maintained on a job, the camp revenue usually enters the books through the camp manager's report (Figure 14-10) and is collected by payroll deduction. Mess hall receipts may be collected in several ways. Meals may be charged at a flat rate per day or per week and then would enter the records and be collected in the same way as the camp charges. If meal tickets are sold, it is usually possible to collect for most of them by payroll deductions. However, some employees prefer to pay cash for them, and such cash will be listed in the report. The camp manager should have a prenumbered receipt book (Figure 7-5) or combination receipt and charge ticket book. The latter may also be issued on a billing machine. When a meal ticket is sold for cash, a "cash ticket" or receipt would be issued. If the price were to be deducted from the employee's paycheck, a charge ticket (Figure 14-6) would be issued.

Commissary Revenue and Receipts

Commissary and company store sales differ from camp and mess hall sales in that the former call for more over-the-counter selling for cash.

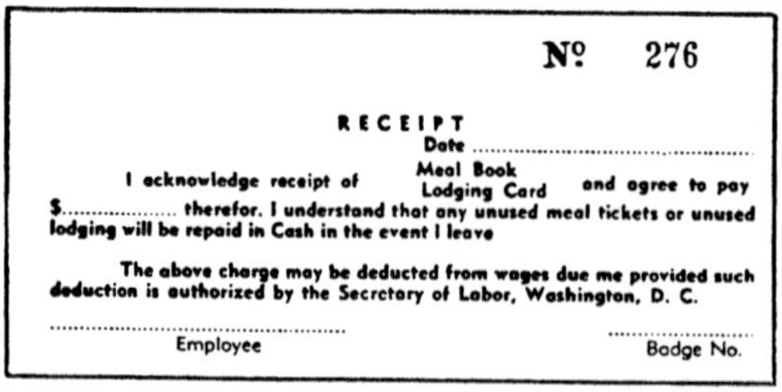

№ 276

RECEIPT

Date ....

I acknowledge receipt of Meal Book / Lodging Card and agree to pay $.................. therefor. I understand that any unused meal tickets or unused lodging will be repaid in Cash in the event I leave

The above charge may be deducted from wages due me provided such deduction is authorized by the Secretary of Labor, Washington, D. C.

.... Employee .... Badge No.

Figure 7-5 **Receipt for Meal Books and Lodging Cards.**

If cash sales are substantial, a cash register is the most efficient and economical way to account for both the sales and the incoming cash. Charge sales would, of course, be handled on prenumbered charge tickets. These all form a part of the camp manager's daily report.

Some contractors follow the practice of subcontracting the camp, mess hall, and commissary activities, and it is true that this procedure limits the paperwork to a minimum. However, it also may defeat its own ends. Such activities can contribute substantially to the efficiency and morale of the personnel on a remote job if properly run, but they can also have the opposite effect. If they are to be run at all, it is better to run them for what they can contribute to the job rather than for the profit they can make. If a contractor runs these operations himself, he can control them better than if he were to sublet them. On the other hand, it is sometimes possible to use the facilities of some contractor already on the site for a fair price, and thereby achieve both objectives.

Equipment and Material Transfers

When books are kept in the general office, a report (Figure 7-6) covering any transfer or sale of equipment, material, supplies, or tools should be made. Once a month these reports are summarized. They form the supporting documents for a journal voucher (a general journal entry) through which they enter the general ledger. To keep the cost ledger up to date from day to day, these reports may be posted directly to the cost ledger as they come in from the jobs.

Change and Extra Work Orders

The use of the field work order (Figure 2-7) has already been described. Essentially it is a means of authorizing extra work to be done. As the work is actually done, its cost must be accounted for and accumulated on a daily extra work report (Figure 7-7) for billing purposes.

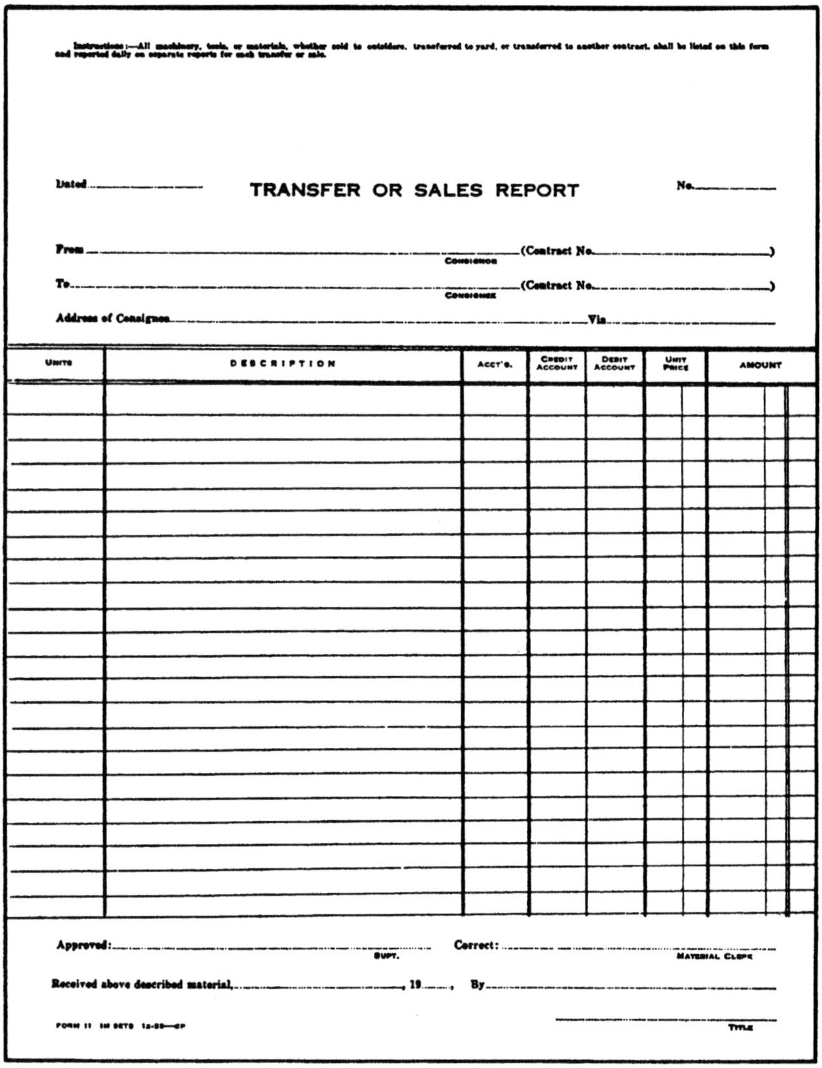

Instructions:—All machinery, tools, or materials, whether sold to outsiders, transferred to yard, or transferred to another contract, shall be listed on this form and reported daily on separate reports for each transfer or sale.

Dated ____________ **TRANSFER OR SALES REPORT** No. ____________

From ______________________ (Contract No. ______________)
Consignor

To ______________________ (Contract No. ______________)
Consignee

Address of Consignee ______________________ Via ______________

Units	Description	Acct's.	Credit Account	Debit Account	Unit Price	Amount

Approved: ______________________ Supt. Correct: ______________________ Material Clerk

Received above described material, ______________________, 19____, By ______________________

FORM 11 1M SETS 12-33—GP ______________________ Title

Figure 7-6 **Transfer or Sales Report.** This is a combination form for use in reporting and recording either transfers of materials, supplies, or equipment from one job to another, or sales of such items. Five different-colored copies are prepared.

A change order that alters the amount of a contract has a direct effect on the contractor's revenue. Chapter 2 deals with change orders in detail, and the illustrations given there are applicable here. In accounting for contract revenue, it is necessary to be sure that progress billings reflect all change orders that have been made final at the date of the billing. Usually this is a function of the engi-

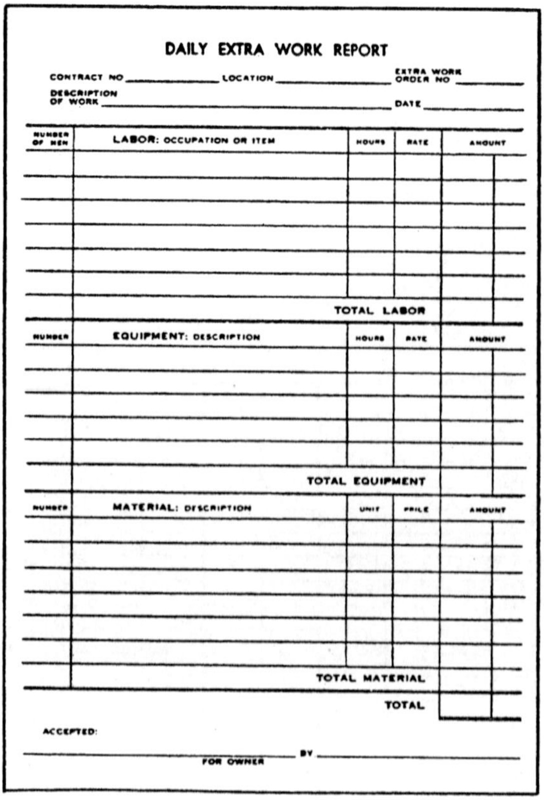

DAILY EXTRA WORK REPORT

CONTRACT NO ______ LOCATION ______ EXTRA WORK ORDER NO ______

DESCRIPTION OF WORK ______ DATE ______

NUMBER OF MEN	LABOR: OCCUPATION OR ITEM	HOURS	RATE	AMOUNT
	TOTAL LABOR			

NUMBER	EQUIPMENT: DESCRIPTION	HOURS	RATE	AMOUNT
	TOTAL EQUIPMENT			

NUMBER	MATERIAL: DESCRIPTION	UNIT	PRICE	AMOUNT
	TOTAL MATERIAL			
	TOTAL			

ACCEPTED:

______ FOR OWNER ______ BY ______

Figure 7-7 **Daily Extra Work Report.** This form provides data for billing clients and others for work not specified in the main contract.

neering or operations personnel, but it is ordinarily up to the accounting and office personnel to see that the change order files are up to date.

On cost-plus jobs, as already noted, there may be an additional fee due when change orders raise the cost beyond an agreed amount. Owners will sometimes seek to avoid such a result by furnishing materials, equipment, insurance, bond, and so on. It is thus possible to increase the scope of the work substantially without increasing the fee, unless the contractor makes provision for such a contingency in the contract. However, even if such provision is made, there is still need for a summary report (Figure 7-8). This report should be prepared and incorporated into the periodic billings.

SUMMARY OF COST ITEMS FURNISHED BY OWNER

Job name. . . . . .Job No. . . . . .Date. . . . .

Items	This month		Job to date	
LABOR Supervisory				
Clerical				
Engineering				
Trades				
Other				
TOTAL LABOR				
MATERIALS Lumber				
Cement				
Sand and aggregate				
Steel				
Hardware and miscellaneous iron				
Other				
EQUIPMENT				
OTHER Bond				
Insurance				
Sundry				
GRAND TOTAL FURNISHED				

Compiled by. . . . . .Approved by. . . . .

Figure 7-8 Summary of Cost Items Furnished by Owner of Cost-Plus Jobs.

Miscellaneous Receipts

Occasionally money will be received from miscellaneous sources, such as the refund of bid or plan deposits, insurance premium refunds, insurance recoveries, recovery of items of expense, or overpayment of an account. When such amounts are received, the credit should go to the account originally charged, if such treatment is not unduly difficult. This procedure is elementary to a trained accountant, but the rule is violated often enough to make it worthy of mention. An outstanding example is the crediting of insurance recoveries to "Miscellaneous income." An equally common, though somewhat less obvious, violation of the same rule is to credit the "Equipment" account (or other fixed asset) with the cash received on the sale of a piece of equipment. This happens because the bookkeeper overlooks the fact that the "Equipment" account is charged with the full cost of the item and that the "Reserve for depreciation" account contains credit equal to the total depreciation since the item was acquired. Therefore the entry is:

Dr. Cash (or Accounts receivable)
Dr. Reserve for depreciation
Dr. Loss on the sale (if book value is more than selling price)

Cr. Equipment (for the total cost)
Cr. Gain on the sale (if book value is less than selling price)

Sometimes, when there is recovery of cost on a job closed out in a prior year, such as insurance refunds affecting premiums of prior years, tax refunds, or bad debt recoveries, it is proper to credit the receipt to "Miscellaneous income."

Refunds or recovery of cost on cost-plus contracts closed out in prior periods require moral as well as accounting judgment. If the cost was reimbursable then it would seem that the contractor has an obligation to pass such savings on to his client.

Cash Receipts

Accounting for actual receipts of cash is much the same in any business. Cash should be controlled when it comes into the company's possession; deposited in a bank account as soon as practicable; and reported on a suitable form (Figure 7-2). Unless the business is so small that the contractor does all the office work personally, it is a good idea to have the mail opened and the checks taken out and listed by someone who does not handle the cash or keep the books. If, as in many smaller firms, the office is run by a combination bookkeeper-office manager, the contractor may find it worthwhile to open all mail personally. Then he knows not only about incoming checks, but also about a great many other things that would otherwise escape his attention. There is one exception to this suggested procedure. Any proprietor who has a tendency to hold out small checks (or any checks) and cash them for pocket money should be kept as far as possible from the incoming mail. Additional bank statements should be mailed to and reconciled by an employee independent of all other cash or related accounting functions.

Cash coming in from scrap or surplus materials sales should, as a matter of standard procedure, be accounted for on a prenumbered receipt which must be made at the time of the transaction and at no other time. If there are numerous over-the-counter sales, a cash register may be useful to a contractor. Otherwise, a sales ticket should be made at the time of sale, showing the amount of cash collected. Often the most convenient way to prepare such a ticket is on a counter-type billing machine (Figure 7-3).

Theft, Misappropriation, and Similar Irregularities

There is no absolute guarantee against dishonesty, but the nearest thing to it is a good system of internal controls coupled with a fidelity bond. Perhaps the best internal control for the larger contractor is an internal audit staff. These measures make it as difficult as possible for an employee to steal, and make ultimate discovery of any theft reasonably certain. In the small one-man firms, the opportunities for use of internal controls are fewer than in a larger firm. On the other hand, the opportunities for management control are usually greater. This statement is true for two reasons. First, in the small firm the contractor either sees or handles most cash transactions personally and if he is alert he can usually detect any dishonesty. Second, there are so few transactions that one person would probably know about all of them and remember the details on most of them.

Dishonesty in handling cash receipts may assume one simple form—pocketing incoming cash instead of recording it. It is the function of internal auditing procedures to look for the likely source of cash, to find out if it actually came in, and then to check the records to see that it was recorded and accounted for correctly. It is the function of internal control to design the accounting procedures in such a way that failure to record incoming cash will cause the cash deposited in the bank (and so controlled by an outside agency) to fail to balance with some record kept by an individual who has access neither to the cash itself nor to the cash records. It is the function of fidelity bonds not only to reimburse the employer for theft losses but to create in the minds of employees handling cash the certainty that once theft is detected it will be prosecuted.

In the one-man contracting firm, the contractor himself usually knows all the transactions which should bring in cash. He also knows how much is coming in. If it does not come in, he knows or will find out why. Such conduct, of course, is internal auditing in its most effective form, and with it the need for internal control disappears. The chance that anyone can steal cash and avoid getting caught is virtually nil.

This theory breaks down, however, when the contractor gets too busy with other things to think about (and follow up on) incoming cash. It also breaks down if he himself takes the incoming cash and makes no record of it. If the contractor follows this practice, he encourages his employees to follow his example, counting on the protective assumption that all will make, that undoubtedly the proprietor took the money himself as usual, but forgot about it.

In larger concerns, the auditing is done either by someone in the accounting department or by special internal auditors. The internal control is obtained by having someone outside the accounting department handle the money. In that way the person who controls the money does not control the cash records and vice versa. When such a division of work exists, it takes collusion among two or more people to conceal a theft by juggling the books. This necessity, in itself, makes stealing harder and less profitable. Coupled with compulsory vacations and occasional transfers, the separation of cash handling from cash accounting can be an effective device. Adding an adequate fidelity bond provides as much protection as is possible against cash losses by defalcation.

Liquidation Problems

After every large job, and after many medium-sized and small jobs, there is a certain amount of surplus material to be sold. This situation results in something more than the usual problem of cash control. Normal routine may be nonexistent, sales may be made on a forced-sale basis, and there is very little comparison between liquidating prices and normal prices for the same items. Often project managers and superintendents will make deals during the job based on promises of preferences to be given in liquidation when the job is closing. By the time the sales are actually being made, the individual who made the deals may be gone.

One way to achieve some control in these postjob liquidations is to send an inventory team around the job, as it is approaching completion, to count and estimate the value of the items to be sold. If possible, it is a good idea to accumulate in one place the things to be sold. Then the liquidator must account for either these items or the price set by the inventory team. Variations from the inventory price would be permitted only when justified. If this system is practicable, it can prevent any major diversion of assets or misappropriation of funds. Each sale would be evidenced by a prenumbered invoice. Receipts would be deposited in a special liquidation bank account on which withdrawals could be made through the head office only.

Liquidations of equipment are usually less difficult than liquidations of materials, supplies, and scrap. The units are larger and usually bear identifying numbers, and there will be an equipment subledger providing an inventory. Depending on the nature of the equipment, prices may be more easily evaluated. Here the inventory and price list technique is definitely practical and should be followed. The prenumbered invoice would also be used in addition to whatever type of bill of sale might be required by local law. The proceeds would still go into a liquidation bank account. Sometimes it is possible to find a finance company or bank that will buy conditional sales contracts "without recourse," and on such a basis, equipment should bring in enough additional proceeds to make credit sales profitable. However, lacking such an arrangement, liquidation sales should be made for cash only. Checks should be accepted with care, and deposited immediately. It may be that some of the bad ones can be caught before the goods are completely out of reach. When equipment is sold on credit in states having the Uniform Commercial Code, a financing statement as illustrated in Figure 7-9 should be filed with the appropriate state authorities.

Lien Laws

Any discussion of contractor income must include a general consideration of lien laws and bond acts. It has already been pointed out that most of a contractor's work consists in improving the real property of others. Therefore, if the contractor is to have any security for his work, there must be some statutory provision that gives him the power to obtain an interest in the real property if he is not paid. Work on privately owned real property is protected by lien laws in all states.

When work is done on real property owned by federal, state, or local governments, it is contrary to public policy to permit private individuals to obtain an interest in the realty through liens. Therefore, the various states and the federal government have provided for payment of workers, material suppliers, and sometimes others, through laws requiring prime contractors to post payment bonds.

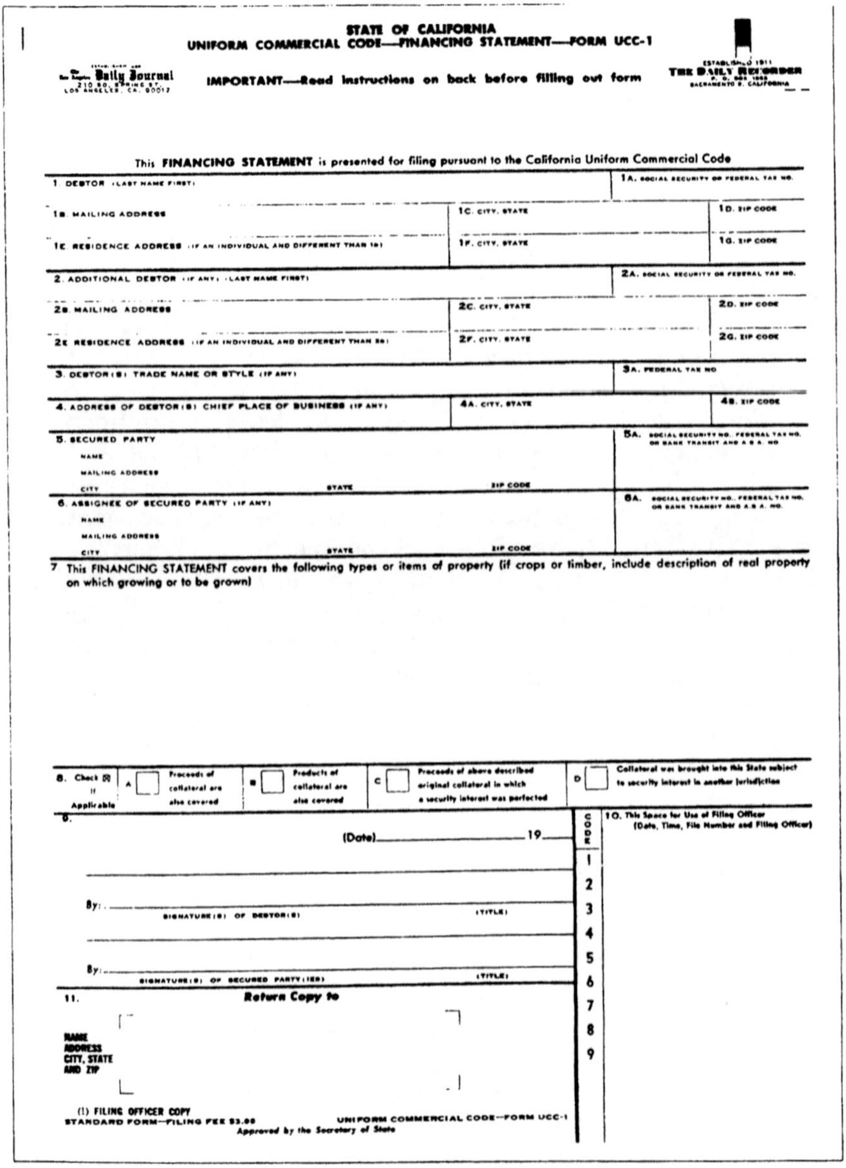

STATE OF CALIFORNIA
UNIFORM COMMERCIAL CODE—FINANCING STATEMENT—FORM UCC-1

Daily Journal 210 SO. SPRING ST. LOS ANGELES, CA. 90012

IMPORTANT—Read instructions on back before filling out form

ESTABLISHED 1911 THE DAILY RECORDER SACRAMENTO, CALIFORNIA

This **FINANCING STATEMENT** is presented for filing pursuant to the California Uniform Commercial Code

1. DEBTOR (LAST NAME FIRST) | 1A. SOCIAL SECURITY OR FEDERAL TAX NO.
1B. MAILING ADDRESS | 1C. CITY, STATE | 1D. ZIP CODE
1E. RESIDENCE ADDRESS (IF AN INDIVIDUAL AND DIFFERENT THAN 1B) | 1F. CITY, STATE | 1G. ZIP CODE
2. ADDITIONAL DEBTOR (IF ANY) (LAST NAME FIRST) | 2A. SOCIAL SECURITY OR FEDERAL TAX NO.
2B. MAILING ADDRESS | 2C. CITY, STATE | 2D. ZIP CODE
2E. RESIDENCE ADDRESS (IF AN INDIVIDUAL AND DIFFERENT THAN 2B) | 2F. CITY, STATE | 2G. ZIP CODE
3. DEBTOR(S) TRADE NAME OR STYLE (IF ANY) | 3A. FEDERAL TAX NO
4. ADDRESS OF DEBTOR(S) CHIEF PLACE OF BUSINESS (IF ANY) | 4A. CITY, STATE | 4B. ZIP CODE
5. SECURED PARTY NAME MAILING ADDRESS CITY STATE ZIP CODE | 5A. SOCIAL SECURITY NO., FEDERAL TAX NO. OR BANK TRANSIT AND A.B.A. NO.
6. ASSIGNEE OF SECURED PARTY (IF ANY) NAME MAILING ADDRESS CITY STATE ZIP CODE | 6A. SOCIAL SECURITY NO., FEDERAL TAX NO. OR BANK TRANSIT AND A.B.A. NO.

7 This FINANCING STATEMENT covers the following types or items of property (if crops or timber, include description of real property on which growing or to be grown)

8. Check ☒ If Applicable | A ☐ Proceeds of collateral are also covered | B ☐ Products of collateral are also covered | C ☐ Proceeds of above described original collateral in which a security interest was perfected | D ☐ Collateral was brought into this State subject to security interest in another jurisdiction

9. (Date) ______ 19__

By: ______ SIGNATURE(S) OF DEBTOR(S) (TITLE)

By: ______ SIGNATURE(S) OF SECURED PARTY(IES) (TITLE)

CODE 1 2 3 4 5 6 7 8 9

10. This Space for Use of Filing Officer (Date, Time, File Number and Filing Officer)

11. Return Copy to

NAME ADDRESS CITY, STATE AND ZIP

(1) FILING OFFICER COPY
STANDARD FORM—FILING FEE $3.00 UNIFORM COMMERCIAL CODE—FORM UCC-1
Approved by the Secretary of State

Figure 7-9 **Uniform Commercial Code—Financing Statement.**

Also, many state laws provide for so-called "stop notices" under which the government owner will withhold payments to the prime contractor until the claimant is paid or his claim secured by a special bond.

Any rights a contractor or subcontractor may have under any of these laws are strictly statutory, and the statute must be strictly complied with or the right will be permanently lost. Most of these laws require formal notice within a relatively short time after the last work is done, followed by the filing of a lawsuit to recover the money within a stated time after notice. A typical provision is one that calls for the filing of a lien within 30 days after notice of completion (or within 90 days after the last work on the job if there has been no notice of completion) followed by the filing of suit within 90 days after notice if payment is not received. Therefore, if payment is not received when due, it is vital that the contractor consult his attorney immediately. Otherwise his action may come too late and the rights may be lost.

The statutes differ in every state, and the state statutes differ from the Federal Bond Act of 1935 (Miller Act), which controls work for the United States government. Also, changes are frequent both in the statutes themselves and in the court decisions that interpret them. The differences include not only the time within which action must be taken but also the specific procedures followed and the extent of the protection afforded. This means that it is unsafe for the average contractor to proceed without the help of his attorney and that preferably, in choosing his attorney, he should be careful to select one who has had experience with construction contractors' problems.

Chapter

8

Factors Affecting Job Financial Results

Introduction

Construction job management is an all-encompassing topic that embodies many technical and managerial aspects of construction contracting. A generally accepted definition of job management is that it is the art of directing and coordinating human and material resources throughout the life of a construction project by using modern techniques to achieve predetermined objectives of scope, cost, time, and quality.

Of course, most contractors realize that managing a project entails more than just sharing the work that is to be performed in a timely and cost-effective manner. This chapter undertakes the task of identifying factors and setting guidelines that should be followed by the construction manager in his attempt to ensure that a construction job is properly planned prior to the commencement of actual fieldwork.

Most contractors require a system of management controls and information weighted toward emphasizing job management and contract cost controls. Of critical importance is the requirement for a set of controls designed to alert the project manager immediately when management attention is needed.

The nature of the job, i.e., commercial, industrial, institutional, or utility, will greatly impact the way a project is organized and staffed. Likewise, whether the project involves new construction versus renovation work will change the complexities in selecting proper management procedures.

Lack of understanding of the contract and failure to perform a careful review of its contents are major shortcomings that often prevent the contractor from making a responsible assessment of the risk involved.

What Is Risk?

Inevitably, there will always be construction disputes, primarily because there is always uncertainty involved in construction. When problems arise on a project, it is usually a combination of the owner's, designer's, and contractor's faults. In construction, as in most businesses, there is risk. In fact, most contractors are in the business because they are risk takers.

Risk is defined in Webster's Dictionary as the chance of injury or damage or loss. While this obviously is a general definition of generic risk, construction risk can also be related to the chance of loss by association with three primary constraints: time, cost, and quality. Only two of these constraints are within the contractor's power to control—time and cost.

The third constraint of quality is determined by the designer through conformity with the profession's acceptable standards and specifications. The constraint of time is not only how fast a project can be completed but also the likelihood of finishing a project according to the agreed-on schedule. The risk constraint of cost is defined as the variability of the actual costs to complete a project as compared to the budgeted costs. A contractor who is to manage risk—and make a profit—must concentrate on managing these two risk constraints.

Pitfalls in Construction

The reason why the construction industry has an increased propensity for risky situations, when compared with other business industry segments, is the numerous pitfalls involved with this industry. These pitfalls represent recurring problems that significantly affect both cost and schedules for almost any type of construction project. Exhibit 8-1 presents an overview of the types of problems that can be related to the major parties of a construction project—owner, designer, and contractor. How these pitfalls affect the success or failure of a project is discussed in more detail below.

Failure to fund. Adequate funding for construction projects, in the current economic period, is becoming much more difficult for the majority of the construction industry. There is significant aversion by financial institutions to adequately fund small- to medium-sized construction projects. Lenders are now enforcing stringent guidelines for construction owners. The banking industry is becoming much more selective in the kinds of projects in which they are willing to invest. The construc-

EXHIBIT 8-1 Pitfalls in Construction—Owner, Designer, and Contractor Perspectives

Owner	Designer	Contractor
■ Failure to fund ■ Owner-furnished materials not available ■ Major changes in requirements ■ Failure to make progress payments ■ Interference	■ Defective plans and specs ■ Shop drawing review and material approval ■ Improper or delayed change orders ■ Failure to coordinate between primes ■ Inadequate information	■ Slow to mobilize ■ Failure to man project ■ Failure to provide sufficient equipment ■ Failure to coordinate ■ Inadequate project management controls

tion contractor must be concerned about whether the owner will be able to reimburse or pay progress payments in a timely manner.

By far the most common cause of contractor bankruptcy is not failure to earn a profit over the long term, but not being paid in a timely manner. Most contractors go bankrupt because they cannot meet short-term obligations, resulting from being unable to manage their current obligations.

Owner-furnished materials. Owners typically want to procure and furnish their own equipment for unique projects, such as educational facilities, hospitals, and research facilities, because of the special needs associated with such projects. Most owners fail to understand that contractors have specific materials management plans for the project. This entails a plan for when certain material will be installed and where it will be stored once it arrives at the site.

Most owners have the mistaken notion that whenever owner-supplied material arrives on the site, the contractor can efficiently store or install the materials. Contractors usually have strict installation schedules that they must follow, and this type of misunderstanding leads to significant material management problems.

Major scope changes. The owner is ultimately responsible for defining the scope of a project. During the conceptual phase of the project, the scope will probably change quite frequently without causing undue hardship. However, changes that occur during the construction phase, or even the detailed design phase, inevitably make life difficult for the construction contractor. Scope changes result from a number of situations, such as project owner desires, design reviews, constructability issues, safety requirements, and operations and maintenance concerns.

Defective plans and specs. There have been few instances where an engineering design was so complete that a project could be built to the exact specifications contained in the original design documents. Errors or omissions in the design documents will always result in some modifications to the originally planned construction process.

Slow to mobilize. Many contractors do not perform sufficient planning in advance to effectively mobilize the project. Often the owner or his representative informs the contractor that the site is available and ready for mobilization to begin. Most contractors do not move onto a site in an efficient or timely manner because of insufficient planning regarding the equipment and labor needs that are required.

Failure to allocate resources. Earning a profit is obviously essential to any contractor's long-term strategy. How does a contractor make a profit? A profit is normally earned by managing a project with the least amount of equipment on site. The more men and equipment situated on the site, the more cash outflow is required. Managing limited amounts of manpower and equipment efficiently and cost-effectively is easier said than done, especially if a contractor is dealing with several projects at once.

What frequently happens is that a contractor either does not have the right mix of people or lacks the right equipment on a particular project because he uses these limited resources on other projects. Labor and equipment availability is an important consideration to take into account when preparing to bid on various types of construction projects.

Inadequate project management controls. Project management control areas that cause the greatest concern include:

- Lack of meaningful cost control information
- Project decisions that are not made in a timely manner or not promptly implemented
- Decisions affecting the total project that are made by one group unilaterally
- Confusion regarding the apparent authority and responsibility of project personnel
- An apparent lack of integrated planning

Insufficient work definition. This has a far-reaching effect because a project's scope must be defined in enough detail for comprehensive scheduling, estimating, and resource allocation to be reasonably accomplished. Whenever the owner is unwilling to make a concerted effort to define the construction project in the early stages, problems will undoubtedly occur throughout the entire project life cycle.

Unrealistic schedules and underestimated costs. Whenever time schedules and/or project costs are based on optimistic predictions, the project will be burdened with missed milestone dates, cost overruns, and even low morale among project personnel. Sufficient time must be allocated in the planning stage of a project to realistically estimate time and cost requirements.

Inadequate cost control and accounting practices. These problem areas are associated with imperfect management planning and control. Possible explanations of why estimated costs for project activities are so inaccurate when compared with actual costs include:

- Lack of planned information systems
- Poor correlation of work definitions
- Improper control of schedules and cost segregation
- Inaccurate cost allocation methods

Identifying Construction Risk Areas

To properly plan and organize the execution of any construction project, a concerted effort must be undertaken to identify risk areas that can affect the project's success or failure. No two construction projects are alike. Thus, assessing the risk areas must be performed for each new undertaking to fully understand which situations may cause undue hardships in completing a project on time and within budget.

Exhibit 8-2 summarizes the type of risk factors and their variable levels of risk that are significant when assessing the likelihood of critical problems occurring on any given construction project. The major components of construction risk are categorized into the areas of:

- Timing
- Project nature
- Approvals
- Owner structure
- Project delivery
- Project controls

Timing. The risk area of timing applies not only to the amount of up-front time that is spent in planning a construction project but also to how fast a project progresses during physical completion of the

EXHIBIT 8-2 Identifying Risk Factors and Level of Risk

Risk Factor	High	Risk Factor Moderate	Low
1. *Timing*			
■ Preconstruction time	Weeks	Months	Years
■ Construction pace	Fast-track	Selected overtime	Normal
2. *Project nature*			
■ Type of construction	State-of-the-art	Government work	Standard office
■ Relative size	Large	Normal	Small
■ Complexity	Rehab; process	Moderate	Spec. building
■ Location	New geography	Regional	Local
■ Status of plans and specs	Insufficient for bid	Essential components shown	Complete
3. *Approvals*			
■ Regulations and codes	Many agencies and rules	Standard safety codes	Basic building codes
■ Politics	Adverse community reactions	Conformity standards	Community support
4. *Owner structure*			
■ Decision process	Building boom	Planned growth	Undeveloped
■ Scope changes	Many committees and users	One small committee	One person
■ Design and construction experience	Many preliminary designs	General program known	Direct approach
■ Monitoring talent	Nobody	Use of consultant	Construction staff
■ Fund resources	No money for overruns	Need to borrow for overruns	In-house funding
5. *Project delivery*			
■ Contract type	Fast-track	GMP	Fixed-price
■ Project management	Owner acts as contractor	Construction manager	Design and build
6. *Project controls*			
■ Owner's representative	None	Architect/engineer	Project management oversight
■ Schedule	No formal schedule	Bar charts	Critical path method
■ Cost control	Schedule of values	Job cost system	Cost loaded schedule
■ Contracts	None	Standards forms	Custom
■ Change order management	Resolve at end of job	Resolve postjob	Resolve prejob

work. If virtually no preplanning has been undertaken for an upcoming project, and the work is contemplated to be done in a fast-track manner (construction is begun before a complete design is finished), the level of risk is sufficiently high that problems will undoubtedly occur on this project.

Project nature. The category of project nature relates to the overall definition of the anticipated project, which includes such variables as type, size, complexity, location, and completeness of design plans. Typically, commercial buildings that are designed to normal specifications are less prone to risk of problems and disputes than complex industrial process plants that are built to unique design criteria.

Each risk factor obviously can have a varying degree of risk depending on the circumstances—a standard office building can be quite risky if it is built in a new geographical location and the plans and design specifications are incomplete prior to beginning construction.

Approvals. All construction projects have various levels of approvals that must be secured before starting construction fieldwork. Typically, projects that must meet stringent environmental regulations (and be reviewed and approved by numerous levels of government agencies) are prone to time delays and cost overruns. Likewise, projects that are adverse to the general public perception, such as nuclear power plants, will encounter significantly more approval problems than publicly supported projects, such as the building of a public park or recreational facility.

Owner. The type and experience of the owner greatly impacts the variability of risk associated with a given construction project. The owner is the single most important entity determining the likelihood of future problems on a planned project. An owner organization that has little construction experience or does not engage qualified construction management consultants creates significantly more risk for a project than an organization that has a defined set of policies and procedures for monitoring the construction work.

Most sophisticated owners fully understand that changes to the scope of work create havoc among construction contractors that try to complete the job on time and within budget. In addition, owners that have not adequately budgeted, with an appropriate level of planned contingency funds, create significant risk hazards.

Project delivery. The category of project delivery is related to the type of contractual relationship between the parties and the organizational makeup for overseeing the work. Fixed-price or lump-sum types of contracts are inherently less risky than fast-track construction that requires a cost-plus type of contract arrangement. In addition, an owner that attempts to fulfill the responsibility of acting as its own general contractor, by subcontracting all or a significant portion of the work, is more likely to have costly problems occurring during physical completion of the project.

Project management controls. The risk area of project management controls has critical impact on ensuring that problems are discovered on a project before they become significant cash outflows. The control mechanism with the lowest risk level is one in which the schedule and cost control components are linked together by a cost-loaded schedule. This type of control ensures that ongoing work activities are monitored accurately by determining the timing of their physical completion, as well as direct costs associated with the individual work items.

In addition, when change orders are dealt with up-front during the ongoing construction of the project (instead of negotiating a disputed amount after the project is finished), risk associated with the planned undertaking is reduced.

Finally, all risk factors should be critically evaluated to determine the overall uncertainty of the construction project. Some factors may be more prone to risk than others, such as owner structure and project controls. Once the risk factors have been identified and evaluated, an appropriate course of action can be developed to ensure that the risk for the project is reduced to an appropriate level.

Minimizing Risks to Avoid Disputes

The periods of initial planning and the periods before and immediately following the award of the contract are extremely important in preventing disputed claim situations altogether. These periods also represent the ideal time to provide for systematic collection and retention of information needed to manage the project—or to support a future claim, if necessary.

Effective management begins with the planning of the project. It requires a coordinated effort between owners, design engineers, and experienced construction personnel, with heavy emphasis on the involvement of construction expertise in the design stages. Three areas warrant special attention:

1. Selecting and installing management systems
2. Careful review of the contract and subcontracts
3. Establishing proper records and filing systems

Selection and installation of management systems

The first step toward preventing a major overrun and related claims disputes on a project is to assure from the start that it will be properly managed and controlled. In past years, with fewer government regulations and other external influences, and with projects being generally smaller, better defined, and less technically complex, establishing effective controls generally was not the problem it is today. Now, with increased emphasis on larger projects, "fast track" approaches, and increased use of negotiated or cost-plus types of contracts, establishing controls has become more difficult and complex—and an even more essential requirement.

The construction industry generally has been slow to recognize the importance of effective integrated management systems, although such systems have been shown to be the first step in preventing major overruns and related claim disputes. By nature, most contractors seem to avoid heavily structured management approaches, priding themselves on being entrepreneurial personalities with the ability to "scramble" through a myriad of adversities. While this attitude can be highly effective, such "shirt pocket," "seat-of-the-pants," and other rudimentary management systems based on the abilities of one or a few individuals are totally inadequate for major construction projects.

Key components of an effective management system include:

1. *Comprehensive schedules.* A study by the Business Roundtable concluded that on average, the duration of most construction projects could be reduced by 10 percent through effective scheduling practices, with the resulting cost savings of approximately 3 percent of the total project cost. At the same time, the study found a strong reluctance by contractors to use and maintain effective scheduling systems, particularly network systems.

Contracts for all major construction projects should require some form of network scheduling system that is actually used to direct and control the project. Moreover, the requirement for such a system should be a provision of the contract. Owners should assure themselves that contractors selected for major portions of the work have the capability to provide and operate under this and other key management systems. Day-to-day management is the responsibility of both the owner and the various contractors hired to perform the work. The owner is generally held responsible for ensuring the coordination of contractors so as not to impede the progress of any individual contractor.

Failure to provide for appropriate controls, or to ensure the competence of any contractor hired, could raise serious questions regarding fulfillment of this responsibility by the owner or a general contractor.

2. *Effective estimating and cost control systems.* Proper scoping of the job and its risks and effective budgeting and cost-accounting systems can have a dramatic effect on performance of the job and on subsequent claims. Yet these areas remain substantially neglected in many, if not most, construction projects. In fact, the combined allocation to all of these functions typically ranges from one to two percent of project costs. Poor scoping of a project alone is estimated to increase costs by two to four percent.

3. *Bid estimates.* In preparing to bid or estimate a package and initial project schedules, contractors should recognize that each element may be critical later in proving that claimed damages did not result from unreasonable estimates or schedules, or from the contractor's own negligence. Where anticipated project costs are sufficient to warrant concern, the contractor should always ensure that estimates and schedules are based on detailed work papers supporting their reasonableness.

Standard preparation and management review procedures should be used to ensure that all estimates have been developed from a thorough investigation of available facts, that any productivity and cost standards are reasonable, and that calculations are accurate.

The contractor must ensure that all specifications used in the estimate either match those provided by the owner or are otherwise specifically justified. Management should also ensure that jobsite conditions represented by the owner are more accurate.

4. *Accounting systems.* Accounting systems appropriate to the particular job's complexities are essential. Most important to potential claim situations is a system that ensures that every extra work item that is not the fault of the contractor be subject to a separate account that identifies all labor, equipment, material, and overhead costs. Important tools in this regard are labor and equipment daily time cards, purchase order and receiving reports, and the cost accounting system.

Every contractor performing a major construction project should have cost accounting systems and procedures that permit separate tracking costs for any work outside the scope of the contract, or for work which results from problems imposed by the owner. The chart of accounts, therefore, should be sufficiently flexible to provide this capability.

During the initial planning phase, and later during mobilization, each area of the job should focus on high-cost items and on items that are otherwise considered to be high risk. Attention should be given at this point to methods for measuring the impact of problems occurring in any area and the documentation needed to support a future claim for each cost category.

Recognizing these potential areas beforehand will help to establish a suitable accounting and reporting system to deal with them when they occur.

The cost accounting system should be tailored specifically to accommodate these potential problems and accounting requirements. At a minimum, the system should provide for a separate cost account or subaccount for labor, material, and equipment on each disputed or extra work activity that occurs.

5. *Materials management systems.* Neglect of materials management systems is one of the most common weaknesses found in the planning and execution of most construction projects. Manufacturing and construction both require the same basic resources: materials, physical space, equipment and tools, trained labor, and usually some form of energy. The absence of any one of these elements can stop production completely. An inadequate supply of appropriate resources when needed can delay production and increase costs substantially.

The construction industry is generally far behind manufacturing operations in recognizing the importance of materials management as part of production planning and control. Considering that materials account for approximately 60 percent of all construction costs, it is surprising that more attention is not given to this function. According to industry estimates, improved materials handling systems could reduce average labor costs (currently 30 to 40 percent of total costs) by approximately six percent. Yet this function consistently remains a stepchild in the organization and management of most jobs.

Every construction contract that involves substantial handling of materials should require a detailed plan and schedule for materials management from the contractor (who should be required to demonstrate the ability to establish and operate an effective materials management system before being awarded the contract). The materials management plan should be integrated with the overall project schedule and provide for the following:

- Assignment of qualified personnel, reporting relationships within the project organizational structure, and any necessary onsite training
- Procedures for vendor selection, purchasing, receiving and inspection, materials storage, accounting, security, and scheduling for timely delivery to the jobsite
- Space and facility requirements for storage
- Lead items, order points, and expediting

Review of the contract and subcontracts

Lack of understanding of the contract and failure to perform a careful review of its contents are major shortcomings that often prevent the contractor from making a reasonable assessment of risk involved, and are substantial contributors to future claims situations.

For example, does the contract contain a "Changes" clause? Does it establish the right of the contractor to collect for additional costs under changed conditions (no clause—no recovery)? Is there a "Termination for Convenience" clause that may preclude the contractor from collecting lost and anticipated profits? Are there exculpatory clauses, that is, no damages for delay (most are considered basically valid)? What are the notification requirements and procedures? What items are incorporated by reference in the contract? Are outside agency regulations (such as EPA general conditions) included? And perhaps most importantly, how are the courts where the work will be performed likely to interpret the contract language and clauses? All of these factors should be matters of concern and careful review by the owner, the contractor, and their respective legal counsels during the bidding or negotiation stage and after the contract is received for signing.

After receiving the awarded contract, contractors and their legal counsel should conduct a thorough review and analysis of its contents to ensure that the scope for the work described accurately reflects the work that was bid. Particular attention should also be given to all areas incorporated by reference, so that all rights and obligations are thoroughly understood. Several articles typically are of special significance with respect to future claims:

- Changes
- Changed conditions
- Variations in quantities (unit cost)
- Terminations for default
- Damages for delay
- Time extensions
- Escalation of labor or material

The prime contractor should also perform a careful review of the forms and contracts provided to subcontractors. Work in these documents must encompass all work required by the owner's contract, and all that is expected from each subcontractor is described completely and clearly. For example, what is the meaning of "installation"? Keep in mind that the contract will be interpreted against the party that drafted it.

Does each subcontract incorporate all the terms and conditions of the contract? Is the incorporation by a general inclusion statement or by specific reference? Are the courts in the area where the work will be performed likely to accept incorporation by reference, or will they require specification of the terms and conditions? All of these legitimate concerns with respect to future claims are matters that should be discussed with the contractor's legal counsel before the subcontract forms are prepared and distributed.

Records and filing systems

Standard record and filing systems should be established and maintained for each contract, along with standard reporting procedures, to provide required information. Good project management, preparation of claims, and defense against claims require systematic collection, preservation, and retrieval of at least the following:

- Request for quotation or bid, specifications, test reports, and other documents provided by the owner
- The contract and all amendments
- Meeting notes and correspondence with the owner/engineer
- Original estimate and estimate work papers plus any revisions
- Project schedule and budgets, including all changes
- Subcontractor files containing contracts, bids, bonding information, and correspondence
- Progress reports and time-phased photographs of construction progress and job conditions

In establishing policies and procedures governing the items that will be retained, contractors and owners alike should bear in mind that project file contents most likely will be subject to discovery by the opposition in future litigation. Potential embarrassment or damage to a future case should be a guiding consideration in preparing written materials and decisions regarding their retention.

Conclusion

Minimizing construction risk requires a multifaceted approach that must be shared by all parties associated with a construction project. All projects contain some degree of uncertainty; thus risk is inherent in every decision made on the project. With proper identification and evaluation of significant risk factors, many troublesome areas can be discovered and action taken prior to costly mishaps.

Contract disputes can be reduced when sufficient efforts are made by the principal parties to the contract to understand the magnitude of risk that is associated with a given construction project.

Chapter

9

General and Subsidiary Ledgers

A record of any one transaction, standing alone, is not likely to mean a great deal. Its meaning increases when it is grouped with other transactions of the same kind. For example, a voucher showing the purchase of a load of lumber shows when it was bought, from whom, and its cost. Grouped with all other purchases of material, the same voucher gains significance as a part of the total cost of materials. Broken down to show the job on which the lumber was used, the voucher gains still more meaning as part of the material cost of a specific job. Broken down further, and classified by the actual use to which the lumber was put on the specific job, it becomes a part of the cost detail that tells the contractor whether his original estimate was right, and whether some aspect of a job is being run as efficiently as it should be.

One of the principal functions of accounting records is to group transactions in ways that will provide useful information to those who manage the business. Stated another way, every figure needed by management should appear as the balance of some ledger account. For example, it is important to know the total amount you can expect to receive from your customers in the near future, and so you have a general ledger account called "Progress payments currently due." Or, if retained percentages are not significant, you may just have an account called "Trade accounts receivable" or "Amounts due from customers." It is equally (if not more) important to know specifically who owes you the money, so a record is also kept of the amount each customer owes. The total of the individual customers' accounts must, of course, equal the amount which the general ledger shows as being due from all customers.

To the experienced bookkeeper or accountant the controlling account and the subsidiary ledger just described are almost too elementary to mention, and the accounts receivable-customer's ledger relationship is the most elementary of the lot. Still, it is not uncommon, among construction companies, to find this elementary relationship being ignored.

For example, it is fairly common to find cash, advanced to jobs in progress, treated as accounts receivable and thrown in with amounts due from customers. This treatment is especially common if job accounting is decentralized. Again, it is not uncommon to find, in the general ledger, an account called "Equipment," which has been charged with a wide variety of capital expenditures. There may be an equipment ledger showing individual pieces of equipment and what part of the general ledger account balance applies to each one. In a given situation such treatment may be entirely proper, but there is a good chance that it is costing the contractor money on his income tax because of his failure to provide the information necessary to claim the proper amount for depreciation.

Every record kept and every figure recorded should have a purpose. If it has no purpose, the cost of putting it down is wasted. Therefore, when deciding what accounts to have in the general ledger and which of the general ledger accounts is to control a subsidiary ledger, it is necessary to determine the use to which the information will be put. Keeping records or making analyses that are not needed just because "somebody might ask for it" is an expensive precaution.

For that reason, whenever an accounting system is being designed or reviewed, the need for each general ledger account and each subsidiary ledger must be determined in terms of the four following questions:

1. What are we trying to accomplish by keeping this account (or this subsidiary ledger)?
2. How much work is going to be required; and is that work justified in view of the objective?
3. Do we have any well-defined accounting policies and, if so, does this account (or subsidiary ledger) contribute anything to carrying out these policies?
4. Are there any peculiar job conditions that require the particular accounting treatment that this account (or subsidiary ledger) provides?

In addition to deciding what accounts to have in the general ledger (chart of accounts) and what subsidiary ledgers are necessary, it is well to consider how to group or sequence the accounts. To do this requires studying the situation based on the end-product needs. For example, a contractor should ask himself, "Do I want to know what it costs periodically to run my plane, equipment, warehouse, etc.?" If the answer is yes, then a lot of analysis time can be saved by grouping accounts to achieve this. The following is an example of how the accounts might be grouped for aircraft operations:

Account No.		*Name*
150		Aircraft operations
	151	Pilot salaries and benefits
	152	Maintenance salaries and benefits
	153	Depreciation—aircraft
	154	Depreciation—other equipment
	155	Hangar rent
	156	Fuel, oil, and lubricants
	157	Maintenance and inspections
	158	Miscellaneous
159		Aircraft cost recovery
	160	Charged to jobs
	161	Third party charters
	162	Selling and bidding

Note that cost recoveries are set out separately in this case so that the contractor sees what the gross cost of the operation is. If costs and recoveries are netted in an account without the detail shown above, there will be much less incentive by those in charge of the department to control cost, and excesses will be hidden by increasing charges to jobs.

As a practical matter construction executives often do without useful information which general and subsidiary ledgers could provide because, when kept by hand, such records are costly and are rarely available soon enough to help much. Applications of computers (see Chapter 14) to the keeping of general and subsidiary ledgers is becoming increasingly practical even for medium-sized and small contractors.

Account Classification

Because the needs of companies in the same industry tend to be similar, it is usual to find more or less standard classifications of accounts, and similar terminology in use. In fact, many trade associations have promoted the use of standard account classifications for their industry in order to obtain comparability of industrywide statistics. In public utilities and other closely regulated industries, standard classifications and terminology are even required by law or regulation. Unfortunately, no standard account classification has gained general acceptance throughout the construction industry, although some specialty trades have designed classifications to fit their special needs. However, when building for public utilities and similar clients, it may be necessary to give the client cost figures to fit the required classifications.

In recent years, a number of publishers of accounting forms have offered so-called "ledgerless" systems for small business firms. Such systems are rarely suitable for any but the very smallest operators, and rarely make adequate provision for keeping track of the cost of each separate job. They usually provide the equivalent of a cash receipts and disbursement record with a column for each account. At the end of each month the totals of the columns are carried to the financial statement without passing through a ledger. If the operation is small enough to minimize the need for accounting as a tool of management, this type of system is often adequate. When applicable, most ledgerless systems meet the minimum requirements for federal income tax reporting. The best way for a contractor to determine when he has outgrown such a system is to ask himself two questions: (1) How often do I need financial information that I cannot get from the books? (2) How often do I have transactions for which the books make no provision?

If the answer to both questions is "rarely" or "never" then the system is still adequate.

Schedules 9-1 to 9-4 give several systems of classifications of accounts that may prove useful in illustrating the arrangement, numbering, and terminology commonly used by construction companies in their general ledgers.

A *medium-sized general contractor's classification* is shown in Schedule 9-1. Note that in Account 70, provision is made for unbalancing the bid or the payment schedule without distorting job income (Chapter 1). Note tool that this classification may be expanded or contracted to meet the needs of almost any size company.

In a business large enough to require such an account classification as the one shown in Schedule 9-1, it would be usual to expect subsidiary records of some sort on accounts 21, 22, 23, 24, 33, 33R, 34, 34R, 35, 35R, 52, 57, 58, 100, 101, and 103. All of these might not be required; on the other hand, others might also be needed.

A *small contractor's classification* is shown in Schedule 9-2. This system would serve the purposes of a small general building contractor.

Note that in Schedule 9-2, provision is made to distribute all overhead to jobs. This procedure need not be followed unless it is particularly desired. If accounting were on a cash basis, receivables or payables would not be included.

A *large contractor's classification* is shown in Schedule 9-3. This classification of accounts is more extensive than that used by many very large contractors. It is useful for a large, diversified general contractor, and provides the skeleton for a very comprehensive and

SCHEDULE 9-1 CLASSIFICATION OF ACCOUNTS FOR MEDIUM-SIZED CONTRACTOR

Account No.	*Assets* Account name
10	Petty cash
11	Cash in bank—general disbursement account
12	Cash in bank—payroll account
13–19	Any special bank accounts
20	Notes receivable for contracts completed
21	Accounts receivable for contracts completed (exclusive of retention)
22	Amounts due on uncompleted contracts
	(a) Approved by owner's representative
	(b) Subject to approval by owner's representative
	(c) Retention
23	Retention due on completed contracts
24	Costs and estimated earnings in excess of billings
25	Inventory of materials and supplies not assigned to jobs in progress
26	Deposits for bids and other guarantees
27	Stocks, bonds, and special assessment obligations
28	Other security investments
29	Other receivables
30	Land not used in business
31	Land used in business
32	Buildings
32R	Accumulated depreciation of buildings
33	Construction machinery and equipment
33R	Accumulated depreciation of construction machinery and equipment
34	Trucks and trailers
34R	Accumulated depreciation of trucks and trailers
35	Passenger automobiles, pickups, and jeeps
35R	Accumulated depreciation of passenger automobiles, pickups, and jeeps
36	Tools (other than small tools)
36R	Accumulated depreciation of tools (other than small tools)
37–39	Special fixed asset groupings and related reserves
40	Prepaid insurance and bonds
41	Prepaid taxes, licenses, and fees
42	Expense advances to employees
43	Expendable tools and supplies
44–45	Other prepaid expenses
46	Deferred prejob expense
47–48	Other deferred charges
49	Other assets
	Liabilities
50	Notes payable to banks
	(a) For bid deposits
	(b) For specific jobs
	(c) General borrowing
51	Accounts payable to general trade creditors
52	Accounts, notes, and contracts payable for equipment—current portion
53	Amounts withheld from employees for taxes
	(a) Income taxes withheld

SCHEDULE 9-1 CLASSIFICATION OF ACCOUNTS FOR MEDIUM-SIZED CONTRACTOR *(Cont.)*

Account No.	Account name
	Liabilities
	(b) FICA taxes
	(c) State unemployment taxes
	(d) Other withholdings
54	Accrued taxes
	(a) Payroll taxes
	(b) Sales and use taxes
	(c) Property taxes
	(d) Income taxes (for use by corporations only)
	(e) Deferred Federal Income Tax
	(f) Other taxes
55	Accrued insurance
56	Due to subcontractors on uncompleted contracts
	(a) Current billings
	(b) Retentions
57	Due to subcontractors on completed contracts
	(a) Current billings
	(b) Retentions
58	Currently due to officers and employees (or related individuals and firms)
59	Other current liabilities
60	Noncurrent portion of equipment contracts
61	Long-term loans, mortgages, trust deeds, and so on
62	Other noncurrent liabilities
70	Billings in excess of cost and estimated earnings
80	Contingent liabilities (memo only)
	Net Worth
90	Capital stock (if corporation)
	Partners' investments (if partnership)
	Owner's investment (if single proprietorship)
95	Retained earnings
99	Personal accounts (if partnership or proprietorship)
	Income and Expense
100	Job revenue taken into income (individual jobs designated by 100.1, 100.2, and so on)
101	Job costs incurred producing job revenue (individual jobs designated by 101.1, 101.2, and so on)
	(a) Labor
	(b) Material
	(c) Subcontracts
	(d) Equipment rental
	(e) Overhead
102	Equipment rentals earned
103	Cost of equipment ownership
	(a) Depreciation
	(b) Maintenance
	(c) Taxes and licenses

SCHEDULE 9-1 CLASSIFICATION OF ACCOUNTS FOR MEDIUM-SIZED CONTRACTOR *(Cont.)*

Account No.	*Income and Expense* Account name
	(d) Insurance
	(e) Rental paid to others
	(f) Other costs
105	Income from joint ventures
106	Fees earned on cost-plus-fixed-fee jobs
107	Discounts earned
108	Income from surplus and salvage sales (net)
109–119	Other income accounts as needed
120–199	General and administrative expenses
121	Executive salaries
122	Office salaries and wages
123	Insurance—general
124	Taxes and licenses
125	Repairs and maintenance
126	Supplies and expendables
127	Postage
128	Telephone and telegraph
129	Depreciation on office and engineering equipment
130	Rent (or comparable ownership costs) and utilities
131	Automobile expense—general office
132	Dues and subscriptions
133	Advertising
134	Travel and promotion
135	Contributions
136	Loss on bad debts
137	Interest
198	Available for other operations as need (asphalt, plant, etc.)
199	Other expenses
200	Provision for income taxes—current
201	Provision for income taxes—deferred

SCHEDULE 9-2 CLASSIFICATION OF ACCOUNTS FOR SMALL UNINCORPORATED CONTRACTOR

Account No.	*Assets* Account name
1	Petty cash
2	Cash in bank
3	Due from customers—current
4	Due from customers—retention
5	Other receivables
6	Prepaid insurance
7	Small tools and supplies
8	Other prepaid and deferred items
9	Machinery, equipment, automobiles, and trucks
9R	Reserve for depreciation
10	Other assets

SCHEDULE 9-2 CLASSIFICATION OF ACCOUNTS FOR SMALL UNINCORPORATED CONTRACTOR *(Cont.)*

Account No.	Account name
	Liabilities
20	Notes payable
21	Trade accounts payable
22	Due to subcontractors
23	Withheld from employees for taxes (with analysis column for detail)
24	Payroll taxes accrued (with analysis columns for detail)
25	Insurance accrued
26	Other accrued expenses
27	Equipment contracts payable—current portion
29	Other liabilities—current
30	Long-term liabilities
	Net Worth
40	Owner's investment
41	Owner's drawing account
	Income
50	Contract revenue on completed jobs (detail by jobs)
51	Cost of jobs completed (detail by jobs)
	(a) Labor
	(b) Materials
	(c) Subcontracts
	(d) Other costs
53	Discounts earned
54	Other income
60	Office salary
61	Office supplies and postage
62	Telephone and telegraph
63	Taxes and licenses
64	Insurance
65	Repairs and maintenance
66	Dues and subscriptions
67	Contributions
68	Advertising and promotion
69	Automobile and truck expense
70	Rent and utilities
71	Depreciation on office equipment
72	Interest
79	Other expense
99	Overhead distributed on jobs (credit)

SCHEDULE 9-3 CLASSIFICATION OF ACCOUNTS FOR LARGE CONTRACTOR

A. *General ledger accounts*

100 *Current assets and prepaid expenses*

- 101 Petty cash
- 102 General disbursement account
- 103 Payroll account
- 104–109 Other special bank accounts
- 110 Billings on construction contracts
  - 110.1 Current billings on lump sum contracts
  - 110.2 Fees billed on CPFF contracts
  - 110.3 Costs billed on CPFF contracts
  - 110.4 Retained percentages
- 112 Other accounts receivable
  - 112.1 Due from employees
  - 112.2 Due from subcontractors
  - 112.9 Due from others
- 115 Allowance for doubtful accounts
- 116 Costs and estimated earnings in excess of billing
- 117 Inventory of materials and supplies
  - 117.1 Warehouse stock
  - 117.2 Machine shop parts and supplies
  - 117.3 Mess hall inventory
  - 117.4 Commissary inventory
  - 117.5 Camp inventory
- 118 Credits to inventory for outside sales
  - 118.1 Sales of warehouse stock
  - 118.2 Sales of machine shop parts and supplies
  - 118.3 Sales of mess hall inventory
  - 118.4 Sales of commissary inventory
  - 118.5 Sales of camp inventory
- 130 Other current assets
- 150 Prepaid insurance
  (Subaccounts for each policy and for compensation insurance deposits)
- 151 Employees' expense advances
- 160 Deferred charges—mobilization costs
  - 160.1 Forms
  - 160.2 Access roads and bridges
  - 160.3 Temporary buildings
  - 160.4 Temporary plant installations
  - 160.5 Temporary utility installations
  - 160.6 Incoming freight on plant and equipment
  - 160.7 Unloading, assembling, and preparation of equipment for use
  - 160.9 Other move-in costs
- 161 Deferred charges—move-out costs
  - 161.1 Disposal of forms
  - 161.2 Restoration of access roads and bridges
  - 161.3 Removal of temporary buildings
  - 161.4 Dismantling temporary plant
  - 161.5 Removing temporary utility installations
  - 161.6 Outgoing freight on plant and equipment
  - 161.7 Preparation and loading out equipment
  - 161.9 Other move-out costs

SCHEDULE 9-3 CLASSIFICATION OF ACCOUNTS FOR LARGE CONTRACTOR (Cont.)

	162	Deferred charges—performance and payment bonds	
	170	Mobilization costs amortized	
		170.1	Forms
		170.2	Access roads and bridges
		170.3	Temporary buildings
		170.4	Temporary plant installations
		170.5	Temporary utility installations
		170.6	Incoming freight on plant and equipment
		170.7	Unloading, assembling, and preparation of equipment for use
		170.8	Performance bonds
		170.9	Other move-in costs
	171	Move-out costs amortized	
		171.1	Disposal of forms
		171.2	Restoration of access roads and bridges
		171.3	Removal of temporary buildings
		171.4	Dismantling temporary plants
		171.5	Removing temporary utility installations
		171.6	Outgoing freight on plant and equipment
		171.7	Preparation and loading out of equipment
		171.9	Other move-out costs
	180	Expendable tools and equipment	
		180.1	Construction tools
		180.2	Office equipment
		180.3	Engineering equipment
200	*Fixed and other assets*		
	250	Equipment	
		250.1	Construction equipment
		250.2	Automobiles, jeeps, and pickup trucks
		250.3	Office equipment
		250.4	Engineering equipment
	251	Accumulated depreciation	
		251.1	Construction equipment
		251.2	Automobiles, jeeps, and pickup trucks
		251.3	Office equipment
		251.4	Engineering equipment
	290	Other assets	
		290.1	Utility, transportation, and tax deposits
		290.2	Returnable container deposits
300	*Liabilities and proprietorship equities*		
	305	Trade accounts payable	
	306	Subcontractor retained percentages	
	308	Accrued payroll	
	310	Payroll taxes accrued	
		310.1	FICA employer's tax
		310.2	FUI employer's tax
		310.3	SUI employer's tax
	311	Amounts withheld from employees for taxes and insurance	
		311.1	FICA employee's tax
		311.2	Federal income tax withheld
		311.3	State income taxes withheld
		311.4	H&A insurance
	312	Accrued insurance based on payrolls	
		312.1	Workers' compensation

SCHEDULE 9-3 CLASSIFICATION OF ACCOUNTS FOR LARGE CONTRACTOR (Cont.)

		312.2	General liability
		312.3	Excess liability
	313	Other state and local taxes accrued	
		313.1	Sales and use taxes
		313.2	City and county business taxes
		313.3	Real and personal property taxes
	320	Accrued income taxes—federal	
		320.1	Federal—current
		320.2	Federal—deferred
	321	Accrued income taxes—state	
		321.1	State—current
		321.2	State—deferred
	322	Accrued income taxes—foreign	
		322.1	Current
		322.2	Deferred
	330	Other accrued liabilities	
	340	CPFF revolving fund	
	350	Unearned revenue deferred	
	380	Capital stock (or partners' capitals or proprietor's equity)	
	385	Retained earnings (unallocated)	
	390	Advances to jobs	
		390.1	Cash
		390.2	Other assets
	391	Interest and overhead reimbursement from jobs	
		391.1	Interest
		391.2	Overhead
	393	Accumulated earnings on jobs	
	393	Withdrawals from jobs	
		393.1	Advances
		393.2	Interest and overhead
		393.3	Accumulated earnings
500	*Income and costs*		
	501	Contract revenue	
		501.1	Progress billings submitted
		501.2	Estimated contract revenue earned but unbilled
		501.9	Revenue deferred to compensate for unbalanced bid (debit)
	505	Fees earned on CPFF contracts	
	506	Revenue from reimbursable costs on CPFF contracts	
	509	Revenue from services	
		509.1	Repairs and maintenance of equipment for subcontractors
		509.2	Repairs to rented equipment chargeable to owners
		509.9	Other income from services
	510	Equipment rentals earned	
	512	Allocated materials and equipment furnished by owner on CPFF jobs	
	520	Income from sales of fixed assets	
	540	Other income	
	551	Construction costs—work items	
		551.1	Labor
		551.2	Materials
		551.3	Subcontracts
		551.4	Equipment costs
		551.5	Other direct costs
		551.7	Bonds—performance and payment

SCHEDULE 9-3 CLASSIFICATION OF ACCOUNTS FOR LARGE CONTRACTOR (Cont.)

		551.8	Allocated work items
		551.9	Overhead allocated
	555	Nonreimbursable costs—CPFF jobs	
	559	Cost of services performed for others	
		559.1	Cost of repair and maintenance of equipment for subcontractors
		559.2	Repairs to rented equipment chargeable to owners
		559.9	Cost of other services
	560	Costs applicable to equipment rented to others	
	562	Reasonable value of materials and equipment furnished by owner on CPFF jobs	
	563	Reimbursable costs on CPFF jobs	
	570	Equipment costs	
		570.1	Rentals paid to others
		570.2	Depreciation on owned equipment
		570.3	Repairs and maintenance—labor
		570.4	Repairs and maintenance—materials and supplies
		570.5	Fuel and lubrication
		570.6	Taxes and licenses
		570.7	Insurance
		570.8	Distributable costs
		570.9	Other equipment costs
	571	Equipment costs charged to work items	
600	*Bidding expenses*		
	601	Estimating salaries	
	605	Bidding expense charged by joint ventures and others	
800	*Overhead*		
	801	Salaries	
		801.1	Executive and supervisory
		801.2	Engineering
		801.3	Accounting and clerical
		801.4	Warehouse and yard
	802	Stationery supplies and expendable equipment	
	803	Depreciation on office and engineering equipment	
		803.1	Office equipment
		803.2	Engineering equipment
	804	Automobile expense	
		804.1	Gas, oil, grease, etc.
		804.2	Tires and repairs
		804.3	Depreciation
		804.4	Insurance
		804.5	Taxes and licenses
		804.6	Other auto expenses
	805	Executive and employee expense accounts	
	806	Professional services	
	807	Rent	
		807.1	Office, etc.
		807.2	Equipment
	808	Telephone and telegraph	
	809	Taxes	
		809.1	Payroll taxes on employer
		809.2	Property taxes
		809.3	Business licenses, etc.

SCHEDULE 9-3 CLASSIFICATION OF ACCOUNTS FOR LARGE CONTRACTOR *(Cont.)*

	809.9	Other taxes
810	Insurance and bonds	
	810.1	Workers' compensation
	810.2	Public liability and property damage
	810.3	Equipment floater
	810.4	Fidelity, position, and forgery bonds
	810.5	Performance, payment, and bid bonds
	810.6	Fire and comprehensive
	810.7	Builder's risk
	810.9	Other
812	Repairs and maintenance	
	812.1	Office and engineering equipment
	812.2	Building and grounds
	812.3	General cleanup
813	Postage	
814	Dues, subscriptions, contributions, and local advertising	
	814.1	Dues and subscriptions
	814.2	Contributions
	814.3	Local advertising
815	Loss on bad debts	
830	Contributions to Employee Retirement Plan	
831	Costs of administering Employee Retirement Plan	
849	Miscellaneous general expense	
850	Commissary	
	850.1	Sales (credit)
	850.4	Other commissary revenue (credit)
	850.5	Cost of sales (debit)
	850.6	Salaries and wages (debit)
	850.9	Depreciation—commissary equipment (debit)
	850.10	All other commissary expense (debit)
851	Mess hall	
	851.1	Sales (credit)
	851.3	Received from subcontractors (credit)
	851.4	Other mess hall revenue (credit)
	851.5	Cost of food sold (debit)
	851.6	Salaries and wages (debit)
	851.7	Laundry and cleaning (debit)
	851.8	Expendable supplies—amortization (debit)
	851.9	Depreciation—mess hall equipment (debit)
	851.10	All other mess hall expense (debit)
852	Camp	
	852.2	Bunk rentals (credit)
	852.3	Received from subcontractors (credit)
	852.4	Other revenue (credit)
	852.6	Camp salaries and wages (debit)
	852.7	Laundry and cleaning (debit)
	852.8	Amortization—linen, bedding, etc. (debit)
	852.9	Depreciation—camp equipment (debit)
	852.10	All other camp expenses (debit)
853	First aid	
	853.3	Received from subcontractors (credit)
	853.4	Other revenue (credit)
	853.6	Salaries and wages (debit)

SCHEDULE 9-3 CLASSIFICATION OF ACCOUNTS FOR LARGE CONTRACTOR (*Cont.*)

	853.7	Laundry and cleaning (debit)
	853.8	Expendable supplies (debit)
	853.9	Depreciation—first aid equipment (debit)
	853.10	All other first aid expenses (debit)
854	Residences	
	854.2	Rentals of residences furnished employees (credit)
	854.3	Received from subcontractors (credit)
	854.4	Other revenue from residences (credit)
	854.6	Salaries and wages (debit)
	854.7	Laundry and cleaning (debit)
	854.8	Amortization of linens, dishes, etc. (debit)
	854.9	Depreciation—residences (debit)
	854.10	All other residence expense including rent (debit)
855	Protection—guards and firemen	
	855.3	Received from subcontractors (credit)
	855.6	Salaries and wages of guards, firemen, etc. (debit)
	855.8	Expendable equipment (debit)
	855.9	Depreciation—protection equipment (debit)
	855.10	All other protection expense (debit)
890	All other income and expense	
899	Overhead allocated to work items	
900	Provision for income taxes	
	900.1	Current
	900.2	Deferred

B. *Cost Accounts*
- 01 Direct labor
- 02 Materials
- 03 Subcontracts
- 04 Equipment costs
- 05 Allocated work items
- 06 Overhead allocated

SCHEDULE 9-4 CLASSIFICATION OF ACCOUNTS FOR DECENTRALIZED, MEDIUM-SIZED TO LARGE JOBS

A. Cash
- 100 Office cash (petty cash)
- 101 Job payroll bank account
- 102 Job general bank account

B. Receivables
- 111 Progress billings approved
- 112 Progress billings awaiting approval
- 113 Amounts earned but unbilled
- 114 Due from employees
- 115 Miscellaneous receivables
- 116 Fees (on CPFF jobs) receivable
- 117 Reimbursable costs billed
- 118 Reimbursable costs unbilled
- 119 Reimbursable costs accrued

C. Inventories
- 120 Warehouse inventories

SCHEDULE 9-4 CLASSIFICATION OF ACCOUNTS FOR DECENTRALIZED, MEDIUM-SIZED TO LARGE JOBS *(Cont.)*

D. Operations of jobsite shops and plants
  130 Shop operations cost
    .1 Labor
    .2 Materials
    .3 Cost of providing job-owned equipment
    .4 Equipment rentals
    .5 Supplies and small tools
    .6 Subcontracts
    .7 Distributable expense (general overhead)
  140 Shop operations charged to job costs (credit)
  150 Production plant operations cost
    .1 Labor
    .2 Materials
    .3 Cost of providing job-owned equipment
    .4 Equipment rentals
    .5 Supplies and small tools
    .6 Subcontracts
    .7 Distributable expense
  160 Cost of materials produced and charged to job costs (credit)
E. Deferred charges and other assets
  170 Cost of temporary structures and special plant charges
  180 Cost of temporary structures charged to job costs (credit)
  190 General expense
  199 General expense distributed to job costs (credit)
  200 Refundable deposits
  210 Returnable containers
  220 Debit items held in temporary suspense
F. Equipment owned by the job
  250 Cost of major equipment (items costing over $1,000 each)
  252 Depreciation on major equipment charged to job costs (credit)
  260 Salvage value (15 percent) of cost of minor equipment
  261 Depreciable cost of minor equipment
  262 Depreciation on minor equipment charged to job costs (credit)
G. Current accounts payable
  300 Vouchers payable—general
  301 Amounts currently due to subcontractors
  302 Retained percentages due to subcontractors
  303 Advances received from owner
  306 Group insurance premiums deducted from payroll
  307 Charitable contributions deducted from payroll
  308 Payroll savings bond deductions
H. Payroll taxes—employee deductions and employer taxes
  310 Federal income tax withheld
  311 State and local income tax withheld
  312 Federal Insurance Contributions Act (Social Security)
  313 FICA tax—employer's tax
  314 State unemployment insurance tax withheld from employees
  315 State unemployment tax—employer's tax
  316 Federal unemployment excise tax
I. Accrued taxes and insurance
  320 Real and personal property taxes
  321 State and local sales, use, and excise taxes
  322 State and local excise taxes, licenses, and fees

SCHEDULE 9-4 CLASSIFICATION OF ACCOUNTS FOR DECENTRALIZED, MEDIUM-SIZED TO LARGE JOBS ***(Cont.)***

325 Special state and local personal taxes (poll taxes, etc.)
330 Workers' compensation insurance
331 Public liability and property damage insurance
332 Employer's liability and occupational disease insurance
333 Builder's risk insurance

J. Other accrued liabilities
340 Accrued payroll
341 Accrued utilities and power bills
345 Accrued union dues deducted from payroll
346 Accrued union health and welfare fund payments
347 Accrued medical and hospitalization insurance
349 Unclaimed wages

K. Reserves
350 Reserve for dismantling and move-out
351 Reserve for major repairs and reconditioning of equipment
352 Reserve for seasonal shut-downs
353 Reserve for credit items held temporarily in suspense

L. Deferred credits to income
360 Amounts billed to owners for uncompleted work

M. Head office equity in job
370 Cash advances to job by head office
371 Excess of billings for completed work over related costs (job operating profit)
375 Cash repaid to head office (debit)
376 Losses chargeable to extraordinary conditions (debit)
377 Amounts charged to job for interest on advances and for head office overhead (debit)

N. Operating revenue and costs—lump sum and unit price jobs
400 Progress billings approved
401 Progress billings awaiting approval
402 Amounts earned but not yet billed
405 Cost of contract work billed or billable
409 Miscellaneous contract revenue

O. Operating revenue and costs—cost-plus-fixed-fee jobs
410 Contractor's fee
411 Reimbursable costs
412 Allocated materials and equipment furnished by client
415 Nonreimbursable costs

P. Nonoperating revenue and costs
420 Equipment rentals
421 Costs chargeable to equipment rental income
425 Gain or loss on sales of equipment
426 Gain or loss on sales of surplus materials and scrap
427 Undistributed equipment operations cost
.1 Operating labor
.2 Repair and servicing labor
.3 Repair parts
.5 Outside repairs and other repair costs
.6 Fuel, oil, grease, power
.7 Other supplies and costs
(NOTE: All equipment operating costs are charged to Account 427 and are allocated to appropriate cost accounts on an hourly basis. Any undistributed balance may remain in Account 427 or may be transferred to general expense.)
450 Miscellaneous gains or losses

detailed accounting system. Furthermore, it is designed so that it can be expanded or contracted as the need arises.

On each job there is a cost ledger with a columnar sheet for each work item. Each of these jobsite ledgers has a summary controlling account with columns for each of these items. The summary of these job ledger controlling accounts balances with the corresponding accounts in the main office general ledger. If the job accounts were centralized the situation would be about the same except that the job cost ledgers could be controlled directly if desired.

A *decentralized accounting classification,* for a medium-sized to large job, is shown in Schedule 9-4. If the jobs keep their own accounts, it is a good idea, wherever possible, to have the general ledger account numbers in the job ledger identical with those in the company's general ledger. This makes the task of consolidating the job books with the home office books much easier. Note that in the decentralized classification there are no accounts for accumulated depreciation as such. Instead, the account which is credited contra to the charge to the cost accounts for equipment depreciation is called "Depreciation charged to job." Note too, that this classification is applicable to lump sum, unit price, or cost-plus-fixed-fee jobs. Use of letters combined with numbers is intended to reduce errors in account designations when coding accounting data.

Integration of Subsidiary Records

Using the general ledger account classification in Schedule 9-4 as a basis, the integration of subsidiary records is illustrated below.

The subsidiary ledger for account C120, "Warehouse inventory," might contain such items as the following:

C1201 Acoustical tile
1202 Aggregate and sand—purchased
1203 Blocks—concrete and pumice
1204 Brick—red, fire, concrete
1205 Bolts, nuts, and screws
1206 Cable
1207 Camp supplies and equipment
1208 Cement
1209 Chimney tile
1210 Conduit and fittings
1211 Doors, frames, and hardware
1212 Drilling supplies and equipment
1213 Electrical supplies and equipment
1214 Fuel, oil and grease
1215 Hardware—form, rough, miscellaneous
1216 Lumber—rough, finish, and moulding
1217 Masonry and plastering supplies
1218 Miscellaneous stores
1219 Nails

1220 Painting and glazing materials

1221 Plumbing, heating, and air conditioning materials and supplies

1222 Powder, caps and blasting supplies

1223 Roofing and sheet metal materials and supplies

1224 Small tools

1225 Steel—reinforcing

1226 Steel—structural, shop repair, tool, etc.

1227 Tile—asphalt, glazed, clay, cement

1228 Tires and tubes

1229 Welding supplies

1230 Window sash, frames and fittings

1231 Commissary supplies

1232 Mess hall supplies and equipment

1233 Camp and bunk house supplies and equipment

1298 Repair parts (with subaccounts for each part)

1299 Permanent materials (not shown above)

The job may not be large enough to require more than one or two shops, so that Account 130 would need no subsidiary accounts. On the other hand there might be, on a particular job, any or all of the following:

D1301 Automobile and truck repair shop

1302 Blacksmith shop

1303 Concrete testing laboratory

1304 Electrical shop

1305 Machine and equipment repair shop

1306 Paint shop

1307 Planing mill and woodworking shop

1308 Plumbing, heating, and sheet metal shop

1309 Tire and battery shop

Similarly, a large job might produce its own aggregate, or batch its own concrete, but it is unlikely that more than one or two plants would be covered by Account 150. If there were only two or three plants they could be covered by Accounts 150, 151, and 152. This method could work for as many as ten plants, which would be more than even the average large job would require. However, if more were needed there could be a subsidiary ledger controlled by account number 150 with the numbers starting at 1501 and going through 1599.

Account 170, "Cost of temporary structures," may or may not require a subsidiary ledger, depending upon the number of temporary structures and the amount of their cost in relation to the size of the job. If some temporary structures are a charge against the job as a whole, whereas others affect only certain work items, it may be necessary to have a subsidiary ledger controlled by Account 170. Some of the accounts such a ledger might be expected to contain are:

E1701 Job office building
1702 Warehouse
1703 Auto and truck repair shop
1704 Blacksmith shop
1705 Laboratory
1706 Electrical shop
1707 Machine and equipment repair shop
1708 Paint shop
1709 Planing mill and woodworking shop
1710 Plumbing, heating, and air conditioning shop
1711 Tire and battery shop
1712 Commissary
1713 Mess hall
1714 Bunk houses
1715 Residences
1716 Hospital and first aid buildings
1717 Garages
1718 Gasoline station
1719 School
1720 Powder house
1721 Electrical generator house
1722 Compressor building
1723 Time shacks
1724 Trailer camp
1725 Site grading
1726 Access and camp roads and bridges
1727 Maintenance of access and camp roads and bridges
1728 Air production plant
1729 Crushing and separation plant
1730 Aggregate plant
1731 Asphalt plant
1732 Asphalt storage
1733 Concrete batch plant
1734 Cement storage
1735 Crane and cableways
1736 Concrete placing structures
1737 Refrigerator plant
1738 Dewatering system
1739 Water storage and distribution
1740 Electrical generators and distribution system
1790 Plant move-in and erection
1791 Plant move-out and dismantling

Each of these accounts would have the following breakdown:

.1 Labor
.2 Materials
.3 Cost of providing job-owned equipment
.4 Equipment rentals
.5 Supplies and small tools
.6 Subcontracts
.7 Distributable expense

The next account that would be likely to require a subsidiary ledger is Account 190, "General expense." There are a good many ways in which the accounts in this ledger might be shown and segregated, for example:

E1901 Salaries and wages
1902 Supplies and expendable equipment
1903 Rent
1904 Maintenance and repairs
1905 Photography and reproduction
1906 Automobile and truck expense
1907 Depreciation (or equipment rental)
1908 Surety bond premiums
1909 Outside professional services
1910 Radio communications
1921 Insurance (other than payroll)
1922 Telephone and telegraph
1923 Heat, light, fuel, and janitor service
1924 Travel expense—operational
1925 Mobilization travel
1926 Employee transportation
1927 Contributions
1928 Promotion, entertainment, and advertising
1929 Bank charges
1930 Head office interest and overhead
1931 Licenses, taxes, permits, and fees
1941 Gain or loss on commissary
1942 Gain or loss on mess hall
1943 Gain or loss on camp operation
1944 Gain or loss on residence and trailer camp
1945 Cost of free meals furnished in mess halls
1980 Cash discounts earned (credit)
1990 Miscellaneous general expense

Some of the larger companies that keep more extensive segregations will have further breakdowns within the accounts. For example, Accounts E1901 to E1919 might have the following breakdown on the ledger sheet:

.1 Management and supervision
.2 Engineering
.3 Accounting, warehouse, purchasing, and payroll
.4 Labor relations and safety
.5 Plant protection—guards and firemen
.6 All other

Accounts E1941 to E1949 might have the following breakdown:

.1 Operating labor
.2 Supplies and expense

.3 Plant amortization

.4 Equipment depreciation or rental

.5 Cost of commodity or service sold

.6 Revenue from sales of commodity or service

.7 Gain or loss transferred to control account

Account F250, "Cost of major equipment," and Accounts F260 and F261 covering the cost of minor equipment, would have subsidiary ledgers according to individual units or types of assets depending upon the system in use. Of course the related reserves for depreciation in Accounts F252 and F262 would appear on the same ledger. Visible-index ledgers are useful for the equipment ledgers.

Normally, the open file of vouchers payable gives adequate support to the vouchers payable account. Subcontractors' accounts, on the other hand, if there are more than two or three, should be kept in a subcontractors' ledger (Figure 2-6). In this account classification, the amount in Account G301 would equal the total of all the amounts currently due subcontractors as shown on the individual ledger sheets. In the same way, Account G302 would control the retained percentages shown by the individual subcontractor accounts.

Account G308, "Withheld for payroll savings bonds," would require a subsidiary ledger in the name of each employee from whom money had been withheld and for whom bonds had not yet been purchased.

The most important subsidiary ledger of all is the job cost ledger, which is made up of the various work items and controlled by Account N405. Illustrations of the types of accounts which might be found in this ledger are given in Chapter 1. Often the job cost ledger will contain a column to show how much of the contract price has been taken into income for each work item. This income might be broken down, as it is in the general ledger, among Accounts N400, N401, and N402. As a practical matter, it is better to use only the account with approved progress billings (in this classification, Account N400) as the controlling account, and to establish the unapproved and unearned balances by journal entry at the end of each month. At the first of the next month, they are reversed and posted in the usual way and there is no duplication. The figures shown on the combined statement of costs and revenues as earnings for the month will usually consist of the billings awaiting approval, plus the amounts earned but unbilled, plus (or minus) any adjustments made when the unapproved billings from the preceding month were approved. The "job-to-date" figures of the statement of costs and revenue will be the total of all three accounts (in this illustration, Accounts N400, N401, and N402) broken down into the individual work items to which they apply.

Location of records. In a fully centralized system all the accounts, including both the general and subsidiary ledgers, will be kept in the general office. In a fully decentralized system the general ledger for the job and all the subsidiary records would be kept in the job office, and the company's general ledger kept in the head office will show only the company's net investment in the job, represented in Schedule 9-4 by accounts in Section M (370, 371, 375, 376, and 377). In a hybrid system it is not uncommon to find the general ledger kept in the head office, and the subsidiary ledgers kept at the jobsite. If a job is large enough to require any substantial amount of jobsite accounting, this method makes fairly good sense because it is at the jobsite that the detail is most likely to be used.

Investment. Often the investment in the job will be treated as a receivable in the head office accounting. This treatment is based on the theory that improvements to real property belong to the landowners. Hence the company's asset is more like a receivable, secured by lien rights, than an inventory item.

Treatment of investments in jobs as an inventory item is more in keeping with the general accounting concept that the company is selling a completed job to the owner. This treatment has the advantage of representing the position of a going concern somewhat more accurately than the receivable treatment, although admittedly not quite so accurately from a liquidating viewpoint. Actually it makes little difference which theory is followed so long as the nature of the item is clearly described in the account title.

A third method is to treat the investment in jobs as an investment. This is rarely an appropriate treatment. The very term "investment" implies that the item is unrelated to operations, so that the result of such treatment is often misleading. One exception to this characterization might be cash contributed to a joint venture that is expected to continue for a considerable period.

Forms. Illustrated are a number of sample forms for the general ledger and subsidiary ledgers (Figures 9-1 to 9-7) with comments on when and how they are used. The ones shown here are all designed to be hand-kept, as are most of the ledgers used by construction companies.

A ledger need not take the form of a book. It can be a tub of cards like those used with bookkeeping machines, a box or tray of cards (Figures 4-3 and 9-5), or a computer printout. The cards illustrate a type of subsidiary ledger that reports units without specific cash reference. Supported by occasional physical inventories, they provide important management information without the cost usually associated with perpetual inventories. If there are as many as 100 accounts

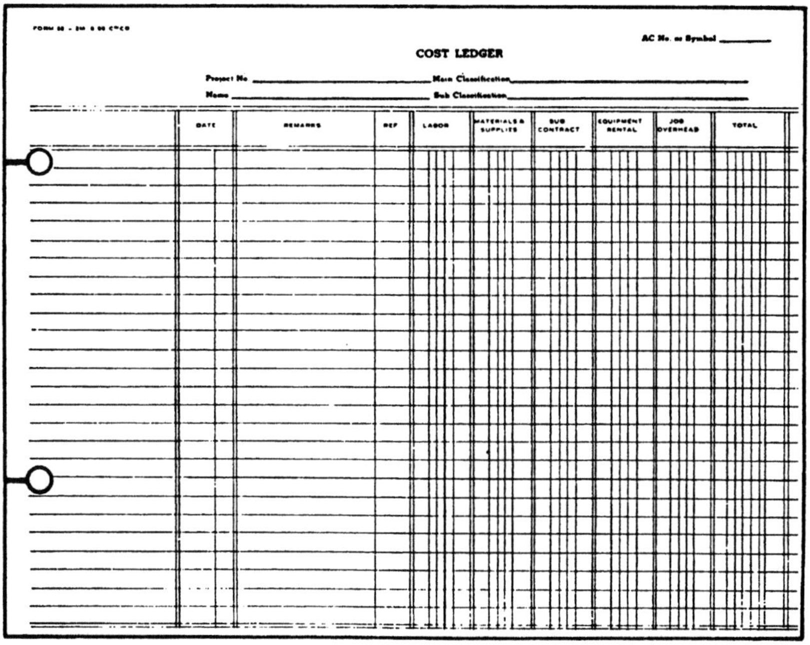
COST LEDGER

AC No. or Symbol ______

Project No ______ Main Classification ______

Name ______ Sub Classification ______

DATE	REMARKS	REF	LABOR	MATERIALS & SUPPLIES	SUB CONTRACT	EQUIPMENT RENTAL	JOB OVERHEAD	TOTAL

Figure 9-1 **Cost Ledger Sheet.** This form is designed for hand bookkeeping, but a similar form may be used in machine systems.

in a ledger, particularly a subsidiary ledger, it may pay to use some visible-index system (Figures 9-6 and 9-7).

Computerizing the Ledger System

Until the middle to late 1970s, computers were beyond the reach of most medium-sized to small contractors because the cost exceeded the benefit. With the advent of minicomputers, however, many companies have begun to take a second look at the advantages of computerizing their accounting and ledger systems.

Even with the very low purchase price or rental rates, other significant applications must also be computerized; otherwise, it will be difficult to justify the cost of what will be simply a high-priced bookkeeping machine. Too often contractors forget that it requires a fairly high-priced clerk to run the computer and a clerk that is fairly well versed in bookkeeping procedures to ensure that the material being input into the computer is prepared properly and to review the material that comes out of the computer as well. There is an expression common among people knowledgeable in computers that summarizes the situation well: "garbage in; garbage out."

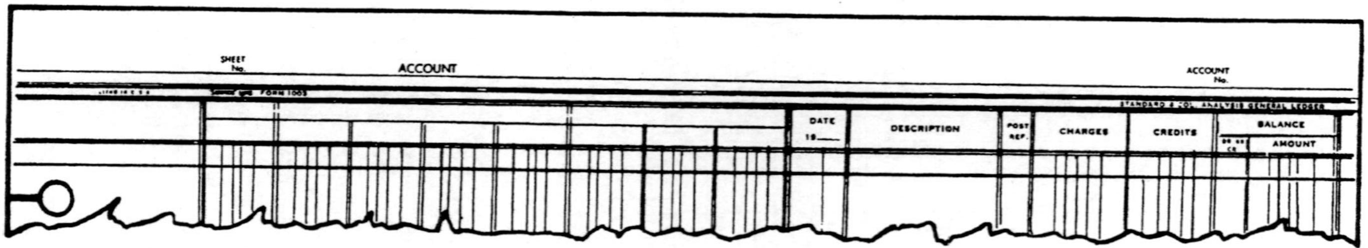

Figure 9-2 General Ledger Sheet. This is a standard 11 × 14 in, 3-column sheet with eight analysis columns. It can also be used as a cost ledger with breakdowns for labor, materials and supplies, subcontracts, equipment rental, job overhead, and so on.

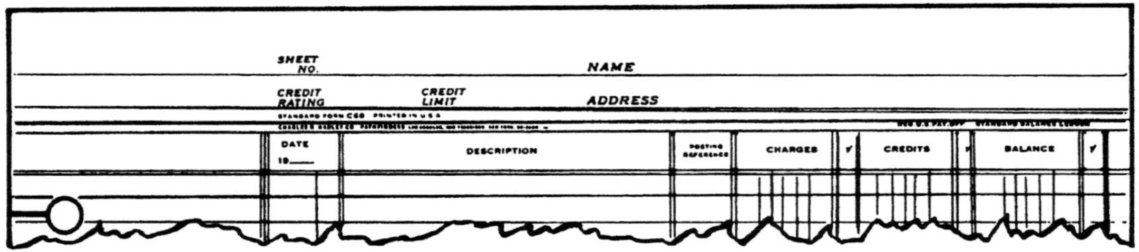

Figure 9-3 **Accounts Receivable Ledger Sheet.** This is a standard 11 × 11 in form.

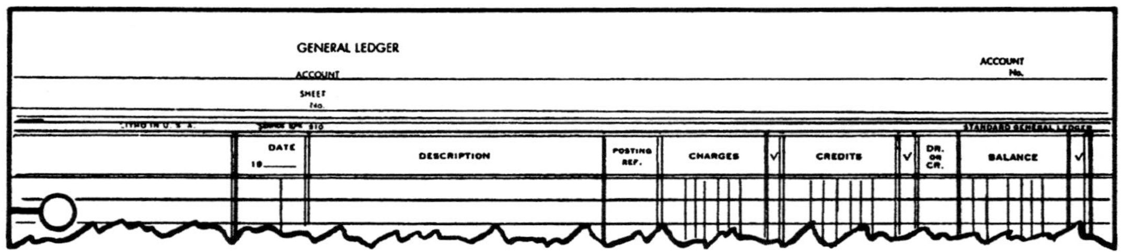

Figure 9-4 **General Ledger Sheet.** This is a standard 11 × 14 in, 3-column sheet. Similar rulings are available in various sizes. This size permits a considerable amount of explanation.

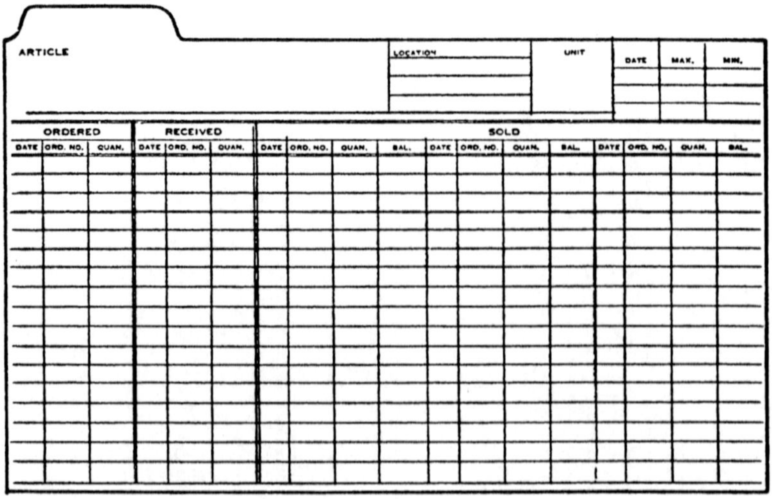

ARTICLE															LOCATION		UNIT	DATE	MAX.	MIN.
ORDERED			RECEIVED			SOLD														
DATE	ORD. NO.	QUAN.	DATE	ORD. NO.	QUAN.	DATE	ORD. NO.	QUAN.	BAL.	DATE	ORD. NO.	QUAN.	BAL.	DATE	ORD. NO.	QUAN.	BAL.			

Figure 9-5 **Inventory Card.**

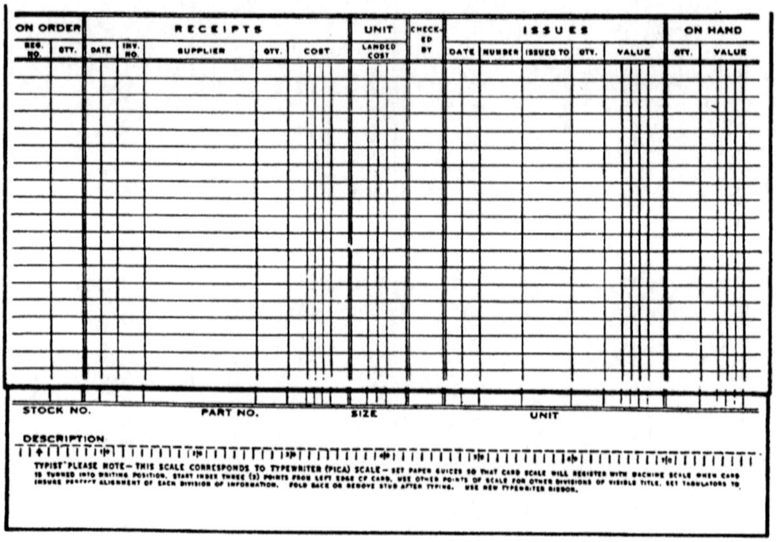

ON ORDER		RECEIPTS					UNIT	CHECKED BY	ISSUES					ON HAND	
REQ. NO.	QTY.	DATE	INV. NO.	SUPPLIER	QTY.	COST	LANDED COST		DATE	NUMBER	ISSUED TO	QTY.	VALUE	QTY.	VALUE

STOCK NO. PART NO. SIZE UNIT

DESCRIPTION

Figure 9-6 **Stock Card for Visible-Index Use.**

Other systems and functions that can be considered for computerization to justify the cost of computerizing the accounting and ledger systems are:

Payroll. The cost of the computer should be considered since it may be cheaper to have a service bureau or bank process the payroll.

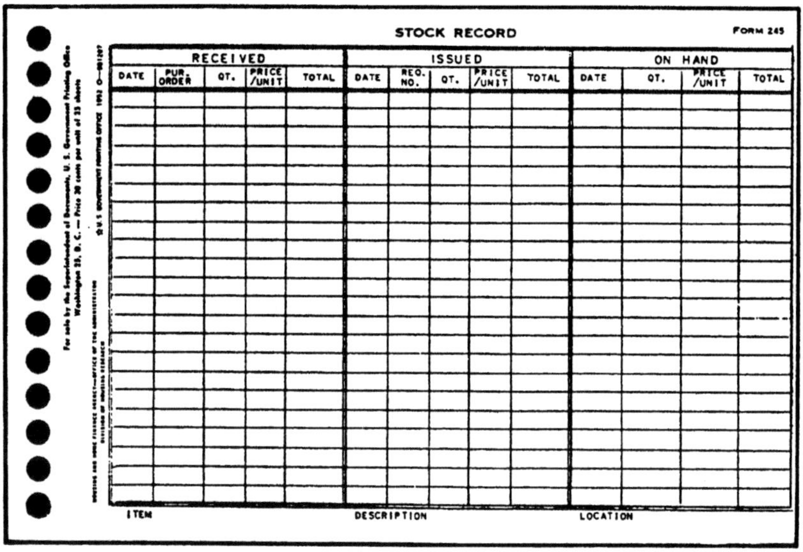

STOCK RECORD FORM 245

RECEIVED					ISSUED					ON HAND			
DATE	PUR. ORDER	QT.	PRICE /UNIT	TOTAL	DATE	REQ. NO.	QT.	PRICE /UNIT	TOTAL	DATE	QT.	PRICE /UNIT	TOTAL

ITEM DESCRIPTION LOCATION

Figure 9-7 **Stock Card for Prong Binder.**

Cash voucher system. The contractor will want to look at the number of checks written monthly to pay for purchases of materials and to pay subcontractors. In most systems it takes something over 200 transactions a month to justify the cost.

Equipment records. Equipment location, depreciation, maintenance, property taxes, etc., can be easily computerized if the volume is significant enough to justify the cost.

Billings and accounts receivable. Generally, most contractors will not have more than ten to twenty jobs going at any one time. The billings required for each job are so customized and the volume is so minor that computerizing this function is usually not feasible.

Construction costs. Any discussion of computerizing subsidiary ledgers would be incomplete without considering the job cost ledger. Compiling job costs by computer is a well-established application, and a complete and up-to-date job cost ledger is an essential element in most systems.

Scheduling and estimating. This function is not usually feasible for smaller contractors since most of the minicomputers will not be able to handle estimating in a form that will be satisfactory to the estimator. Estimating tends to be somewhat of an art and every estimator has his own way of racking up the take-off quantities.

In summary, whether or not to computerize the accounting and ledger systems as well as other functions requires careful analysis. Contractors should seek the advice of consultants before making the decision to purchase or lease computer hardware. Once committed to a computer, contractors tend to be a little like boat and airplane owners, wanting to up-size periodically. Without careful controls, the number of people working in the computer operation will tend to grow at an unnecessarily accelerated rate.

Chapter

10

Financial and Cost Reporting

It is the purpose of financial and cost reporting to group significant facts in such a way that the person reading them can form an accurate judgment of the financial condition of the company and the results of its operations or can make management decisions. In the past, there has been a tendency to group the figures on construction company statements in the same way that they are grouped on the statements of a manufacturing company, and, in general, to interpret the results as if they were the results of a manufacturing process. To some extent, construction is like manufacturing, but the differences between the two are important enough so that accounting statements prepared and interpreted by manufacturing standards are seriously misleading.

Attention has already been directed (Chapter 7) to an important difference in the treatment of inventories, particularly "work in process." As pointed out there, in manufacturing, inventories are owned by the company and are a tangible asset with a definite dollar value. Construction work-in-process inventories are generally more intangible in nature in that they generally consist of improvement to someone else's real property and therefore are not owned by the construction company. Additionally, if the contractor is using the percentage-of-completion method of accounting, work in process will be stated in terms of a dollar value resulting from estimates of cost to complete and cost to date. Therefore, if a contractor's financial statement is being read in terms of values on liquidation, jobs in process could easily be a liability rather than an asset. On the other hand, once completed, the value, even on liquidation, is more firmly fixed because its value is secured by lien laws (private work) and bonding laws (public work).

Perhaps the most significant difference in the financial reports of a manufacturing company and a construction company is found in the income statement. In manufacturing, a completed item is sold with a known cost at a known price and title passes generally at the time of sale. In construction companies with long-term contracts, and where the percentage-of-completion method of accounting is used, the major components of the income statement are based on "best estimates" of final results on jobs in progress at the income statement date.

Financial statement ratios, used by analysts to evaluate manufacturing companies (current ratio, net quick assets, etc.), are of less use when attempting to evaluate a construction company. Perhaps the most important financial indicator of a construction company's financial position is the amount of backlog (value of uncompleted work) and the amount of estimated profit in that backlog, whereas in a manufacturing company's statement, firm sales commitments would be interesting but not vital.

Management

It is to be expected that the most detailed use of a company's financial statements will be made by the management of the company itself. The use of accounting figures in management is increasing now that computers (see Chapter 12) are making those figures available in a more timely and usable form. For example, aggressive management will use the financial statements as a starting point for planning the company's operation many months in advance. Individual jobs will be analyzed to see when the money, the personnel, and the equipment committed to them may be released. On the basis of that analysis, management will look at the jobs being offered for bid to see which ones should be bid, and how small a margin of profit the company is willing to accept to win them.

Management will also consider the effect of the statements on borrowing and bonding capacity, and determine what its policies will be on those aspects of its financing. Management will be seeking answers, in the financial statements, to questions of the effects of the company's policies on such matters as subcontracting and the ownership of equipment.

Whenever past performance provides a basis for present policies and future planning, sound management will start with its financial statements. A number of examples of how this procedure is followed are given in Chapter 1. In this chapter emphasis is given to the form that the financial reports should take.

Some contractors still express more than a mild form of contempt for the "paper mill" (the office) and its products. These are the contractors who boast that they can "guess closer than those clerks can figure." They are also the contractors who curse the "stupidity" of their estimators when they lose money on a job and who call their bankers and bonding companies "shortsighted" when they refuse to underwrite the contractor's less fortunate guesses.

In this chapter several typical statement forms and their supporting schedules are illustrated. Of particular interest is the form released by Robert Morris Associates (Figure 10-1). There are very few contractors who could not improve their statements by incorporating into their own financial reporting some of the principles of statement structure illustrated in that form.

Bankers and Bonding Companies

Contracting is, at best, a risky business, and it usually involves large sums of money. For these reasons alone, many banks and some surety companies have been reluctant to lend money to contractors or to bond their jobs. As one banker puts it: "There are too many things that can go wrong; too few contractors who know anything about banking; and too few accountants who know anything about either construction or banking."

In recent years there has been a considerable effort on the part of the professional and trade associations serving the accounting profession, the banks, the surety companies, and the contractors to reach a common understanding. Typical of these efforts are the following:

1. Release by the AICPA of its *Accounting Research Bulletin 43* (dealing with accounting for cost-plus-fixed-fee construction contracts) and *Bulletin 45* (dealing with methods of accounting for long-term contracts). These bulletins are quoted in Chapter 1.
2. Release by the AICPA of its *Audit and Accounting Guide—Construction Contractors* which provides guidance for and clarification of *Bulletins 43* and *45*.
3. Release by the AICPA of Statement of Position (SOP) 81-1 which provides guidance in the application of contract accounting.
4. The release by Robert Morris Associates of a standard form (Figure 10-1)—"Contractor's Financial Statement and Supporting Information."

For any contractor who wants to receive maximum credit consideration, Figure 10-1 is required reading. Among the items to be noted particularly are the following:

1. There is no attempt to break down the assets and liabilities into current and fixed categories in the traditional way.

CONTRACTOR'S FINANCIAL STATEMENT
AND SUPPORTING INFORMATION

INDIVIDUAL ☐
PARTNERSHIP ☐
CORPORATION ☐

For the purpose of procuring and maintaining credit from time to time in any form whatsoever with the below named Bank, for claims and demands against the undersigned, the undersigned submits the following as being a true and accurate statement of its financial condition on the following date, and agrees that if any change occurs that materially reduces the means or ability of the undersigned to pay all claims or demands against it, the undersigned will immediately and without delay notify the said Bank, and unless the Bank is so notified it may continue to rely upon the statement herein given as a true and accurate statement of the financial condition of the undersigned as of the close of business on the date set forth in the following line.

Statement Date ________ 19__ Submitted To ________

Submitted By ________

Date Statement Signed ________ 19__ Signature ________ Title ________

EXHIBIT A: BALANCE SHEET (OMIT CENTS IN ALL EXHIBITS AND SCHEDULES)		ASSETS
1. Cash (Excluding Items in Line 8)	SCH. (B)	$
2. Notes Receivable for Contracts Completed	SCH. (A)	
3. Accounts Receivable for Contracts Completed (Excl. Retainage)	SCH. (A)	
4. Amounts Due on Uncompleted Contracts:		
(a) Approved by Engineers or Architects	EXH. (C)	
(b) Subject to Approval of Engineers or Architects	EXH. (C)	
(c) Retainage	EXH. (C)	
5. Retainage Due on Contracts Completed		
6. Labor, Material and Direct Overhead Charged to Jobs (Not Included Above)		
7. Inventory Not Applicable to Contracts in Progress		
8. Deposits for Bids and Other Guarantees		
9. Stocks, Bonds and Special Assessment Obligations Received in Payment of Contracts Completed or Uncompleted	SCH. (C)	
10. Other Security Investments	SCH. (C)	
11. Other Receivables		
12. Land and Buildings—Not Used in the Business—*Less Reserve For Depreciation*	SCH. (D)	
—Used in the Business—*Less Reserve For Depreciation*	SCH. (D)	
13. Machinery, Equipment, Vehicles, etc. (Net)	SCH. (E)	
14. Deferred Charges and Prepaid Expense		
15. Other Assets		
16.		
TOTAL ASSETS		$
		LIABILITIES
17. Notes Payable to Banks—For Bid Checks	SCH. (B)	$
—Other	SCH. (B)	
18. Accounts and/or Notes Payable for Equipment	SCH. (E)	
19. Advances or Loans From Officers, Partners, Friends or Relatives		
20. Accounts Payable for Material Purchases		
21. Accrued Expenses Other Than Federal Income Taxes		
22. Federal Income Taxes—Due $ ________ Accrued $ ________		
23. Due to Sub-Contractors—On Uncompleted Contracts	EXH. (C)	
—On Completed Contracts		
24. Mortgage on Real Estate (Amount Due in Next 12 Mos. $ ________)	SCH. (D)	
25. Other Liabilities*		
26.		
TOTAL Lines 17 Thru 26 —— $		
27. Reserve for Unearned Income *OR* Profit or Loss to Date on Uncompleted Contracts	EXH. (C)	
28. Preferred Stock Outstanding (Par Value $ ________)		
29. Common Stock Outstanding (Par Value $ ________)		
30. Earned Surplus (Net Worth If Not a Corporation)	EXH. (B)	
31. Capital Surplus		
TOTAL Lines 28 Thru 31 —— $		
TOTAL LIABILITIES		$

*See SCH. (H) For Contingent Liabilities.

Page One

Carried in Stock by Cadwallader & Johnson, Chicago.
Robert Morris Associates—Form No. C111H

Figure 10-1 **Contractor's Financial Statement and Supporting Information.** Robert Morris Associates form can be used by most contractors.

2. The language in the captions is precise enough to let the person reading it know, at least in general, what the figures mean.
3. Earned-in-excess-of-billed revenue is included in "Amounts due on uncompleted contracts" in either subcategory *a* or *b*. Presumably if progress billings were in excess of earned revenue, the amount would be a subcaption to Line 25 "Other Liabilities" and called

EXHIBIT B: PROFIT AND LOSS STATEMENT From________19___ to________19___

1.	Contracts Completed During Period	$
2.	Less—Direct Costs Including Material, Labor, Job Overhead and Sub-Contracts	
3.	Gross Profit on Contracts Completed	
4.	Other Income	
5.	Total Income	
6.	Less—General and Administrative Expense	
7.	Interest Paid $______ Other Expense $______	
8.	Provision for Federal Income Taxes	
9.	Net Profit or Loss	
10.	Add—Earned Surplus (Net Worth if Not a Corporation) at Beginning of Period	
11.	Adjustments (Net) $______ Dividends or Withdrawals $______	
12.	Earned Surplus (Net Worth if Not a Corporation) at Close of Period	$
13.	Memo: Depreciation Included in Above Expenses	$

EXHIBIT C: CONTRACTS IN PROGRESS { Items (1) thru (4) Based on Actual Figures. Items (5) thru (8) Based on Estimated Figures.

Note: Page Three contains a spread of the individual uncompleted contracts making up the grand total in the column to the right.

		GRAND TOTALS ON UNCOMPLETED CONTRACTS TO DATE______ 19___
1.	TOTAL CONTRACT PRICE	$
	COSTS TO DATE:	
	(A) Labor	
	(B) Material	
	(C) Direct Overhead Charged to Contract to Date	
	(D) Paid to Sub-Contractors to Date	
	(E) Owing to Sub-Contractors	
	(F)	
2.	Total Items (A) Thru (F)	$
	ESTIMATES RECEIVED OR DUE:	
	(G) Approved Draws Received	
	(H) Due and Subject to Draw Pending Architect's or Engineer's Approval	
	(I) Retainage on (G) and (H)	
	(J)	
3.	Total Items (G) Thru (J)	$
4.	PROFIT OR LOSS TO DATE (Total of Line 2 Minus Total of Line 3)	$
5.	CONTRACT BALANCE TO BE COMPLETED AFTER STATEMENT DATE	$
	ESTIMATED COST TO COMPLETE CONTRACT:	
	(K) Labor	
	(L) Material	
	(M) Direct Overhead	
	(N) Let or to Be Let to Sub-Contractors	
	(O)	
6.	Total Items (K) Thru (O)	$
7.	ESTIMATED DRAWS ON CONTRACT BALANCE (Including Retainage)	$
8.	ESTIMATED PROFIT OR LOSS ON CONTRACT BALANCE (Total of Lines 6 Minus 7)	$
9.	COST OF EQUIPMENT ACQUIRED TO DATE	$
10.	ESTIMATE OF ADDITIONAL EQUIPMENT NEEDED TO COMPLETE CONTRACT	$
11.	EQUIPMENT RENTALS PAID TO DATE AND INCLUDED IN LINE (C)	$
12.	LOCATION OF JOB	
13.	NATURE OF WORK	
14.	CONTRACT COMPLETION DATE	
15.	ORIGINAL CONTRACT PRICE (BID)	$
16.	CONTRACT PRICE OF NEXT LOWEST BIDDER	$
17.	ADDITIONS OR EXTRAS TO DATE	$
18.	CREDITS OR OTHER DEDUCTIONS TO DATE	$
19.	INCREASE OR DECREASE IN ORIGINAL PROFIT ESTIMATE BASED ON EXPERIENCE TO DATE	$
20.	State Contract Provisions for Liquidated Damages, Penalties or Bonuses with Respect to Completion Within Specified Time.	

Page Two

Figure 10-1 *continued.*

"Billings in excess of earned revenue." See Chapter 7 for a discussion of earned revenue.

4. While the statement still uses the term "Earned surplus," instead of the more modern terminology "Retained earnings," the meaning is sufficiently clear for the purpose of the banker or bonding company.

5. Exhibit C calls for a separate "profit-or-loss" statement on each job and this, plus the absence of any accounting for fees on cost-plus-fixed-fee work, would appear to indicate a preference for the "profit-

SCH. A. ACCOUNTS & NOTES RECEIVABLE FOR COMPLETED CONTRACTS (List 4 Largest—Group Others in Line 5)				
DUE FROM	Original Contract Price	Amount Due Excl. Unapproved Claims	Maturity Date	Claims Not Approved For Payment
1.				
2.				
3.				
4.				
5.				

EXHIBIT C: (Continued): (If you have more than six contracts in progress list the five largest contracts in Columns 1 to 5 and lump all the smaller ones in Column 6).

Col. FOR WHOM CONTRACT PERFORMED	NAME OF SURETY COMPANY	NAME OF PRINCIPAL SUB-CONTRACTOR
COL. (1)		
COL. (2)		
COL. (3)		
COL. (4)		
COL. (5)		
COL. (6)		

	COL. (1)	COL. (2)	COL. (3)	COL. (4)	COL. (5)	COL. (6)
1						
A						
B						
C						
D						
E						
F						
2						
G						
H						
I						
J						
3						
4						
5						
K						
L						
M						
N						
O						
6						
7						
8						
9						
10						
11						
12						
13						
14						
15						
16						
17						
18						
19						
20						

Page Three

Figure 10-1 *continued.*

and-loss" approach over the "owner's equity" approach. Presumably, too, if there were a guaranteed maximum or "upset" price on a fee contract, that fact would be disclosed in Schedule H.

6. Items 7 to 20 on Exhibit C are peculiar to construction contractors and are of prime importance in any analysis of the position of each job.

SCH. B. CASH AND NOTES PAYABLE BANKS: (Cash Not in Deposit Column Below $________).

NAME OF BANK	CASH ON DEPOSIT	AMOUNT OWING	PORTION COLLATERALIZED	MAXIMUM BANK LOANS DURING PAST 12 MONTHS

SCH. C. STOCKS, BONDS, ETC.:

DESCRIPTION	UNITS	BOOK VALUE	MARKET VALUE	AMOUNT DEPOSITED AS COLLATERAL OR IN ESCROW (STATE WHICH)

SCH. D. LAND AND BUILDINGS: (Use asterisk if not used in business). (Use hundreds of dollars).

LOCATION	IMPROVEMENTS	BOOK VALUE	ENCUMBRANCE Amount	ENCUMBRANCE Maturity	TITLE RECORDED IN NAME OF

SCH. E. MACHINERY, EQUIPMENT, VEHICLES, ETC.: (Use separate line for each general classification).

No. of Items	DESCRIPTION	AGE Note (1)	PURCHASE PRICE	PRESENT BOOK VALUE	ENCUMBRANCE Note (2)	DATE DUE (Final)

Note (1) If two or more items are lumped give sum of ages. Note (2) Use asterisk if payable monthly.

SCH. F. Principal Contracts Completed Past 3 Years.

FOR WHOM PERFORMED	AMOUNT (Omit 000)

SCH. G. PRINCIPAL MATERIAL SUPPLIERS AT PRESENT:

NAME AND ADDRESS

SCH. H. CONTINGENT LIABILITIES:

	YES	NO
1. Have any notes or accounts receivable been sold or discounted?		
2. Are there any liens for labor or material filed on your work anywhere?		
3. Are there any disputes over payment of labor or materials on any contract?		
4. Are there any suits or claims or judgments pending or unsettled?		
5. Are you endorsor or guarantor on any contracts, accounts or notes of others?		
6. Do you have any financial interest in a joint venture with other contractors?		
7. Are there any other contingent liabilities?		

If any known contingent liabilities exist of a nature not set forth in the above questions one to seven inclusive, please attach supplemental schedule giving full particulars. The bank is correct in assuming no other contingent liabilities exist if no such schedule is attached.

Page Four

Figure 10-1 *continued.*

Prospective Clients

An insolvent or defaulting contractor causes nothing but trouble and expense for his client. First, there are delays in completing the work. Second, he may tend to try to save money by using inferior materials and workmanship. Third, unpaid mechanics and material suppliers will doubtless file liens (or other similar statutory claims) against the real property, and if the contractor has been paid for the work represented by the lien, the client may be required to pay for it a second time. Even if the contractor has provided the partial protection of a performance and payment bond, the client's loss in terms of time, annoyance, and legal fees can be substantial. To reduce such risks, before they accept a bid on proposed construction work, clients often require that contractors submit financial statements. The client's interest, in many respects, parallels that of the banker and the bondsman, but the emphasis is different. The banker and the bondsman want to know if the contractor has the money, experience, personnel, and the equipment to complete the jobs he is working on. The prospective client wants to know whether the contractor will be able to finance, staff, and equip the client's job in addition to keeping up his other work. He also wants to see if the contractor has in progress one or more losing jobs that will drain off the progress payment money that the client pays to finance his own job.

For the contractor who is faced with the problem of getting new work while completing some losing jobs, the best way to overcome the client's objection is to offer to take the job on the basis of cost-plus-fixed-fee with the bid price as the guaranteed maximum (sometimes called an "upset" price). In that way the client keeps control of his money until he knows it was spent on his job. He can cover the risk that the contractor might go over the upset price by requiring bond, and he stands a chance to gain if the contractor does the job below his expected cost. Usually it is possible for the contractor to negotiate an agreement giving him a participation in savings if the work is done for an amount below the guaranteed maximum.

Suppliers and Subcontractors

Suppliers and subcontractors are neither buying the contractor's services nor providing him directly with money or credit. Therefore they are not in a position to ask for a great deal of financial information from the contractor. Usually they get their information from credit reporting agencies and from other suppliers and subcontractors. Building material dealers usually keep rather complete credit files and exchange information on their contractor customers. For the contractor in good financial condition, it is often good business to take a few selected suppliers and subcontractors into his confidence and show them his statements. The advantage can be twofold. First, the news travels in the trade. Second, the contractor may be able to get better prices from suppliers and subcontractors who feel their payment is secure.

Stockholders and Partners

Stockholders, partners, and others who have a proprietary interest in a construction company use the company's financial statements to find out four things. First, they want to know whether their investment is being soundly managed, and that question can be answered, in part, by review of the company's financial condition, results of its operations, and changes in financial position. Unless they know something about the status of the jobs and the accounting methods used in reporting on their operation, their understanding is likely to be limited and their judgment may be wrong.

Second, as individuals, they want to know how much income (in dividends) they can expect to receive currently.

Third, they want to know the long-run prospects for income from their investments.

Fourth, they want to know if their investment is increasing in value.

Even the individual proprietor, who personally makes all the management decisions, will want to know the answers to these questions. For him it is a matter of checking his own thinking, rather than that of someone else. Still, the questions are the same and the facts needed to get the answers are the same. The only difference lies in the extent to which the individual proprietor or partner can rely on his own knowledge instead of that provided by the statements. Actually, the form and content of the statements should be the same regardless, because of the other uses to which the statement will be put.

Usually, a joint venture is itself a contractor, and the members of the venture use the statements as if they were partners. The sponsor would, of course, also use the statements of the venture as one of its tools in managing the venture.

Statement Requirements

Thus far the discussion has dealt principally with the form in which financial information is presented for outsiders, and the uses to which it is put. Here the emphasis is on content. Running down the balance sheet, the items appear as follows:

Cash should consist principally of demand deposits in solvent banks. Office and job petty cash funds may be included if they have been replenished at the end of the accounting period. In a decentralized accounting system, cash may include the job bank accounts of wholly owned jobs, but it may not include the company's share in joint venture bank accounts unless the cash is under the company's control. Cash should not include bid deposits or other cash deposits to guarantee performance, nor should it include any expense advances to employees or deposits subject to substantial limitations on withdrawal. Cash in foreign banks subject to currency restrictions may be treated as cash so long as it does not exceed the company's need for cash in the country whose currency is represented. In establishing the schedule showing cash in the bank, it is a good idea to relate the cash on deposit with any loans that may be outstanding with the same bank, the collateral pledged, the maximum credit extended by that bank, and the average balance during the past year. Every report should include a cash forecast, at least in the copies issued to management.

Receivables should be broken down to show the following:

1. Progress billings approved and currently due.
2. Progress billings made but not yet approved.
3. Amounts earned and billable but not actually billed at the statement date.
4. Retentions on contracts in progress.
5. Retentions on completed jobs.
6. If advances to decentralized jobs and joint ventures are treated as receivables, they should be set apart in a separate schedule and should appear on the face of the balance sheet under a separate caption. Whether an equity in such a job arising from job profits should be treated as a receivable is open to question, but when such treatment is used that fact should be disclosed. Generally, the best practice is to show advances and undistributed earnings of a joint venture as "Investments in joint ventures," with adequate footnote disclosure as to the nature of the investment and capsulized financial data on the joint venture.
7. Cost of jobs in process. When these items are treated as receivables, only the excess of earned revenue over progress billings on any particular job should be shown. If there is an excess of progress billings over earned revenue, they should be treated as a liability and presented in the current section if the balance sheet is classified into current and noncurrent sections.
8. Amounts due from miscellaneous sales and services and from insurance claims.
9. Amounts due from officers and employees.

Supporting schedules should show these items in detail, and when they arise from or are related to any particular job, they should be identified with that job in the schedules.

Inventories, technically speaking, are limited to unappropriated stocks of materials, supplies, and spare parts. Sometimes small tools are classified with supplies, and sometimes they are shown separately. Either way, they should be separately identified, either on the face of the balance sheet, or in the supporting schedules. Purchased inventory should be valued at cost or market value, whichever is lower.

There are no hard and fast rules for used materials of a capital nature (plywood and metal forms) and small tools (jackhammers, drills, etc.) returned to central storage from jobs. One standard treatment is to "sell" new items to jobs at a 10 percent markup on cost. When returned to central storage, the job is credited for 50 percent of the original selling price and the inventory account is charged for a like amount. On the next "sale" to a job, 10 percent is added to the return value from the previous job. On return, the process is again repeated and so on until the item is useless, lost, or the remaining value is too minor to bother with. The 10 percent markup factor is intended to recover the cost of minor repairs and warehousing. A variation of this approach is to assign an estimated value on return rather than a consistently applied percentage. This approach, however, requires the time of someone knowledgeable enough to appraise the value on each return and may be more trouble than the procedure is worth. However, when "selling" such items to joint ventures or to cost-plus-jobs, it is a good practice to be sure that the price fairly represents the value of the materials or small tools. Care should also be taken that the transaction is not so documented that the state wants to treat it as a sale subject to sales tax.

Investments—in the sense of stocks, bonds, or mortgages used as a temporary reservoir for surplus funds—are not often found in the balance sheet of an active construction company. On the contrary, an active contractor is often short of cash. It is not uncommon to see the contractor take all or part of his payment in special assessment bonds or similar obligations. It is also common to see funds advanced to joint ventures, or even to large, wholly owned jobs with decentralized accounting, referred to as investments.

The three types of items are totally unrelated, and each should be treated separately, and shown separately, on the balance sheet. The true investments should be stated at the lower of cost or market value. The securities taken as a part of the contract price of work done should be taken into account at market value on the date received. The Tax Reform Act of 1986, for the first time, made the interest on certain state and local bonds subject to federal income tax; if the statement shows any such bonds, their interest status should be disclosed. The advances to large jobs should be stated as actual cash advanced less any amounts refunded during the life of the job. Generally speaking, it is not good practice to carry the portion of the contractor's equity which represents profit from such jobs in an investment account.

All items designated as investments should be classified and shown in detail on the supporting schedules with enough description to clarify their source and nature. If they have been pledged as collateral for any liability, that fact should be shown.

Fixed assets, in addition to the customary segregations of land, buildings, machinery and equipment, trucks and trailers, automobiles, and so on, should also be classified according to whether or not they are used in the business. Thus, land held for a future equipment yard and warehouse would be segregated from land taken in as part payment on a job and held for sale; and both would be segregated from the land on which the present yard and warehouses are situated.

Equipment and vehicles are the life blood of some contracting firms. Others own little or none, and there should be some note on the statements to show why. For firms that do own equipment, a schedule showing all of it is best, if not too long. If a detailed list is impractical, then a list by types (dump trucks, power shovels, compactors, and so on) should be made. The schedule should show cost, depreciation (or amortization), book value, amount owed, and net equity. If accelerated depreciation or amortization has been claimed, that fact should be noted in the equipment schedules.

Notes and accounts payable should be shown in detail in a schedule whenever practical. The schedule of notes payable should show in detail the payee, due date, interest rate, collateral given, and amount. If some notes are given to finance bid deposits, that fact should be indicated, and those notes segregated from regular loans. Notes given for the purchase of equipment and secured by equipment should be separated in the schedule, and the encumbered equipment should be described as accurately as possible. In the equipment schedule, of course, the encumbrance will be shown against the encumbered equipment. Notes to officers, employees, friends, and relatives should be scheduled separately, and if any have been subordinated to other claims, that fact should be noted, and the claims to which they have been subordinated should be indicated.

For internal management purposes, a list of significant trade accounts payable and amounts due to subcontractors should accompany the report. In the reports issued to the bank and bonding company, the ten or fifteen largest accounts may be shown in detail and the others lumped together. The "Trade accounts payable" schedule should show as a footnote any disputed payable or other item which might affect the contractor's credit, such as a very large advance purchase made to secure scarce materials or to obtain bargain prices.

Taxes payable include not only the routine payroll tax, property tax, and sales and use tax accruals, but may also include federal and state income taxes. The income taxes should never be lumped with the other taxes and, in fact, the current and deferred portions should be shown as separate items on the face of the balance sheet. There are some technical objections to showing on the balance sheet of a proprietorship or a partnership the income tax liabilities of the individual owner or owners. However, if the business is going to be called upon to pay the money, the technical objections must give way to the practical considerations, and the technical exception may be taken in a footnote. A supporting schedule may show the estimated tax, the prepayments, and the net amount of cash the business is going to be called upon to supply.

If, for any reason, taxes withheld from employees' wages become delinquent, it is essential that the delinquency be shown and the reason noted. Withheld taxes are trust funds, and willful failure to pay them is a very serious offense. In fact, the individual responsible for the failure to pay may be held personally responsible for the tax plus heavy penalties.

Contingent liabilities may include a number of items. Following are some of the better-known examples:

1. Notes or accounts receivable discounted or sold "with recourse"
2. Liens filed or threatened on any job
3. Disputes over amounts due to laborers, material suppliers, or subcontractors (including labor disputes involving retroactive pay or union health and welfare fund contributions)
4. Suits, claims, or judgments pending or unsettled
5. Endorsements or guarantees on the notes or contracts of others
6. Possible liability for a share of joint venture losses
7. Possible liabilities for refund of progress payments made, should jobs not be finished
8. Possible loss on fixed price or cost-plus contracts with a guaranteed maximum
9. Possible penalties for alleged violations of wage and hour laws and other similar statutory or contractual penalties
10. Possible liquidated damages for failure to complete jobs on time
11. Possible additional tax assessments
12. Asserted income tax payment deficiencies from prior years

Proper financial disclosure calls for an "accounting policy" footnote. For example, in addition to stating that "the percentage-of-completion method of accounting is used to recognize revenue for the period on contracts in progress," the note will also include a description of the basis for determining percentage of completion and the accounting procedures for "Loss jobs." Other policy matters discussed in the note might be fixed asset depreciation methods and usable basis, inventory pricing policies, consolidation procedures where there are subsidiary companies, joint venture accounting procedures, and other matters which might be of importance to the financial statement users.

Job revenue and job costs should be supported by a schedule showing, in summary, an operating statement on each job together with the following information about each job (Schedule 7-4 can be expanded to provide a convenient way to present this information):

1. Original contract price (estimated in terms of expected units on unit price jobs)
2. Number and amount of additive change orders
3. Number and amount of deductive change orders
4. Total contract
5. Balance of contract to be completed after the statement date
6. Estimated cost to complete, broken down into the estimated cost of:
   a. Labor
   b. Materials
   c. Subcontracts
   d. Equipment rentals
   e. Job overhead
   f. Total
7. Estimated cash, including retention, still to be received
8. Estimated profit (or loss) still to be earned
9. Cost of job-owned equipment to date
10. Equipment rentals paid to date
11. Location of job
12. Name and contract number (if any) by which the job is known
13. Date presently scheduled for completion
14. Date presently estimated for completion
15. Liquidated damages, and completion bonuses (if any)
16. Kind of work
17. Original bid
18. Next lowest bid

19. Amount of bond (if any)
20. Bonding company (if any)

If a blanket job order number is assigned to a large number of small jobs, the total should be reported as a single job.

When subcontractors are doing an important part of the work, it is not uncommon to see prepared for the bank a so-called "exposure schedule" which deducts, from the remaining work to be done on each job, the amount that is to be done by subcontractors. The function of this schedule is to show the banker what part of the risk and the financial load is being carried by subcontractors.

Sample Financial Statements

Audit and Accounting Guide—Construction Contractors, prepared by the Construction Contractor Guide Committee of the AICPA and published in 1981, explains and illustrates the recommended format for presentation of audited financial statements prepared on the percentage-of-completion method of accounting and statements presented on the completed-contract method of accounting. Schedules 10-1 and 10-2 present the guide's recommended format for audited financial statements that might be used by the management of a building contractor using the percentage-of-completion method of accounting. Schedule 10-3 presents the audited financial statements which might be used by the management of the same type of contractor if it were using the completed-contract method of accounting. In addition to internal usage by company management, these statements would be suitable for presentation to bonding companies, bankers, and others. However, when completed-contract statements are prepared, the contractor may find that bonding companies, banks, and others will demand more detail than readily can be gotten from books maintained on the completed-contract method of accounting.

Schedule 10-4 presents the financial statement of an individual job where the accounting function has been decentralized. These statements would be prepared at the construction site, forwarded to the head office where they would be used for purposes of analysis and evaluation of job progress, and then consolidated with financial statements of other jobs and the head office to present overall company financial position and results of operations. When job accounting is decentralized, the statements are somewhat like those of a separate construction company engaged in only one job. The primary and most significant difference between job financials and company financials lies in the equity section of the balance sheet where, in lieu of the normal equity classification such as common stock or retained earnings,

SCHEDULE 10-1 SAMPLE FINANCIAL STATEMENTS PREPARED ACCORDING TO THE PERCENTAGE-OF-COMPLETION METHOD OF ACCOUNTING

Independent Accountants' Report

The Shareholders and Board of Directors
Percentage Contractors, Inc.

We have examined the consolidated balance sheets of Percentage Contractors, Inc., and subsidiaries as of December 31, 19X8 and 19X7, and the related consolidated statements of income and retained earnings and changes in financial position for the years then ended. Our examinations were made in accordance with generally accepted auditing standards and, accordingly, included such tests of the accounting records and such other auditing procedures as we considered necessary in the circumstances.

In our opinion, the financial statements referred to above present fairly the financial position of Percentage Contractors, Inc., and subsidiaries at December 31, 19X8 and 19X7, and the results of their operations and the changes in their financial position for the years then ended, in conformity with generally accepted accounting principles applied on a consistent basis.

(*Firm Signature*)
Certified Public Accountants

City, State
February 18, 19X9

Percentage Contractors, Inc.
Consolidated Balance Sheets
December 31, 19X8 and 19X7*

Assets

	19X8	*19X7*
Cash	$ 264,100	$ 221,300
Certificates of deposit	40,300	
Contract receivables (Note 2)	3,789,200	3,334,100
Costs and estimated earnings in excess of billings on uncompleted contracts (Note 3)	80,200	100,600
Inventory, at lower of cost, on a first-in, first-out basis, or market	89,700	99,100
Prepaid charges and other assets	118,400	83,200
Advances to and equity in joint venture (Note 4)	205,600	130,700
Note receivable, related company (Note 5)	175,000	150,000
Property and equipment, net of accumulated depreciation and amortization (Note 6)	976,400	1,019,200
	$5,738,900	$5,138,200

*The accompanying notes are an integral part of these financial statements.

SCHEDULE 10-1 SAMPLE FINANCIAL STATEMENTS PREPARED ACCORDING TO THE PERCENTAGE-OF-COMPLETION METHOD OF ACCOUNTING *(Cont.)*

Liabilities and Shareholders' Equity

	19X8	*19X7*
Notes payable (Note 8)	$ 468,100	$ 578,400
Lease obligations payable (Note 9)	197,600	251,300
Accounts payable (Note 7)	2,543,100	2,588,500
Billings in excess of costs and estimated earnings on uncompleted contracts (Note 3)	242,000	221,700
Accrued income taxes payable	52,000	78,600
Other accrued liabilities	36,600	36,000
Due to consolidated joint venture minority interests	154,200	26,200
Deferred income taxes (Note 13)	619,200	408,000
	4,312,800	4,188,700
Contingent liability (Note 10)		
Shareholders' equity		
Common stock—$1 par value, 500,000 authorized shares, 300,000 issued and outstanding shares	300,000	300,000
Retained earnings	1,126,100	649,500
Total shareholders' equity	1,426,100	949,500
	$5,738,900	$5,138,200

Percentage Contractors, Inc.
Consolidated Statements of Income and Retained Earnings
Years Ended December 31, 19X8 and 19X7*

	19X8	*19X7*
Contract revenues earned	$22,554,100	$16,225,400
Cost of revenues earned	20,359,400	14,951,300
Gross profit	2,194,700	1,274,100
Selling, general, and administrative expense	895,600	755,600
Income from operations	1,299,100	518,500
Other income (expense)		
Equity in earnings from unconsolidated joint venture	49,900	5,700
Gain on sale of equipment	10,000	2,000
Interest expense (net of interest income of $8,800 in 19X8 and $6,300 in 19X7)	(69,500)	(70,800)
	(9,600)	(63,100)
Income before taxes	1,289,500	455,400
Provision for income taxes (Note 13)	662,900	225,000
Net income (per share, $2.09 (19X8); $.77 (19X7))	626,600	230,400
Retained earnings, beginning of year	649,500	569,100
	1,276,100	799,500

*The accompanying notes are an integral part of these financial statements.

SCHEDULE 10-1 SAMPLE FINANCIAL STATEMENTS PREPARED ACCORDING TO THE PERCENTAGE-OF-COMPLETION METHOD OF ACCOUNTING *(Cont.)*

	19X8	*19X7*
Less: Dividends paid (per share, $.50 (19X8); $.50 (19X7)	150,000	150,000
Retained earnings, end of year	$ 1,126,100	$ 649,500

Percentage Contractors, Inc.
Consolidated Statements of Changes in Financial Position*
Years Ended December 31, 19X8 and 19X7

	19X8	*19X7*
Source of funds		
From operations		
Net income	$ 626,600	$ 230,400
Charges (credits) to income not involving cash and cash equivalents		
Depreciation and amortization	167,800	153,500
Deferred income taxes	211,200	(75,900)
Gain on sale of equipment	(10,000)	(2,000)
	995,600	306,000
Proceeds from equipment sold	25,000	5,000
Net increase in billings related to costs and estimated earnings on uncompleted contracts	40,700	10,500
Decrease in inventory	9,400	
Decrease in prepaid charges and other assets		16,100
Increase in accounts payable		113,200
Increase in other accrued liabilities	600	21,200
Increase in amount due to consolidated joint venture minority interests	128,000	26,200
Total	1,199,300	498,200
Use of funds		
Acquisition of equipment		
Shop and construction equipment	100,000	155,000
Automobile and trucks	40,000	20,000
Dividends paid	150,000	150,000
Increase in contract receivables	455,100	9,100
Increase in inventory		3,600
Increase in advances to and equity in joint venture	74,900	15,400
Increase in notes receivable, related company	25,000	50,000
Increase in prepaid charges and other assets	35,200	
Decrease in notes payable	110,300	90,300
Decrease in lease obligations payable	53,700	9,700
Decrease in accounts payable	45,400	
Decrease in accrued income taxes payable	26,600	2,400
Total	1,116,200	505,500

*On statements after July 31, 1988, the "Statement of Changes in Financial Position" is no longer required. It has been replaced by a "Statement of Cash Flow."

SCHEDULE 10-1 SAMPLE FINANCIAL STATEMENTS PREPARED ACCORDING TO THE PERCENTAGE-OF-COMPLETION METHOD OF ACCOUNTING *(Cont.)*

	19X8	*19X7*
Increase (decrease) in cash and certificates of deposit for year	83,100	(7,300)
Cash and certificates of deposit		
Beginning of year	221,300	228,600
End of year	$ 304,400	$221,300

Percentage Contractors, Inc.
Notes to Consolidated Financial Statements
December 31, 19X8 and 19X7

1. Significant Accounting Policies

Company's activities and operating cycle. The company is engaged in a single industry: the construction of industrial and commercial buildings. The work is performed under cost-plus-fee contracts, fixed-price contracts, and fixed-price contracts modified by incentive and penalty provisions. These contracts are undertaken by the company or its wholly owned subsidiary alone or in partnership with other contractors through joint ventures. The company also manages, for a fee, construction projects of others.

The length of the company's contracts varies but is typically about two years. Therefore, assets and liabilities are not classified as current and noncurrent because the contract-related items in the balance sheet have realization and liquidation periods extending beyond one year.

Principles of consolidation. The consolidated financial statements include the company's majority-owned entities, a wholly owned corporate subsidiary and a 75 percent-owned joint venture (a partnership). All significant intercompany transactions are eliminated. The company has a minority interest in a joint venture (partnership), which is reported on the equity method.

Revenue and cost recognition. Revenues from fixed-price and modified fixed-price construction contracts are recognized on the percentage-of-completion method, measured by the percentage of labor hours incurred to date to estimated total labor hours for each contract.* This method is used because management considers expended labor hours to be the best available measure of progress on these contracts. Revenues from cost-plus-fee contracts are recognized on the basis of costs incurred during the period plus the fee earned, measured by the cost-to-cost method.

Contracts to manage, supervise, or coordinate the construction activity of others are recognized only to the extent of the fee revenue. The revenue earned in a period is based on the ratio of hours incurred to the total estimated hours required by the contract.

Contract costs include all direct material and labor costs and those indirect costs related to contract performance, such as indirect labor, supplies, tools, repairs, and depreciation costs. Selling, general, and administrative costs are charged to expense as incurred. Provisions for estimated losses on uncompleted contracts are made in the period in which such losses are determined. Changes in job performance, job conditions, and estimated profitability, including those arising from contract penalty provisions, and final contract settlements may result in revisions to costs and income and are recognized in the period in which the revisions are determined. Profit incentives

*There are various other alternatives to the percentage of labor hours method for measuring percentage of completion, which, in many cases, may be more appropriate in measuring the extent of progress toward completion of the contract (labor dollars, units of output, and the cost-to-cost method and its variations).

SCHEDULE 10-1 SAMPLE FINANCIAL STATEMENTS PREPARED ACCORDING TO THE PERCENTAGE-OF-COMPLETION METHOD OF ACCOUNTING *(Cont.)*

are included in revenues when their realization is reasonably assured. An amount equal to contract costs attributable to claims is included in revenues when realization is probable and the amount can be reliably estimated.

The asset, "Cost and estimated earnings in excess of billings on uncompleted contracts," represents revenues recognized in excess of amounts billed. The liability, "Billings in excess of costs and estimated earnings on uncompleted contracts," represents billings in excess of revenues recognized.

Property and equipment. Depreciation and amortization are provided principally on the straight-line method over the estimated useful lives of the assets. Amortization of leased equipment under capital leases is included in depreciation and amortization.

Pension plan. The company has a pension plan covering substantially all employees not covered by union-sponsored plans. Pension costs charged to earnings include current-year costs and the amortization of prior-service costs over 30 years. The company's policy is to fund the costs accrued.

Income taxes. Deferred income taxes are provided for differences in timing in reporting income for financial statement and tax purposes arising from differences in the methods of accounting for construction contracts and depreciation.

Construction contracts are reported for tax purposes on the completed-contract method and for financial statement purposes on the percentage-of-completion method. Accelerated depreciation is used for tax reporting, and straight-line depreciation is used for financial statement reporting.

Investment tax credits are applied as a reduction to the current provision for federal income taxes using the flow-through method.

2. Contract Receivables

	December 31, 19X8	*December 31, 19X7*
Contract receivables		
Billed		
Complete contracts	$ 621,100	$ 500,600
Contracts in progress	2,146,100	1,931,500
Retained	976,300	866,200
Unbilled	121,600	105,400
	3,865,100	3,403,700
Less: Allowances for doubtful collections	75,900	69,600
	$3,789,200	$3,334,100

Contract receivables at December 31, 19X8, include a claim, expected to be collected within one year, for $290,600 arising from a dispute with the owner over design and specification changes in a building currently under construction. The changes were made at the request of the owner to improve the thermal characteristics of the building and, in the opinion of counsel, gave rise to a valid claim against the owner.

The retained and unbilled contract receivables at December 31, 19X8, included $38,600 that was not expected to be collected within one year.

SCHEDULE 10-1 SAMPLE FINANCIAL STATEMENTS PREPARED ACCORDING TO THE PERCENTAGE-OF-COMPLETION METHOD OF ACCOUNTING *(Cont.)*

3. Costs and Estimated Earnings on Uncompleted Contracts

	December 31, 19X8	*December 31, 19X7*
Costs incurred on uncompleted contracts	$15,771,500	$12,165,400
Estimated earnings	1,685,900	1,246,800
	17,457,400	13,412,200
Less: Billings to date	17,619,200	13,533,300
	$ (161,800)	$ (121,100)
Included in accompanying balance sheets under the following captions:		
Costs and estimated earnings in excess of billings on uncompleted contracts	$ 80,200	$ 100,600
Billings in excess of costs and estimated earnings on uncompleted contracts	(242,000)	(221,700)
	$ (161,800)	$ (121,100)

4. Advances to and Equity in Joint Venture

The company has a minority interest (one-third) in a general partnership joint venture formed to construct an office building. All of the partners participate in construction, which is under the general management of the company. Summary information on the joint venture follows:

	December 31, 19X8	*December 31, 19X7*
Current assets	$ 483,100	$280,300
Construction and other assets	220,500	190,800
	703,600	471,100
Less: Liabilities	236,800	154,000
Net assets	$ 466,800	$317,100
Revenue	$3,442,700	$299,400
Net income	$ 149,700	$ 17,100
Company's interest		
Share of net income	$ 49,900	$ 5,700
Advances to joint venture	$ 50,000	$ 25,000
Equity in net assets	155,600	105,700
Total advances and equity	$ 205,600	130,700

(For the purposes of illustrative financial statements, the one-line equity method of presentation is used in both the balance sheet and the income statement. However, the pro rata consolidation method is acceptable if the investment is deemed to represent an undivided interest.)

SCHEDULE 10-1 SAMPLE FINANCIAL STATEMENTS PREPARED ACCORDING TO THE PERCENTAGE-OF-COMPLETION METHOD OF ACCOUNTING *(Cont.)*

5. Transactions with Related Party

The note receivable, related company, is an installment note bearing annual interest at 9¼%, payable quarterly, with the principal payable in annual installments of $25,000, commencing October 1, 19Y0.

The major stockholder of Percentage Contractors, Inc., owns the majority of the outstanding common stock of this related company, whose principal activity is leasing land and buildings. Percentage Contractors, Inc., rents land and office facilities from the related company on a ten-year lease ending September 30, 19Y6, for an annual rental of $19,000.

6. Property and Equipment

	December 31, 19X8	*December 31, 19X7*
Assets		
Land	$ 57,500	$ 57,500
Buildings	262,500	262,500
Shop and construction equipment	827,600	727,600
Automobiles and trucks	104,400	89,100
Leased equipment under capital leases	300,000	300,000
	1,552,000	1,436,700
Accumulated depreciation and amortization		
Buildings	140,000	130,000
Shop and construction equipment	265,600	195,500
Automobiles and trucks	70,000	42,000
Leased equipment under capital leases	100,000	50,000
	575,600	417,500
Net property and equipment	$ 976,400	$1,019,200

7. Accounts Payable

Accounts payable include amounts due to subcontractors, totaling $634,900 at December 31, 19X8, and $560,400 at December 31, 19X7, which have been retained pending completion and customer acceptance of jobs. Accounts payable at December 31, 19X8, include $6,500 that are not expected to be paid within one year.

8. Notes Payable

	December 31, 19X8	*December 31, 19X7*
Unsecured note payable to bank, due in quarterly installments of $22,575 plus interest at 1% over prime	$388,100	$478,400
Note payable to bank, collateralized by equipment, due in monthly installments of $1,667 plus interest at 10% through January, 19Y3	80,000	100,000
	$468,100	$578,400

At December 31, 19X8, the payments due within one year totaled $110,300.

SCHEDULE 10-1 SAMPLE FINANCIAL STATEMENTS PREPARED ACCORDING TO THE PERCENTAGE-OF-COMPLETION METHOD OF ACCOUNTING *(Cont.)*

9. Lease Obligations Payable

The company leases certain specialized construction equipment under leases classified as capital leases. The following is a schedule showing the future minimum lease payments under capital leases by years and the present value of the minimum lease payments as of December 31, 19X8:

Year ending December 31	
19X9	$ 76,500
19Y0	76,500
19Y1	76,500
Total minimum lease payments	229,500
Less: Amount representing interest	31,900
Present value of minimum lease payments	$197,600

At December 31, 19X8, the present value of minimum lease payments due within one year is $92,250.

Total rental expense, excluding payments on capital leases, totaled $86,300 in 19X8 and $74,400 in 19X7.

10. Contingent Liability

A claim for $180,000 has been filed against the company and its bonding company arising out of the failure of a subcontractor of the company to pay its suppliers. In the opinion of counsel and management, the outcome of this claim will not have a material effect on the company's financial position or results of operations.

11. Pension Plan

Pension costs charged to earnings were $61,400 in 19X8 and $57,300 in 19X7. At December 31, 19X8, the estimated actuarial value of vested benefits exceeded the fund assets (at market) and contribution accruals by $197,600.

12. Management Contracts

The company manages or supervises commercial and industrial building contracts of others for a fee. These fees totaled $121,600 in 19X8 and $1,700 in 19X7 and are included in contract revenues earned.

13. Income Taxes and Deferred Income Taxes

The provision for taxes on income consists of the following:

	December 31, 19X8	*December 31, 19X7*
Currently payable, net of investment credits of $9,400 and $13,800	$451,700	$300,900
Deferred		
Contract related	204,200	(80,900)
Property and equipment related	7,000	5,000
	$662,900	$225,000

At December 31 of the respective years, the components of the balance of deferred income taxes were:

SCHEDULE 10-1 SAMPLE FINANCIAL STATEMENTS PREPARED ACCORDING TO THE PERCENTAGE-OF-COMPLETION METHOD OF ACCOUNTING *(Cont.)*

	December 31, 19X8	*December 31, 19X7*
Contract related	$594,000	$389,800
Property and equipment related	25,200	18,200
	$619,200	$408,000

14. Backlog

The following schedule shows a reconciliation of backlog representing signed contracts, excluding fees from management contracts, in existence at December 31, 19X7 and 19X8.*

Balance, December 31, 19X7	$24,142,600
Contract adjustments	1,067,100
New contracts, 19X8	3,690,600
	28,900,300
Less: Contract revenue earned, 19X8	22,432,500
Balance, December 31, 19X8	$ 6,467,800

In addition, between January 1, 19X9 and February 18, 19X9, the company entered into additional construction contracts with revenues of $5,332,800.

*The presentation of backlog information, although encouraged, is not a required disclosure.

there will usually be one account entitled "intercompany" or a similar title. Another difference is that a condensed form of the income statement may be presented in the equity section rather than a more detailed presentation on a separate statement. The income statement, if presented separately, will be much different from the statement presented in the company's overall financial statements. It will usually contain considerably more detail in the revenue and cost sections and may be so detailed as to present costs and revenues by bid or work items. Quite often another difference will be that the balance sheet will be much more like a detailed trial balance so that it can be more readily consolidated with the other job trial balance and head office trial balance. Also common in jobsite financials is the fact that there will be no general and administrative expenses. These costs are generally deferred as part of the various bid and work items on the job.

If the contractor has numerous decentralized jobs in progress, jobsite statements should be designed to allow for "pegboarding" or keypunching when computers are used to facilitate consolidating the job data for overall company financial statement purposes.

SCHEDULE 10-2 SAMPLE ADDITIONAL INFORMATION TO AUDITED FINANCIAL STATEMENTS (SCHEDULE 10-1). THIS INFORMATION IS USUALLY REQUIRED BY BONDING COMPANIES AND BANKS.

Independent Accountants' Report on Additional Information

The Shareholders and Board of Directors
Percentage Contractors, Inc.

Our examinations of the basic financial statements presented in the preceding section of this report were made primarily to form an opinion on such financial statements taken as a whole. The additional information, contained in the following pages, is not considered essential for the fair presentation of the financial position of Percentage Contractors, Inc., the results of its operations, or the changes in its financial position in conformity with generally accepted accounting principles. However, the following data were subjected to the audit procedures applied in the examinations of the basic financial statements and, in our opinion, are fairly stated in all material respects in relation to the basic financial statements taken as a whole.

(Firm Signature)
Certified Public Accountants

City, State
February 18, 19X9

Percentage Contractors, Inc.
Schedule 1
Earnings from Contracts
Year Ended December 31, 19X8

	19X8			*19X7*
	Revenues earned	*Cost of revenues earned*	*Gross profit (loss)*	*Gross profit (loss)*
Contracts completed during the year	$ 6,290,800	$ 5,334,000	$ 956,800	$ 415,300
Contracts in progress at year end	16,141,700	14,636,900	1,504,800	921,400
Management contract fees earned	121,600	51,800	69,800	1,700
Unallocated indirect and warranty costs		46,700	(46,700)	(38,100)
Minority interest in joint venture		128,000	(128,000)	(26,200)
Charges on prior year contracts		162,000	(162,000)	
	$22,554,100	$20,359,400	$2,194,700	$1,274,100

SCHEDULE 10-2 SAMPLE ADDITIONAL INFORMATION TO AUDITED FINANCIAL STATEMENTS (SCHEDULE 10-1). THIS INFORMATION IS USUALLY REQUIRED BY BONDING COMPANIES AND BANKS.

Percentage Contractors, Inc.
Schedule 2
Contracts Completed
Year Ended December 31, 19X8

Contract		*Contract totals*			*Before January 1, 19X8*			*During the year ended December 31, 19X8*		
Number	*Type**	*Revenues earned*	*Cost of revenues*	*Gross profit (loss)*	*Revenues earned*	*Cost of revenues*	*Gross profit) (loss)*	*Revenues earned*	*Cost of revenues*	*Gross profit (loss)*
1511	B	$ 5,475,300	$ 4,802,500	$ 672,800	$3,223,400	$2,932,700	$290,700	$2,251,900	$1,869,800	$382,100
1605	A	695,000	880,900	(185,900)	596,100	558,100	38,000	98,900	322,800	(223,900)
1624	A	140,700	150,700	(10,000)	29,600	31,800	(2,200)	111,100	118,900	(7,800)
1711	A	2,725,100	2,391,700	333,400	1,654,100	1,510,000	144,100	1,071,000	881,700	189,300
1791	B	4,770,100	4,288,900	481,200	3,028,500	2,929,600	98,900	1,741,600	1,359,300	382,300
1792	A	635,000	457,900	177,100				635,000	457,900	177,100
Small contracts		413,400	349,500	63,900	32,100	25,900	6,200	381,300	323,600	57,700
		$14,854,600	$13,322,100	$1,532,500	$8,563,800	$7,988,100	$575,700	$6,290,800	$5,334,000	$956,800

*Contract types:
A—Fixed-price.
B—Cost-plus-fee.

SCHEDULE 10-2 SAMPLE ADDITIONAL INFORMATION TO AUDITED FINANCIAL STATEMENTS (SCHEDULE 10-1). THIS INFORMATION IS USUALLY REQUIRED BY BONDING COMPANIES AND BANKS. ***(Cont.)***

Percentage Contractors, Inc.
Schedule 3
Contracts in Progress
December 31, 19X8

Contract		*Total contract*		*From inception to December 31, 19X8*						*At December 31, 19X8*		*For the year ended December 31, 19X8*		
Number	*Type**	*Revenues*	*Estimated gross profit (loss)*	*Revenues earned*	*Total costs incurred*	*Cost of revenues*	*Gross profit (loss)*	*Billed to date*	*Estimated cost to complete*	*Costs and estimated earnings in excess of billings*	*Billings in excess of costs and estimated earnings*	*Revenues earned*	*Cost of revenues*	*Gross profit (loss)*
1845	A	$ 6,750,200	$ 877,000	$ 5,890,500	$ 5,244,500	$ 5,143,900	$ 746,600	$ 5,976,000	$ 628,700	$15,100		$ 5,664,200	$ 4,984,500	$ 679,700
1847	B	1,471,800	127,100	1,250,400	1,139,800	1,139,800	110,600	1,195,800	204,900	54,600		962,800	899,000	63,800
1912	A	451,800	(130,100)	108,600	238,700	238,700	(130,100)	98,100	343,200	10,500		98,600	191,500	(92,900)
1937	B	11,125,000	847,900	7,337,900	7,045,500	6,721,100	616,800	7,808,000	3,231,600		$145,700	6,981,900	6,469,900	512,000
1945	A	3,650,100	497,000	2,395,200	2,061,300	2,061,300	333,900	2,491,500	1,091,800		96,300	2,395,200	2,061,300	333,900
Small contracts		51,300	8,400	49,800	41,700	41,700	8,100	49,800	1,200			39,000	30,700	8,300
		$23,500,200	$2,227,300	$17,032,400	$15,771,500	$15,346,500	$1,685,900	$17,619,200	$5,501,400	$80,200	$242,000	$16,141,700	$14,636,900	$1,504,800

*Contract types
A—Fixed-price.
B—Cost-plus-fee.

It would be impractical to attempt here to illustrate the great variety of forms used to report decentralized jobsite activities. The degree of decentralization varies so widely that jobsite statements and forms that would be suitable for one job might be completely inadequate for the next. If complete decentralization exists, any documents like delivery tickets or time cards are simply sent to the main office. If payroll and vouchers are made, and paychecks and accounts payable checks prepared and distributed at the jobsite, copies of all these and of the voucher register should be sent to the main office. Suitable forms for payroll, voucher system, and other documents have been illustrated in preceding chapters.

A joint venture in the construction industry is usually a special partnership of two or more companies formed to do a particular job. When the job is complete, the joint venture is liquidated. The finan-

SCHEDULE 10-3 SAMPLE FINANCIAL STATEMENTS PREPARED ON THE COMPLETED-CONTRACT METHOD OF ACCOUNTING. WHEN STATEMENTS ARE PREPARED ON THIS BASIS, BONDING COMPANIES, BANKS, AND OTHERS MAY REQUIRE THE ADDITIONAL INFORMATION SET FORTH IN SCHEDULE 10-2.

Sample Financial Statements
Completed Contractors, Inc.*

Independent Accountants' Report

The Stockholders and Board of Directors
Completed Contractors, Inc.

We have examined the balance sheets of Completed Contractors, Inc., as of December 31, 19X8 and 19X7, and the related statements of income and retained earnings and changes in financial position for the years then ended. Our examinations were made in accordance with generally accepted auditing standards and, accordingly, included such tests of the accounting records and such other auditing procedures as we considered necessary in the circumstances.

In our opinion, the financial statements referred to above present fairly the financial position of Completed Contractors, Inc., at December 31, 19X8 and 19X7, and the results of its operations and the changes in its financial position for the years then ended, in conformity with generally accepted accounting principles applied on a consistent basis.

(Firm Signature)
Certified Public Accountants

City, State
February 18, 19X9

*These statements were prepared prior to the passage of the Tax Equity and Fiscal Responsibility Act and so do not reflect any of the concepts of that act or the regulation changes it mandates.

SCHEDULE 10-3 SAMPLE FINANCIAL STATEMENTS PREPARED ON THE COMPLETED-CONTRACT METHOD OF ACCOUNTING. WHEN STATEMENTS ARE PREPARED ON THIS BASIS, BONDING COMPANIES, BANKS, AND OTHERS MAY REQUIRE THE ADDITIONAL INFORMATION SET FORTH IN SCHEDULE 10-2. *(Cont.)*

Completed Contractors, Inc.
Balance Sheets
December 31, 19X8 and 19X7*

Assets

	19X8	*19X7*
Current assets		
Cash	$ 242,700	$ 185,300
Contract receivables (less allowance for doubtful accounts of $10,000 and $8,000) (Note 2)	893,900	723,600
Costs in excess of billings on uncompleted contracts (Note 3)	418,700	437,100
Inventories, at lower of cost or realizable value on first-in, first-out basis (Note 4)	463,600	491,300
Prepaid expenses	89,900	53,900
Total current assets	2,108,800	1,891,200
Cash value of life insurance	35,800	32,900
Property and equipment, at cost		
Building	110,000	110,000
Equipment	178,000	163,000
Trucks and autos	220,000	200,000
	508,000	473,000
Less: Accumulated depreciation	218,000	203,200
	290,000	269,800
Land	21,500	21,500
	311,500	291,300
	$2,456,100	$2,215,400

*The accompanying notes are an integral part of these financial statements.

cial statements shown in Schedule 10-4 are generally adequate for transmittal to venture members for inclusion in their own financial statements. In larger construction joint ventures the financial statements will usually be audited and, if the percentage-of-completion method is being used, will take the form of Schedule 10-1. If the completed-contract method of accounting is used, they will take the form of Schedule 10-4. However, on joint venture if it were necessary to show in the net worth section of the balance sheet the details regarding each venturer's equity (cash advanced, share of earnings, special

SCHEDULE 10-3 SAMPLE FINANCIAL STATEMENTS PREPARED ON THE COMPLETED-CONTRACT METHOD OF ACCOUNTING. WHEN STATEMENTS ARE PREPARED ON THIS BASIS, BONDING COMPANIES, BANKS, AND OTHERS MAY REQUIRE THE ADDITIONAL INFORMATION SET FORTH IN SCHEDULE 10-2. *(Cont.)*

Liabilities and Stockholders' Equity

	19X8	*19X7*
Current liabilities		
Current maturities, long-term debt (Note 5)	$ 37,000	$ 30,600
Accounts payable	904,900	821,200
Accrued salaries and wages	138,300	155,100
Accrued income taxes	53,000	36,200
Accrued and other liabilities	116,400	55,550
Billings in excess of costs on uncompleted contracts (Note 3)	34,500	43,700
Total current liabilities	1,284,100	1,142,350
Long-term debt, less current maturities (Note 5)	245,000	241,000
	1,529,100	1,383,350
Stockholders' equity		
Common stock—$10 par value, 50,000 authorized shares, 23,500 issued and outstanding shares	235,000	235,000
Additional paid-in capital	65,000	65,000
Retained earnings	627,000	532,050
	927,000	832,050
	$2,456,100	$2,215,400

Completed Contractors, Inc.
Statements of Income and Retained Earnings
Years Ended December 31, 19X8 and 19X7*

	19X8	*19X7*
Contract revenues	$9,487,000	$8,123,400
Costs and expenses		
Cost of contracts completed	8,458,500	7,392,300
General and administrative	684,300	588,900
Interest expense	26,500	23,000
	9,169,300	8,004,200
Income before income taxes	317,700	119,200
Income taxes	164,000	54,200
Net income ($6.54 and $2.77 per share)	153,700	65,000
Retained earnings		
Balance, beginning of year	532,050	525,800
	685,750	590,800
Dividends paid ($2.50 per share)	58,750	58,750
Balance, end of year	$ 627,000	$ 532,050

*The accompanying notes are an integral part of these financial statements.

SCHEDULE 10-3 SAMPLE FINANCIAL STATEMENTS PREPARED ON THE COMPLETED-CONTRACT METHOD OF ACCOUNTING. WHEN STATEMENTS ARE PREPARED ON THIS BASIS, BONDING COMPANIES, BANKS, AND OTHERS MAY REQUIRE THE ADDITIONAL INFORMATION SET FORTH IN SCHEDULE 10-2. *(Cont.)*

Completed Contractors, Inc.
Statements of Changes in Financial Position*
Years Ended December 31, 19X8 and 19X7**

	19X8	*19X7*
Source of working capital		
Net income	$153,700	$ 65,000
Charge to income not requiring outlay of working capital—depreciation	54,800	50,300
Working capital from operations	208,500	115,300
Proceeds of notes payable	44,000	68,000
	252,200	183,300
Use of working capital		
Purchase of property and equipment	75,000	53,500
Reduction of long-term debt	40,000	28,000
Payment of dividends	58,750	58,750
Increase in cash value of life insurance	2,900	2,685
	176,650	142,935
Increase in working capital	$ 75,850	$ 40,365
Changes in components of working capital		
Increase (decrease) in current assets		
Cash	$ 57,400	$ (26,435)
Contract receivables	170,300	36,500
Costs in excess of billings on uncompleted contracts	(18,400)	49,100
Inventories	(27,700)	3,400
Prepaid expenses	36,000	(16,500)
	217,600	46,065
Decrease (increase) in current liabilities		
Current maturities, long-term debt		
Notes payable, bank	(6,000)	(12,000)
Mortgage payable	(400)	(500)
Accounts payable	(83,700)	(24,600)
Accrued salaries and wages	16,800	(24,300)
Accrued income taxes	(16,800)	6,300
Accrued and other liabilities	(60,850)	33,100
Billings in excess of costs on uncompleted contracts	9,200	16,300
	(141,750)	(5,700)
Increase in working capital	$ 75,850	$ 40,365

*On statements after July 31, 1988, the "Statement of Changes in Financial Position" is no longer required. It has been replaced by a "Statement of Cash Flow."

**The accompanying notes are an integral part of these financial statements.

SCHEDULE 10-3 SAMPLE FINANCIAL STATEMENTS PREPARED ON THE COMPLETED-CONTRACT METHOD OF ACCOUNTING. WHEN STATEMENTS ARE PREPARED ON THIS BASIS, BONDING COMPANIES, BANKS, AND OTHERS MAY REQUIRE THE ADDITIONAL INFORMATION SET FORTH IN SCHEDULE 10-2. *(Cont.)*

Completed Contractors, Inc.
Notes to Financial Statements
December 31, 19X8 and 19X7

---

1. Significant Accounting Policies

Company's activities. The company is a heating and air-conditioning contractor for residential and commercial properties. Work on new structures is performed primarily under fixed-price contracts. Work on existing structures is performed under fixed-price or time-and-material contracts.

Revenue and cost recognition. Revenues from fixed-price construction contracts are recognized on the completed-contract method. This method is used because the typical contract is completed in two months or less and financial position and results of operations do not vary significantly from those which would result from use of the percentage-of-completion method. A contract is considered complete when all costs except insignificant items have been incurred and the installation is operating according to specifications or has been accepted by the customer.

Revenues from time-and-material contracts are recognized currently as the work is performed.

Contract costs include all direct material and labor costs and those indirect costs related to contract performance, such as indirect labor, supplies, tools, repairs, and depreciation costs. General and administrative costs are charged to expense as incurred. Provisions for estimated losses on uncompleted contracts are made in the period in which such losses are determined. Claims are included in revenues when received.

Costs in excess of amounts billed are classified as current assets under costs in excess of billings on uncompleted contracts. Billings in excess of costs are classified under current liabilities as billings in excess of costs on uncompleted contracts. Contract retentions are included in accounts receivable.

Inventories. Inventories are stated at cost on the first-in, first-out basis using unit cost for furnace and air-conditioning components and average cost for parts and supplies. The carrying value of furnace and air-conditioning component units is reduced to realizable value when such values are less than cost.

Property and equipment. Depreciation is provided over the estimated lives of the assets principally on the declining-balance method, except on the building where the straight-line method is used.

Pension plan. The company has a pension plan covering all employees not covered by union-sponsored plans. Pension costs charged to income include current-year costs and the amortization of prior-service costs over 30 years. The company's policy is to fund the costs accrued.

Investment tax credit. Investment tax credits are applied as a reduction to the current provision for federal income taxes using the flow-through method.

SCHEDULE 10-3 SAMPLE FINANCIAL STATEMENTS PREPARED ON THE COMPLETED-CONTRACT METHOD OF ACCOUNTING. WHEN STATEMENTS ARE PREPARED ON THIS BASIS, BONDING COMPANIES, BANKS, AND OTHERS MAY REQUIRE THE ADDITIONAL INFORMATION SET FORTH IN SCHEDULE 10-2. *(Cont.)*

2. Contract Receivables

	December 31, 19X8	*December 31, 19X7*
Completed contracts, including retentions	$438,300	$408,600
Contracts in progress		
Current accounts	386,900	276,400
Retentions	78,700	46,600
	903,900	731,600
Less: Allowance for doubtful accounts	10,000	8,000
	$893,900	$723,600

Retentions include $10,300 in 19X8, which are expected to be collected after 12 months.

3. Costs and Billings on Uncompleted Contracts

	December 31, 19X8	*December 31, 19X7*
Costs incurred on uncompleted contracts	$2,140,400	$1,966,900
Billings on uncompleted contracts	1,756,200	1,573,500
	$ 384,200	$ 393,400
Included in accompanying balance sheets under the following captions:		
Costs in excess of billings on uncompleted contracts	$ 418,700	$ 437,100
Billings in excess of costs on uncompleted contracts	(34,500)	(43,700)
	$ 384,200	$ 393,400

4. Inventories

	December 31, 19X8	*December 31, 19X7*
Furnace and air-conditioning components	$303,200	$308,700
Parts and supplies	160,400	182,600
	$463,600	$491,300

Furnace and air-conditioning components include used items of $78,400 in 19X8 and $71,900 in 19X7 that are carried at the lower of cost or realizable value.

SCHEDULE 10-3 SAMPLE FINANCIAL STATEMENTS PREPARED ON THE COMPLETED-CONTRACT METHOD OF ACCOUNTING. WHEN STATEMENTS ARE PREPARED ON THIS BASIS, BONDING COMPANIES, BANKS, AND OTHERS MAY REQUIRE THE ADDITIONAL INFORMATION SET FORTH IN SCHEDULE 10-2. *(Cont.)*

5. Long-Term Debt

	December 31, 19X8	*December 31, 19X7*
Notes payable, bank		
Notes due in quarterly installments of $2,500, plus interest at 8%	$140,000	$150,000
Notes due in monthly installments of $1,500, plus interest at prime plus 1½%	87,000	58,000
Mortgage payable		
Due in quarterly payments of $3,500, including interest at 9%	55,000	63,600
	282,000	271,600
Less: Current maturities	37,000	30,600
	$245,000	$241,000

6. Pension Plans

The total pension expenses for the years 19X8 and 19X7 were $31,200 and $27,300, including contributions to union-sponsored plans.

At December 31, 19X8, the estimated actuarial value of vested benefits of the company's plan exceeded the fund assets (at market) and contribution accruals by $48,000.

7. Backlog

The estimated gross revenue on work to be performed on signed contracts was $4,691,000 at December 31, 19X8, and $3,617,400 at December 31, 19X7. In addition to the backlog of work to be performed, there was gross revenue, to be reported in future periods under the completed-contract method used by the company of $2,460,000 at December 31, 19X8, and $2,170,000 at December 31, 19X7.*

*The presentation of backlog information, although encouraged, is not a required disclosure.

allowances for interest or overhead, withdrawals, and so on), it would be necessary to prepare the operating statement separately.

Some contractors whose jobs are decentralized issue the reports on a comparative basis, or show separately the operating figures for the current period and for the job to date. Others prefer to make a comparative analysis after the statements come in from the job. The summary of cost and revenue for a cost-plus-fixed-fee job (Schedule 10-3) is an example of the comparative type of statement.

To make this type of statement for a cost-plus-fixed-fee job, the records should be kept by the "profit-and-loss" rather than by the "owner's equity" method of cost-plus-fixed-fee accounting.

SCHEDULE 10-4 FINANCIAL STATEMENTS FOR A DECENTRALIZED JOB

Ecks and Wye Construction Co.
General Ledger—Contract 51
Mountain City, Calif.
At close of business for month of February, 19—

Assets

Cash:			
Petty cash		$ 50.00	
First National Bank—General account		2,150.00	
Security Bank—Payroll account		1,000.00	
			$ 3,200.00
Accounts receivable:			
Approved progress billings		$28,500.00	
Retained percentages		15,000.00	
Progress billings not yet approved		14,500.00	
Employees' accounts		75.00	
Miscellaneous receivables (including back charges)		2,100,00	
Fees receivable			
Reimbursable costs			
Reimbursable costs unbilled			
Reimbursable costs accrued			
			60,175.00
Materials and supplies:			
Purchases for warehouse		$21,500.00	
Credits for warehouse issues	$ 8,100,00		
Credits for warehouse sales	900.00	9,000.00	
			12,500.00
Mess hall			
Food		$ 1,150.00	
Commissary merchandise		200.00	
Machine shop operation	$ 2,875.00		
Credits for machine shop work	2,350.00	525.00	
			1,875.00
Deferred charges and other assets:			
Plant and moving in	$37,750.00		
Credits for amortization	18,900.00	$18,850.00	
General expense	14,200.00		
Credits for amortization	12,550.00	1,650.00	
Refundable deposits		180.00	
Debits (unallocated)			
Equipment rental suspense			
Returnable containers		20.00	
			20,700.00
Equipment (job owned):			
Major equipment—cost		$ 4,800.00	
Depreciated cost transferred	$ 500.00		
Depreciation charged to job	2,400.00	2,900.00	
			1,900.00

SCHEDULE 10-4 FINANCIAL STATEMENTS FOR A DECENTRALIZED JOB *(Cont.)*

Minor equipment—salvage value		
Cost to be depreciated		
Depreciation charged to job		
TOTAL ASSETS		$100,350.00

Ecks and Wye Construction Co.
General Ledger—Contract 51
Mountain City, Calif.
At close of business for month of February, 19—

Liabilities

Accounts payable:		
Accounts payable—general	$12,750.00	
Contract advances received		
		$12,750.00
Due subcontractors	2,600.00	
Due subcontractors—retained percentages	950.00	
		3,550.00
Group hospitalization insurance		100.00
Accrued payroll		1,250.00
Accrued S.S. and withholding taxes:		
Federal withholding tax	$ 1,200.00	
State withholding tax		
Payroll deductions (FICA)	200.00	
Company (FICA)	200.00	
State payroll deductions	200.00	
State unemployment—company	500.00	
Federal unemployment tax	150.00	
		2,450.00
Other accrued expenses:		
State and county property tax	200.00	
Accrued insurance—compensation	1,400.00	
Accrued insurance—P.L. and P.D.	200.00	
Accrued insurance—E.L. and O.D.		
Accrued power bill		
Accrued rental—outside equipment		
Unclaimed wages	150.00	
		1,950.00
Reserves:		
Dismantling and moving out	2,700.00	
Uncompleted work		
Unallocated credits	500.00	
		3,200.00
TOTAL LIABILITIES		$25,250.00
Due to head office		53,300.00

SCHEDULE 10-4 FINANCIAL STATEMENTS FOR A DECENTRALIZED JOB *(Cont.)*

Operations

Contract revenue:	$175,500.00		
Cost of contract:	155,000.00		
		$20,500.00	
Other operations:		1,300.00	
			21,800.00
TOTAL LIABILITIES AND OPERATIONS			$100,350.00

Ecks and Wye Construction Co.
Expense Ledger—Contract 51
Mountain City, Calif.
At close of business for month of February, 19—

	Additions for current month	*Balance for job to date*	*Balance unamortized to date*
Plant and moving in:			
Temporary buildings	$ —	$ 5,000.00	$ 2,500.00
Utilities installations	—	2,000.00	1,000.00
Built-in construction plant	5,000.00	23,000.00	11,500.00
Unloading, assembling, and erection	—	2,000.00	1,000.00
Transportation	—	3,000.00	1,500.00
Dismantling and moving out	—	2,750.00	1,400.00
Miscellaneous			
TOTAL	$5,000.00	$37,750.00	$18,900.00
General expense			
Surety bond premium	$ —	$ 3,000.00	$ 1,500.00
Management and supervision salaries	200.00	1,500.00	—
Office salaries	300.00	1,000.00	—
Engineering salaries	200.00	500.00	—
Warehouse salaries			
Insurance—other than payroll insurance		200.00	150.00
Taxes—other than payroll taxes	100.00	200.00	—
Auto expense			
Warehouse expense	50.00	100.00	—
Home office—interest	100.00	200.00	
Home office—general expense	900.00	4,500.00	
Interest—other than home office			
Losses from bad accounts			
Telephone and telegraph	50.00	200.00	—
Light and power			
Legal expense			
Travel	100.00	200.00	—
Stationery and office supplies	50.00	100.00	—
Meals—mess hall	50.00	100.00	—

SCHEDULE 10-4 FINANCIAL STATEMENTS FOR A DECENTRALIZED JOB *(Cont.)*

Safety expense—including hospital gain or loss			
Engineering supplies and expense			
Rent—office			
Depreciation—office equipment			
Camp, commissary, and mess hall—net gain or loss	$1,500.00	$ 2,500.00	—
Cash discounts earned—credit	(25.00)	(100.00)	—
Miscellaneous general expense			
TOTAL	$3,575.00	$14,200.00	$ 1,650.00

The balance sheet with the operating statement included in the net worth section can be applied without much difficulty to the "owner's equity" type of cost-plus-fixed-fee accounting by showing reimbursed costs and earned fees in the net worth section. The cost statement can be applied without difficulty to "cost-plus" contracts, as can the mess hall report.

The summary of cost and revenue can be easily applied to any kind of contract by changing the account titles either to suit the individual job or to reflect the company's standard account classification.

The summary of volume and uncompleted work should be used to supply management with the facts not shown by the financial statements as such.

Cost Reports and Job Management

Probably the most important of all the cost reports is the daily labor cost report (Figure 3-20). Through it, job management can keep a day-to-day check on the most elusive element in the whole field of construction costs. If this one report is received daily, and is accurate, there is no better guide for the application of management control. An experienced construction man can compare the labor cost to date, the total estimated labor cost, and the physical work still to be done on any particular work item, and tell with great accuracy how well that part of the job is progressing. In a company with a number of smaller jobs, or on a large job with numerous work items, it is on this basis that the person in charge decides where to spend his own time and that of his supervisory staff.

Similar reports may be made daily on materials consumed or placed and on equipment time. Both serve a purpose similar to that of the labor cost report. When time and staff permit, a daily cost sum-

SCHEDULE 10-4 FINANCIAL STATEMENTS FOR A DECENTRALIZED JOB *(Cont.)*

Ecks and Wye Construction Co.
Job Revenue Statement—Contract 51
Mountain City, Calif.

Week of month ending February 28, 19—

	Quantity		*Amounts ($)*			*Unit costs*		
Item	*Total contract*	*To date*	*Actual cost to date*	*Estimated cost to date*	*Contract revenue to date*	*Actual*	*Estimated*	*Contract price*
1. Excavation—earth, cu yd	336,000	200,000	$ 94,000	$ 96,000	$110,000	$ 0.47	$ 0.48	$ 0.55
2. Excavation—rock, cu yd	4,000	3,500	8,050	8,050	8,750	2.30	2.30	2.50
3. Excavation—hand, cu yd	2,000	1,800	15,300	15,600	14,400	8.50	8.68	8.00
4. Reinforced steel in place, lb	35,000	10,000	1,900	2,400	2,500	0.19	0.24	0.25
5. Concrete in place, cu yd	1,800	528	35,550	35,904	39,600	67.33	68.00	75.00
6. Miscellaneous metal in place, lb	4,000	625	200	225	250	0.32	0.36	0.40
TOTAL			$155,000	$158,179	$175,500			

SCHEDULE 10-4 FINANCIAL STATEMENTS FOR A DECENTRALIZED JOB *(Cont.)*

Ecks and Wye Construction Co.
Job Revenue Statement—Contract 51
Mountain City, Calif.

Week of month ending February 28, 19—

	Contract			*This week*		*To date*		
Item	*Quantity*	*Price*	*Amount*	*Quantity*	*Amount*	*Quantity*	*Amount*	*Percent completion*
1. Excavation—earth, cu yd	336,000	$ .55	$184,000			200,000	$110,000	59.5
2. Excavation—rock, cu yd	4,000	2.50	10,000			3,500	8,750	87.5
3. Excavation—hand, cu yd	2,000	8.00	16,000			1,800	14,400	90.0
4. Reinforcing steel in place, lb	35,000	.25	8,750			10,000	2,500	28.6
5. Concrete in place, cu yd	1,800	75.00	135,000			528	39,600	29.3
6. Miscellaneous metal in place, lb	4,000	.40	1,600			625	250	15.6
TOTAL			$356,150				$175,500	

SCHEDULE 10-4 FINANCIAL STATEMENTS FOR A DECENTRALIZED JOB *(Cont.)*

Ecks and Wye Construction Co.
Accounts Receivable—Contract 51
Mountain City, Calif.
At close of business for month of February, 19—

Name	*Amount*	*Under 60 days*	*Over 60 days*	*Over 90 days*
Estimates				
E.Z. Doe Co., Inc. Progress billing 11	$28,500.00	$28,500.00	$	$
E.Z. Doe Co., Inc. Retained percentage	15,000.00	7,000.00	4,000.00	4,000.00
Progress billing 12 not yet approved	14,500.00	14,500.00		
	$58,000.00	$50,000.00	$4,000.00	$4,000.00
Employees				
W. T. Jones	$ 50.00	$ 50.00	$	
J. S. Smith	25.00		25.00	
	$ 75.00	$ 50.00	$ 25.00	
Miscellaneous				
Zee Equipment Co.	$ 1,900.00	$ 1,900.00	$	
Notany Co.	200.00		200.00	
	$ 2,100.00	$ 1,900.00	$ 200.00	

SCHEDULE 10-4 FINANCIAL STATEMENTS FOR A DECENTRALIZED JOB *(Cont.)*

Ecks and Wye Construction Co.
Subcontract Accounts—Contract 52
Mountain City, Calif.
At close of business for month of February, 19—

Subcontractor	*Amount of subcontract*	*Amount of work performed*	*Due on estimates*	*Due on retained percentages*
ABC Concrete Co. 800 Main Street Chico, Calif.	$20,000.00	$8,500.00	$2,600.00	$950.00

Ecks and Wye Construction Co.
Monthly Report of Camp and Mess Operations—Contract 51
Mountain City, Calif.
At close of business for month of February, 19—

Mess Operations

	Current month		*To date*	
Mess revenue:				
Payroll deductions	$1,600.00		$2,500.00	
Cash meal sales	100.00		200.00	
Executive and guest meals	50.00		100.00	
Other mess revenue				
Total mess revenue		$ 1,750.00		$ 2,800.00
Mess expense				
Beginning inventory	$1,450.00		$1,450.00	
Food purchased	1,200.00		2,100.00	
	$2,650.00		$3,550.00	
Less: Closing inventory	1,150.00		1,150.00	
Cost of food	$1,500.00		$2,400.00	
Fuel and supplies	200.00		300.00	
Labor	1,100.00		1,600.00	
Plant amortization	200.00		500.00	
Total mess costs		$ 3,000.00		4,800.00
NET GAIN (OR LOSS) FROM MESS		$(1,250.00)		$(2,000.00)
Average costs				
Total number meals served	2253		3733	
Cost per meal served	1.33		1.29	
Net loss per meal served	.55		.54	
Food cost per meal served	.666		.645	
Labor cost per meal served	.489		.429	
Supply cost per meal served	.089		.082	
Plant cost per meal served	.089		.134	
Total cost per meal	$1.33		$1.29	

SCHEDULE 10-4 FINANCIAL STATEMENTS FOR A DECENTRALIZED JOB *(Cont.)*

Camp Operations

	Current month		To date	
Camp revenue:				
Payroll deduction	$ 150.00		$ 300.00	
Other revenue				
Total camp revenue		$ 150.00		$ 300.00
Camp expense:				
Fuel, laundry, and supplies	$ 75.00		$ 150.00	
Labor	50.00		100.00	
Plant amortization	275.00		550.00	
Total camp costs		$ 400.00		$ 800.00
NET GAIN (OR LOSS) FROM CAMP		$(250.00)		$ (500.00)
Capacity of camp (number of men)	30			
Average daily occupancy	25			
Average gain (or loss) per man-month	(10.00)		(10.00)	

Residence Operations

Residence revenue		
Residence costs		
GAIN (OR LOSS) FROM RESIDENCES		

Ecks and Wye Construction Co.
Summary of Cost and Revenue
(Cost-Plus-Fixed-Fee Contracts)
Contract 52, Mountain City, Calif., month ended February 28, 19—

	Job to date	Current year to date	Current month
Revenue as shown by job books:			
Work done to date			
Reimbursable costs billed to client	$30,000.00	$22,000.00	$ 8,500.00
Reimbursable costs unbilled or accrued	12,000.00	9,000.00	3,500.00
Fee earned	3,500.00	2,700.00	850.00
Equipment rental and other revenue from client	7,000.00	5,200.00	1,600.00
TOTAL REVENUE	$52,500.00	$38,900.00	$14,450.00
Cost of work done:			
Reimbursable costs	$42,000.00	$30,200.00	$11,600.00
Costs relating to contractor owned equipment	4,000.00	3,000.00	1,000.00
Total direct costs	$46,000.00	$33,200.00	$12,600.00
Net to contractor before non-reimbursable cost	$ 6,500.00	$ 5,700.00	$ 1,850.00

SCHEDULE 10-4 FINANCIAL STATEMENTS FOR A DECENTRALIZED JOB *(Cont.)*

	Job to date	*Current year to date*	*Current month*
Nonreimbursable costs:			
General expense items nonreimbursable per contract	$350.00	$240.00	$120.00
Home office charges (if nonreimbursable)	1,100.00	1,065.00	30.00
Disallowed or suspended costs	50.00	30.00	15.00
Total nonreimbursable costs	$1,500.00	$1,335.00	$165.00
Operating profit	$5,000.00	$4,365.00	$1,685.00
Other operations, excess depreciation, etc.	100.00	60.00	30.00
NET PROFIT AS SHOWN BY JOB BOOKS	$5,100.00	$4,425.00	$1,715.00

Estimated amount of contract, including change orders—$157,500.00
Amount of main office investment in contract (M-1)—$28,780.00
Date set for completion—June 30, 19—
Percent of work done—33.33
Percent of time elapsed—30

	Payrolls	*Company-owned equipment depreciation*	*Company-owned equipment rental to home office*
This month	$8,000.00		$2,000.00
Previous months	9,000.00		2,500.00
TOTALS	$17,000.00		$4,500.00

Superintendent *Office Manager*

Ecks and Wye Construction Co.
Summary of Contract Volume and Uncompleted Work
Contract 52 Mountain City, Calif.
At close of business for month of February, 19—
Client's name: A.B.C. Co.
Type of contract: Unit price of lump sum _____ Bid __________
Cost-plus-fixed-fee __________ Negotiated __________

1. Amount of original contract		$120,000.00
Fee		12,000.00
2. Net change orders previous months	$15,000.00	
Fee	1,500.00	
3. Change orders this month (list):		
C.O. No. 2 extension—south wing	$ 8,181.81	
Fee	818.19	
Total change orders to date		$ 25,500.00

SCHEDULE 10-4 FINANCIAL STATEMENTS FOR A DECENTRALIZED JOB *(Cont.)*

4. Actual variations in contract amount due to overrun (or underrun) of quantities on bid items completed to 19—	$
5. Rental revenue from contract owner not included above	
6. Estimated total amount of principal contract(s)	$
7. Other contract work not included in principal contract(s)	$
8. Total estimated contract volume	$157,500.00
9. Amount of work completed to date	$ 52,500.00
10 Uncompleted contract volume	$105,000.00
Date contract was physically completed (if completed)	
Expected date of physical completion (if not completed)	June 30, 19—

Instructions for summary of contract volume and uncompleted work

1. *Amount of original contract:* Enter the amount which appears in the contract document, or if no amount is shown, estimate the total contract amount on the basis of the facts known at the time the contract was negotiated. For fixed-fee contracts include the estimated fee in the gross amount of the contract.

2. *Net amount of change orders previous month:* Enter the net amount of additive and deductive change orders reported in detail on previous reports. Include only approved changes.

3. *Change orders current month (list):* Enter each change order that affects the amount of the contract, and the amount (if any) by which the contract is changed. If change order shows no definite amount, enter the best available estimate of the amount of the change. For this purpose, treat extra work orders as changes, but clearly designate them as extra work.

4. *Actual variations in contract amount due to overruns (or underruns) of quantities on bid items completed:* Enter the sum of the variations in contract volume arising from differences in quantities paid for and quantities estimated in the contract, or from change orders on bid items completed and for which final quantities have been determined. (Supporting list by bid item should be available in job files.) Enter differences between estimated costs or fee and actual costs or fee in the case of cost-plus-fixed-fee contracts. This amount must be adjusted each month to agree with actual billings.

5. *Rental revenue from contract owner not included above:* Enter the amount of equipment rentals billed to the contract owner to date, but only to the extent that such equipment rentals are not included in the contract under items 1, 2, or 3 above.

6. *Estimated total amount of principal contract(s):* This amount, the sum of the preceding items, should represent the known volume of work under contract at the current date. If additional overruns or underruns are expected to be substantial, the amount should be included here and supported by a special schedule.

7. *Other contract work not included in principal contract(s):* Enter here the total of work completed by and on the contract for others than the owner. This item will include work for subcontractors and outsiders with contract equipment and labor. It will include bare equipment rental from others than contract owner.

8. *Total estimated contract volume:* Enter the total of items 5 and 6 above.

9. *Amount of work completed to date:* This amount should agree with the total contract revenue as shown on the "Summary of cost and revenue" under "Total revenue."

10. *Uncompleted contract volume:* Enter here the difference between items 7 and 8 above.

Cost and Revenue Detail

Location: Mountain City, Calif. ***Month: February*** ***Contract 52***

Contract items and quantities	*Cost Total*	*Cost Unit*	*Contract price Total*	*Contract price Unit*
Nonreimbursable costs:				
General expense items:				
Travel expense of John Jones, project manager	$ 200.00			
Donation—American Red Cross	100.00			
Employee entertainment (party)	50.00			
	$ 350.00			
Home office charges:				
Interest on investment	$ 50.00			
Engineering salaries and expenses	1,050.00			
	$1,100.00			
Disallowed costs:				
Replacement of defective concrete in foundation	$ 50.00			
	$ 50.00			

mary (Figure 10-2) can be extremely useful. Otherwise, the labor summary is usually the most effective tool in helping job management direct its efforts most effectively.

On some types of jobs, the most effective unit of measurement is the amount of material moved or placed. One illustration of that type of report can be found in the grade foreman's report (Figure 5-12). On pile driving jobs the daily report might cover the number of feet of piling driven and the unit cost per foot. On concrete jobs the number of cubic yards placed, and on pier or dock jobs the number of MBF (thousand "board feet," or feet board measure) of lumber placed might be reported. On this type of report the unit cost of the work done should be watched closely. Any substantial deviation from the estimated unit cost would constitute a warning to management, as would a comparable deviation in labor cost. The important points are these:

1. Job management is a day-to-day and hour-to-hour matter, and it must have progress data on that basis. If it is to be done properly the information on which it acts must be current and correct.
2. The exact information needed to control job progress varies according to the nature of the jobs and of the people who manage them. It is better, therefore, to determine exactly what information is needed and to design a form to present it, rather than to try to force all job progress information to conform to a theoretical ideal.

ECKS & WYE CONSTRUCTION CO.

Daily Cost Summary

Job No. . . . .

Prepared by . . . .
Checked by . . . .
Approved by . . . . Date . . . . 19 . . . .

Work item	Today					Job to date					Total budget	Balance unspent	Per cent complete
	Labor	Mat'ls.	Equip.	Other	Total	Labor	Mat'ls.	Equip.	Other	Total			
Totals													

Figure 10-2 **Daily Cost Summary.**

3. The only reason for daily cost figures is to help job management direct its efforts where they will do the most good. It is only when this purpose is accomplished that the cost of compiling daily figures can be justified.

Job management often needs a detailed analysis of cost on a particular work item. If the job cost ledger (Figure 9-1) has been kept, the information should be available there. If the daily cost summary (Figure 10-2) has been kept, the answer would appear in that report. If neither has been kept, then it may be necessary to extract the needed information from such basic data as time cards and purchase invoices.

Such information might be needed for a number of purposes. For example, price negotiations for a change order may be in progress; daily cost reports may have indicated differences between the job estimate and the actual operating results that require further analysis; or current results on this job may be needed for a guide in bidding new work or justifying a claim in renegotiation or an estimate for financing purposes. Whether any or all of these possibilities justify the cost of a job cost ledger is, of course, a problem peculiar to each contractor.

Cumulative cost figures, by individual work items and for the entire job, are meaningful to job management because they may be compared to the total contract price to show how much money is left in the contract and in the budget for each work item to complete the work. A cumulative form for labor or equipment (Figure 10-3) provides such information. This information, combined with unit costs and elapsed time figures, indicates to job management what it must do in order to earn the estimated profit on the job. For example, if unit costs are below the estimate, but time is running short, some overtime to bring the work up to schedule may be justified. On the other hand, if unit costs are high and the job is lagging it means more and better supervision is needed quickly to bring the troublesome work item back into line to prevent loss. It may mean that work is being done that should be compensated as an "extra" and, if so, the sooner the "extra" is claimed, the better. Usually it means that either the work is not being done efficiently, or the job estimate is wrong. In either case, prompt attention from job management is required.

Cost Reports and General Management

There are two indicators of satisfactory job progress—time and money. Up to a certain point the faster the work progresses, the less it costs. This is true for several reasons. Rapid progress means plan-

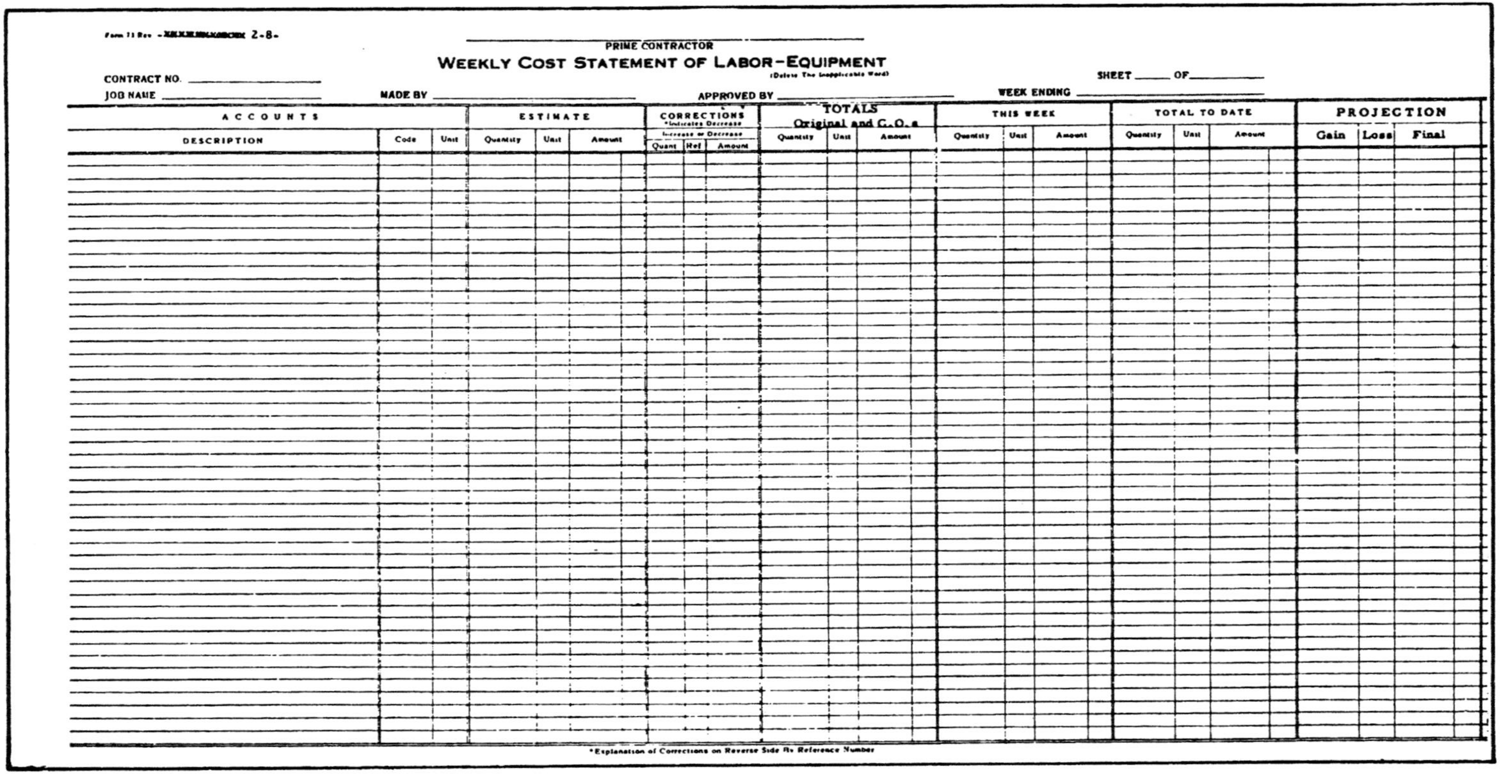

Form 11 Rev. 2-8-

PRIME CONTRACTOR

WEEKLY COST STATEMENT OF LABOR-EQUIPMENT

(Delete The Inapplicable Word)

CONTRACT NO. ______ SHEET ______ OF ______

JOB NAME ______ MADE BY ______ APPROVED BY ______ WEEK ENDING ______

ACCOUNTS			ESTIMATE			CORRECTIONS *Indicates Decrease Increase or Decrease			TOTALS Original and C.O.s			THIS WEEK			TOTAL TO DATE			PROJECTION		
DESCRIPTION	Code	Unit	Quantity	Unit	Amount	Quant	Ref	Amount	Quantity	Unit	Amount	Quantity	Unit	Amount	Quantity	Unit	Amount	Gain	Loss	Final

*Explanation of Corrections on Reverse Side By Reference Number

Figure 10-3 **Weekly Cost Statement of Labor or Equipment.** Note that unit costs for each work item are compared with those estimated for that item.

ning has been effective, and there are few time-consuming errors to correct. Also, the faster the job progresses, the lower the amount of overhead that each unit of production has to carry.

Beyond a certain rate, however, each unit of production increases in cost (Ricardo's theory). Overtime, inadequate space, and mistakes resulting from speed and pressure all contribute to this loss of efficiency. To determine how closely any particular job can or should be run to this point of diminishing returns, and to see that it keeps running at the predetermined rate, is the job of general management. One of the tools which general management uses to accomplish this purpose is the job progress report (Figure 10.3). This form shows one week's progress on one job. It breaks the job down into work items, and compares the key variables (labor and equipment costs) on each work item with the totals estimated for that item, the total labor or equipment cost of the work accomplished for the week and for the job to date, and the unit costs according to the estimate and according to the actual operations. These data, when combined with the more complete figures on the monthly statement, give most key executives the current money figures they need.

But money figures alone rarely tell the whole story. To supplement them it is usually necessary to have an additional weekly report (Figure 10-4). This report covers weather, work force, special job conditions, time progress, and financial forecast to complete the analysis of job operations already developed from the financial statements and the weekly cost reports.

One of the most commonly discussed and least understood uses of cost reports is in bidding other work. It is easy to think that, having done a certain operation before under a variety of circumstances and having kept accurate cost records on it, the unit costs might be helpful in bidding.

Actually, this practice is dangerous because it tends to lead to bidding at a fixed price per unit of work. For example, a sewer contractor might find his trenching costs are running \$8.00 per lineal foot. So long as he is digging the same type of trench through the same type of material, he may use this figure in bidding, but suppose in the next job trenching is through loose, sandy soil that will not hold a bank. Or, suppose the cost figures were developed in an area where the soil is loam overlying a loose shale at about 4 feet, whereas the area where the new work is to be done has patches of caliche that have to be blasted before it can be dug. Or, suppose the previous jobs had been in open areas, permitting machine digging, whereas the job being bid is in close quarters and will require a large proportion of hand excavation. Geographical weather and union labor conditions are also important factors in determining unit costs.

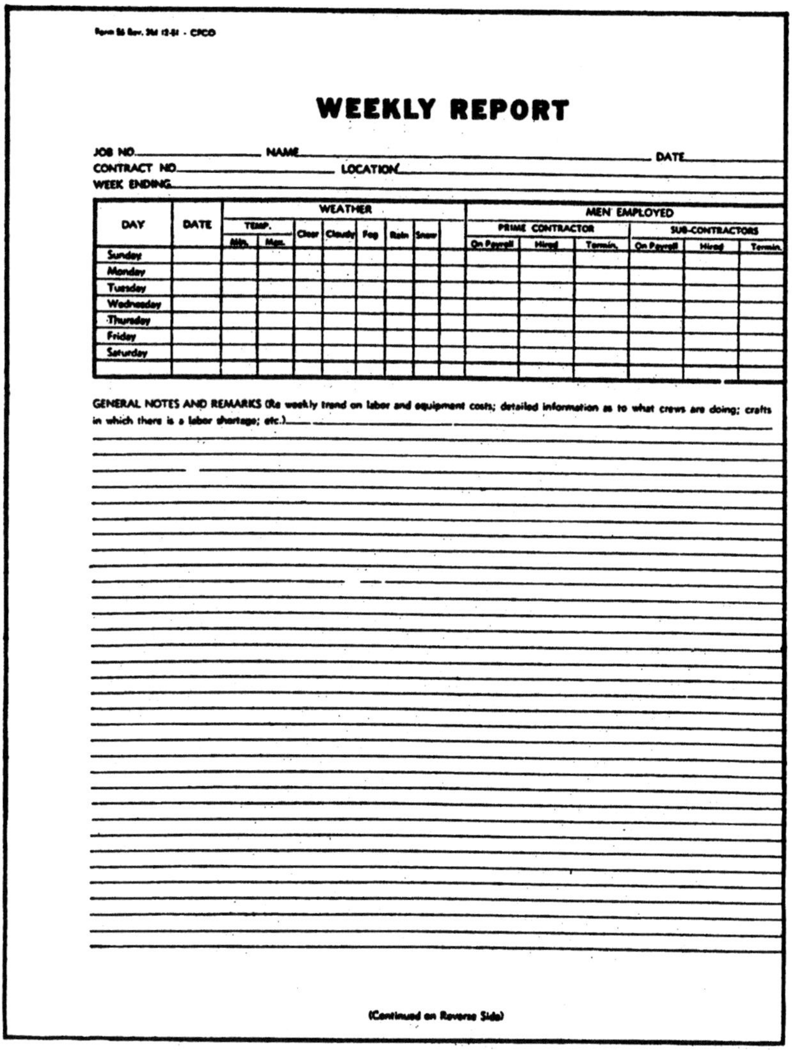

WEEKLY REPORT

JOB NO.________ NAME________ DATE________
CONTRACT NO.________ LOCATION________
WEEK ENDING________

DAY	DATE	WEATHER								MEN EMPLOYED					
		TEMP.		Clear	Cloudy	Fog	Rain	Snow		PRIME CONTRACTOR			SUB-CONTRACTORS		
		Min.	Max.							On Payroll	Hired	Termin.	On Payroll	Hired	Termin.
Sunday															
Monday															
Tuesday															
Wednesday															
Thursday															
Friday															
Saturday															

GENERAL NOTES AND REMARKS (Re weekly trend on labor and equipment costs; detailed information as to what crews are doing; crafts in which there is a labor shortage; etc.)

(Continued on Reverse Side)

Figure 10-4a **Weekly Report—Face.**

The risk in using cost figures in bidding lies in the ease with which an estimator may assume that the same kind of work will cost the same on all jobs. If unit prices are available from similar jobs, it is easy for an estimator, working against a bid-opening deadline, to use them without considering the individual problems of the job being bid.

On the other hand, if averages may be safely used, good unit cost figures can make it possible for an estimator to prepare more bids and do a faster and better job. Thus, on estimating costs in tract

DELAYS (State steps taken to stop these delays)______________________

PROGRESS	WORK COMPLETED %	ELAPSED TIME %
PREVIOUS—WEEK ENDING		
THIS PERIOD—WEEK ENDING		
TOTAL TO DATE		

Original Contract . . . . $__________

Change Orders & Supplemental Agreements issued, signed and received $__________

Change Orders & Supplemental Agreements approved, in process of being issued and signed, but NOT received . . . . $__________

TOTAL CURRENT CONTRACT . . . . $__________

Change Estimates for work proceeding without Change Orders or Supplemental Agreements . . . . $__________

Change Estimates quoted, which will probably be approved (Omit all dead or abandoned Change Estimates) . . . . $__________

TOTAL CHANGE ESTIMATES PENDING . . . . $__________

TOTAL ANTICIPATED CONTRACT . . . . $__________

Dates on which Partial Payment Requests or Progress Estimates are to be made______________

Net Progress Payments paid to date of this report . . . . $__________

Retention on above Progress Payments . . . . $__________

Gross Partial Payment Requests or Progress Estimates submitted but not paid to date of this report:

No.______submitted______195__for work completed to______195__ $__________

No.______submitted______195__for work completed to______195__ $__________

No.______submitted______195__for work completed to______195__ $__________

Total gross Partial Payment Requests or Progress Estimates due but not paid . . . . . $__________

TOTAL PROGRESS PAYMENT REQUESTS or PROGRESS ESTIMATES TO DATE OF THIS REPORT $__________

CASH PAID OUT	THIS WEEK	TO DATE	ESTIMATED FOR NEXT WEEK
PAYROLLS	$______	$______	$______
SUB-CONTRACTORS . .	$______	$______	$______
VENDORS	$______	$______	$______
MISCELLANEOUS . .	$______	$______	$______
TOTALS	$______	$______	$______

CASH IN BANK . . . . $__________

COLLECTIONS ESTIMATED FOR NEXT WEEK

PENDING PARTIAL PAYMENTS . . . . . . $__________

MISCELLANEOUS ACCOUNTS RECEIVABLE . . . . $__________

TOTAL ESTIMATED COLLECTIONS . . . . $__________

REMARKS (Concerning Changes, Progress Payments, Disbursements and Collections)______________________

______________________ Project Manager
Superintendent

Figure 10-4b **Weekly Report—Reverse.**

housing developments, a degree of uniformity is assured, and the estimator can assume, for example, that certain plumbing assemblies may safely be prefabricated in the shop, or that a given number of foundations can be poured in a day.

Probably the safest way to use cost figures in bidding is as a check on the estimator's thinking. If he is estimating new work and the difference between the bid and the former actual performance is substantial, then he should find the reasons for the difference. If he is analyzing operations to find the reasons for a difference between unit costs estimated and unit costs actually incurred on a current job, deviations in the unit cost of the same type of work on other jobs may provide the answers.

The most useful application of cost figures to bidding is the analysis of each job after it is complete, to account for differences between the estimate and the actual result. In this way errors in the methods used in preparing bid estimates can be found and corrected. Equally important, methods of operation assumed by the estimator can be compared with those followed on the job, and the results of such study can be considered on the next job involving similar work.

Chapter 11

Internal Auditing

In the United States, internal auditing has become a widely accepted institution in most industries with the exception of the construction industry. In recent years, some of the larger turnkey and engineering type contractors have instituted internal auditing on an extensive and sophisticated basis. Many of these internal audit groups primarily use "operational" auditing in which the principal emphasis is on engineering, procurement, planning and scheduling, the accuracy of cost forecasting, and computation of percentage completion. As a result, these audit teams are staffed with specialists in the subject matter which they audit.

For the great majority of contractors, however, any auditing that may be done is performed by the contractor-owner or by someone in his office on a "hit or miss" basis. In most of the smaller construction companies, the only auditing done is performed by the public accountant employed to prepare the tax return and review the year-end financial statements.

Field Auditing

The general attitude of most smaller contractors toward internal auditing is pretty well summed up by one California contractor who recently said: "What is there to audit? I see all the bills before they are paid and I sign all the checks. Besides, my bookkeeper is so honest he wouldn't touch a penny."

That type of reasoning is sound as long as the company is small enough so that one man can know everything that needs to be known about the jobs under his control. He needs to know these jobs well enough to identify the invoices and the employees, and to determine whether the invoices are correct and whether the employees shown on the payroll were actually on the job. Because he is managing the job, the efficiency is usually only as good as his management ability, since he may be reluctant to accept advice from lower-level employees on how the work should be done. A regular system of internal auditing can and often does work well in operations where the auditing functions are delegated to the project manager or superintendent on large decentralized jobs. One word of caution, however. When done on this basis, the auditing will be only as good as the project manager and of course will not provide for regular reporting to management.

There is one major reason why a hard-headed, fraud-conscious industry like the construction industry has seen fit to leave what auditing is performed to job management. The reason is that accounting and accounting departments within the construction industry generally are not held in high esteem by construction and engineering people. In fact, one construction company president recently described his accounting department as "mere clerks." This reasoning comes from the fact that relatively few accountants have the technical knowledge of construction work necessary to function in any capacity other than accountants or auditors. The major accounting function is cost forecasting, and that is performed by engineers or construction supervisors.

Also, top management of most construction companies is made up of engineers and operating personnel whose experience with internal auditors often tends to support the view that their principal function is that of mere routine figure checking such as recomputing payrolls and matching invoices with purchase orders. They do not look upon their own function in monitoring the progress of the work, reviewing payrolls and vouchers, and signing checks, as auditing. They regard this function as far too important to fit their idea of the audit function.

Nevertheless, among some of the larger construction companies, the use of special internal auditors has become an accepted practice, largely because the demands on the time of the project managers and superintendents has become so great that the division of work has proved necessary. Also, the larger construction companies tend to be more accounting-minded, and there is increased recognition of the need for more of the accounting type auditing tempered by a great deal of technical construction knowledge.

In order to be fully effective in doing the internal auditing in a construction company, the auditor should be able to review the plans, observe the physical progress of the job, and know what the records should show. He should be able to go to the records and know whether they reflect the physical progress he has observed. He must also know the sequence in which the physical work is to be accomplished. Payroll padding and other similar irregularities have often been detected simply because the labor charges on a work item were out of line with physical progress, or were charged to a work item which was not in sequence with the rest of the job.

To summarize then, most of the auditing done within the construction industry today is performed by engineering-trained management people who are in a so-called "online" function and who are reviewing the physical progress of the work, the payrolls, and the vouchers preparatory to signing checks. Accounting personnel, more often than not, are responsible for the comparison between final vouchers and supporting documents and for mathematical accuracy. The project manager, superintendent, or job engineer reviews the documentation for overall reasonableness and propriety before signing the checks. Any additional internal auditing is performed by head office personnel who call at the jobs to see if the routines called for by the company's procedure manual are being followed (i.e., compliance auditing).

General Office Auditing

When auditing is done in the home office, it follows, in many respects, the pattern described for the field. However, the further auditing is removed from the point of actual operations, the more it tends to take on a character of clerical and accounting checks and to lose the characteristics of critical analysis which it normally has when performed in the field. There are two important reasons for this tendency:

1. When a centralized system is in use, the demands on the time of the principals are so great that there is a tendency to treat internal auditing lightly. The day-to-day construction and engineering functions of the principal take priority over internal auditing, which too often is considered a necessary evil, at best.
2. If the person doing the auditing is an accountant or office engineer who lacks detailed knowledge of the job or jobs in progress, he has no basis for comparison beyond the form followed and its apparent compliance with the system.

Accordingly, when a contractor's accounting is centralized, the internal auditing needs to be done by someone who has direct personal knowledge of the job.

No internal audit program is more effective than the procedures in effect for following up on its findings. If auditing is carried out by job management personnel and errors are corrected at once, then there is no need for further follow-up. A number of the larger and more progressive companies, however, are finding that something more is needed, particularly if general management is separated from job management. High-level management should take a thorough, informed, and unbiased look at what is happening on each job and then, after taking time to think about it, analyze it and make suggestions on how the operations can be improved. This step cannot be performed by the individual job managers themselves because they have the primary function of completing the project. It cannot be adequately performed by the ordinary accountant or auditor because he usually lacks the construction know-how. It cannot be done by the ordinary superintendent or engineer because he lacks knowledge of accounting, taxation, auditing, and financial management. Individuals who already have this combined knowledge are hard to find and, if available, carry a salary equal to their abilities. Since such salaries are overhead expense, they are often beyond the reach of the smaller companies.

Once an employee is prepared by training and experience to audit construction jobs, he still requires two important tools. First, he needs auditable records. In other words, he needs adequate documents and support for the transactions he is reviewing. A good example is cost-forecasting, which in smaller companies is a function too often performed on a "seat of the pants" basis. Another example is the analysis of transactions recorded in the general journal. Complete explanations are often difficult in the customary two-column journal, particularly if the entry is based on a large worksheet analysis by engineering or operations people. Cross references to the worksheet should be adequate as long as the sheet is attached to the journal voucher or is otherwise filed where it can be found. However, if the worksheet is lost or misplaced, it is next to impossible to reconstruct the basis for the accounting entry. For this reason the more progressive construction firms are standardizing their cost-forecasting worksheets and other analysis work performed by engineering and operating people, properly filing them and using journal vouchers which can readily be checked back to the engineering and operating workpapers as needed.

Second, the auditor needs an audit program or checklist to remind him of the things that he needs to accomplish during his audit. In other words, the audit program consists of an inventory of key steps to be performed. More sophisticated internal audit groups in the larger companies will tailor their audit program for each audit. However, they will have a skeletal audit program (Schedule 11-1) which is fleshed out after they have determined the overall objectives of the audit and have obtained an adequate understanding of the project and tasks to be performed. Other internal audit groups will have standard checklists designed to fit all situations within the company and which are completed through a series of questions and answers. One such checklist is shown in Schedule 11-2. Here, the emphasis is on accounting, but it need not be confined to that. Each company would of course prepare its own standardized checklist.

SCHEDULE 11-1 SYSTEMS APPROACH AUDIT PROGRAM FOR INTERNAL AUDIT

1. Determine audit objectives, which generally fall into the following three areas:
   a. Financial or compliance
   b. Operating procedures and efficiencies
   c. Program results
2. Obtain an overview of the project, department, or other unit to be audited.
   a. How does it fit into the overall company organization?
   b. What are the backgrounds and personalities of the key management?
   c. What are the objectives of the project, department, or other unit being audited?
   d. What are the contractual terms—financial and operational?
   e. What are the known problems?
   f. Where are the "soft spots" or what should we look out for?
3. Select the staff for the assignment. Match individual abilities of the members of the internal audit group with the audit objectives and information obtained in step 2.
4. By inquiry and reference to policy manuals and similar sources, obtain a detailed understanding of the systems and procedures in effect at or in the project, department, or other unit to be audited.
5. By use of flowcharts and written narratives, record the data obtained in step 4.
6. Select representative transactions for each function or system cycle recorded in step 5, and, by taking the transaction through the system, determine that the underlying facts are correctly understood.
7. Identify key financial or operating controls (depending on audit objectives). Financial controls are those that help ensure the safeguarding of company assets and the accurate accumulation of data for financial reporting. Operating controls are those that show whether company assets are efficiently used or employed and whether data are accurately gathered and reported for management decision making.
8. Test the operation of the key controls (identified in step 2). Where the volume of data is great, use statistical sampling of documents to test the control.
9. Prepare a report on any control weaknesses noted during previous steps. For a maximum impact with management, state the results in terms of dollars: that is, show what it might cost the company if a control is lacking or not operating.
10. Based on steps 1 through 9, determine extent of additional testing needed to validate the results of the examination and to achieve overall assignment objectives.
11. Prepare final report.
12. The internal audit department manager should review audit procedures and findings to ensure adequate documentation to support findings.

In contrast to the skeleton audit program shown in Schedule 11-1, Schedule 11-2 illustrates a program based on a comprehensive series of questions.

SCHEDULE 11-2 CHECKLIST FOR INTERNAL AUDITOR

Job Audit Report

Job:
Location:
Dates Visited:

	Yes	*No*	*Not applicable*	*See comment*
A. *Cash on hand*				
1. Did you count it?				
2. Did you reconcile it with the books?				
3. Did you consider all possible sources?				
4. Do you feel certain that all cash from miscellaneous sources is being recorded?				
5. What did you do to satisfy yourself that it is (or is not) being recorded?				
6. Do we cash personal checks from cash on hand?				
B. *Cash in bank*				
1. Did you reconcile all bank accounts?				
2. Did you trace out the reconciling items in last month's reconciliations?				
3. Are any items from last month still open?				
4. If so, have you satisfied yourself that they are proper?				
5. Did you test-check the signatures on the checks?				
6. In reconciling the payroll account, did you compare the signatures and endorsements on termination checks with those on the two preceding paychecks?				
7. Did you make a test-check of the endorsements on the checks?				
8. Any exceptions not satisfactorily explained?				
9. Did you make a test comparison of the checks against the supporting payrolls and vouchers? What steps were followed in the test?				
10. Did you compare the deposits as shown on the statements with the books, the detail of the duplicate deposit slips, and the underlying estimates, invoices, cash reports, and so on?				
11. Are receipts from extra work being properly accounted for?				
12. Did all duplicate deposit slips bear the bank's stamp and receipt?				
13. Did any duplicate deposit slips show signs of alteration?				

SCHEDULE 11-2 CHECKLIST FOR INTERNAL AUDITOR (*Cont.*)

	Yes	*No*	*Not applicable*	*See comment*
14. Was there any substantial lag between receipt and deposit of funds?				
15. Did any part of your examination give indication of signature irregularities, such as checks going through on only one signature?				
16. Did you account for all check numbers?				
17. Is the name of the payee visible to the signer when checks are signed?				
18. Did you notice any weakness in the control of cash that should be strengthened?				
C. *Purchasing and vouchering*				
1. Do you have a list of persons authorized to sign requisitions?				
2. Were any requisitions not signed by an authorized signer?				
3. Were requisitions coded?				
4. Did you test the coding for reasonableness?				
5. Were any purchase orders not covered by requisitions?				
6. If there were any, was the omission satisfactorily explained?				
7. Are purchase orders priced and coded?				
8. If not, was omission satisfactorily explained?				
9. Do you have a list of individuals authorized to sign purchase orders?				
10. Did you test-check the purchase orders for authorized signatures?				
11. Any exceptions?				
12. Are purchase orders issued at the time of purchase?				
13. If not, are confirming orders sent out before receipt of the invoice?				
14. Did you investigate the procedure for handling back orders?				
15. Is it adequate to prevent duplicate shipments?				
16. Are purchase orders checked, before being issued, by someone not under the purchasing agent's control?				
17. Is numerical control maintained on purchase orders?				
18. Are blanket purchase orders used?				
19. If so, are there proper safeguards to prevent misuse?				
20. Are receiving memoranda prepared at the time goods are received?				

SCHEDULE 11-2 CHECKLIST FOR INTERNAL AUDITOR (*Cont.*)

	Yes	*No*	*Not applicable*	*See comment*
21. Are they compared with the purchase orders by someone not under the control of the individuals controlling the warehouse or the purchasing?				
22. Are vendors' invoices received by the accounting department before anyone else?				
23. Are vendors' invoices compared with the purchase order and receiving memoranda?				
24. Did you test-check vouchers to see that the system is being followed?				
25. Did your test-check of vouchers include a test of pricing, extensions, and discounts claimed?				
26. Did you see evidence of adequate follow-up on vendor's invoices bearing a discount to assure the taking of all discounts?				
27. Is there any preaudit of vouchers and checks before checks are released?				
28. If so, is it effective?				
29. Does the project manager or superintendent see all vouchers?				
30. Does he actually review them or just sign them?				
31. Do you have a list of authorized check signers?				
32. Did you see any evidence of presigning of checks?				
33. Are documents supporting paid vouchers stamped to prevent reuse?				
D. *Receiving and warehousing*				
1. Is responsibility for receipt of materials and supplies fixed?				
2. Is an adequate system in effect to check incoming materials and supplies?				
3. Are adequate receiving memoranda prepared?				
4. Is there any procedure for checking on the person who receives materials and supplies?				
5. Is there any attempt to keep a perpetual warehouse inventory?				
6. If so, is the account kept in terms of units rather than money?				
7. If not, is there a satisfactory control on warehouse stocks by other means? (If other means are used, describe in comments.)				

SCHEDULE 11-2 CHECKLIST FOR INTERNAL AUDITOR (*Cont.*)

	Yes	*No*	*Not applicable*	*See comment*
8. If no warehouse account is maintained, is there an adequate system for controlling charges for purchased materials and supplies to work items?				
9. Is the system being followed? What did you do to check this point?				
10. On the basis of your review of the procedures, personnel, and records, do you believe that all incoming materials and supplies are being properly checked and accounted for?				
11. Do you believe that materials and supplies being taken from the warehouse are being accounted for correctly?				
E. Timekeeping and payroll				
1. Have you watched the operation of the system for reporting field time—brass, cards, foreman's report, or some combination of the three? Describe briefly the system in use and its operation.				
2. Did you check at least one payroll?				
3. Did you supervise the distribution of at least one week's paychecks?				
4. Were there any complaints from the workers on the accuracy of their checks for the last two pay periods?				
5. If so, did you secure satisfactory explanations of the claimed differences?				
6. Are the personnel records in satisfactory condition?				
7. Did you make test-checks on the hiring and termination procedures?				
8. If the job uses field timekeepers, did you check the terminations on their reports? If not, explain how attendance of employees is checked by the job.				
9. Is such a check effectively maintained?				
10. Did you check the rates of pay against the union agreement?				
11. If such a check is made, does it include variable wage rates?				
12. When a worker is transferred from one crew to another, does the second foreman turn in his time for the entire day?				
13. Is there any possibility that both foremen might report his time?				
14. Is there an adequate control on identification badges?				

SCHEDULE 11-2 CHECKLIST FOR INTERNAL AUDITOR (*Cont.*)

	Yes	*No*	*Not applicable*	*See comment*
15. When time comes in from the field, is it subjected to any check before being entered in the records?				
16. Would there be any way of detecting duplications or omissions?				
17. If the job is subject to the Davis-Bacon Act, have you checked the hourly rates with the specifications?				
18. Did you note any possible violations of the Copeland antikickback law?				
19. Did you check the specifications for requirements regarding overtime?				
20. If there are any, are they being observed?				
21. Did you trace any checks to the employee's earnings record?				
22. Did you secure satisfactory explanations for all split checks (when the employee was paid with two or more checks for the same period)?				
23. If facsimile signatures are used, is a satisfactory control kept on the machines and plates?				
24. Is there any postaudit on payrolls?				
25. Is it made by someone outside the payroll department?				
26. Does it include a check of all documents supporting earnings of all employees paid off during the week and test-checks of comparable documents of other employees?				
27. Have you looked into the procedure for handling garnishments and attachments?				
28. Have you any suggestions for strengthening internal control of payroll and timekeeping?				
F. Cost accounting				
1. Has the original estimate been revised to reflect changes in the manner of doing the job?				
2. Are labor costs reported daily?				
3. Are equipment costs reported daily?				
4. Did you look into the possibilities that costs or quantities of materials, or both, might exceed estimates?				
5. Is the cost of extra work segregated?				
6. Is extra work incurred only on written order signed by or for the client?				

SCHEDULE 11-2 CHECKLIST FOR INTERNAL AUDITOR (*Cont.*)

	Yes	*No*	*Not applicable*	*See comment*
7. If the job is using ready-mixed concrete, dry mix, or similar materials, are comparisons being made between quantities shown on vendors' billings and quantities actually used?				
8. Are labor, materials, and equipment charges coded by the engineering department and checked by the accounting department?				
9. If so, is the check effective?				
10. If any attempt were made to divert costs from one work item to another, would present procedures disclose and correct the practice?				
11. Have you test-checked the cost codings and traced them to the cost ledger?				
12. Is the cost ledger up to date?				
13. How often are cost ledger postings made?				
14. If unit costs as of last night were required on any particular work item, could they be provided sometime today?				
15. If not, how long would it take and why?				
16. Can purchase commitments be determined at the close of any particular day?				
17. Can the cost per hour to operate any particular piece of equipment or any particular type of equipment be determined quickly and accurately?				
18. If the job is a joint venture, did you check the rentals paid to each joint venturer against equipment time records and the rental schedule in the joint venture agreement?				
19. Can repair and maintenance costs on all equipment be determined quickly from the records?				
20. Can repair and maintenance costs be determined by individual units and by types of equipment?				
21. Did you check equipment time records against rental schedules?				
22. If so, was there any evidence of idle equipment that should be terminated?				
23. If there is a mess hall or a camp, did you check the costs per meal and the costs per worker-day?				
24. Did you check the computations of percentages of completion?				

SCHEDULE 11-2 CHECKLIST FOR INTERNAL AUDITOR (*Cont.*)

	Yes	*No*	*Not applicable*	*See comment*
25. If so, what method was used? Do you consider it sound in view of the nature of the job?				
26. Did you inspect the work visually and compare what you saw with the percentages of completion shown for the various work items?				
27. Did you encounter any confusion about which work item should be charged with certain costs?				
28. How were such items being handled?				
29. Would it be feasible to record worker-hours and equipment hours as well as dollar costs?				
30. If you were the project manager, would you feel safe in making decisions based on our cost figures? If not, why?				
31. If you were an estimator, would you feel safe in relying on our cost figures in bidding a comparable job? If not, why?				
32. Have you any suggestions for improving our job cost accounting?				
G. *General accounting*				
1. Did you make test-checks of the accuracy of coding vouchers?				
2. Did you trace the entries to the books of original entry?				
3. Did you test the footings of the books of original entry?				
4. Did you test-check postings and footings in the general ledger?				
5. Did you check the figures in the last financial statements with the general ledger?				
6. Did they agree?				
7. Did you check all general journal entries?				
8. If so, were they proper?				
9. Were they properly authorized?				
10. Were explanations complete and accurate ?				
11. Did you analyze all deferred income and deferred cost accounts?				
12. If so, were the deferred items properly supported?				
13. Does the job maintain an equipment ledger?				
14. Does the job maintain a subcontract ledger?				

SCHEDULE 11-2 CHECKLIST FOR INTERNAL AUDITOR (*Cont.*)

	Yes	*No*	*Not applicable*	*See comment*
15. Did you review subcontract retentions?				
16. Were any released in advance without consent of the subcontractor's bonding company?				
17. Are any being held that should be released?				
18. Did you reconcile all subsidiary ledgers with the related general ledger control accounts?				
19. If any inventory accounts are carried, did you see the actual inventory to see if it looked reasonable in relation to the book amount?				
20. Did you review receivables and payables for questionable items?				
21. Do the income accounts and the related receivables reflect all items of sundry income and extra work?				
22. Are you satisfied that the receivables are in order?				
23. What did you do to satisfy yourself on the point?				
24. Did you review the correspondence file for indications of unusual or questionable items?				
H. Office procedure				
1. Are files orderly and complete?				
2. Is filing kept up to date?				
3. Do employees keep desk files?				
4. Are papers put away and desks cleared at night?				
5. Are blank checks controlled?				
6. Is the office work being properly scheduled?				
7. Are insurance reports made promptly at the end of the month?				
8. Are financial statements prepared promptly?				
9. Is the work so scheduled as to allow ample time for the preparation of tax returns required to be filed and tax data required by the general office?				

Comments

Question

No.

Chapter 12

Automated Systems for Construction Contractors

Historically, the construction industry has lagged behind most others in automating its manual accounting, bidding, job cost, and other systems. The "mainframe" environment of the 1960s required capital investments that were prohibitive to all but the largest contractors. In the 1970s, a change began to occur. The trend toward smaller "minicomputers" and their reduced cost and increased capabilities expanded the market of potential users among contractors. Although costs were still excessive to most, the market was sufficient to attract a number of software vendors, and a variety of "canned" software applications were developed to meet the general system needs of construction companies. Customers had few choices, both in software and the types of equipment available to run the applications, however, and conversion to automated systems remained limited essentially to the larger and medium-sized companies.

The trend of the 1980s has brought a major change to the computer industry and to construction company accounting and operational systems and capabilities. Microcomputer technology and economy have created both a vast market of personal and small-business users and an intense competition among manufacturers. Customers are now offered a wide variety of equipment and configurations which, in many instances, exceeds the capabilities of expensive mainframe and minicomputer predecessors. Sophisticated capabilities which once would cost hundreds of thousands may now be acquired within a cost range of $10,000 to $50,000. More simple equipment and applications are available at a fraction of these costs.

The explosion of the microcomputer industry has brought with it a huge increase in software development firms, each seeking a profitable niche in the marketplace. Specialty applications now abound, and the construction industry is considered a significant and relatively untapped market. Today, there exists a virtual myriad of construction software applications and vendors, offering a wide choice of capabilities and compatibility with many brands of equipment and with the operating systems to run them.

Available applications range from general accounting packages to more sophisticated integrated systems, including digital takeoff and estimating, scheduling, and job cost accounting and reporting modules. Costs typically range from as little as $2,000 upward to more than $25,000.

Impact on the Construction Industry

Along with everyone else in the business world, the construction industry is now a part of an ever-increasing information age. Computer utilization and "computer literacy" are becoming commonplace now and will become more so in the future. Traditional concerns and considerations about converting to automated systems are no longer applicable to those businesses intending to grow and maintain a competitive edge. The question to the contractor is no longer whether to utilize computerized systems. Rather, it is when to develop the capability and how to structure the conversion in a manner which best meets current requirements and resources, which minimizes disruption, and which provides for orderly growth and development.

What Can a Computer Be Expected to Do?

A large number of construction companies, ranging in size from local firms doing less than $1 million per year to the huge multinational firms, use computerized records in some or all of the following applications:

1. Payroll, paychecks, labor cost distributions, payroll tax returns, insurance reports based on payrolls, fringe benefit reports to unions, certified payroll to owners, pension and profit sharing trust fund analyses, and any other payroll or labor cost-oriented reports
2. Accounts payable, including subcontractor accounts and check and voucher controls
3. Job costs, including weekly or monthly summaries by work item of labor, material and equipment costs, and, where needed, overhead allocations
4. Equipment costs, including analysis by hours of owned and rented equipment as well as repair costs by unit, equipment location, depreciation, rentals, and idle time

5. Accounts receivable, including aging, discount, status, billings, statements, and sales reports
6. General ledger listing all transactions by account number, trial balances, and yearly summary of income, expense, and retained earnings
7. Bidding, i.e., takeoff and detailed computations involved in bid preparation, materials lists, materials and supplies budgets, and labor computations
8. Inventories and small tools controls
9. Overhead distributions and allocations
10. Critical path scheduling

Overview of a Typical Contractor Information System

Characteristics of construction accounting systems may vary according to the requirements and resource capabilities of a particular company, its equipment configuration, and the applications the company has selected for its use. Generally, the basic applications package is similar to that illustrated in Figure 12-1. It consists of the following subsystems: general ledger and job cost; payroll; accounts payable–subcontract; accounts payable–material; equipment; and accounts receivable. The general ledger and job cost module(s) interacts with all other subsystems to form the basis for cost control. More sophisticated systems may also contain an estimating module and a scheduling module.

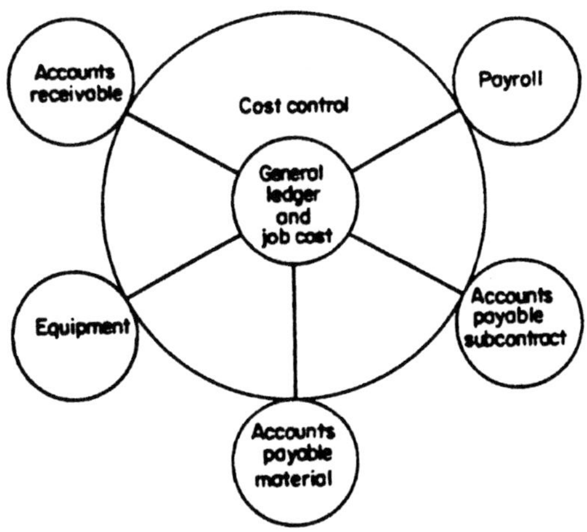

Figure 12-1 Typical Contractor Information System.

Usually the logic of the system and the output provided will be similar to that displayed by the flow diagram in Figure 12-2. Inputs are made using one or more direct entry terminals, either at a central headquarters office or, in some cases, from the field using remote job entry (RJE) terminals. The outputs, or reports, produced typically include the following:

- Trial balance and general ledger
- Accounts receivable and accounts receivable aging
- Accounts payable, accounts payable aging, and period cash requirements
- Payroll register, labor distribution report, payroll tax reports, union reports, and workers' compensation reports
- Customer, vendor, and subcontractor lists
- Job cost report, job cost summary (LEMS), and subcontractor status reports
- Monthly financial statements
- Jobs in progress

Reporting formats differ depending on the specific applications in use; however, the following descriptions illustrate typical characteristics.

General ledger and job cost subsidiary ledger. The general ledger, with a built-in job cost subsidiary, is the heart of the construction information system. While each of the system modules typically is designed to operate independently, job-related data generated by other subsystems is integrated into the files of the general ledger and job cost module. This integration of data allows all cost information to be recorded in one place, and it provides a link with all other systems while maintaining system independence and flexibility. A detail report can be printed at any time for each project showing current contract amount, detail job cost data, billings, and retention receivables.

ECKS & WYE COMPANY JOB COST SUBSIDIARY LEDGER RUN DATE 12/18/7-

CUSTOMER BOARD OF WATER COMMISSIONER'S OF THE CITY OF DETROIT CUST. NO. DET 21 JOB NO. 26075

DESCRIPTION	REF.	DATE	CODE	LABOR	MATERIAL	SUB-CONTR	CONTRACT	BILLING	RETENTION
INS. LABOR 09/7-	0911	09/30/7-		.00	9,638.27	.00	.00	.00	.00
A/PM 09/7-	0912	10/09/7-		.00	12,996.05	.00	.00	.00	.00
AP/S 09/7-	0913	09/30/7-		.00	.00	755,570.99	.00	.00	.00
BM-CM 09/7-	0915	09/30/7-		.00	56.50	.00	.00	.00	.00
ACCOUNTS RECEIVABLE	0916	09/30/7-		.00	.00	.00	.00	431,176.85-	43,117.69
AV 17 09/7-	0924	09/30/7-		.00	2,088.75	.00	.00	.00	.00
** TO-DATE BALANCES **				380,202.76	690,879.18	5,018,455.39	12,825,465.50	6,039,509.29-	603,950.93
PAYROLL 10/03/7-	0950	10/03/7-		6,434.89	.00	.00	.00	.00	.00
PAYROLL 10/10/7-	0951	10/10/7-		6,584.67	.00	.00	.00	.00	.00
PAYROLL 10/17/7-	0952	10/17/7-		6,809.07	.00	.00	.00	.00	.00

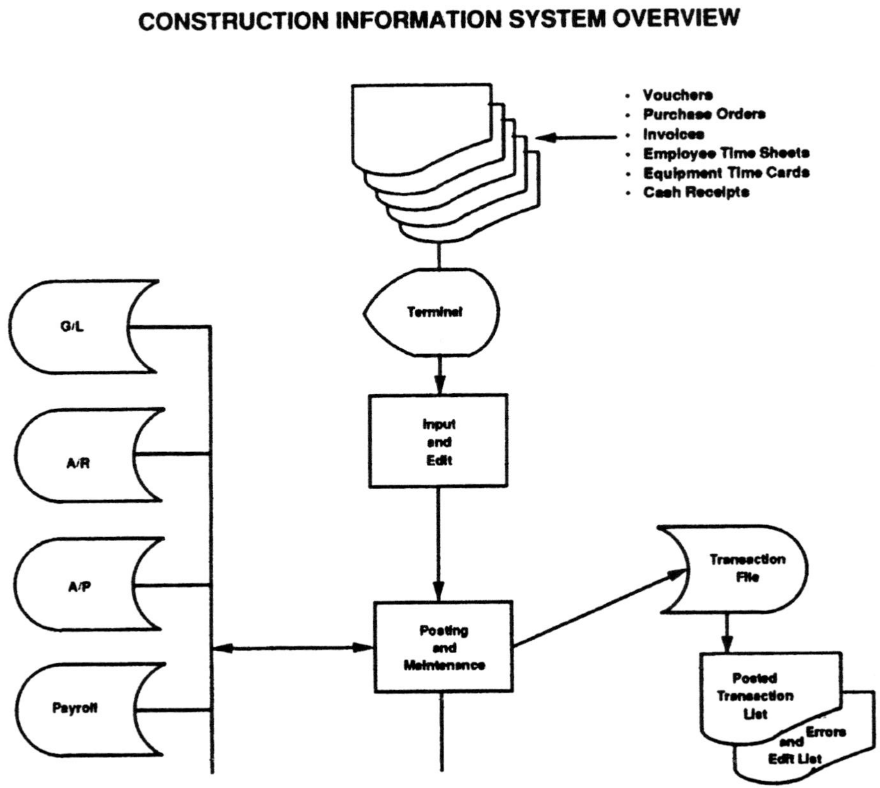

Figure 12-2 **Typical Flow Diagram.**

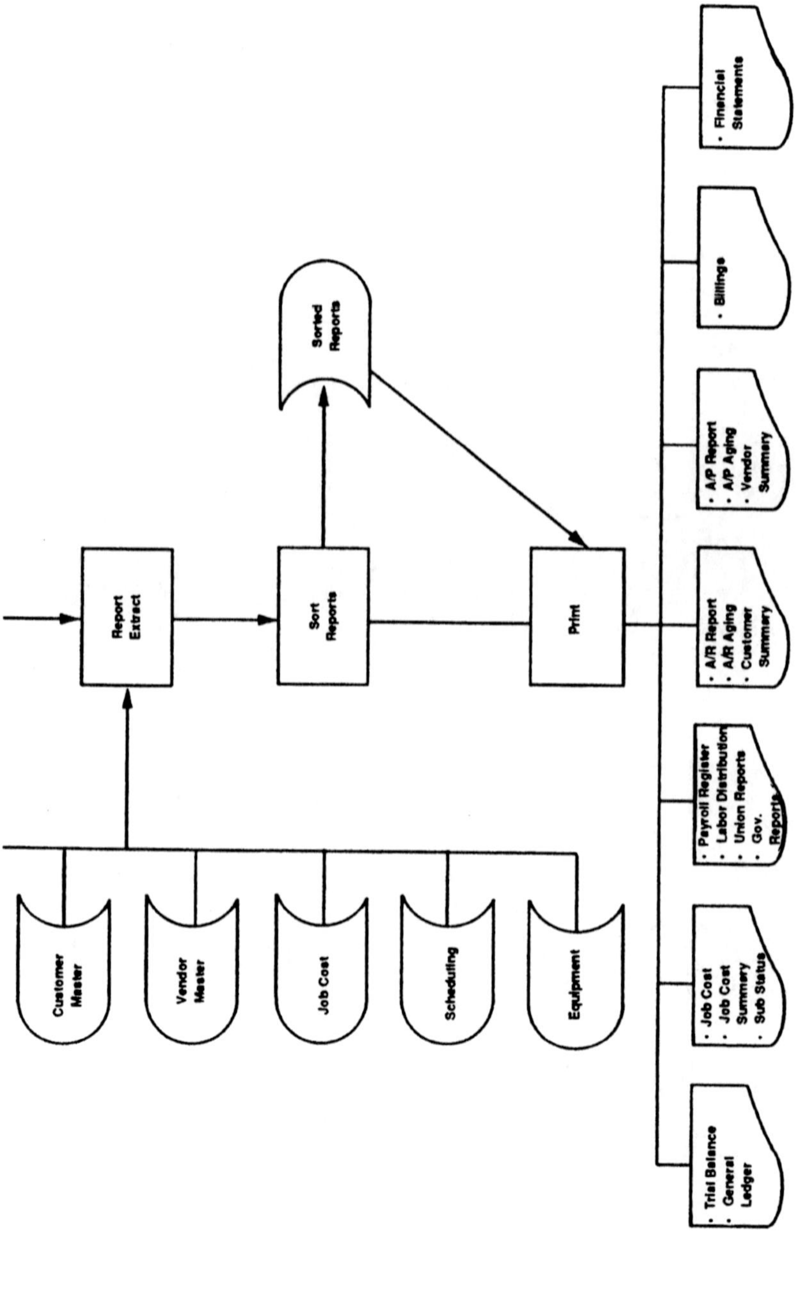

Figure 12-2 (*Cont.*).

Detail trial balance. Complete details of each general ledger account provide management with debit and credit activity, net, and net change.

DETAIL TRIAL BALANCE AT 11/30/7- PAGE 006

ACCT.NO.	ACCT. DESCRIPTION	DEBIT	CREDIT	NET	NET CHANGE
5000	SELLING, PROMOTION	3,371.03	414.00	2,957.03	475.40
5010	OFFICE PAYROLL	14,877.73	1,285.00	13,592.73	2,146.20
5020	MAINTENANCE OFFICE CARS	44.20	.00	44.20	
5050	OFFICE RENT EXPENSE	3,500.00	500.00	3,000.00	500.00
5060	OFFICE TELEPHONE EXPENSE	491.36	.00	491.36	104.48
5070	OFFICE SUPPLIES INCLUDING DIRECT	1,170.58	.00	1,170.58	29.34
5080	DEPRECIATION OFFICE EQUIPMENT	537.53	.00	537.53	97.81
5090	LEGAL EXPENSE	200.00	.00	200.00	200.00
5100	AUDIT EXPENSE	2,995.00	.00	2,995.00	
5140	ADVERTISING EXPENSE	2,818.60	.00	2,818.60	701.85
5160	SHOP AND OFFICE HEAT AND LIGHT	213.59	.00	213.59	100.46
5280	OFFICE SUPPLIES INCLUDING DIRECT	423.95	.00	170.35	184.56
5290	SUNDRY EXPENSE	187.35	.00	187.35	
5600	NOT LISTED IN CHART OF ACCOUNTS	.00	3,901.02	3,901.02-	
	BALANCE	49,689.14	13,728.84	35,960.30	7,027.58

Summary trial balance. This report summarizes the detail trial balance and indicates the status of general ledger accounts by major classification.

ECKS & WYE COMPANY

SUMMARY TRIAL BALANCE AT 01/31/7- PAGE 006

CHART OF ACCOUNTS	DEBIT	CREDIT	NET	NET CHANGE
** TOTAL AMOUNTS **	55,652,160.79	55,652,160.79	44,704,832.92-	1,575,087.88-
** ASSETS **	52,620,574.92	4,648,796.24		
** LIABILITIES **	2,639,851,83	50,887,907.50		
** TOTAL **	55,260,426.75	55,536,703.74	276,276.99-	
INCOME, JOB COSTS AND DEDUCTIONS	108,984.38	110,284.05		
** YARD EXPENSE	4,783.30	.00		
** GENERAL EXPENSE	277,966.36	5,173.00		
** TOTAL **	391,734.04	115,457.05	276,276.99	

NET CHANGE FIGURES FOR ENTRY NUMBERS 0029 TO 0034

Job Cost Progress Reports

Job cost report. This report is a monthly summary of job progress versus the budget, usually separated by home office costs, general conditions, direct labor, materials, equipment, and subcontractor costs. Each cost account (e.g., job area) represents a separate line item on the report, and a summary of project costs versus budget concludes the report. Figure 12-3 illustrates a typical format. Variations of this report may show planned hours versus actual hours.

LABOR		PROJECT COST REPORT		Period: June 30, 198x		
	EXPENDITURES			UNITS		
COST CODE	BUDGET	ACTUAL-TO-DATE	ECAC	ESTIMATE	ACTUAL-TO-DATE	PROJECTED
00-01000	100,000	75,000	100,000	1,000 HRS	750	1,000
00-01010	20,000	10,000	20,000	200	100	200
040-01000	300,000	200,000	400,000	150,000 SF	75,000	150,000
040-01010	150,000	75,000	150,000	50,000 SF	25,000	50,000
040-01050	400,000	50,000	400,000	100,000 SF	12,500	100,000
040-01100				150,000 SF	75,000	
040-0				37,500 SF	15,000	

Figure 12-3 **Typical Format for Project Cost Report.**

Percentage-of-completion statement. Management can evaluate the relationships between costs, billings, retentions, etc., because this report provides an overview of the entire job. If problems exist, detail entries are examined and appropriate action is initiated.

This report is also used where invoices (payment applications) set forth values which are progress billed on the basis of the percentage of completion for each work item.

PERCENTAGE OF COMPLETION STATEMENT [illegible]

[illegible] WASTE TREATMENT PLT. ORIG. COMPLETION DATE 01/31/7-

PROJECT MGR. D.STOUT ESTIMATOR JIM [illegible] SUPT. D.STOUT EST. COMPLETION DATE 01/31/7-

	LABOR	MATERIAL	SUB-CONTRACT	TOTAL
1. ESTIMATED COST	1,[illegible],534.00	1,[illegible],910.52	13,104,995.60	15,[illegible].12
2. VARIANCE TO-DATE	55,140.00-	32,132.19	5,031.56-	[illegible].17-
3. ADJUSTED EST.COST (1+2)	[illegible],394.00	1,633,042.71	13,159,162.34	15,770,590.95
4. CONTRACT AMOUNT				16,[illegible],013.52
	LAST MONTH	NET CHANGE		THIS MONTH
5. ADJ. O.H. - P. (4-3)	752,731.19	34,516.02-		[illegible],214.57
6. O.H. AT 2.5 PERCENT - ADJ.COST	394,102.05	562.92		394,464.97
		34,079.74-		

Labor Equipment Materials Subcontractor Summary (LEMS)

This report is a summary printout of costs incurred on each job, broken out by labor, equipment, and subcontractor costs. Usually, each job is printed as a separate line entry on the report.

Payroll

The profitability of a project depends on the effective control of labor costs. Controlling labor costs begins with the preparation of time sheets used for processing the payroll. Depending on the size of the project, time sheets are filled in daily or weekly by the superintendent, who enters a code indicating the type of work performed (wall forms, foundations, etc.) and the related hours. The sheets typically are forwarded to the home office to serve as source documents for data entry. The data are then processed by the computer to provide the following payroll records: payroll journal, employee pay statement, and employee check. Other reports are also provided, such as: labor distribution report, labor union reports, and government reports. The payroll data provide information needed for job cost reports.

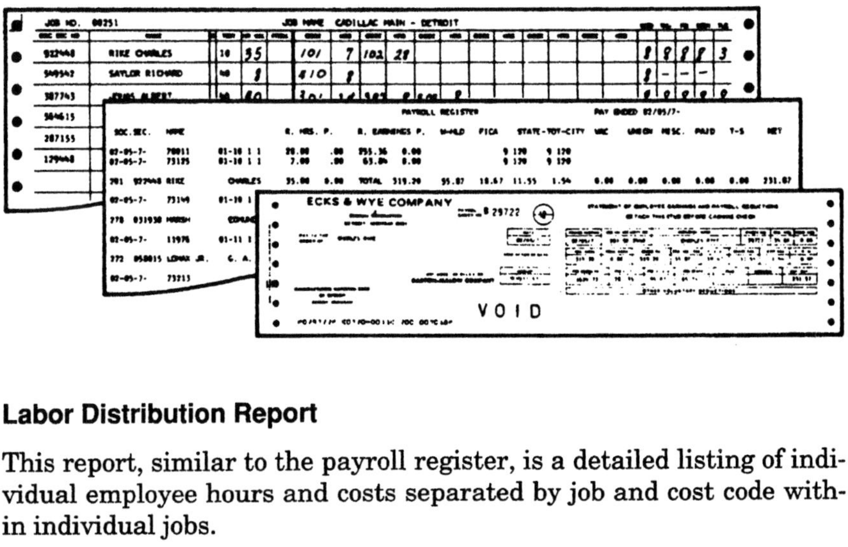

Labor Distribution Report

This report, similar to the payroll register, is a detailed listing of individual employee hours and costs separated by job and cost code within individual jobs.

Labor job charge report. This report shows the total weekly labor charges and the job to which they are charged.

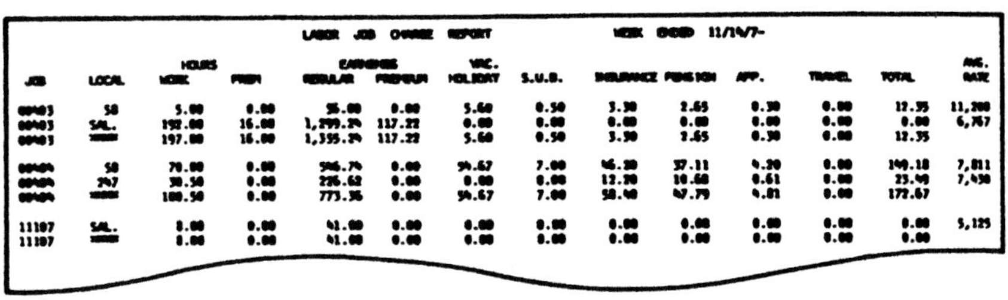

LABOR JOB CHARGE REPORT WEEK ENDED 11/14/7-

JOB	LOCAL	HOURS WORK	PREM	EARNINGS REGULAR	PREMIUM	VAC. HOLIDAY	S.U.B.	INSURANCE	PENSION	APP.	TRAVEL	TOTAL	AVG. RATE
00403	58	5.00	0.00	55.00	0.00	5.60	0.50	3.30	2.65	0.30	0.00	12.35	11,200
00403	SAL.	192.00	16.00	1,299.24	117.22	0.00	0.00	0.00	0.00	0.00	0.00	0.00	6,767
00403	[illegible]	197.00	16.00	1,355.24	117.22	5.60	0.50	3.30	2.65	0.30	0.00	12.35	
00404	58	70.00	0.00	546.74	0.00	54.67	7.00	46.20	37.11	4.20	0.00	149.18	7,811
00404	247	30.50	0.00	226.62	0.00	0.00	0.00	12.20	10.68	0.61	0.00	23.49	7,430
00404	[illegible]	100.50	0.00	773.36	0.00	54.67	7.00	58.40	47.79	4.81	0.00	172.67	
11107	SAL.	8.00	0.00	41.00	0.00	0.00	0.00	0.00	0.00	0.00	0.00	0.00	5,125
11107	[illegible]	8.00	0.00	41.00	0.00	0.00	0.00	0.00	0.00	0.00	0.00	0.00	

Union reporting. Certain information needed for union reporting is extracted from the payroll master file to produce the fringe benefit report.

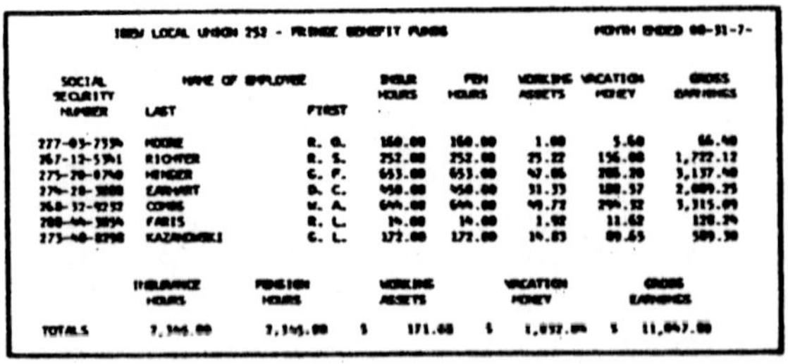

IBEW LOCAL UNION 252 - FRINGE BENEFIT FUNDS — MONTH ENDED 08-31-7-

SOCIAL SECURITY NUMBER	NAME OF EMPLOYEE LAST	FIRST	INSUR HOURS	PEN HOURS	WORKING ASSETS	VACATION MONEY	GROSS EARNINGS
277-03-7334	MOORE	R. G.	160.00	160.00	1.00	5.60	66.40
367-12-5341	RICHTER	R. S.	252.00	252.00	25.22	156.00	1,722.12
275-20-0740	HENGER	G. F.	653.00	653.00	47.06	285.20	3,137.40
374-20-3000	EARHART	D. C.	450.00	450.00	31.33	180.57	2,089.25
368-32-9232	COMBS	W. A.	644.00	644.00	49.72	294.32	3,315.09
280-44-3034	FARIS	R. L.	14.00	14.00	1.92	11.62	120.24
273-40-8290	KAZANOWSKI	G. L.	172.00	172.00	14.03	89.65	509.50

	INSURANCE HOURS	PENSION HOURS	WORKING ASSETS	VACATION MONEY	GROSS EARNINGS
TOTALS	2,345.00	2,345.00	$ 171.08	$ 1,032.04	$ 11,047.80

Certified payroll report. The system may also provide payroll reports required by government projects.

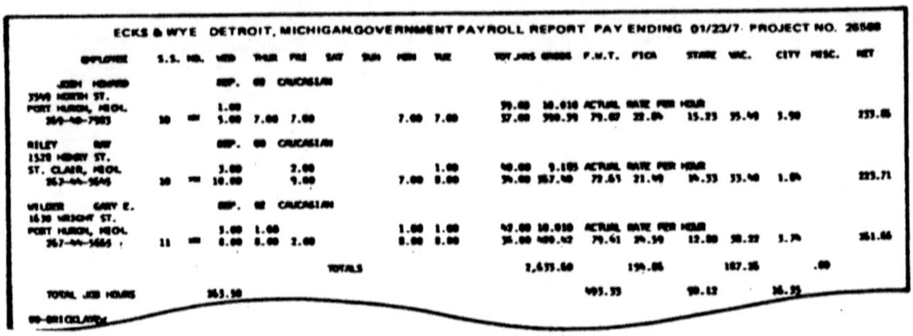

ECKS & WYE DETROIT, MICHIGAN GOVERNMENT PAYROLL REPORT PAY ENDING 01/23/7- PROJECT NO. 26500

EMPLOYEE	S.S. NO.		WED	THUR	FRI	SAT	SUN	MON	TUE	TOT.HRS	GROSS	F.W.T.	FICA	STATE	VAC.	CITY	MISC.	NET
JOHN HOWARD / 3540 NORTH ST. / PORT HURON, MICH. / 369-40-7303	10	—	EMP. 00 CAUCASIAN 1.00 5.00	7.00	7.00			7.00	7.00	39.00 37.00	10.010 ACTUAL RATE PER HOUR 390.39	79.07	22.04	15.25	35.40	3.90		233.05
RILEY ROY / 1520 HENRY ST. / ST. CLAIR, MICH. / 367-44-5645	10	—	EMP. 00 CAUCASIAN 3.00 10.00		2.00 9.00			7.00	1.00 8.00	40.00 34.00	9.105 ACTUAL RATE PER HOUR 357.40	72.63	21.40	14.33	33.40	1.04		223.71
WILDER GARY E. / 1630 WRIGHT ST. / PORT HURON, MICH. / 367-44-5665	11	—	EMP. 02 CAUCASIAN 3.00 8.00	1.00 8.00	2.00			1.00 8.00	1.00 8.00	42.00 35.00	10.010 ACTUAL RATE PER HOUR 400.42	79.41	24.50	12.00	30.22	3.74		251.66
TOTALS											2,633.60		194.06		107.26		.00	
TOTAL JOB HOURS			363.50									493.33		90.12		36.35		

00-BRICKLAYER

Employee distribution. This optional report, available in some systems, lists the number of employees in each ethnic group by employee classification.

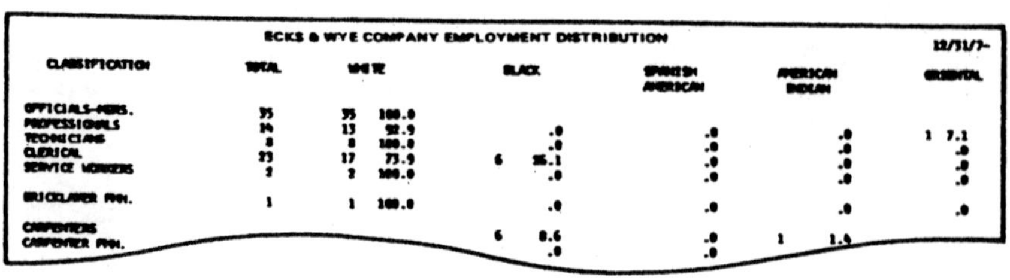

ECKS & WYE COMPANY EMPLOYMENT DISTRIBUTION — 12/31/7-

CLASSIFICATION	TOTAL	WHITE		BLACK		SPANISH AMERICAN	AMERICAN INDIAN		ORIENTAL	
OFFICIALS-MGRS.	35	35	100.0							
PROFESSIONALS	14	13	92.9		.0	.0		.0	1	7.1
TECHNICIANS	8	8	100.0		.0	.0		.0		.0
CLERICAL	23	17	73.9	6	26.1	.0		.0		.0
SERVICE WORKERS	2	2	100.0		.0	.0		.0		.0
BRICKLAYER FMN.	1	1	100.0		.0	.0		.0		.0
CARPENTERS				6	8.6	.0	1	1.4		
CARPENTER FMN.					.0	.0				

Accounts Payable Purchases and Warehouse Inventory

One of the most common applications of computers to construction accounting is a combined application to purchases, accounts payable, and material cost. Two further extensions of this application can include subcontractor payables and warehouse inventories.

In these applications it is common to see the voucher system used as the basis. Starting from there, it would be normal to see the computer produce, at a minimum, the following:

1. Accounts payable (and, if applicable, subcontractor) checks
2. Check register—monthly
3. Voucher register—monthly
4. Register of open accounts payable
5. Reciprocity report and other vendor-oriented reports

If the system is extended to include, again at a minimum, subcontractor payments and warehouse inventory, then these further statements may be expected:

1. A summary of subcontractor accounts, including:
   a. Total earned to date
   b. Amount retained
   c. Balance to be paid
   d. Amount paid to date
   e. Amount payable this month
2. A monthly summary, by inventory numbers, of warehouse quantities on hand

If the system is extended to include subcontractor payables, the subcontractor progress billings must be compared with the subcontract progress shown on the job progress reports. If warehouse inventories are to be tied in with computerized processing of payables, the warehouse is treated just as another job in coding vendor invoices, and warehouse requisitions are coded as if they were vendor invoices charged to the job.

The material cost should be posted to the job cost ledger at least weekly, and the posting should be by work items within the jobs. There should be a weekly statement of materials, supplies, and vendor services charged to jobs as a part of a general cost statement, which would also include the cost of labor, equipment, subcontracts, materials, supplies, and vendor services.

Sometimes the accounts payable application will be expanded not only to reflect vendor-purchased materials and supplies and subcontractor payables but also to include warehouse requisitions so that the job cost statements reflect this source of costs too.

One of the most important by-products of a computer system applied to accounts payable is the ease with which it can take up a lag that has always plagued engineers computing the "cost to complete" a job. The lag between purchasing (commitment) and receipt of the various items of construction cost has always been a factor in estimating the cost to complete. With the computer picking up the materials and supplies commitments along with other elements of cost, the lag factor can be minimized and the promptness and accuracy of the cost to complete are improved.

Typical reporting formats are illustrated below.

Accounts Payable—Subcontractors

Accounts payable subcontract journal. Billings, payments, and retentions for the current period are listed by project, by subcontractor.

PROC.NO.122-01 — ACCOUNTS PAYABLE SUB-ORDER JOURNAL — DATE 12/19/7- — PAGE NO 001

SUPP NO	NAME	INV.	ORDER	JOB	REFERENCE	POSTED	PAYABLE	PAID	RETENTION	ADD.RETENTION
ACM 04	ACME WIRE - IRON WORKS	37365	504516	93100		.00	30.75	.00	30.75-	.00
ALU 01	ALUMINUM - ARCHITECTURAL M	8690	505005	02600		2,605.00	2,605.34	.00	.00	.00
ALU 04	ALUM-A-WALL COMPANY	2154	506904	26075		9,334.00	8,400.60	.00	933.40	.00
AME 02	AMERICAN ROOFING COMPANY	5152	507479	26300		1,443.00	1,443.00	.00	.00	.00
AME 23	AMERICAN RAILROAD CONSTRUC	72601	507063	26524		7,600.00	6,840.00	.00	760.00	.00
AME 35	AMERICAN FURNITURE - FIXTU	26972	506750	26072		.00	107.95	.00	107.95-	.00
APC 01	APCON PORCELAIN, INCORPORA	26716	505539	26116		.00	856.00	.00	856.00-	.00
ARM 03	J.D.ARMSTRONG LANDSCAPING	77509	507931	77500		.00	620.00	.00	620.00-	.00

PROC.NO.122-01 — ACCOUNTS PAYABLE SUB-ORDER JOURNAL — DATE 12/19/7- — PAGE NO 007

SUPP NO	NAME	INV.	ORDER	JOB	REFERENCE	POSTED	PAYABLE	PAID	RETENTION	ADD.RETENTION
					TOTAL	1,048,783.37		.00		4,380.60-

Accounts payable balance report. This report provides a history record of accounts payable transactions for each subcontractor.

ACCOUNTS PAYABLE SUB-ORDER BALANCE REPORT — 11/30/7- — PAGE 010

JOB NO.	MEMO	INV.NO.	PROC. NO.	DATE	POSTED	PAYABLE	PAID	RETENTION	ADD-RETENTION	P	
77500		270750	092 - 01	09-30-7-	109.75	109.75	.00	.00	.00	D	D
77500	PAYMENT		-	10-11-7-	.00	109.75-	109.75	.00	.00	D	D
				TOTAL	109.75	.00	109.75	.00	.00		
AME 10	AMERICAN STEEL PRODUCTS CORPORAT			ORDER NO A-46787		ORDER AMT. $	.00				
25995		25995	032 - 00	03-21-7-	20,700.00	20,700.00	.00	.00	.00	D	
25995	PAYMENT				.00	20,700.00-	20,700.00	.00			
25995						19,214.00	.00				
				TOTAL					.00		
AME 21	AMERICAN BRIDGE DIVISION			ORDER NO S-07032		ORDER AMT. $ 636,000.00					
26525		438890	112 - 02	11-30-7-	127,200.00	114,480.00	.00	12,720.00	.00	D	
26525	PAYMENT		-	12-10-7-	.00	114,480.00-	114,480.00	.00	.00	D	
				TOTAL	127,200.00	.00	114,480.00	12,720.00	.00		

Subcontractor trial balance. Produced for each job, this report reflects the status of current accounts payable balances of all subcontractors.

SUB - CONTRACT TRIAL BALANCE FOR 26604 DATE RUN 11/30/7-

NO.	** SUB-CONTRACTOR		ORDER	POSTED	PAYABLE	PAID	RETENTION	ADD-RETENTION
BEA 03 CODE 0211	BEAVER LUMBER - WRECKING CO. TOTAL CONTR.	1,500.00	S-07167	0.00	0.00	0.00	0.00	0.00
REW 01 CODE 1500	W. J.				0.00			
	CONTR.	42,914.91	S-07168	28,314.			2,831.49	0.00
WAY 02 CODE 0360	WAYNE STEEL ERECTORS, INC. TOTAL CONTR.	52,885.30	S-07165	14,536.50	0.00	13,082.85	1,453.65	0.00
WES 02 CODE 0710	WESTERN WATERPROOFING TOTAL CONTR.	10,960.00	S-07163	500.00	0.00	450.00	50.00	0.00
** JOB					0.00			

Accounts payable subcontract distribution. This report details distribution of subcontract charges to each job.

ECKS & WYE COMPANY 10/31/7-

* DETAIL * A/P SUBCONTRACT DISTRIBUTION

JOB NO.	PROC. NO.		INV. NO.	POSTED
26524	10200	ACORN IRON WORKS, INC.	4672	25,672.00
26524	10200	AMERICAN RAILROAD CONSTRUCTION	72791	23,400.00
26524	10200	COLAGANTI FLOOR CO.	4132	1,616.00
26524	10200	GIZZI METAL ERECTORS CORP.	26624	20,214.00
26524	10201	INLAND RYERSON CONSTR. PROD.	926562	24,946.00
26524	10200	EAGLE PAVING COMPANY	36524	2,500.00
26524	10200	GORDON SEL-WAY, INC.	4636	24,890.00
26524	10200	WAYNE STEEL ERECTORS, INC.	3529	628.40
26524	10200	WESTWOOD CARPENTRY COMPANY	26624	2,400.00
		A/C 26524 TOTALS		126,268.40

Accounts Payable—Material

Accounts payable journal. An accounts payable journal is provided to serve as an audit trail of all invoices entering the system.

RUN NO 03/52
RUN NO 03-52

ECKS & WYE COMPANY

ACCOUNTS PAYABLE JOURNAL

DATE 03/25/7-
PAGE 1 TO TOTALS

SUPP. NO	SUPPLIER	INV. NO	DATE INV.	DATE DUE	INV. AMT.	PER	DISCOUNT AVAILABLE	*** DEBIT *** ACCT.	*** DEBIT *** AMOUNT	TAX CODE	*** CREDIT *** USE TAX	*** CREDIT *** ACCT. PAY
BET- 01	BETHLEHEM STEEL CO.	142473	03/03/7-	03/25/7	2,875.00		14.08	77200	2,875.00			2,875.00
BET- 01	BETHLEHEM STEEL CO.	145301	03/13/7-	03/25/7-	175.00		.88	77500	175.00			175.00
BET- 01	BETHLEHEM STEEL CO.	145302	03/13/7-	03/25/7-	54.00		.27	77500	54.00			54.00
BET- 01	BETHLEHEM STEEL CO.	145303	03/13/7-	03/25/7-	128.00		.64	77500	128.00			128.00
BET- 01	BETHLEHEM STEEL CO.	247002	03/03/7-	03/25/7-	695.00		3.41	77918	695.00			695.00
UNI- 06	UNITED PARCEL SERVICE	A	03/07/7-	03/25/7-	25.09			05270	25.09			25.09
VIC- 01	VICTORY RE-STEEL INC.	13153	03/11/7-	03/25/7-	236.00	.5	1.18	93300	236.00			236.00

DEBIT AMT 9,519.55
USE TAX

TOTAL INVOICE AMT. 9,519.55 - TOTAL DEBIT AMT. 9,519.55 - TOTAL CREDIT AMT. 9,519.55

Accounts payable material distribution. This report shows the distribution of material cost to the job by classification and vendor.

ACCOUNTS PAYABLE MATERIAL DISTRIBUTION 01/02/7-

VENDOR	JOB	CODE	NO.	AMT.
GORDON BARBER	26605	0107	12	22.50
RICHARD LAWITZKE	26605	0107	12	22.50
JOHN R. WIELAND	26605	0107	12	32.89
	MONTH SUBTOTAL			77.89
				77.89
		CODE		
CLOVERDALE RENTAL CO.	26605	0151	12	38.44
	MONTH SUBTOTAL			38.44
	CODE SUBTOTAL			38.44
SYMONS MANUFACTURING COMPANY	26605	0353	12	981.59
USM FASTENER COMPANY	26605	0353	12	169.80
	MONTH SUBTOTAL			1,151.39
	CODE SUBTOTAL			1,151.39
	JOB TOTAL			6,153.12

Voucher and check. The system provides for a check on all vouchers by entering them (with checks attached) in a voucher register. Then, when the checks are released, they would be entered in a cash disbursements journal.

MEMO	DATE	REF.	AMOUNT	DEDUCTIONS	NET AMOUNT
82600	11-03-7-	A	22.28		22.28
26054	10-02-7-	B	40.00		62.28
26054	10-02-7-	C	78.86		141.14

ECKS & WYE COMPANY
GENERAL CONTRACTORS
DETROIT, MICHIGAN

MANUFACTURERS NATIONAL BANK
OF DETROIT
DETROIT, MICHIGAN

CHECK NO. 55030 11/15/7-

PAY ONE HUNDRED FORTY ONE AND 14 CENTS AMT. $141.14

TO THE ORDER OF
DET. BD. OF WATER COMM.
735 RANDOLPH ST.
DETROIT, MICHIGAN 48226

ECKS & WYE COMPANY

NON-NEGOTIABLE

⑈055030⑈ ⑆0720–0033⑆ 200 009384⑈

Equipment

When equipment represents a major asset, the company should establish tight control over equipment inventory. Each time equipment is moved from yard to site or from one site to another, a transfer slip is prepared to record the quantity and type of equipment being moved. The transfer data are input to the computer system to generate a detailed equipment location report and cost distribution. Similar records are also prepared when equipment is rented to other contractors.

Equipment rental invoice. At the end of the month, equipment invoices are automatically prepared by the computer system.

ECKS & WYE CO. PERIOD ENDING 11/20/7-

JOB NO. 26605 EQUIPMENT RENTAL INVOICE INV. NO. E05096

WM. BEAUMONT HOSPITAL, ROYAL OAK, MICHIGAN

CLASS	DESCRIPTION	HOURS OR QUANTITY	DAYS	RATE	AMOUNT
1099					9.86
	CORNER	40	21		
	4FT OUTSID CORNER	6	19	.26	
1100	SYMONS 2FTX3FT PANEL	6	11	.62 /WK.	8.18
1103	SYMONS 18INX3FT PANEL	4	11	.59 /WK.	5.19
1125	SYMON SCAFFOLD BRACKET	37	21	.99 /MO.	56.63
1125	SYMON SCAFFOLD BRACKET	25	11	.37 /WK.	20.35
1126	SYMONS			4.25 /MO.	34.00
1127					40.56

Equipment cost report. This report shows current and inception to date income and cost for each piece of equipment. Costs are automatically calculated and distributed by each project.

ECKS & WYE COMPANY EQUIPMENT COST REPORT DATE RUN 10-30-7-

PURCHASE DATE		INCOME	TOTAL EXPENSE	OPERATOR LABOR	MAINTENANCE LABOR	MAINTENANCE PARTS	MAINTENANCE SUBLET	OTHER EXPENSE	DEPRECIATION EXPENSE
04-7-	OLIVER BULLDOZER CLASS NO. 0119								
	CUR	1,230.00	350.80	34.60	80.00	170.60	.00	.00	123.60
	YTD	4,920.00	1,326.40	176.00	460.00	464.00	100.00	.00	1,250.60
	ACQ	5,650.00	7,055.00	1,194.30	1,225.00	1,275.00	100.00	.00	3,260.70
03-7-	TROJAN ENDLOADER								
	CUR	4,650.00	2,616.50	56.20	140.00	600.00	510.00	400.00	870.30
	YTD	18,600.00	9,264.60	2,204.30	1,100.00	1,370.00	510.00	600.00	3,400.30
	ACQ	18,600.00	9,264.60	2,204.30	1,100.00	1,370.00	510.00	600.00	3,300.30
TOTAL	CUR	5,880.00	3,015.30	120.00	220.00	770.60	510.00	400.00	993.90
TOTAL	YTD	23,520.00	10,591.00	2,381.10	1,560.00	1,834.00	610.00	600.00	4,730.90
TOTAL	ACQ	24,250.00				2,645.00	610.00		

Equipment depreciation. Periodically, or as required, a depreciation report is automatically produced and updated to reflect the depreciation status of each piece of equipment.

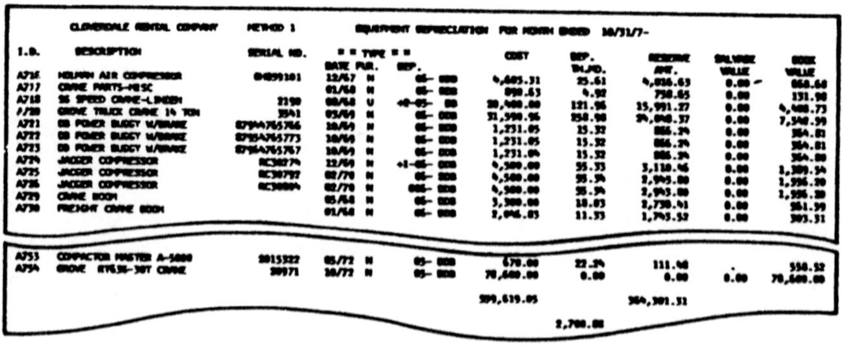

CLOVERDALE RENTAL COMPANY METHOD 1 EQUIPMENT DEPRECIATION FOR MONTH ENDED 10/31/7-

I.D.	DESCRIPTION	SERIAL NO.	DATE PUR.	TYPE	DEP.	COST	DEP. THIS MO.	RESERVE AMT.	SALVAGE VALUE	BOOK VALUE
A716	HOLMAN AIR COMPRESSOR	04059101	12/67	N	05- 000	4,605.31	25.61	4,036.63	0.00	568.68
A717	CRANE PARTS-MISC		01/68	N	05- 000	890.63	4.92	758.65	0.00	131.98
A718	55 SPEED CRANE-LINDEN	2290	08/68	U	+0-05- 00	20,400.00	121.96	15,991.27	0.00	4,408.73
A720	GROVE TRUCK CRANE 14 TON	3541	05/69	N	05- 000	31,590.96	258.98	24,040.37	0.00	7,540.59
A721	DD POWER BUGGY W/BRAKE	8799A765766	10/69	N	05- 000	1,231.05	15.32	866.24	0.00	364.81
A722	DD POWER BUGGY W/BRAKE	8795A765773	10/69	N	05- 000	1,231.05	15.32	866.24	0.00	364.81
A723	DD POWER BUGGY W/BRAKE	8796A765767	10/69	N	05- 000	1,231.04	15.32	866.24	0.00	364.80
A724	JAEGER COMPRESSOR	RC30274	12/69	N	+1-05- 000	4,500.00	55.33	3,110.46	0.00	1,389.54
A725	JAEGER COMPRESSOR	RC30792	02/70	N	05- 000	4,500.00	55.54	2,945.00	0.00	1,556.20
A726	JAEGER COMPRESSOR	RC30804	02/70	N	005- 000	4,500.00	55.54	2,943.00	0.00	1,556.20
A729	CRANE BOOM		05/68	N	05- 000	3,300.00	18.03	2,738.41	0.00	561.59
A730	FREIGHT CRANE BOOM		01/68	N	05- 000	2,046.83	11.33	1,743.52	0.00	303.31
A753	COMPACTOR MASTER A-5000	2015322	05/72	N	05- 000	670.00	22.24	111.48	.	558.52
A754	GROVE RT63S-30T CRANE	30971	10/72	N	05- 000	70,600.00	0.00	0.00	0.00	70,600.00
						599,639.05		364,301.31		
							2,700.08			

Accounts Receivable

The company's accounts receivable system is designed to record and maintain sales, cash receipts, and retentions. When these data are recorded, management reports can then be produced as an aid to improve collections and increase cash flow.

Accounts receivable sales journal. The sales journal reflects all charges to accounts receivable for the current month.

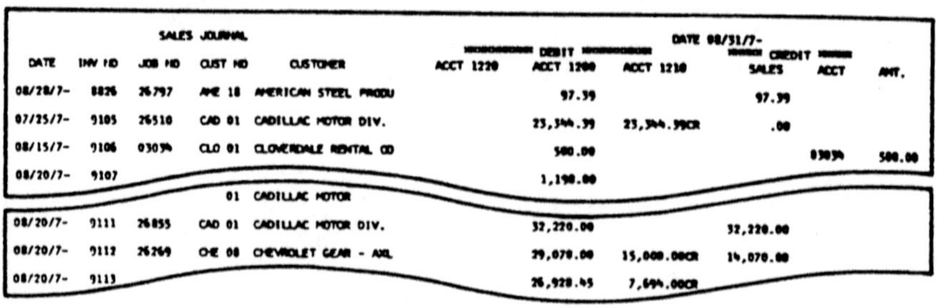

SALES JOURNAL DATE 08/31/7-

DATE	INV NO	JOB NO	CUST NO	CUSTOMER	DEBIT ACCT 1220	DEBIT ACCT 1200	DEBIT ACCT 1210	CREDIT SALES	CREDIT ACCT	CREDIT AMT.
08/28/7-	8826	26797	AME 18	AMERICAN STEEL PRODU		97.39		97.39		
07/25/7-	9105	26510	CAD 01	CADILLAC MOTOR DIV.		23,344.39	23,344.39CR	.00		
08/15/7-	9106	03034	CLO 01	CLOVERDALE RENTAL CO		500.00			03034	500.00
08/20/7-	9107					1,190.00				
			01	CADILLAC MOTOR						
08/20/7-	9111	26855	CAD 01	CADILLAC MOTOR DIV.		32,220.00		32,220.00		
08/20/7-	9112	26269	CHE 08	CHEVROLET GEAR - AXL		29,070.00	15,000.00CR	14,070.00		
08/20/7-	9113					26,920.45	7,694.00CR			

Balance report. This report provides a historical record of all transactions affecting each customer.

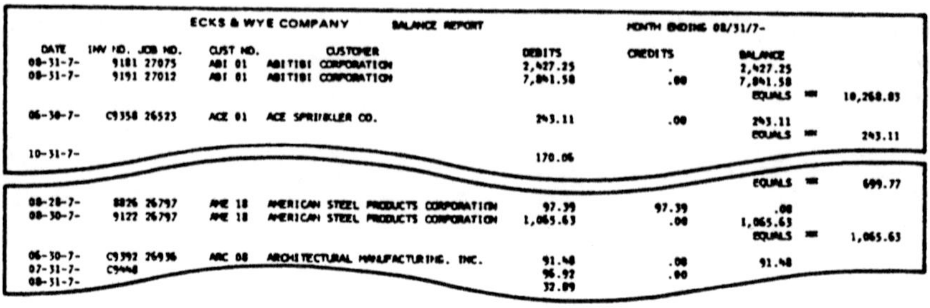

ECKS & WYE COMPANY BALANCE REPORT MONTH ENDING 08/31/7-

DATE	INV NO.	JOB NO.	CUST NO.	CUSTOMER	DEBITS	CREDITS	BALANCE		
08-31-7-	9181	27075	ABI 01	ABITIBI CORPORATION	2,427.25	.	2,427.25		
08-31-7-	9191	27012	ABI 01	ABITIBI CORPORATION	7,841.58	.00	7,841.58		
							EQUALS	==	10,268.83
06-30-7-	C9358	26523	ACE 01	ACE SPRINKLER CO.	243.11	.00	243.11		
							EQUALS	==	243.11
10-31-7-					170.06				
							EQUALS	==	699.77
08-28-7-	8826	26797	AME 18	AMERICAN STEEL PRODUCTS CORPORATION	97.39	97.39	.00		
08-30-7-	9122	26797	AME 18	AMERICAN STEEL PRODUCTS CORPORATION	1,065.63	.00	1,065.63		
							EQUALS	==	1,065.63
06-30-7-	C9392	26936	ARC 08	ARCHITECTURAL MANUFACTURING, INC.	91.48	.00	91.48		
07-31-7-	C9448				96.92	.00			
08-31-7-					32.09				

Accounts receivable trial balance. The trial balance provides aging of all customer accounts and serves as a tool for collection follow-up.

ECKS & WYE COMPANY ACCOUNTS RECEIVABLE TRIAL BALANCE MONTH ENDING 08/31/7-

CUST.NO.	CUSTOMER	PRIOR TO 07/7-	MONTH 07/7-	MONTH 08/7-	BALANCE AT 09/14/7-
ABI 01	ABITIBI CORPORATION			10,268.83	10,268.83
ACE 01	ACE SPRINKLER CO.	243.11	.00	.00	243.11
AFF 01	AFFILIATED INCINERATOR CORP.	170.06	.00	.00	170.06
AME 11	AMER. PRESTRESSED CONC., INC.	.00	116.44	583.33	699.77
AME 18	AMERICAN STEEL PRODUCTS CORPORATION	.00	.00	1,065.63	1,065.63
ARC 08	ARCHITECTURAL MANUFACTURING, INC.	91.48	96.92	32.89	221.29
AUS 01	AUSTINS PAINTERS	.00	185.53	.00	185.53

Statement. A statement of all open items is automatically prepared at the end of each month.

Monthly Financial Reports

Most systems provide for monthly balance sheet and income statement generation following the closing of the files. Income statements usually report results for the period and cumulative results to date.

When Is a Contractor Ready for Computerized Accounting?

As a general rule, the need for computerized information systems is a function of the volume of information being processed on a regular basis. When adequate manual systems are in place along with competent accounting personnel, yet a contractor finds that his present system still is not providing the information needed to support his business, or is providing it too late to be useful, automation may be a reasonable consideration.

The principal consideration in assessing whether or not to convert to automated systems should be the perceived benefits to the operation of the business in the near and longer term; e.g., better estimating capability, improved management reporting, and better cost controls. Expectations of savings in the accounting function may be realized; however, it is not unusual to find that information processing costs actually increase following conversion to automated systems. This is especially true during conversion periods when parallel systems are being maintained.

Another important consideration is how well prepared the management and staff are to implement and operate an automated system.

Are good manual systems currently in place? Do company personnel have an acceptable level of computer literacy as opposed to complete apprehension of computers? Are the existing working environment and controls sufficiently structured to support systematic data entry and controls? If these aspects of the system are deficient, a computerized information system, although possible, will be difficult to implement and the results may be extremely disappointing.

Selecting and Implementing Automated Information Systems

Selection and implementation of a good automated system should be a structured and well-thought-out process. The initial focus should be on identifying software which best meets the company's current and projected needs. Specific equipment decisions should follow identification of the desired software. Implementation should be under the supervision of a top-management executive who knows the company's problems thoroughly and understands the use of accounting information in solving them.

One of the better publications we have found to assist in this process is the recent book, *How to Find and Buy Good Software,* Curry & Bonner, Prentice-Hall. The authors describe an eight-step process consisting of the activities shown in Figure 12-4.

The purpose of the approach is to assure the following prior to any purchase of software or equipment:

- A thorough understanding of your own requirements and desired outputs
- A good familiarity with the software available
- A thorough demonstration of any systems under consideration
- A good understanding of the questions which should be asked
- A level of comfort in the automated systems environment

Key aspects of the approach are highlighted below. However, careful study of the Curry & Bonner text or similar publications is strongly recommended to anyone about to embark on the development of an automated accounting system.

Identify your needs. The objective in this step is to conduct a thorough review of your operations, to identify and prioritize the functions which may warrant automation, and to identify the specific types of products desired from the system. The analysis should result in a list of processes and outputs, including both "must" items and those "nice to have." The lists may be pared later in analyzing what is obtainable or practical given cost considerations.

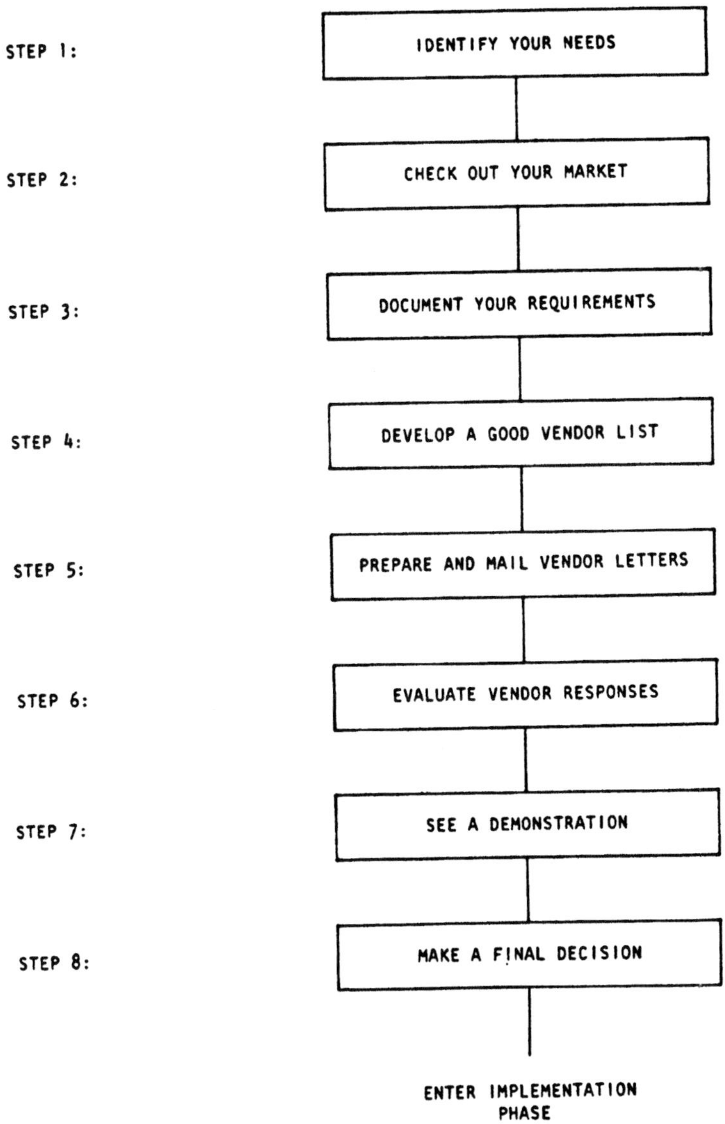

Figure 12-4 **The Eight-Step Process: An Overview.**

Key questions in this analysis are: what, when, where, how much, and how critical to the business?

Survey the marketplace. Having established a wish list, the next step is to begin identifying what software exists to meet your needs and what the related costs are. Initial sources of potential vendors might include:

- Construction Computer Applications Directory, Construction Industry Press
- Data Pro
- Local computer retail stores
- Other contractors and industry association representatives

Visits to several local retail stores will also assist you in developing a general familiarity with the computer environment, will offer an opportunity to view demonstrations, and will provide a general feeling for how well available systems meet your own specific needs.

Document your requirements. The next step is to outline what the system must be able to do in order to meet your requirements. Add additional remarks to each item outlined to make it clear to yourself and the vendors exactly what you want the system to do. A clear description here will pay dividends in subsequent steps.

Develop a vendor list and prepare and mail vendor letters. From initial investigation of the marketplace, a group of vendors potentially meeting your application requirements and budget objectives should be obtained. Vendor letters should then be prepared and mailed. The letters should include:

- A cover letter identifying the company and type of business, automated objectives and cost limits, and the time line established for selection
- The rating criteria to be used: e.g., meeting the requirements, system documentation, cost, conversion assistance, and customer support
- A description of system requirements
- A request for a proposal from the vendor

Evaluate the vendors' proposals. Once proposals are received, each should be evaluated against the established rating criteria and requirements documentation. Rating sheets should include notes regarding particular strong or weak areas and rating points assigned to each area. Individual points should then be summed and the proposals ranked in order of those best meeting the requirements.

In making the selections, a key consideration other than strict conformance to requirements should be the types and brands of equipment on which the applications can be executed. For example, if only one brand of equipment will support a particular package, rejection may be in order even though the system meets defined requirements.

Arrange for a full demonstration. After rating the proposals, two or three vendors should be selected for further consideration. Arrangements should be made for a full demonstration of the system operation, allowing *at least* an hour to test its operation and to resolve any questions remaining from the evaluation. Key testing should include running several transactions against new master files in each system component and testing automated interfaces between systems. This will provide a level of assurance that the system design works and will provide an understanding of the difficulty involved in its operation.

Select the system best for the company. By this point, the desirability of each system or of the components of individual systems should be reasonably apparent. However, some care will still be necessary in coordinating the choice around equipment, which now should take an important role in the selection decision. The system should be supported by dependable hardware that meets the company's specific needs and price range and is backed by reliable customer support.

Glossary of Computer Terminology

The following glossary, while far from complete, contains some of the more common terms which may be found in discussions regarding computers:

Accumulate To bring data together from separate sources.

Application Any task you tell the computer to perform.

Bit The term is short for "binary digit." It is the smallest unit of data a computer can process.

Byte Eight bits constitute 1 byte. Each byte makes up one of the 256 characters that most computers can produce. In other words, there are twenty-six lowercase letters, twenty-six uppercase letters, and numerous symbols: plus (+), minus (−), multiply (×), equals (=), asterisk (*), and so forth.

Chip The name used to designate an integrated circuit (or circuits) embedded in a silicon wafer.

Collate To arrange in proper order.

Compatibility Exists when input from one type of hardware can be used in another type.

Computer language Computers can only do what they are told to do, and the computer "language" consists of the words or symbols put into the machine for that purpose. There are four principal computer languages: BASIC, which stands for *b*eginner's *a*ll-purpose *s*ymbolic *i*nstruction *c*ode; COBOL, which stands for *co*mmon *b*usiness *o*riented *l*anguage; FORTRAN, which stands for *for*mula *tran*slation; and Pascal, named after the inventor.

CPU Stands for a microprocessor and a memory, or a *c*entral *p*rocessing *u*nit.

Daisy-wheel printer One type of "impact" printer which produces business-letter-quality type. Another type of impact printer is the "dot matrix" type. There are several varieties of nonimpact printers.

Database A particular file or set of files of information, such as a mailing list.

Database management system (DBMS) A program which will organize, query, search, and retrieve records from files.

Disk storage Data may be stored either on disks or on magnetic tape. There are two kinds of disks. The so-called "floppy disk" is a flexible plastic. The so-called "hard disk" is aluminum or stiff plastic covered with a magnetic oxide.

DOS An acronym for *d*isk *o*perating *s*ystem. It controls the operation of the computer. Other operating systems include MS, Unix, and Apple DOS.

Electronic spreadsheets An application which analyzes accounting data and classifies it on a spreadsheet.

Graphics An application which converts numbers and text into charts.

Hardware This is the machine itself as distinguished from the software, which is the programming.

Integration Takes place when the parts of a system are interdependent.

Interface A liaison point or person. The term is also used to mean a device that makes equipment from different manufacturers compatible.

Internal storage The capacity of the central processing unit to store instructions. It is to be distinguished from external storage on tapes or disks.

I/O An input-output device.

Modularity Exists when a system can be expanded or contracted by adding or subtracting modules.

Mainframe This is the largest, most powerful, and fastest type of computer.

Mass storage Either magnetic tapes or disks on which data are stored.

Microcomputer A small desktop machine normally used by only one person at a time.

Microprocessor A large integrated circuit which combines numerous chips in one package.

Minicomputer A computer which is large and more versatile than a microcomputer and which can be used by more than one person at a time. It is smaller and slower than a mainframe.

On line A term used to indicate a computer tied in with input equipment.

Peripherals I/O devices that get data into a processor and also see and record results.

Printout An actual printed record prepared by a computer.

RAM An acronym for *r*andom-*a*ccess *m*emory. It is the machine's internal memory and it can be altered at will so long as it is in the machine. It can be removed and stored. Through it, any bit can be extracted in the same time regardless of where it is stored physically.

ROM An acronym for *r*ead-*o*nly *m*emory. This is the permanent memory which is built into the machine when it is manufactured.

Software The programming used in any particular system.

Chapter

13

Tax Considerations in Construction Management

At the outset it should be understood that the purpose of this discussion is to call attention to the issues and not to take the place of an up-to-date tax service.

Taxation, at all levels, affects the construction industry in unusual and unexpected ways because in many respects the construction industry is unique. For example, in manufacturing the work flows through a factory and production tends to fall into a fixed routine. In construction, a new mobilization of work force, materials, equipment, and facilities must be carried out for each new job. New routines must be established quickly. They must operate at top speed and at top efficiency in a matter of days, and when the job is done they must be demobilized just as quickly.

Because of these peculiar features, the unit for accounting is the job, and a job may last only a few days or it may last several years. Also, most construction contractors do not have inventories as such. Instead, as pointed out elsewhere in this book, the materials put on a job and the work done are more in the nature of receivables secured by lien rights or public works bonds.

The result is a set of tax rules that are applicable primarily to construction work and only incidentally to other industries. As with all tax rules, those applicable to construction change regularly and sometimes rapidly. For that reason, this discussion is aimed more at general principles rather than specific applications and procedures.

One of the best known and least understood of the special tax rules for the construction industry deals with acceptable methods of accounting. For a good many years the Internal Revenue Service has been trying to find a way to require construction companies to use, for income tax purposes, the same method they use for financial accounting. So far the attempts have been only partially successful, as the objectives of tax accounting and financial accounting are different.

The accounting methods described in this chapter are strictly tax accounting methods. They are similar in many respects to financial accounting methods and sometimes go by the same names. Nevertheless, tax accounting is not required, either by law or common sense, to follow financial accounting and there is no good reason why it should.

Selecting Income Tax Accounting Policies

Generally the soundest tax policy is to defer income wherever possible consistent with a reasonable degree of uniformity in the flow of income. There are several reasons, the most obvious of which is the use of the money for the length of time payment is deferred. Another advantage which is somewhat less obvious is the possible loss of a deduction by having it assigned by an Internal Revenue Agent to a period already outlawed under the statute of limitations.

For smaller corporations it is worthwhile to plan, wherever possible, to keep the maximum amount of income within the lowest tax bracket available under whatever rate schedule that is in effect at any given time.

Effective for years after 1986, the reserve method of computing and deducting bad-debt losses may no longer be used by construction contractors.

Depreciation policies must be set with equipment replacement policies and the hazards of recapture of depreciation in mind. Where recapture is overlooked it can bring about some unpleasant surprises. Recapture problems usually arise in connection with the more or less standard depreciation methods and also under the accelerated cost recovery system enacted in 1981. However, in the construction industry there are some alternatives that need to be considered. For example, on a large joint venture that may run for several years there may be advantages to buying new equipment, charging it to the job, and selling it at the end of the job. For another example, trucks that might last several years on over-the-road hauls might be expendable equipment if used to haul gravel out of a deep pit or make their hauls over roads that would wear the truck out in a year or less.

Inventories, in the true sense of tangible personal property (and sometimes land) held for sale to customers in the ordinary course of business, are not common in the construction industry. However, inventories of tools and supplies exist and, in some instances at least, need to be taken into account. As a practical matter, inventories, even where technically required for tax purposes, are frequently ignored when their size, in relation to total costs, is not great.

General Income Tax Rules

Accounting procedures must produce results which, in the commissioner's opinion, fairly reflect income. In arriving at that opinion, consistency is a major consideration. Generally accepted accounting principles in the industry will normally be deemed to fairly reflect income. Where a taxpayer deviates from generally accepted accounting principles he must be prepared to show that such deviations do not produce distorted results. If the taxpayer's method does not fairly reflect income, the commissioner may adjust it to the point where it does.

Where a taxpayer has two or more types of operation he may use different accounting methods to account for them. For example, the same contractor may take large public works contracts as one operation; he may perform small paving and curb and gutter jobs as another; and take large cost-plus jobs as a third. He is entitled to separate accounting elections on each type of operation.

A taxpayer filing his first return is entitled to a complete new set of elections. Sometimes, where a taxpayer is committed to a very bad set of elections, he may find it worthwhile to reorganize his business so as to get new elections. If the taxpayer were merely to change his accounting methods, he must have permission from the commissioner to make such a change.

Usually the commissioner looks with disfavor on changes which reduce income in the year of the change or move income backward into outlawed periods. Difficulty may also be encountered where there is a deferral of income for long periods.

Accelerated Cost Recovery System (ACRS)

The ACRS survived the Tax Reform Act of 1986, as the modified accelerated cost recovery system. However, changes in rates and property classifications were so drastic that any contractor with substantial depreciation problems should make a detailed study of the 1986 tax act. Specific attention should be paid to the recapture provisions.

Summary of the Recent Tax Legislation

Recent tax legislation contains many provisions that have a major impact on the construction industry. The Tax Reform Act of 1986 (TRA 86) and the Reconciliation Act of 1987 will have an effect on all contractors in the country. Because of the rapid development of new rules in the area of accounting for purposes of taxation, this discussion will concentrate on general principles of taxation applicable to the construction industry.

The construction industry has a unique form of business from an accounting and taxation perspective in that the unit of measurement is the job. A job can last for several days or several years. Each job may have a different work force and require different materials and equipment. As a result, special tax rules aimed at acceptable methods of accounting have been developed particularly for the construction industry.

The accounting methods discussed in this section are tax accounting methods. They are similar to financial accounting methods, but they are not necessarily the same.

Methods of Accounting

Prior to the Tax Reform Act of 1986 (TRA 86), taxpayers generally could elect to use any method of accounting so long as the method clearly reflected income.

For activities other than the performance of long-term contracts, taxable income of construction contractors may still be computed as for any other business, using the cash receipts and disbursements method, the accrual method, or any other method that clearly reflects income. Limitations on the use of the cash method are discussed below in this section.

TRA 86 provides two methods of computing taxable income for long-term contracts, which are contracts started in one tax year and finished in a later tax year: (1) the percentage-of-completion method and (2) the percentage-of-completion–capitalized-cost method. These methods are required for contracts entered into after February 28, 1986. Prior rules permitted income from long-term contracts to be reported under the percentage-of-completion method, the completed-contract method, or any other method that clearly reflected income.

Cash method. Revenue is included in taxable income under the cash method of accounting in the year of either actual or constructive receipt. "Constructive receipt" means made available to the taxpayer so that he can or could have drawn on that income if he had chosen to do so. Income includes payments received for work done and advances for work to be done.

Expenses generally are deducted from taxable income in the year paid. The cost of assets with a useful life of more than one year, such as equipment or a shop building, must be depreciated in computing taxable income. Even the cost of items such as an insurance policy covering more than one year must be prorated over the years benefited.

The cash method is by far the simplest method of accounting to use. It does not require estimates and complex calculations, but it has the advantage that tax is payable for the year that the cash generating the income for tax purposes is received by the taxpayer.

While the cash method provides the opportunity for a certain amount of control over the amount of taxable income any given year, it is not widely used by major contractors. When financial statements are required for obtaining credit, bonding, or determining the financial condition of a business, other methods of accounting that better reflect the financial position of the company are generally employed.

TRA 86 generally prohibits the use of the cash method of accounting by general business corporations (referred to in tax parlance as C corporations) and partnerships having a C corporation as a partner. An exception is provided for entities having average annual gross receipts of $5 million or less for the three prior taxable years. In applying the gross receipts test, related entities are generally aggregated and intercompany receipts are disregarded.

Accrual method. The accrual method is not widely used in construction accounting, although some companies that say they are using the percentage-of-completion method are, in fact, using the accrual method. Under the accrual method, for tax purposes, income consists of all amounts for which the right to receive the income has become fixed. This is different than the financial accounting concept which would include in income all amounts earned. The right to receive the retained percentage has not been fixed for tax purposes until the job is completed and a period of time (usually equivalent to the lien period) has expired. The same is true for deductions, which should be reduced by the amount retained by the contractor out of billings from subcontractors.

If the retained percentage has historically been included in income when it is earned, the taxpayer has adopted a method of accounting which cannot be changed without the consent of the Commissioner of the Internal Revenue Service.

Commonly, the payment schedule is front-loaded where payments are high for work done early relative to the cost of doing the job. This resulted in an earlier recognition of income under the accrual method than under the "old law" percentage-of-completion method. Also, another disadvantage of the accrual method occurs if final billing is scheduled on completion of the work, because income is recognized before the expense of demobilization and moveout is deductible.

Long-term contracts. A long-term contract is defined as any contract for the manufacture, building, installation, or construction of property if such contract is entered into and not completed within the same taxable year. Under pre-1986 rules a taxpayer could use the cash or accrual method or, at the taxpayer's election, the percentage-of-completion method or the completed-contract method. The definition of a long-term contract was not changed by TRA 86, but only the percentage-of-completion method or the percentage-of-completion–capitalized-cost method may be used. Taxpayers currently using the percentage-of-completion method may not change methods without the consent of the Commissioner of the Internal Revenue Service.

The percentage-of-completion method. The percentage-of-completion method is the method of accounting most commonly used for long-term construction contracts and may provide the clearest reflection of income for financial purposes. For tax purposes, it is more than a variation of the accrual method. It is recognized as a separate and distinct method of accounting and generally must be applied consistently for all long-term contracts.

Under the percentage-of-completion method, revenue is recognized based on the completion of the work under the contract. For contracts entered into after February 28, 1986, the only method allowed for calculating the percentage of completion is based on costs incurred under the contract to total costs estimated to be incurred.

The percentage of completion is the ratio of total costs incurred through the end of the taxable year to the estimated total costs to be incurred under the contract. Prior law permitted any reasonable method of computing the percentage of completion that was consistently used. Thus, the physical completion method based on engineering estimates for determining the percentage of a contract completed during the year is no longer permitted.

The contract costs taken into account in determining the percentage-of-completion ratio include all costs required to be taken into account under the extended period long-term contract regulations of prior law. The percentage-of-completion calculation is done on a cumulative basis. The amount included in gross income for a particular year is the proportion of the expected contract price determined by the percentage-of-completion ratio based on costs incurred through the end of the taxable year, less the amount of the gross contract price included in income in previous taxable years. Costs incurred that are allocable to the contract are generally currently deductible. Taxpayers may elect to not recognize any income or deduct any costs under the percentage-of-completion method until the first year in which at least 10 percent of the total estimated costs under the contract have been incurred in that year. Once this election is made, it is binding for all future contracts entered into by the taxpayer.

Since revenue is recognized based on the estimated total contract costs, the percentage-of-completion method is inherently inaccurate. To compensate for the inaccuracy of the estimates, a new "look-back" rule requires the taxpayer to recompute the tax based on the actual total contract costs incurred and revenue received for each year under the contract. The recomputed tax is computed to the tax paid, and interest is paid to or by the taxpayer on the overpayment or underpayment. The interest rate is the federal rate used to calculate interest on overpayments of tax, compounded daily.

Completed-contract method. Under this method, income or loss from a contract is reported in the year the contract is finally completed and accepted. The advantage of this method is that it is not subject to estimates and generally provides an income deferral. However, the completed-contract method may not provide adequate financial information, so the books are often kept on the percentage-of-completion method.

The completed-contract method, while not subject to estimation, provides a potential controversy as to what a contract is and when it is complete for tax purposes. Generally, a single agreement for an integrated project will be treated as one contract for tax purposes. The taxpayer must follow the form of the transaction in determining what is a single contract. The Internal Revenue Service may sever or aggregate contracts, if necessary, to reflect income clearly.

Whether a single agreement is aggregated with other agreements or severed into several agreements depends on all the facts and circumstances surrounding the particular agreement. If the taxpayer's purpose in having several agreements covering several projects incorporated into a single contract is to delay the recognition of income, the IRS is likely to treat the agreement as more than one contract. If the taxpayer is trying to recognize a loss in the early stages of a project by having separate agreements, the IRS may determine that there is only one contract. Only the IRS may aggregate or separate agreements; contractors must account for them as drawn.

Time of completion of the contract is the year in which it is completed and accepted by the owner. However, a contractor may not deliberately postpone completion in order to defer income recognition into the next year. From a practical standpoint, the contractor generally does not want to delay completion of a project any longer than necessary.

In the case of a subcontractor, the time of completion where the subcontractor performs a portion of a long-term contract is upon acceptance by the general contractor, which may occur prior to completion of the entire contract.

The effect of a dispute related to the performance of a long-term contract is not necessarily to treat the contract as incomplete. If the amount in dispute is so material that the job could result in either a profit or a loss depending on the outcome, income or loss recognition is deferred. Otherwise, a profit is recognized, less the amount reasonably in dispute. The difference between the amount in dispute and the actual settlement is recognized in the year the dispute is resolved. Similarly, if there is a loss on the job, that loss which occurred before the amount in dispute occurred is recognized currently, and when the dispute is resolved, any additional loss is recognized.

Percentage-of-completion–capitalized-cost method. The second method of accounting for long-term contracts permitted under TRA 86 for contracts entered into before July 11, 1989, is the percentage-of-completion–capitalized-cost method. Under this method, a percentage of the revenues and expenditures is accounted for under the percentage-of-completion method, and the remaining portion is taken into account under the taxpayer's normal method of accounting. Upon completion of the contract, the look-back rule described above is applied to the portion computed under the percentage-of-completion method and the taxpayer will either pay or receive interest. For contracts entered into on or after February 28, 1986, but on or before October 13, 1987, 40 percent is recognized under the percentage-of-completion method. The remaining 60 percent of the items under the contract is accounted for under the taxpayer's normal method of accounting. For contracts entered into after October 13, 1987, but before June 21, 1988, the percentage recognized under the percentage-of-completion method has been raised to 70 percent, and 30 percent is accounted for under the normal method of accounting. For contracts entered into after June 20, 1988, but before July 11, 1989, the percentage recognized under the percentage-of-completion method had been raised to 90 percent. The completed-contract method is most commonly used to compute the portion not accounted for under the percentage-of-completion method. All contracts entered into after July 10, 1989, must be accounted for by using the percentage-of-completion method.

Exception for small construction contracts. The new tax rules for long-term contracts do not apply to certain small construction contracts. The contract must be for the construction or improvement of real property and must have an expected duration of two years or less at the commencement date of the contract. The taxpayer must also have average annual gross receipts for the three prior years not exceeding $10 million, including receipts of certain commonly controlled or related businesses. If the above requirements are met, the contract may be accounted for under the previously allowed methods, such as the cash, accrual, or completed-contract methods.

Overhead Expense Allocation

The Tax Reform Act of 1986 broadened the scope of costs required to be capitalized. Generally, the capitalization rules established by the regulation applicable to extended period long-term contracts as set out below *will be applied.* Exceptions for certain costs that may continue to be deducted include independent research and development costs, unsuccessful bid and proposal costs, and marketing expenses.

> Treas. Reg. 1.451-3(d)(6). Allocation of costs to extended period long-term contracts. Except as provided in paragraph (g) of this section, this paragraph (d)(6) applies to taxable years beginning after December 31, 1982. The following rules shall apply in determining what costs are properly allocable to an extended period long-term contract for tax purposes. These rules may also apply to certain extended period long-term contracts accounted for under a method of accounting that uses inventories (other than a long-term contract method). See paragraph (d)(6)(v) of this section.
>
> (i) Direct costs. Direct material costs and direct labor costs must be treated as costs properly allocable to an extended period long-term contract. "Direct material costs" include the costs of those materials which become an integral part of the subject matter of the extended period long-term contract and those materials which are consumed in the ordinary course of building, constructing, installing or manufacturing the subject matter of an extended period long-term contract. See § 1.471-3(b) for the elements of direct material costs. "Direct labor costs" include the cost of labor which can be identified or associated with a particular extended period long-term contract. The elements of direct labor costs include such items as basic compensation, overtime pay, vacation and holiday pay, sick leave pay (other than payments pursuant to a wage continuation plan under section 105(d) as it existed prior to its repeal in 1983), shift differential, payroll taxes and payments to a supplemental unemployment benefit plan paid or incurred on behalf of employees engaged in direct labor.
>
> (ii) Indirect costs allocated to extended period long-term contracts. The term "indirect costs" includes all costs other than direct material costs and direct labor costs. In determining what indirect costs are properly allocable to an extended period long-term contract, all such costs that directly benefit the performance of extended period long-term contracts, or are incurred by reason of the performance of

extended period long-term contracts, must be allocated to extended period long-term contract unless otherwise provided in paragraph (d)(6)(iii) of this section. Certain types of costs may directly benefit, or be incurred by reason of the performance of extended period long-term contracts of the taxpayer even though the same type of costs also benefits other activities of the taxpayer. Accordingly, such costs require a reasonable allocation between the portion of such costs that are attributable to extended period long-term contracts and the portion attributable to the other activities of the taxpayer. Indirect costs that must be allocated to extended period long-term contracts include:

(A) Repair expenses of equipment or facilities used in the performance of particular extended period long-term contracts,

(B) Maintenance of equipment or facilities used in the performance of particular extended period long-term contracts,

(C) Utilities, such as heat, light, and power, relating to equipment or facilities used in the performance of particular extended period long-term contracts,

(D) Rent of equipment or facilities used in the performance of particular extended period long-term contracts,

(E) Indirect labor and contract supervisory wages, including basic compensation, overtime pay, vacation and holiday pay, sick leave pay (other than payment pursuant to a wage continuation plan under section 105(d) as it existed prior to its repeal in 1983), shift differential, payroll taxes and contributions to a supplemental unemployment benefit plan incurred in the performance of particular extended period long-term contracts,

(F) Indirect materials and supplies used in the performance of particular extended period long-term contracts,

(G) Tools and equipment not capitalized used in the performance of particular extended period long-term contracts,

(H) Costs of quality control and inspection incurred in the performance of particular extended period long-term contracts,

(I) Taxes otherwise allowable as a deduction under section 164 (other than state and local, and foreign income taxes) to the extent such taxes are attributable to labor, materials, supplies, equipment or facilities used in the performance of particular extended period long-term contracts,

(J) Depreciation, amortization and cost recovery allowances on equipment and facilities (to the extent allowable as deductions under Chapter I of the Code) used in the performance of particular extended period long-term contracts,

(K) Depletion (whether or not in excess of cost) incurred in the performance of particular extended period long-term contracts,

(L) Administrative costs (whether or not performed on a job-site) directly attributable to the performance of particular extended period long-term contracts (but not including any cost of selling, or any return on capital),

(M) Direct and indirect costs incurred by any administrative, service, or support function or department to the extent such costs are allocable to particular extended period long-term contracts pursuant to paragraph (d)(9) of this section,

(N) Compensation paid to officers attributable to services performed on particular extended period long-term contracts (but not including any cost of selling),

(O) Costs of insurance incurred in the performance of particular extended period long-term contracts, such as insurance on machinery and equipment used in the construction of the subject matter of an extended period long-term contract,

(P) Contributions paid to or under a stock bonus, pension, profit-sharing or annuity plan or other plan deferring the receipt of compensation whether or not the plan qualifies under section 401(a) (except for amounts described in paragraph (d)(6)(iii)(I) of this section), and other employee benefit expenses paid or accrued on behalf of labor, to the extent such contributions or expenses are otherwise allowable as deductions under Chapter 1 of the Code. "Other employee benefit expenses" include (but are not limited to): workers' compensation; amounts deductible or for whose payment reduction in earnings of profits is allowed under section 404A and the regulations thereunder; payments pursuant to a wage continuation plan under section 105(d) as it existed prior to its repeal in 1983; amounts includible in the gross income of employees under a method or arrangement of employer contributions or compensation which has the effect of a stock bonus, pension, profit-sharing, or annuity plan, or other plan deferring the receipt of compensation or providing deferred benefits; premiums on life and health insurance; and miscellaneous benefits provided for employees such as safety, medical treatment, recreational and eating facilities, membership dues, etc.

(Q) Research and experimental expenses (described in section 174 and the regulations thereunder) directly attributable to particular extended period long-term contracts in existence at the time such expenses are incurred, or incurred under an agreement to perform research or experimentation,

(R) Rework labor, scrap and spoilage to the extent incurred in the performance of particular extended period long-term contracts, and

(S) Bidding expenses incurred in the solicitation of particular extended period long-term contracts ultimately awarded to the taxpayer. For purposes of this section, the term "bidding expenses" does not include any research and experimental expenses described in section 174 and the regulations thereunder. The taxpayer shall defer all bidding expenses paid or incurred in the solicitation of a particular extended period long-term contract until the contract is awarded. If the contract is awarded to the taxpayer, the bidding costs become part of the indirect costs assigned to the contract. If the contract is not awarded to the taxpayer, bidding costs become deductible in the taxable year the contract is awarded, or the taxable year the taxpayer is notified in writing that no contract will be awarded and that the contract (or similar or related contract) will not be re-bid, or in the taxable year that the taxpayer abandons its bid or proposal, whichever occurs first. Abandoning a bid does not include modifying, supplementing, or changing the original bid or proposal. If the taxpayer is awarded only part of the bid (for example, the taxpayer submitted one bid to build each of two different types of bridges and the taxpayer was awarded a contract to build only one of the two bridges), the taxpayer shall deduct the portion of the bidding expenses related to the portion of the bid not awarded to the taxpayer; in the case of a bid or proposal for a multi-unit contract, however, all the bidding expenses shall be allocated to a contract awarded to the taxpayer to produce any of such units (for example, where the taxpayer submitted one bid to produce three similar turbines and the taxpayer was awarded a contract to produce only two of the three turbines).

Although the above regulation refers to extended period long-term contracts, it is to be applied in capitalizing costs under the current rules for all long-term contracts except contracts which fall under the exception for small construction contracts. Under the current rules, there are no "extended period" long-term contracts.

The Tax Reform Act of 1986 made substantial changes in recovery periods and property classifications under the pre-1986 Accelerated Cost Recovery System (ACRS). The details of the Modified Accelerated Cost Recovery System (MACRS) are outside the scope of this discussion. However, construction companies with substantial depreciation problems should study the changes in relation to their individual problems.

Alternative Minimum Tax

The alternative minimum tax (AMT) was designed to ensure that no taxpayer with substantial economic income can avoid a significant tax liability through legitimate exclusions, deductions, and credits. All taxpayers, including corporations, are now subject to AMT. For partnerships and S corporations, the AMT is computed at the partner and shareholder levels, respectively. The tax is imposed at the rate of 20 percent of alternative minimum taxable income (AMTI) for corporations.

AMTI is computed for corporations by adding to regular taxable income adjustments, preferences, and an adjustment for current earnings and profits (ACE). Adjustments can be positive or negative and are items in the nature of timing differences under the alternative minimum taxable income system. Preferences are items where the amount of the income or deduction, rather than the timing, gives rise to the treatment as a preference. The ACE adjustment is equal to 75 percent of the difference between the corporation's ACE and AMTI computed without regard to ACE.

The details of ACE are beyond the scope of this discussion. However, one component is an entirely different computation of depreciation expense.

In computing AMTI, long-term contracts must be computed under the percentage-of-completion method for all contracts entered into after February 28, 1986. The adjustment may be positive or negative since it is in the nature of a timing difference.

A minimum tax credit is provided to offset the regular tax liability to the extent that the minimum tax was payable in a prior year due to deferral preferences because the intent of Congress was to accelerate the timing of the payment of tax. No credit is permitted if the items of tax preference were exclusion preferences which are included in AMTI in order to permanently tax the items.

Corporations are required to make estimated tax payments based on their tax liability including the AMT. In addition, all corporations including S corporations must pay the Superfund tax. The tax is 0.12 percent of modified AMTI in excess of $2 million. Modified AMTI is AMTI computed without regard to net operating losses or the deduction for the Superfund tax, which will be allowed for computing both regular taxable income and AMTI.

State and Local Income Taxes and Franchise Taxes

Most states in the United States impose some sort of income tax or franchise tax based on net income. The rates and rules for computing the tax vary widely and the tax can be material in many states. Opportunities for significant tax savings exist in the area of state taxation, but because the rates were relatively low in comparison to the federal rates in the past, state tax planning was a very low priority. For any state in which a company currently pays a material amount of tax or plans to expand such that income taxes may increase significantly, tax planning and a thorough review of recent tax returns is essential.

Franchise taxes are imposed on the privilege or license of a corporation to engage in corporate activity. The tax base varies by state. The tax may be imposed on capital stock, capital employed in the state, net worth, net taxable income earned within the state, or gross receipts.

Forty-three states and the District of Columbia impose a franchise tax measured by net income or a direct net income tax, or both. Nevada, South Dakota, Washington, and Wyoming currently do not impose such a tax. In Michigan, a single business tax is imposed on the privilege of doing business in the state that is based on federal taxable income. In New Hampshire and California, a tax is imposed on profits of business organizations allocated to the state.

Payroll Taxes

Here, again, is a type of taxation which needs to be investigated for each new location. There are two types of payroll taxes involved. First is the withholding of income tax. Virtually all employers are familiar with these taxes at the federal level. Most are familiar with them at the state level. However, these taxes have been showing up at the city and county levels where local income taxes are levied. Second is the social security and unemployment type of payroll tax. Any time a contractor gets a job or establishes a field office in a new location, he may become liable for this type of payroll tax.

Another matter which requires mention at this point is the 100 percent penalty levied on persons who have the power to pay over taxes withheld from employees but who fail or refuse to make such payments or prefer other creditors by paying them ahead of the Internal Revenue Service. This law is strictly enforced and can result in catastrophic liability.

Also important to construction contractors are the increasingly strict rules for distinguishing between independent contractors and employees. Essentially, any person who is treated as a subcontractor is not subject to withholding tax or employment taxes. When such a person is held to be an employee, the employer is liable for all payroll taxes not withheld and for tax on any payments on which income tax was owed but not paid by the recipient. The safest procedure is to use only licensed subcontractors and to use a standard form of subcontract in connection with the work to be subcontracted. Obviously, cash payments under the table are, per se, fraudulent and will normally be dealt with as such.

State and Local Sales and Use and Gross Receipts Taxes

A majority of the states and many local governments have retail sales and use taxes. They are levied at widely varying rates from 3 to 7 percent at the state level. A local tax may be imposed in addition to the state tax. Contractors usually must pay sales tax to their vendors on the materials, supplies, and equipment they use or consume. Gross income or gross receipts taxes are imposed on income derived from all sources unless they are imposed as occupation taxes from specific sources. When a contractor buys materials, supplies,or equipment from a source not subject to sales tax, he must usually pay a use tax at the same rate he would pay if he were subject to sales tax.

Sales taxes are generally intended to tax the privilege of making the sale and not the articles sold. However, if the contractor buys tangible personal property for resale he may avoid tax at the time of purchase by qualifying as a seller and giving a resale certificate, but when he does this he must be prepared to collect and pay the tax when the items bought under resale certificate are sold or used. An example would be a plumbing contractor who sells appliances from his showroom. He would buy the appliances for resale without sales tax and would collect and pay tax on the sales when they are sold. If he took one of the appliances for installation on a job, he would become liable for use tax.

When he sells equipment or trades it in, he should get either the tax or, if he sells or trades to an equipment dealer, a resale certificate. Equipment rentals are also subject to sales tax in certain circumstances.

A contractor needs to be aware of the attitude of the sales tax authorities on "fabrication labor." There are cases where tax on fabrication labor was levied on the construction of door and window frames on the job, on gravel produced and sold, on the product of a machine which mixes road oil and native material, and, of course, on ready-mixed concrete and asphalt road mix.

As with most other forms of state and local taxation, variations in laws and rulings make anything but a general statement impossible except in relation to a given taxing jurisdiction.

State and Local Property Taxes

Of major importance to contractors who own equipment and who have large quantities of materials, supplies, or inventory either in the warehouse or at the jobsite are ad valorem taxes on personal property. Usually the tax applies to the value of personal property which is present in a given jurisdiction on a given day. Often substantial savings may be made by timing the acceptance of delivery on such property. On equipment, assessors frequently ignore book value and substitute a "percent good" concept based on the continuing value of the equipment in the business.

Property taxes are sometimes imposed on intangible personal property. The capital stock tax and the capital employed tax are often considered types of property taxes.

Contractors' Licenses

Contractors' license laws can be either regulator or income producing, or both. In some states, such as California, local license laws cannot be regulatory in nature. However, unless they are given careful *and detailed attention,* they can prove very costly. Licensing considerations should be on every bidding checklist.

Summary

This discussion is only for the purpose of highlighting the existence and general application of the principal taxes affecting the construction industry. The tax laws are increasingly complex and constantly changing. The details of any specific tax by any specific governmental body must be left up to the contractor's tax advisers. The point here is to direct attention to the need to find out about the tax liabilities the contractor may face in this particular business.

Chapter

14

Special Problems for the Construction Accountant

Functions of Outside Bookkeepers and Accountants

All too often contractors, and for that matter management in general, fail to understand the definitions and differences between bookkeepers and accountants. In general terms, a bookkeeper is an individual with the ability to record business transactions in a formal set of books. Such a person may or may not fully understand the significance of the transactions, so the function is largely routine. A "full-charge" bookkeeper generally has the ability to recognize the original documents from which entries are made in the books of original entry (i.e., the various journals), can post the journals to the general ledger and various subsidiary ledgers and balance them, and can prepare trial balances of the various ledgers at the end of the month. An accountant, while having all the skills of a full-charge bookkeeper, will also have the ability to prepare financial statements and accounting reports from the general ledger and other books of account and prepare such other analyses as may be required for management decision making.

A "public accountant" is a professional accountant who performs accounting and auditing services for various clients. In all states, certified public accountants and sometimes noncertified public accountants are licensed by the state. A certified public accountant is an accountant who has met high educational requirements and has passed a rigorous examination. He is licensed by the state and holds a certificate as a certified public accountant. In addition to being an expert accountant, he will ordinarily be knowledgeable in tax matters, governmental regulations with respect to financial reporting, and if experienced in the construction industry will have the ability to assist in dealing with bonding companies and banks.

Many small contractors employ a public bookkeeper (who is sometimes an accountant) to "write up" their books each month, to prepare their current payroll tax and sales tax returns, and to prepare their federal and state income tax returns at the end of the year. For very small operations, this is usually a reasonably satisfactory arrangement. As a rule, the cost is quite reasonable. The old adage, however, that you "get what you pay for" holds true for bookkeeping and accounting services as well as for other things. A first-rate certified public accountant, knowledgeable in the construction industry, will usually provide the contractor with a return many times the fee for his services.

If the records or preparation of the income tax returns for the small contractor require the interpretation of a contract or other legal document, or if lien rights or the title to the property are involved, the contractor should trust neither his own knowledge nor that of his bookkeeper or accountant, but should consult his attorney. Usually it is "free" legal or accounting advice that is the most expensive, and this truism is particularly true when applied to tax planning.

If the bookkeeping is being performed by an outside bookkeeper or accountant in the manner described, the contractor should carefully review every statement and tax return, and should not accept it or act on it until he understands and agrees with the essential accuracy of every figure. Regardless of how competent or dependable the bookkeeper or accountant may be, it is the contractor who is financially responsible. Unless the figures make sense, there is a good chance that they are not correct.

Financial statement preparation

When the volume becomes large enough to justify hiring a full-time employee as a bookkeeper, the contractor usually finds that the bookkeeper he is able to hire does not have the knowledge to prepare financial statements and other financial analyses which he may need. Additionally, the bookkeeper may well need expert advice on the preparation of monthly and quarterly reports, the year-end closing of the books, and certainly on preparation of the income tax returns. At this stage, it is usually most economical and satisfactory to employ a thoroughly competent public accountant (preferably a certified public accountant experienced in construction accounting) to perform this service. At this stage, too, the public accountant can and probably should provide an important control for the contractor, in that he can review the accuracy of financial statements and tax returns prepared in the contractor's office. In selecting a public accountant, it is a good idea for the contractor to talk with his banker and his bonding company to seek their recommendations on which accounting firms or individuals have a sound reputation within the construction industry. Even with the additional control which the certified public accountant provides over accuracy of financial statements and tax returns, the contractor should release no statements or tax return until he is personally satisfied with the overall correctness of such statements and returns. Any important legal or tax decision should be checked with a specialist in the field.

As the contractor's business grows in volume, he usually finds it profitable to employ a full-time accountant to supervise his bookkeeper and who, in addition, is competent to prepare periodic financial statements and tax returns. This change in emphasis is accompanied by a shift in the type of service he requires from his public accountant. At this point, the need is even greater to make sure that he employs a public accountant who is thoroughly experienced in the accounting and tax problems of construction companies, in general, and if possible with the particular type of construction in which the contractor is engaged. When he gives an unqualified opinion on financial statements, it is the duty of the public accountant to satisfy himself as to the correctness of the financial statements and tax returns, and he will generally prepare such statements and returns over his own name and on his own letterhead. On the other hand, while he may not perform an audit in accordance with generally accepted auditing standards, he will do enough work to ensure that the statements are prepared in accordance with generally accepted accounting principles before allowing his name to be associated with such statements. When this is done, he will be required to put a letter with the financial statements saying that he has "prepared the accompanying financial statements from the contractor's accounting records and from information supplied by the contractor, without audit, and accordingly expresses no opinion on the fairness of the financial statements." When the auditor finds facts which are unfavorable, he is not permitted to ignore them, but must qualify his opinions accordingly. If the public accountant does his job well, it will be unnecessary to make demands on the contractor's time except to secure decisions on the accounting or tax treatment of important issues that require technical or construction knowledge.

It may be at this stage that the public accountant will carry out an audit in accordance with generally accepted auditing standards and render an audit report and an auditor's opinion on the financial statements. The audit, if required, will often be required by the contractor's bonding company and possibly also the contractor's banker. Nevertheless, the public accountant's principal function is to relieve the contractor of the detailed supervision of record keeping that is necessary under the system in which the outside accountant keeps all or part of the books. At this point too, the contractor begins to enjoy the benefits of an outside check on his own thinking, accounting methods, and record-keeping procedures, as well as the advantages of being able to supply his bank and bonding company with audited financial statements.

Employing a controller or office manager

At some point in the growth of the contractor's business, he will find it necessary to hire a fully trained and experienced accountant as a chief accountant, controller, or office manager. When this has been accomplished, the level of service required of the outside accountant rises to a peak and accordingly calls for engaging an accounting firm with the highest professional skills and qualifications in accounting and income tax knowledge, as well as a thorough understanding of all phases of construction accounting and financing.

At this level of service, the outside accountants, during the year, serve as consultants on technical tax and accounting problems as they arise. At the year's end, the audit is performed, and tax returns are either prepared by the outside accountants or prepared by the controller and reviewed by the outside accountants for technical correctness. When the annual audit is performed, the public accountants will also make a critical analysis of the overall system of internal controls and accounting procedures of the company, and will make suggestions to improve the level of internal control or efficiency of the contractor's operations. These suggestions are a by-product of the public accountant's annual audit. In choosing the outside public accountant, it is good practice for the contractor to find out if it is routine for the outside accountant to supply him at the end of the audit with a so-called "management letter" setting forth his observations during the course of the audit and any recommendations he may have based on those observations.

At this peak level of financial operations, it is not uncommon for the outside accountants to schedule periodic meetings with the company's bankers and surety bondsman to review the company's financial statements. If statements or reports are to be made to regulatory bodies such as the Securities and Exchange Commission, the outside certified public accountants customarily take care of them. Some states require prequalification statements before a contractor may bid on state work. These statements are usually prepared by the outside accountants at the time of their annual audit. Here again, it is well to engage outside accountants who are knowledgeable in such prequalification statements, because these statements can significantly affect the size and total volume of the jobs that the contractor is qualified to bid. A knowledgeable certified public accountant can usually make suggestions which may significantly increase the contractor's bidding capacity with the state. Most states using these prequalification forms have more or less arbitrary guidelines to determine bidding capacity. For example, it is not unusual to find that the rule of thumb is ten times working capital (current assets less current liabilities) or four times net worth.

Internal control and auditing

Systems of internal control and internal auditing are not well developed within the construction industry. This is primarily due to the fact that the overwhelming number of companies making up this industry are relatively small family-owned businesses. Also, not many public accountants have enough knowledge of the operational end of construction to evaluate some of the important relationships between job progress as it exists in the field and what the job records show. To be really effective the auditor must be able to evaluate the figures he sees in terms of what they disclose about the technical construction operations. Therefore, auditing on jobs and other internal cost centers is better done either by the employees of the company who are especially trained for the job, or employees of an accounting firm whose background includes substantial experience in construction job offices. A public accounting firm which can furnish such experienced construction auditors can make a very valuable contribution to the auditing of its construction company clients and their financial affairs in general.

There are functions, however, in the field of internal auditing that any competent public accountant can and (given the authority) should do. First, he can make sure that he knows how the data are compiled for each record he examines. With that knowledge, he can evaluate critically the chances of fraud or of error in the accumulation of data escaping discovery. Any weakness in the accounting system he discovers can then be discussed with management to determine whether the risk created by the weakness is worth the cost of installing a control to eliminate the weakness.

For example, if each foreman fills out and turns in the time card for the men on his crew, the opportunity for fraud exists through collusion between the foreman and the workers. The best way to detect or prevent such fraud would be through periodic spot checks and head counts by a field timekeeper who accounts physically for the presence of all employees on the payroll and the fact that they are working. Whether the risk justifies the cost in a given situation is a matter for management to decide but the auditor should call it to their attention.

Secondly, the auditor can provide technical supervision over the accounting, and carry out regular tests to see whether established procedures are being followed. For example, if the procedures call for a field timekeeper, as described above, failure to follow the procedure should be discovered and reported by the public accountant. One of the generally accepted auditing standards requires that an adequate review be made of internal control. However, the controls contemplated by this review are controls which help to preclude defalcations and also provide for accurate gathering of data for financial statement purposes. While the scope of the outside auditors' assignment should be such that they can find weaknesses in internal controls, it should also include, as a by-product, reviews of operational controls (e.g., those controls which ensure efficient use of assets and provide for efficient operations).

For example, if the client-contractor's business is such that the system for timekeeping and payroll could be improved so as to do away with the timekeepers and still retain the same degree of control, then it is up to the outside auditors to call the possible improvement to the attention of management and if it can be made without disturbing other phases of the operation, to supervise putting the improvement into effect.

In summary, it is best right from the start for the contractor to become associated with a first-rate, construction-oriented accounting firm. He also needs a good, construction-oriented lawyer. Between the two of them, they can keep the contractor out of expensive accounting, tax, and legal trouble and let the contractor and his staff get on with running the construction business.

Management Considerations in Selecting an Accounting Method

In Chapter 1 the various accounting methods available to a construction company were compared technically. Included in the comparison were the cash method, the accrual method, and variations of the accrual method known as the completed-contract method and the percentage-of-completion method. In the following subsections the emphasis is not upon the methods themselves but on their application to a number of widely varied situations.

The Tax Reform Act of 1986 made major changes in the accounting which construction companies may use for federal income tax purposes. Sometimes the application of these changes can be incorporated into a construction contractor's general fiscal accounting without harm to the usefulness of the financial statements for management purposes. Sometimes they cannot and it will take the issuance of final regulations and probably several years to form valid judgments regarding their effect.

For example, many small contractors have incorporated their companies to obtain the protection of corporate operation. Of these companies, many use either cash basis accounting or a hybrid method combining cash and accrual methods. The new rule that ordinary business corporations (the so-called "C" corporations) cannot use either cash or hybrid methods will inevitably affect the usefulness of the financial statements to management.

Another example is an apparent attempt to make the use of the completed-contract method, by large contractors, so burdensome as to be impractical for federal income tax purposes. This result is obtained by, in effect, redefining the term "long-term contract" to mean one which (1) is not "expected" to be performed within two years and (2) is performed by a taxpayer whose average gross revenue over the three taxable years immediately preceding the year in question does not exceed $10 million annually. The "average gross revenue" includes that of all companies under common control. The language of the 1986 amendments would appear to include businesses not engaged in the construction business and also nonsponsored joint ventures. Obviously such terms as "common control" and "gross revenue" for the purposes of the Internal Revenue Code are still to be defined by regulations. The same applies to such rules as attribution.

At this writing it appears likely that most small and medium-sized contractors, and possibly some of the larger firms, will continue their present accounting methods and, for federal income tax purposes, make any required year-end adjustments. The comments in this section are based on this assumption except in those cases where the context clearly indicates otherwise.

Many large jobs widely scattered. If there are numerous large jobs spread over a wide area, diversification should cause individual differences to average out, so that the aggregate result should be reasonably accurate regardless of the method of accounting chosen. Individual jobs would normally report enough operating figures to permit adequate jobsite management. Assuming a corporate form of organization, and in the absence of an excess profits tax, the time when profits are realized would seem to make little difference. On that basis the completed-contract method of determining income would appear to be logical, and certainly the most conservative.

On the other hand, operations of this size and extent may involve financing problems, bonding problems, and stockholder problems.

These problems are often simpler when the percentage-of-completion method keeps the flow of book profits somewhat more regular than would normally be possible under the completed-contract method. The AICPA, in its SOP 81-1, generally favors the percentage-of-completion method. In the present state of the art, large companies with widely scattered operations tend to use the percentage-of-completion method to arrive at profits for financial accounting and the completed-contract method for income tax purposes.

Few large jobs in restricted area. The fewer jobs a company has, the greater the variation in income between years that can be occasioned by use of the completed-contract method of determining income. Also, the smaller the territory in which the company operates, the greater will be the variations caused by purely local conditions. If the company is incorporated so that it need not be concerned with high individual surtax rates, and if it is well enough financed so that its borrowing and bonding capacity is above suspicion, the completed-contract method of determining income ensures against spending profits that may fade away before the end of the job.

However, if bonding capacity, credit, and stockholder considerations call for a more or less uniform flow of income on the operating statement, then the percentage-of-completion method is indicated. From a management point of view, the principal hazard inherent in reflecting profits that are still tied up in retained percentages and other job assets lies in the tendency on the part of many contractors to think of job profits as available cash.

They will therefore spend money and make commitments on the basis of estimated profits which, even if they are ultimately realized, have not yet been realized in cash and, in fact, may never be.

If joint ventures are sponsored (and hence operated) by others, there is not much control over the time when their profits will appear on the financial statements. Therefore, a somewhat greater degree of control over the timing of profits from the company's own jobs is required. If the percentage-of-completion method is used, a normal margin for variations in estimating can often provide the necessary control.

One or two large jobs. Companies having one or two large jobs and one or two joint ventures are less likely to be incorporated than the larger companies, and hence the pressure of high individual surtax rates is very strong. It is rare to find a construction company organized as a partnership when the facts favor reporting as a corporation for tax purposes. Therefore, the use of the percentage-of-completion method becomes a virtual economic necessity.

However, the completed-contract method may be practical for income tax purposes if the contractor is able to unbalance his payment schedules so as to free enough cash to pay the taxes in the year they become a liability, and if the contractor's flow of joint venture income is steady enough to smooth out the income curve.

Many large and small jobs widely scattered. A large number of jobs over a wide area, regardless of the size of the individual jobs and regardless of the accounting methods used, will tend to have an equalizing effect on income. Strict interpretation of the income tax laws would seem to require that, at least for jobs running less than one year, either the cash method or the straight accrual method be used. For obvious reasons, the cash method provides little in the way of information for management, although if the jobs are small enough there may be little enough variation for this method to produce a reasonably accurate overall result.

On jobs that run over one year the contractor is entitled to use one of the long-term contract methods, but sometimes it is better to sacrifice the advantage of the long-term methods for the greater economy of a single consistent method. The straight accrual method has the disadvantage of treating as income the amounts currently billable without reference to job progress or earnings. This method becomes increasingly applicable as the jobs become smaller and shorter. If the jobs are small enough to dispense with progress billings and substitute a single payment on completion of the job, the straight accrual method approximates the completed-contract method. In fact, as a practical matter, many small contractors close all jobs on completion regardless of whether there are any progress billings.

Small, short jobs. Companies having small, short jobs almost always collect payment on completion, and even if there are jobs with progress billings, the completed-contract method is most practical even though technically not permissible for tax purposes. Of course, the straight accrual method is preferable when applicable.

Speculative builders. Many small contractors build houses or stores on speculation. The speculative structure is usually owned by the contractor and assumes more of the character of an inventory item than the usual construction job. Thus, the straight accrual method becomes applicable.

Builders of tract housing often receive their costs through a disbursing agent or "joint control." Then an order on the joint control is the equivalent of a check so far as the contractor's accounting is concerned. However, the person to whom the order is issued must present it to the joint control to receive payment. An exception is usually made for payroll checks. The joint control check is issued to reimburse the builder's payroll bank account, and the builder's paychecks are issued to the employees.

Custom builders. It is usual for the custom builder of residences and small commercial buildings to receive progress payments. It is unusual to find a builder who has a job that lasts over a year and who is thus qualified (income tax-wise, at least) to use one of the long-term contract methods of reporting income. Technically, most builders in this group could qualify for the use of the cash method of accounting but, for the most part, they can achieve a result closely approaching the percentage-of-completion method by using straight accrual accounting.

Many builders in this class, however, prefer to take up income on the completed-contract method. When properly applied to work of this type, the completed-contract method actually reflects income more accurately than the other available methods, and is therefore usually accepted without question for income tax purposes.

Income tax effect

As indicated at the beginning of this section, not many valid judgments are possible at this time as to choices of accounting methods for income tax purposes. For the small, unincorporated contractor it will be important to know whether or not the cash method, heretofore acceptable, is to be considered an unacceptable "hybrid" because certain accrual procedures (i.e., depreciation and amortization) have been engrafted on it. Present indications are that it will be acceptable at least until contrary regulations are adopted.

Another unanswered question involves companies whose three-year average of gross receipts falls below $10 million and who have some contracts expected to last less than two years and some which will last more than two years. Again, practicality seems to favor continuing present procedures and making the necessary adjustments (adequately explained in the working papers) for tax return purposes.

Usefulness to management

As was pointed out previously in this discussion, the cash method of accounting is of minimum usefulness to management. However, so long as the business is small enough so that management needs no help from accounting, the cash method enables the contractor to supply the demands of government at a minimum cost in time, effort, and money.

The percentage-of-completion method of accounting is often the method most likely to provide management with usable data, because it takes into account more of the known facts relative to the company's operations.

Bank and bonding credit

Generally the method of accounting used by the contractor has little bearing on his credit with banks and bonding companies. If supplied with all the supplementary data they request, these concerns will make their own adjustments. The banks, of course, are interested in the sources of funds from which their loans are to be repaid, and are often more interested in a cash forecast than in the financial statements themselves. The bonding company is generally more interested in the current position and the ability to finance the work bonded. Both, of course, are interested in the security to which they may look if they incur losses.

Banks and bonding companies may have to adjust the figures in their clients' statements to secure the specific information they need. Beyond their own immediate needs, however, banks and bonding companies are interested in their clients' accounting methods as some indication of the management skill upon which the client depends for success. To the extent that the management of a contracting firm depends upon obsolete or inadequate methods, its credit and bonding capacity may be limited.

Insurance Problems and Job Accounting

It is the responsibility of general management to see that a construction company's operations are adequately covered by insurance and, in meeting that responsibility, management usually makes use of the services of an insurance broker. Because the broker is customarily the agent of the insured, and not of the insurance company, it is important that he be both responsible and competent.

Premiums

Once the insurance coverage is placed, it is usually up to the accounting department to see that the necessary reports are made and that any claims are properly filed and documented.

Premiums on *workers' compensation insurance* are based on payrolls. However, despite the language of some policies and some reporting forms, the base on which premiums are computed is usually not the total dollar amount of the payroll, but rather the amount arrived at by multiplying the total hours (whether straight time or overtime) by the straight time rates. For that reason many contractors now compute "gross pay" by computing total hours at straight time and adding the overtime bonus.

Other types of insurance for which premiums are usually based on payrolls include *public liability and property damage* (other than automobile), *employer's liability and occupational disease,* and *group health and accident.* A few states, like California, have a compulsory state health and accident coverage tied to their unemployment insurance laws. In these states, the coverage is automatic unless, by special permission, it is written by private insurance companies. As a practical matter little, if any, private insurance is written in this field because of onerous legal requirements. When it is privately written, the report is usually a copy of the state unemployment insurance tax form. As for the other reports on payroll for insurance purposes, each company has its own reporting requirements, although the differences tend to be minor. In any event, complete instructions should be obtained from the contractor's insurance broker.

For accurate and timely reporting of payroll data for insurance purposes, the accrual of insurance should be made on each payroll, and the preparation of the insurance reports should be an item on the paymaster's checklist.

Public liability and property damage insurance on automotive equipment is usually covered separately for each piece of equipment reported, under a blanket policy. Each piece must be reported as soon as use starts, and special provisions are made for vehicles not owned by the company, but used on company business ("nonownership" coverage).

The so-called "equipment floater" is a policy that protects the contractor from loss or damage to his equipment from a wide variety of causes including, generally, fire and storm. There are, however, certain risks (especially some marine risks) that are not covered by the general floater policy. To facilitate the preparation of equipment insurance reports, it is important that equipment records show the value, for insurance purposes, of each piece of equipment. At the beginning of the policy year a list of equipment to be covered and the value of each piece is furnished to the insurance company. Thereafter a report of changes in items covered and their values is made. With most equipment this procedure is followed monthly. With automotive equipment it is sometimes necessary to report each piece of equipment as soon as it goes into service, particularly if public liability and property damage insurance is placed on the automotive equipment at the time the floater goes into effect.

Builder's risk insurance is usually placed on individual jobs, and the only reports required are the periodic (usually monthly) statements of the value of materials stored at the jobsite and cost of completed work to date. Of course, any substantial amount of materials stored in a warehouse or elsewhere should have the usual fire and comprehensive coverage. If the amounts vary, a monthly report is usually required on this type of insurance.

Fidelity bonds may be set on the basis of individual employees, or they may cover any employee who occupies a given position. To cover individuals, a copy of the hiring slip on all bonded employees should go to the bonding company. To cover jobs, a periodic report showing the names of the employees covered and their positions is customarily required (usually quarterly or semiannually) by the bonding company.

"All risk" insurance is sometimes used in lieu of some of the other types of coverage. This type of coverage has many advantages but should be used with full knowledge of its limitations.

Other types of insurance—such as comprehensive, fire, burglary, robbery, business records, and business continuation—are on an annual basis and only their expiration dates need to be checked.

It is a good idea to charge one person with full responsibility for all insurance matters. That person should maintain a checklist of all coverage, the due dates of reports and returns, the existence of coinsurance clauses in reporting-type policies, and similar matters. It should be his responsibility to follow up on all reports, claims, cancellations and renewals, and all retrospective and cancellation refunds. To help top management keep abreast of the insurance situation, a number of large and medium-sized firms, and some small firms, have an insurance summary included in their monthly financial statements showing all policies and their current status.

Claims

As soon as an employee is injured, entirely apart from the reports required by the Occupational Safety and Health Act (OSHA), a complete and accurate accident report must be made to the insurance company. All companies writing workers' compensation insurance require such reports, and check to see that they are made. Usually the questions asked by the accident report required by the insurance company will provide a checklist of all the necessary information.

It must be remembered, however, that injuries due to "serious and willful negligence" on the part of the employer are not normally covered by workers' compensation insurance. It must also be remembered that a great many innocent details can add up to a circumstantial case of "serious and willful" negligence. For that reason, any injury that may be serious is worthy of a special report of conditions, including, if possible, pictures of the site of the accident and the surrounding area; names, addresses, and telephone numbers of witnesses; memoranda regarding safety measures in effect at the time; and anything else that might indicate the exact cause of the accident. Compensation insurance rates tend to be in direct proportion to losses and, since workers' compensation insurance is an essential part of job cost, a low rate of losses can result in a bidding advantage. The reverse is also true. A high accident rate can force bidding prices up and place the bidder at a severe disadvantage in a highly competitive market.

If a person injured on the job is the employee of a subcontractor, or is merely a business visitor, the prime contractor must obtain complete details in provable form. Otherwise he may find himself faced with a lawsuit months after the event, and have nothing with which to determine his liability or on which to base a defense against an accusation of negligence. Therefore, no matter who is injured at the jobsite, the information and evidence assembled should always be the same. If the injured person is not an employee, the accident should be reported as completely as possible to the company carrying the contractor's public liability insurance.

If job offices are maintained, it is usually the job accountant or timekeeper who makes the accident reports. If there is no job office, the foreman usually gives the rough data for any accident reports to the general office, and someone in the accounting department makes them up. Accident reports are too often looked upon as an annoying bit of clerical routine to be passed on to anyone available. Of course, if one person has responsibility for all insurance reports, he should either prepare or review the reports.

Automobile accidents must be reported on special forms provided by the insurer and in most states either the state or local police require additional reports to be made on their forms. Many states place liability on the owner of the vehicle as well as the driver if an accident is caused by negligence, and in all states the general rules of law place responsibility on an employer for acts of his employees in the course of their employment. Because of this fact, most contractors carry insurance against liability for accidents their employees may have while driving personal cars on company business. If "nonownership" coverage is carried, reports of on-duty accidents to employees should be reported as completely and carefully as those in company vehicles.

Claims against common carriers or insurance companies for shortage and damage to goods in transit are usually first made against the carrier. In companies that maintain a warehouseman, it is usually his job to see that such claims are made. In smaller companies, the losses on materials and supplies in transit are usually reported to the office by a foreman, and the claims against the carrier are made by the bookkeeper.

Claims under marine insurance policies are probably the least well understood by contractors, and any contractor undertaking waterfront work should employ an insurance broker who is well versed in marine insurance matters. Many of the rules of negligence in maritime law are the same as those applicable on land, but the tests used in applying the rules often differ substantially. It can only be stated here that such differences do exist and that they call for the services of a specialist.

Storm and fire damage is reported on an affidavit customarily referred to as a "proof of loss." As soon as possible after the storm or fire, the damage is surveyed and estimated. Sometimes settlement is made from the estimate, but it is not uncommon for the insurance company to insist that the damage be repaired and a detailed schedule of costs submitted. If this procedure is followed, it is necessary to set up a work order to cover the repairs, and charge all the repair costs to the work order. Then the charges to the work order are documented in the same way as for cost-plus-fixed-fee work. As in such work, care must be taken to see that overhead and equipment rates are adequate and that impact cost, if any, is given proper recognition.

If the insurance company pays on the basis of the estimate of damage, the contractor's unit costs often form the basis for computing the total amount to be paid. Unless the unit costs contain all the applicable costs, including appropriate overhead allocations, the contractor stands to lose by their application. There is a tendency on the part of many contractor's accountants to take the attitude that, as long as the insurance company is liable for a loss, the matter is no concern of theirs. Actually, when the insurer admits liability, the accountant's work may be just beginning. The next step is to prove the amount of the loss.

Good cost records and complete records of equipment and inventories are essential to establish what has been lost and what it cost. It is true that most insurance claims are based on "present value" instead of cost. It is also true that, on the type of loss usually covered by builder's risk policies, the number of units lost and their replacement cost will be estimated by engineers. Nevertheless, when differences arise, as they frequently do, between the estimates of the insurance company's engineers and the contractor's engineers, good accounting records provide the means for reconciling them. Lacking such records, the contractor usually has to choose between an unreimbursed loss and a lawsuit that might result in further loss.

Another problem that sometimes arises on damage claims is the allocation of lump sum insurance settlements among the prime contractor and subcontractors. There is often a feeling (sometimes justified) on the part of the subcontractors that the prime contractor is trying to profit at their expense in allocating the loss. If an overcollection has been made, it is better to refund it to the insurance company than to divide it among the prime contractor and subcontractors. There are two good reasons for this policy. First, a contractor who has established a reputation with his insurers for strictly scrupulous dealing has a better chance of getting paid on doubtful claims. Second, excessive payments secured by overstated or fraudulent claims can be recovered by the insurance company. If, under such circumstances, some part of an excessive payment had been passed on to a subcontractor, the prime contractor might find himself forced to repay the insurance company, but unable to recover from the subcontractor.

The interest of the construction accountant in insurance claims cannot end when their amount is determined. He must record the claim as an asset, list it on the income tax returns, and report it on the financial statements. Normally a claim is treated as a current account receivable from nonoperating sources. However, it may not always be so. For example, if overlapping coverage causes a dispute among the insurers over who must pay, collection may be delayed for many months. Or the insurance company may deny liability or try to reduce the amount of the claim because of a coinsurance provision. Collection of a claim under such circumstances may require litigation, and the claim could not be properly treated as a current asset, either on financial statements or in working capital forecasts. If the insurer denies liability, the most conservative balance sheet treatment is to show the claim as an asset, reduced by a reserve for loss that is equal in amount. Normally such losses should be claimed as income tax deductions in the year in which they occur, and any recovery in a subsequent year should be treated as income in that year.

Accounting for insurance costs

Whenever an insurance policy is issued, there should be a covering invoice from the broker. The amount on the invoice should agree with the premium information on the policy itself, or on the endorsement, as the case may be. When the invoice is found to be correct, it should be so marked and sent through to be vouchered and entered in the insurance register. For most contractors the standard columnar insurance register, in which the premium is spread over the life of the policy, is adequate. Credit memoranda for subsequent cancellations can be entered on a separate line when they come in. Each entry would, of course, be identified by policy number.

Workers' compensation premiums would not normally be recorded through the insurance register, but through the payroll vouchers. The deposit that is normally required is applied against the last payment due for the policy year, and the deposit for the new policy would be charged to the "Deposits" account.

Each year, after losses for the preceding year have been determined, a number of the insurance companies make refunds on premiums collected from each company whose losses are below an agreed percentage of the premiums paid. These "retrospective" refunds should, theoretically, be credited back to the cost of compensation insurance and such treatment is sometimes possible. Usually, however, by the time the retrospective refund is paid, the year to which it relates is long since closed. For that reason it is often better and simpler to treat the retrospective refunds as either a reduction in the cost of the company's current accident prevention activities or merely as miscellaneous income.

Subcontractors' insurance certificates

Most subcontracts require that the subcontractor carry certain basic insurance, such as public liability, property damage, and workers' compensation. To assure the prime contractor that such insurance is, in fact, in force, the subcontractor secures from his insurers either copies of the policies, or certificates that the policies are in force, or both. The prime contractor should make sure that the certificates of insurance contain the provision that the prime contractor will be given notice before the policy is cancelled or allowed to terminate for any reason.

In establishing the requirements for subcontractors' insurance, the possible complications from overlapping coverage must be considered. If any possibility exists of dispute over which insurer has the primary liability, that question should be recognized and settled before there is any claim. Usually compensation insurance would be required by law, and public liability and property damage would apply primarily to accidents caused by the subcontractor's employees. If builder's risk is desired, it might well be carried by the prime contractor, with resulting adjustment of the subcontract prices.

Most subcontracts contain a so-called "hold-harmless" clause under which the subcontractor assumes responsibility for any loss or damage caused by his operations. If the size of the operation warrants, this clause can be reinforced by insurance certificates, and on larger operations a control file may be maintained to see that all necessary subcontractor coverage is maintained. However, if a prime contractor wants to protect himself fully, his subcontract forms should stipulate all required insurance coverages and provide that, if they are not maintained by the subcontractor, the prime contractor has the privilege of purchasing the required coverage at the subcontractor's expense.

If the subcontract contains such a clause, it should be simple to decide what to do when a subcontractor's insurance is canceled. Often, however, it is not simple, because a subcontractor whose insurance has lapsed has probably exhausted his credit with the prime contractor, and may have built up liens or potential liens far beyond his capacity to pay and far in excess of his retained percentage.

Insurance premiums are not the sort of expense normally covered by a payment bond, even if the subcontractor is bonded (and frequently he is not). The prime contractor can usually bring a defaulting subcontractor's risks under the coverage of his own insurance by taking over the work. Then, if the subcontract is relet, the risks can again be shifted to the insurers of any new subcontractor who takes over from the old one.

Some question may arise about the protection of the subcontractor against the risks which the prime contractor should cover. Usually the subcontractor's own casualty coverage will hedge the most important risks. However, if the question is one that might, under the circumstances, become important, it should be covered in the provisions of the subcontract document and an attempt should be made to include it in the risks covered by the payment bond.

Automated Accounting for Contractors

Automated accounting has been slower to take hold in the construction industry than in some other industries, for several reasons.

First, very small contractors do not have the volume of business necessary to justify the cost of installing the equipment.

Second, there has been a tendency on the part of many construction executives to minimize the part that accounting plays in their operations. Accounting personnel have been looked upon as "mere clerks," and the memorandum field costs kept by the engineers have been thought by many to be the only records that "mean anything."

Third, the long time which has normally elapsed between the happening of events and the time they have been reflected in the accounting records has detracted heavily from the usefulness of accounting records in day-to-day construction management.

Fourth, until recently the larger firms, which one might expect to be first to use such equipment, have tended to decentralize their accounting to the field offices, where conditions are least favorable to mechanization.

Fifth, it is only lately that computer equipment has achieved the degree of dependability and economy that has made its use practical in a field office, where breakdowns and delays can cause real trouble on a fast-moving job with a large number of employees. Currently, given the increasing cost of keeping time and of making payrolls and cost distributions—and given the ever-increasing number of collateral reports required by government agencies, insurers, bonding agencies, labor unions, and others—more serious and successful attempts to computerize payroll and labor cost routines are being made. Once these routines are computerized, other computer applications follow quickly.

The first innovations in automated construction accounting were the "peg board" and the so-called "write it once" systems. These were followed shortly by bookkeeping machines such as National Cash Register, Burroughs, and other comparable equipment. Figures 14-1 and 14-2 illustrate bookkeeping machine operations which, though they are being phased out by some manufacturers, are still used in the construction industry; while these examples illustrate only a part of a payroll application, they do present a fair sample of a typical application on typical equipment. There are recent developments that seem likely to displace the old-style adding machines and the calculators, which for many years have been standard equipment in construction offices. These are the new breed of electronic calculators. Generally speaking they are smaller in size and have greater speed and a greater range of capabilities. They are available with or without tapes. Like other mechanical equipment they are no more accurate than the people who use them, but they can be used by people who know less and therefore are less expensive to employ. Now computers have moved into the field and give indications of ultimately taking it over completely.

In spite of the fact that mechanization is becoming the rule rather than the exception, the illustrations in prior chapters of this book have been based on hand-kept systems. The reason is simple and practical. The hand-kept forms illustrate information required, and procedural steps that must be taken, before any mechanized or computerized system can be successful.

The savings in time and money made possible by bookkeeping machines fall into essentially two categories:

1. Duplication of work is reduced by preparing two or more records at a single writing.
2. Faster and more accurate machine calculations, together with automatic balancing, combine to speed up the processing of detailed data and reduce errors in calculation.

Computerized records can properly claim these advantages and can add the following:

1. The "data retrieval" capabilities and the so-called "memory" mechanisms of computers make it possible to pull together related facts from numerous diverse sources and thus produce, in an incredibly short time, analyses of data which were either impossible or impractical in hand-kept or even bookkeeping machine accounting systems.
2. The capability of computer systems to feed data from a jobsite office directly to a central processing bureau and get the processed

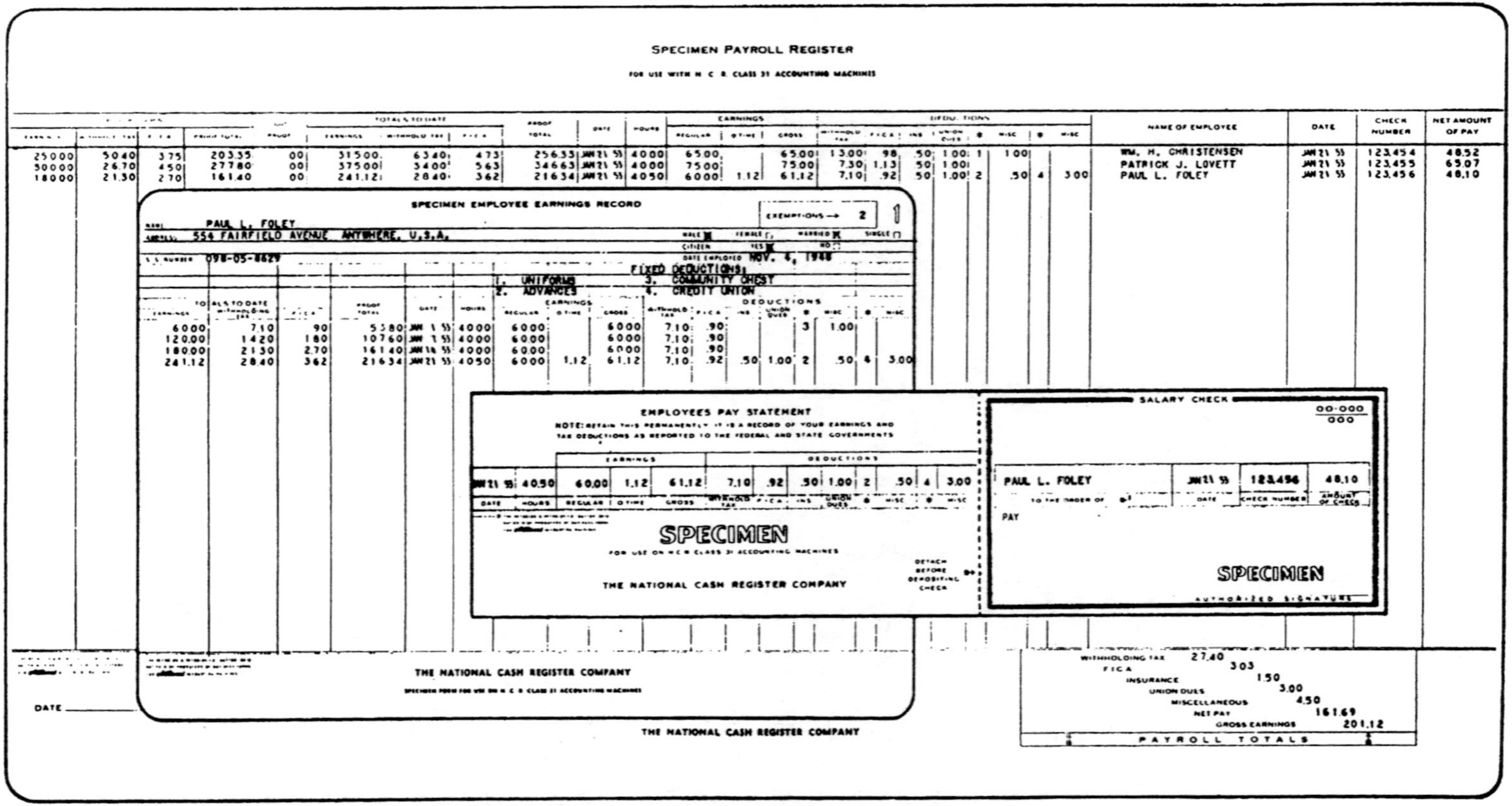

Figure 14-1 Payroll System. This simple system and the one shown in Figure 14-2 were designed for the National Cash Register Company's Model 31 bookkeeping machine.

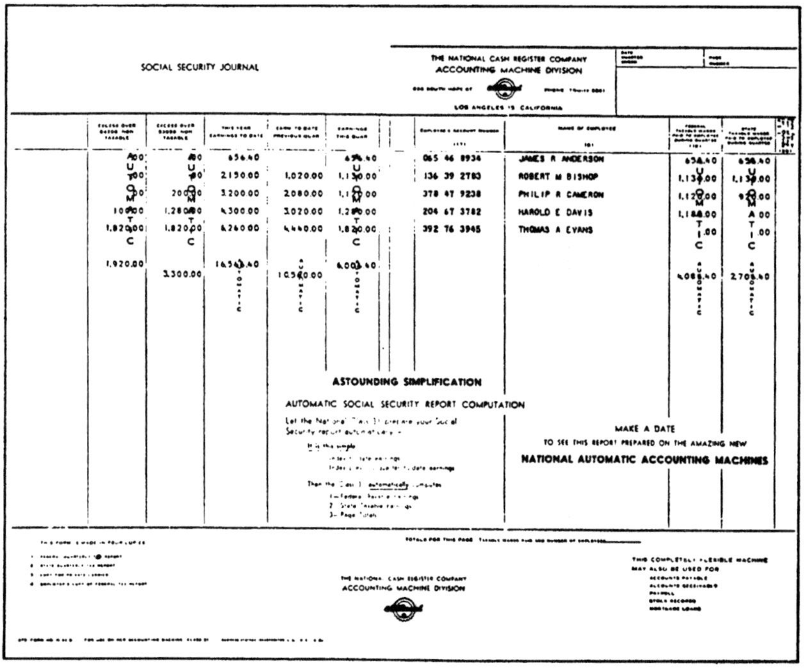
SOCIAL SECURITY JOURNAL

THE NATIONAL CASH REGISTER COMPANY
ACCOUNTING MACHINE DIVISION
LOS ANGELES 15 CALIFORNIA

Employee	Account No.	This Year Earnings to Date	Earnings This Quarter
JAMES R ANDERSON	065 46 8934	656.40	656.40
ROBERT M BISHOP	136 39 2783	2150.00	1,130.00
PHILIP R CAMERON	378 47 9238	3200.00	1,120.00
HAROLD E DAVIS	204 67 3782	4300.00	1,280.00
THOMAS A EVANS	392 76 3945	6260.00	1,820.00

ASTOUNDING SIMPLIFICATION
AUTOMATIC SOCIAL SECURITY REPORT COMPUTATION

MAKE A DATE
TO SEE THIS REPORT PREPARED ON THE AMAZING NEW
NATIONAL AUTOMATIC ACCOUNTING MACHINES

THE NATIONAL CASH REGISTER COMPANY
ACCOUNTING MACHINE DIVISION

Figure 14-2 **Social Security Journal.**

data back at the jobsite has greatly improved the quantity and quality of job progress reporting. More importantly, the computers have proved their capacity to get such reports back to job management and general management in time to be useful in job cost and progress control.

There are pitfalls, too. Some mechanized equipment requires experienced operators. Other equipment may require service from some distant point if it breaks down. On jobs where there is work for only one machine and one operator, a failure on the part of either could delay the entire job. Most construction jobs move too fast to allow for such delays. The records, and particularly the payrolls, must be completed accurately and on time or the job will be faced with heavy additional costs. Often the supply of electric power at jobsites is uncertain or is of a different frequency from that in the cities. Machines built for alternating current may fail when used with direct current. Dust and dirt at jobsites may foul delicate mechanisms and cause malfunctions, and inexperienced operation can cause some equipment to break down. These are some of the reasons, other than cost (which in the past has been very high), that have caused the construction industry to hesitate to adopt mechanized or computerized accounting.

Insofar as cost is concerned, the advent of the so-called "minicomputer" has brought the cost of bookkeeping machines and computers close enough so that some machine manufacturers are phasing out their bookkeeping machines and going to the minicomputer. For example, the installation, including forms and procedures, for the most advanced bookkeeping machine would cost from $10,000 to $15,000. Whereas, for example, National Cash Register Company's Century 100 minicomputer would cost from $15,000 to $25,000 for both hardware and software. A system where data could be fed into a service center by telephone lines could cost even less.

One factor to be considered is the trend toward improvement. The machines themselves (i.e., the "hardware") are constantly being simplified and improved, and mechanical means of catching and correcting operator errors is becoming more dependable. The programming (i.e., the "software") is constantly becoming more sophisticated. Jobsite power and working conditions are tending to improve and the newer types of machines are less vulnerable. Most important of all, the level of accounting know-how in the construction industry has been steadily rising and, as a result, individual construction executives have become more open minded about the desirability of spending money on improved accounting methods. As a result, computers are being used by an increasing number of construction firms when volume is large enough to warrant their use. Even very small contractors make use of computer services offered by banks.

The entire discussion of computerized accounting in this book is based on the assumption that most contractors who are contemplating installing a computerized accounting system already have an accounting system in place and operating. Of all the problems that face the chief accounting officer of a construction company, one of the most complicated and demanding is the conversion of the existing system over to computers. It is a job that requires a combination of skills:

1. A complete and detailed knowledge of the operation of the company, of its accounting system, and of the information needed to control its operations.
2. A complete knowledge of computers and of what they can and cannot do.
3. A good working knowledge of the special accounting problems of the construction industry as well as of what it takes to make an accounting system auditable.

It is rarely possible to find all these qualities in one person; therefore, the conversion of an existing accounting system to a computerized system is almost always a team effort. Finally, it is necessary that top management give the installation its complete backing.

The subject of automated accounting for construction contractors is discussed fully in Chapter 12.

Effect of Personnel Policies on Job Accounting and Reporting

Employees whose jobs are unionized must normally be governed, in their relationships with their employer, by union rules and by the laws governing unions. Except for the effect of the federal wage and hour law and the various state laws, nonunion employees are governed by company personnel policies. This section is principally concerned with the nonunion group.

Schedule 14.1 is a statement of personnel policies established by a construction company of medium size. This statement became a part of the company's procedure manual. The significance of these policies in terms of financial and accounting control is, to a large extent, self-explanatory. It is the duty of the accounting department to recognize, in each policy, the financial and accounting requirements involved, and to make preparations to fulfill those requirements. For example, if the expense account policy is to be adhered to, the procedures followed by the voucher clerk must be so designed that no unauthorized expense voucher will be paid, and those that are properly authorized will be processed and paid without delay. Similar requirements could be cited for each of the illustrative personnel policies in Schedule 14-1.

Recent developments in two special areas deserve special mention. The first is the potential liability which can arise from failure to keep the personnel records required by the Immigration Reform and Control Act of 1986. The law is too new and developments are too rapid to recommend any policy except to make every effort to comply to the letter with the record-keeping requirements. The required forms and instructions are available from the U.S. Immigration and Naturalization Service.

The second area of potential liability is the increasing tendency of the courts to find for the employee in lawsuits involving discrimination, sexual harassment, and wrongful discharge. It is outside the scope of this discussion to go into the details. However, a study of the decisions clearly shows the importance of sound, well-defined policies clearly spelled out and carefully documented by proper personnel records. The policies are the job of management, but the record keeping is usually the responsibility of whoever is in charge of the office.

Jobsite personnel responsibility

It has been pointed out elsewhere in this volume that full control over jobsite personnel must be in the hands of the person in charge of the job. If that person is "accounting-minded" this control creates no real

SCHEDULE 14-1 MEDIUM-SIZED CONTRACTOR'S PERSONNEL POLICIES*

General policies

All employees will be classified for the purposes of establishing policies and creating incentives. The classifications will be as follows:

E—Executive
A—Administrative
O—Operational

The executive group includes top management, department heads, and such other key personnel as top management may wish to classify in this group.

The administrative group consists of those employees who, regardless of their duties, are considered permanent employees, as distinguished from those hired on a job-to-job basis.

The operational group includes all employees not otherwise classified.

To the extent practicable, advancements to the "E" classification will be made from employees in the "A" classification, and advancements to the "A" classification will be made from the "O" group.

Hiring procedures—nonunion employees

This company is an equal opportunity employer. As such, its policy of nondiscrimination in matters of race, creed, color, sex, or national origin covers all hiring and promotion procedures. It is the policy of the company to make promotions from within the organization whenever possible. Accordingly, whenever trainees or others are brought in from the outside they will normally be placed in the operational group. Subject to the limitations of our affirmative action program, jobs in the administrative group will first be offered to qualified employees in the "O" group, and only if none can be found will outside applicants be considered. When an employee is brought into the "A" group, he is covered by a blanket position bond. He will be required to fill out an application blank.

When an employee is needed in the "A" group and there is no qualified person already employed to fill the position, the first reference will be to files of former employees who have requested reemployment. If none is found there, the general application files will be searched and only after these have failed to produce the desired result will recourse be had to such sources as advertisements, employment agencies, and so on.

Occasionally top management may decide to offer a position in the "A" or "E" group to someone outside the company. Such a step will first be thoroughly discussed by top management among themselves and with the employees who might otherwise have been considered for the job the newcomer is to fill.

If an application is received by mail and there appears to be a reasonable chance that someone of the type indicated by the application may be needed, either immediately or in the foreseeable future, one of the company's regular application blanks should be sent with a request that it be completed for early consideration. If the application cannot be considered at once, a postcard should be sent to the applicant bearing the following message:

"This will acknowledge receipt of your inquiry regarding employment with our company. At present we have no opening available requiring experience and qualifications such as you have stated. We will, however, keep your application on file and if anything develops in which we feel you will be interested, we will be pleased to contact you. Ecks & Wye Co., 000 Hoe St. Los Angeles 90017, (213) 681-2345."

When an individual is interviewed for a position, he should be notified as soon as the position is filled even though he may not be selected.

*Those employers who have a retirement plan should include in their personnel policy provision for reference to the Employee Retirement Income Security Act of 1975.

SCHEDULE 14-1 MEDIUM-SIZED CONTRACTOR'S PERSONNEL POLICIES *(Cont.)*

Grievance procedure

One of the most important psychological factors in any business organization is employee morale, and one of the best ways to keep morale good is to establish among the employees the feeling that grievances will be fully heard and fairly dealt with. Most employees are reluctant to take a grievance directly to a department head. Therefore, in each department some senior employee should be designated to bring employee grievances to the department head. In listening to and acting upon such grievances, the department head must remember that upon his fairness and good judgment depends the morale of the people upon whom his own job, in a large measure, depends.

It is the policy of the company to encourage employees to bring their personal problems to the attention of their department head, since we do not have a personnel counselor. The reason for this policy is to discover, and take action to remedy, the problems that disturb the employees and keep their minds from their jobs. It is expected that this policy, if effectively carried out, will create a closer and more personalized relationship between the department heads and their employees.

Absenteeism

It is the responsibility of each department head to eliminate tardiness, leaving the job early, and absence for any but legitimate causes. See policy on time off.

Sick leave

Sick leave, as the term implies, is to be used only when sickness of the employee himself, or of his immediate family, prevents attendance. It is not a form of vacation to be taken as a matter of right.

"E" employees

Employees in the "E" classification may accumulate sick leave at the rate of one day per month up to a maximum of 25 working days. Any sick leave in excess of that accumulated must be specifically approved by top management.

"A" employees

Employees in the "A" classification may accumulate sick leave at the rate of one day per month up to a maximum of 15 working days. At the discretion of his department head an employee in this group may receive as much as 5 additional working days. Any additional sick leave for an employee in the "A" classification must be approved by top management.

"O" employees

As a matter of policy, employees in the "O" classification are not entitled to sick leave. Virtually all such employees will be paid on an hourly basis and will be paid for time in attendance only. An exception to this general rule occurs when an "O" employee is paid on a straight salary and has not been paid for overtime. Under such circumstances the department head or project manager may authorize up to one week of sick leave. Any allowance in excess of one week must be on instructions from top management.

Joint ventures

Employees of this company who are employed on a joint venture job shall be subject to whatever sick leave rules are laid down by the sponsor. However, if those rules result in treatment less favorable than the employee would have received had he been employed on a company job, the company will make up the difference.

SCHEDULE 14-1 MEDIUM-SIZED CONTRACTOR'S PERSONNEL POLICIES *(Cont.)*

Computation and record keeping

A record of accumulated sick leave will be maintained in the general office for all "E" and "A" employees. This will require that any sick leave taken by employees assigned to jobs in the field be reported to the general office.

Once an employee has accumulated the maximum sick leave, no further time will accumulate. However, when such an employee returns to work after taking sick leave, accumulations will start in the following month and continue until the maximum is again reached.

Vacation policy

It is the general policy of the company that all salaried employees having twelve months or more of continuous service shall receive two weeks paid vacation each calendar year. Except as noted below, such vacations must be taken in one continuous period and may not be carried over in whole or in part into a subsequent calendar year. Also, subject to the exceptions noted below, it is not the policy of the company to pay for vacation time not taken during the calendar year.

Salaried employees on construction projects

Salaried employees in job offices will normally take their vacations at the conclusion of a job. Occasionally a salaried employee on a construction project may be needed continuously during one calendar year, in which event, with the written authorization of the department head, a carry-over into the next calendar year will be permitted with a maximum accumulation of four weeks to be taken at the earliest opportunity. If, for company reasons, some part of such an accumulation cannot be taken, payment may be made for the time, provided written permission from the department head is obtained.

New employees' vacations

If, during the first calendar year of employment, an employee has less than six months' continuous service, he shall receive no vacation in that calendar year, and shall not be eligible for his first two-week vacation until completion of twelve months' service.

If, during the first calendar year of employment, an employee has more than six months' but less than twelve months' service, he shall receive a vacation of one week to be taken within the calendar year.

Effect of terminations on vacations

Employees who are qualified for a vacation and who voluntarily leave the employ of the company before taking it shall be entitled to such vacation provided they are in good standing and give the company adequate notice of their intention to resign. Vacation time will not be considered in determining the period of notice.

Management reserves the right to determine to what extent, if any, vacation allowances will be granted to those whose services are terminated under other conditions.

Holidays during vacation

A legal holiday falling within an employee's vacation period will have the effect of extending the vacation by one day.

Vacations with leave of absence

An employee going on leave of absence and otherwise eligible for a vacation shall be granted the vacation prior to the leave. Should a vacation be taken immediately prior

SCHEDULE 14-1 MEDIUM-SIZED CONTRACTOR'S PERSONNEL POLICIES *(Cont.)*

to a leave of absence, such leave of absence shall be considered to begin on the day following the last day of vacation.

Exceptions to vacation policy

It is recognized that under exceptional circumstances some deviation from the established policy may be justified because of company necessity or some equally compelling reason. Under such conditions there may be a deviation from the foregoing, provided that such deviation is authorized by top management.

Salary adjustment and pay schedules

Compensation of employees in the "E" classification will be set by top management.

Compensation of employees in the "A" classification will be set in accordance with a predetermined salary schedule showing rate ranges. This schedule will be reviewed in June of each year. The construction manager and the controller will review the rate ranges for jobs in their respective departments and make any recommendations they see fit to the general manager. The schedule will not be changed without specific authorization from the general manager.

Adjustments of compensation of employees in the "A" classification will arise in two ways: (1) by upgrading or downgrading or (2) by a semiannual review of each employee's record.

During the month preceding the anniversary date of each employee's service with the company, and at the six months point, each department head will review each employee's record for the preceding year. This review will take into account such factors as quantity and quality of work, attitude, initiative, resourcefulness, length of service, potential capacity for progress, relationship of present salary to going rates for comparable jobs both within and outside the company, and any other factors that may be material.

On the basis of this review, the department heads will make recommendations for pay adjustments to the general manager. Any adjustments approved by the general manager will be put into effect for the first pay period beginning after the anniversary or midyear date of the employees affected.

Adjustments due to upgrading or downgrading will be effective on the date of the change in grade.

Employees in the "O" classification will normally be paid according to established hourly rates. If, on occasion, such employees are paid on a weekly or semimonthly basis, their salaries shall be set by applying the applicable hourly rate to the established work week. Semimonthly rates, in such cases, would be computed on the basis of four and one-third weeks per month.

Any adjustment in compensation of employees in this group will normally reflect changes in going rates for such employees or equalize inequities between jobs. However, if, in the opinion of the department head, variations in the going rates are justified, the department head has discretionary power to allow such variations up to 10 percent of the going rate.

Overtime†

Only employees in the "O" classification shall be paid for overtime and then only if called for by an applicable union agreement, and the rate will then be that established by the union agreement.

†The Federal Fair Labor Standards Law, which at one time was not uniformly applicable to all construction work, is now generally applicable. Any attempt to operate outside that law should be cleared with legal counsel.

SCHEDULE 14-1 MEDIUM-SIZED CONTRACTOR'S PERSONNEL POLICIES *(Cont.)*

Bonuses

Bonuses have been discontinued as a matter of company policy except where necessary to honor previous commitments. It is now the company's policy to pay all compensation in the form of salary. For the policy on gifts at Christmas, see the section on employee benefits.

Time off‡

Occasions will arise when an employee, for legitimate personal reasons, may need to be absent during regular working hours. On such occasions, it is within the discretion of the department head to grant reasonable time off without penalty. However, it is also within the discretion of the department head to charge such time off to any vacation which may be due to the employee or, in extreme cases, to charge the time against the employee's current salary.

Normally this problem will arise only in connection with employees in the "A" group or with those in the "O" group who are paid on a weekly or semimonthly basis.

Jury duty

If employees are called for jury duty, it will be the company's policy to have them excused if possible. If all such efforts fail and an employee is forced to serve on a jury, the employee will be paid an amount equal to his usual salary reduced by his jury fees for the usual 30-day term. In the event an employee, while serving a term of jury duty, is selected for a long trial that will run over the 30-day term, the employee's department head will consult with the general manager to determine the policy to be followed for the excess period of service.

Rest periods

There will be no regularly scheduled rest periods. At the discretion of each department head, employees may be permitted such time as may be reasonable for an occasional cup of coffee at a coffee vending machine, or for similar interim relaxation as may be justified by the nature of the work. However, such rest periods as the department head may sanction are a matter of privilege, and not a matter of right. Abuse will justify their withdrawal either on an individual or departmental basis.

Employee benefits

At present, a group health and accident insurance policy is in effect. Premiums are deducted from the employees' paychecks. The company handles the accounting work and the remittances.

During the last week of November each year, the general manager with such assistance from the department heads as he may require, will select an appropriate gift (such as a turkey or a box of fruit) to be given to all employees in the "E" and "A" classifications and to any "O" group employees to whom such gifts may be in order.

Future consideration is to be given, as conditions warrant, to such further benefits as: a pension- or profit-sharing trust; group life insurance; employee activities such as picnics, parties, or a company magazine.

Policies for their administration will be established when and if such benefits are inaugurated.

‡Construction companies frequently find it necessary to include a section regarding the allowance of time off to those employees who have an obligation to the armed forces reserve organizations.

SCHEDULE 14-1 MEDIUM-SIZED CONTRACTOR'S PERSONNEL POLICIES *(Cont.)*

Ideas

It is the policy of the company to stimulate and reward constructive thinking on the part of all employees. It is one of the duties of supervisory employees to encourage the people under their supervision to come to them with ideas, and to assist in the development of any of those ideas that appear practical. Due credit will be given any employee for useful ideas, but the greatest credit, and the most substantial recognition, will be given to those supervisory employees who are most successful in stimulating thought.

Holidays

The general office will be closed on the following holidays:
New Year's Day (office closes at noon December 31)
Independence Day
Labor Day
Veterans Day
Thanksgiving Day
Christmas (office closes at noon December 24)
Should any of these holidays fall on Sunday, the following Monday will be a holiday.

Office hours

The general office will open at 8:15 a.m. and close at 5:00 p.m. Monday through Friday each week and there will be a 45-minute lunch period. All employees, including executives, are expected to observe these hours.

Contributions

The company belongs to a group known as A-I-D (Associated In-Group Donors). Once each year you will be asked to contribute to that group. There will be no other solicitations during the year, and all requests will be referred to A-I-D. Internal solicitation (passing the hat) must be cleared through the office manager who will also have a budget for company contributions to such funds.

Memberships

The company encourages membership in professional societies and in suitable circumstances, to be determined by the general manager, will pay all or part of the dues of an employee active in such societies. Each of the employees in the "E" classification is encouraged to belong to a service club such as Rotary, Lions, Kiwanis, Optimists, or Exchange Club. Dues will be paid by the company. Reasonable allowances of time will be made for organization activities. In engaging in such activities, employees are expected to remember that they are representing the company, and that by promoting its interest at every opportunity they are thereby promoting their own.

Expense accounts

All employees who incur expense on behalf of the company are entitled to reimbursement upon submitting to the controller a properly authorized expense account. To be properly authorized, an expense account must meet the following tests:

Employees in the "E" classification

Expense accounts need not be authorized in advance but they must be authorized for payment by the general manager. Since employees in this group are supplied with company cars and credit cards, only such company car expense as cannot be placed on the

SCHEDULE 14-1 MEDIUM-SIZED CONTRACTOR'S PERSONNEL POLICIES *(Cont.)*

credit card without undue inconvenience will appear on the expense account. Hotel, Diners Club, railway, airline, and telephone credit cards are also furnished to employees in the "E" classification, and are to be used in lieu of cash if possible. In keeping with the standing of the company, it is expected that employees in the "E" classification will travel in first-class accommodations and will stay at first-class hotels.

Cash spent for entertainment of guests will be reimbursed without supporting vouchers. However, it will be necessary for the expense account to show the names of the persons entertained, their business connection, and the date and place of entertainment. The cost of entertaining public officials or public employees is not deductible for income tax purposes.

Tips, cab fare, postage, parking, local telephone calls, and similar costs should be identified and not aggregated into a single item called "miscellaneous." All purchases that cannot be charged to the company should, if possible, be supported by receipted bills or invoices.

Travel advances will be charged against the individual and, if not covered by expense accounts within a reasonable time, will be collected from the employee. Expense accounts are due on or before the fifth of each month for the preceding month, and expense accounts rendered after that date may be disallowed in whole or in part.

Employees in "A" and "O" classification

Employees in the "A" and "O" classifications who are called upon to travel regularly are subject to the same rules as employees in the "E" classification, except that air and rail travel will be tourist class where possible. Hotel accommodations will be in tourist class hotels. Employees in the "A" and "O" classifications will not ordinarily be authorized to spend money for entertainment. When such expenditures are made, they will be subject to very close scrutiny.

Usually, any travel expense incurred by employees in the "A" or "O" classifications must be authorized in advance by the head or assistant head of the department in which the employee works. When the expense account is turned in by the employee, it is the department head or his assistant who must approve the expense account for payment.

Discipline

Normally, the maintenance of discipline in each department is the responsibility of the department head. However, it must be remembered that the company has a substantial investment in each employee and indiscriminate firing can be costly. Any borderline cases should be discussed with the general manager.

There are, however, certain practices and offenses that will merit termination. These include, but are not necessarily limited to, insubordination, bad work habits, inefficiency, substance abuse, dishonesty, conduct detrimental to safety, and conduct detrimental to the company.

In any case, the first formal step in subjecting an employee to discipline is a written notice that certain conduct is not acceptable and can subject the employee to discipline. Second is a written warning calling attention to the first notice and informing the employee that further breaches will subject the employee to a specific form of discipline (i.e., docking, termination, etc.). After a reasonable time following the second notice, if the employee's bad conduct continues, the discipline shall be imposed.

Garnishments, attachments, and assignments

The company is legally bound to honor any garnishments, attachments, or assignments of an employee's wages or salary. We should, nevertheless, make every legitimate effort, without subjecting ourselves to liability, to protect the employees against such

SCHEDULE 14-1 MEDIUM-SIZED CONTRACTOR'S PERSONNEL POLICIES *(Cont.)*

legal levies. There are three reasons for this policy. First, we gain more from securing the loyalty of our own employees than we do from the reputation among loan companies and credit retailers as a "cooperative" employer. Second, an employee who can expect nothing on payday but more attachments and garnishments is rarely a good employee, and sometimes the pressure of debts will turn him into a dishonest one. Third, every payroll deduction or split check increases the cost of operating the payroll department, and to the extent that these special problems can be held to a minimum, so also can the costs.

Therefore, if it should become necessary to fire the employee, a record of having done everything reasonably possible to protect him tends to justify the result in the minds of the other employees. Also, it frequently happens that the employee is being victimized by a racket, and when a dishonest creditor feels he is up against the employer as well as the employee, he frequently abandons his efforts.

When faced with a difficult garnishment problem, the paymaster should consult the head of the employee's department. If the department head feels the employee is valuable enough to the organization to make it worthwhile, he may consult the company's attorneys on ways to assist the employee in controlling the collection efforts of his creditors.

Group insurance

The company has made available to its salaried employees an employee-financed group health and accident insurance policy. This group policy was selected only after a careful survey of all available policies to secure the best coverage. Every employee should have an opportunity to participate. It is not company policy to exert pressure of any kind on any employee to take part in any insurance plan. At this time, a number of group life programs are under study and, as soon as the most suitable is selected, it will be presented to the employees for their consideration.

Joint venture service

Often an employee of this company in the "A" or "O" classifications will be transferred to the payroll of a joint venture. When this company sponsors the venture, there will be no problems of seniority, vacation, insurance, expense accounts, moving expenses, and similar matters, because joint venture policies will conform to company policies. However, if the joint venture is sponsored by one of the other joint venture partners, then the employee will be treated in accordance with joint venture policy. If any joint venture personnel policy is less favorable to the employee than that of this company, an equitable adjustment will be made by this company. This policy does not include working hours, however. Any requirement of the joint venture regarding working hours will be binding on the employee without adjustment.

Lending employees to others

It is contrary to the policy of this company to "lend" its employees to others or to "borrow" the employees of others for its own work. If, with company permission, an employee performs services under the direction and control of another person or firm, it will be company policy to insist that he be placed on the payroll of such other person or firm. Aside from the business aspects of this policy, work compensation insurance and industrial accident considerations make it imperative.

Likewise, employees of subcontractors, or other outside employees should never, under any circumstances, be placed on the payroll, unless the company is prepared to take full responsibility for their work and conduct.

problem. In fact, if jobs are large enough to warrant a field office, it is possible to place on the job an office manager or project controller who can take a great deal of the administrative burden from the project manager or superintendent. It is often entirely possible to give the office manager control, under the project manager, of employees in the accounting, office clerical, payroll, timekeeping, cost, warehouse, purchasing, camp, mess hall, first aid, safety, security guard, and fire protection departments. Any fully qualified construction office manager should have experience in handling all these activities on a large job. The combination of a strong office manager and an accounting-minded project manager make a profitable team that can operate autonomously without any direct line of functional control from the general office to the job accountant.

If the project manager is inclined, as many are, to look upon the job office manager as a "mere clerk," or if the job office manager is a "mere clerk" in fact, some direct control by the head office of the job office is necessary. This control usually is most effective when exercised by the company controller or a field auditor who is an experienced job office manager in his own right. To make split control work at all, the head office representative must have the authority to require, even over the project manager's objection, that the actual accounting be done according to his instructions. The head office representative must thus have the complete confidence and backing of top management. If these cannot be obtained, it is better to centralize the accounting.

Accounting reports

Decentralized job accounting leaves the head office largely dependent upon the job accounting reports for much of its management information. Placing responsibility for such reports is an important personnel decision, and upon it may depend the success of the job. The responsibility for correct accounting principles must lie essentially with the head office and its accounting manual. Applying those principles on the job must be the responsibility of the job accountant or, in a split control system, of his functional supervisor in the head office. Responsibility for percentage-of-completion figures and the overall reasonableness of the reports must necessarily rest with the project manager or superintendent. Many of the more progressive construction companies have both the job office manager and the project manager or superintendent sign the accounting reports.

Forms and Printing in Job Accounting

It has been pointed out (Chapter 1) that the very small construction companies rarely find it economically feasible to use decentralized accounting and, in fact, the smaller the concern the less likely it is to have problems with forms or printing. Usually the standard forms that can be purchased from commercial printers are adequate. However, as a construction business grows larger, and starts to design its own forms, some problems do arise. Some of these problems are created by decentralized accounting.

Decentralized system

It is usual for a decentralized office to have separate payroll and general bank accounts. If these accounts are kept in the same bank with other accounts of the same contractor, the checks should be a different color to prevent charges to the wrong accounts. Sometimes, particularly if the job is some distance from the general office, a bank near the jobsite will be used, especially for payroll. In either event, it will be necessary to have special checks printed for the job, and special printing may take from one to three weeks. Meanwhile, as soon as a contract is awarded there is immediate activity, and checks, as well as all other forms, are needed at once.

One of the simplest solutions to this problem is to have a standard form for voucher checks and a standard form for paychecks. A supply can then be printed up without a bank designation and with a number series (say in the 900,000 series) that will be distinctive. All banks have their own check stickers that can be put on the special series checks until the regular forms can be printed. In that way payments can start immediately from the proper bank accounts. The same special checks may be used at the end of a job if the estimated number of regular checks proves too small and the job runs out of checks so near to the end that a second run would be too expensive. They may also be used if some of the regular paychecks are stolen and it becomes necessary to change accounts in the middle of a job. When the checks are designed, space can be left above and below the company name so that the name of a joint venture partner could be stamped in should it become necessary to use such checks for a joint venture.

Checks should be so designed that if the flap of an ordinary mailing envelope is clipped over the top of the check, the name of the payee will still be visible. This tends to prevent the kinds of defalcations that occur when a dishonest voucher clerk manages to slip a check through the authorized signers without a payee. On voucher checks this safeguard can easily be effected by having the voucher at the top and the check on the bottom.

When a job is completed and the bank accounts are closed out, all the extra checks applicable to those accounts should be destroyed. Omitting this precaution is an open invitation to theft of the check forms and forgery. Needless to say, the special checks should be kept under lock and key for the same reason, and all numbers issued should be accounted for strictly.

Letterheads and return addresses on envelopes need not cause much concern. It is usually possible, as soon as a job starts, to secure a quick "overprint" job on enough letterheads and envelopes to start work, or a rubber stamp can usually serve until the regular supply is ready. It is often practical to have a supply of stock without the specific job address, which can then be stamped or typed in. However, the simplest solution is to have some regular letterheads with the following words printed in red at the top:

ADDRESS REPLY TO:________

The job address can then be inserted in the blank space until the job letterheads are available.

Purchase orders usually show the job number and location and contain a space for delivery instructions (Ship to ____ at ____) and billing instructions. Still, if the job is large enough to warrant decentralized purchasing and a large volume of work is going through the purchasing office, it is practical to have this information printed on the form instead of having to write or type it in on each order. Therefore, as with the checks, it saves time and money to have a supply of the write-in type of form with a distinctive number series to fill in at the start of a job until the regular printing can be completed.

Work orders, statement forms, reports, requisitions, vouchers, receiving reports, personnel and payroll forms, and most other forms are usually made up to show the job number and designation and usually can be taken directly from stock. If the job is being done by a joint venture, and it is necessary to have the joint venture name on the forms, it can usually be inserted with a rubber stamp until the regular printing is received.

Other considerations

It is ordinarily most economical to use stock forms whenever they will fit the needs of the business. The forms that cannot be bought from the stock of the contractor's regular printer or stationer must be specially printed, and it is usually good business to select one responsible firm and let that firm handle all the printing of such forms and stationery. However, occasions arise, particularly if large projects are undertaken some distance from the general office, when it is good business to have as much as possible of the special printing for that job done locally. In smaller towns job printing is often done in the shop of the local newspaper, and it is common for a contractor to find that he needs the support of the local press when his operations cause complaints from the local citizens. If conditions indicate that such a situation might arise, a few extra dollars spent on locally printed forms might prove invaluable in terms of public relations.

Another matter that should be considered, in ordering printing and designing forms, is the time needed for preparation. In a decentralized job the need for such design is multiplied by the lack of the usual general office controls. Thus, snap-out forms cost more initially, but that cost must be weighed against the time required to interleaf the carbon paper and line up the forms at the time they are being prepared. The use of single-use carbon paper in snap-out forms avoids the errors and delays inherent in failure to use new enough carbon to produce legible copies. If forms are to be used by foremen, superintendents, or warehousemen, they will be more promptly filled out and submitted if they are designed to be executed in pencil rather than by typewriter. If several copies of such forms are required, it may be cheaper to reproduce them on one of the numerous photographic duplicators than to try to produce enough copies with carbon paper. Of course, if the forms can be typed, it is possible to produce as many copies as desired by using a duplicating process.

Centralized system

If the office work on all jobs is centralized in the head office, it is possible to centralize the printing of forms and to realize the economy of volume buying from a single reputable source. As one of the printer's better customers, it is easier for a contractor to get prompt delivery dates and occasional "rush job" service. Unless the operation is very small, snapout check and voucher forms can save a great deal of employee time in the office, and can be bought at reasonable prices in volume. It is quite simple, as a rule, to find forms of this kind that are entirely adequate and that can be bought from stock.

If all banking is done at a single bank, it is simple enough to get voucher check and paycheck forms to fit a wide variety of needs. Both shape and color can be used to distinguish checks on the various accounts from each other and, if volume warrants it, the checks can be punched for mechanical sorting.

If separate bank accounts are kept for each job in a centralized system, it is advisable to have them at separate banks. However, if a number of accounts must be kept in a single bank, it is almost imperative to use various sizes, shapes, and colors of checks to prevent intermingling of the accounts.

Purchasing forms in a centralized purchasing system can be the snapout type and still have a maximum of shipping and billing information printed on the form. With a properly designed form and an electric typewriter it is possible to get ten or more copies without duplicating. Moreover, all billing instructions can be uniform, so that additional time is saved in prompting suppliers to provide billings in the right form and number of copies.

If purchasing is decentralized it is sometimes more economical, particularly if the purchasing is done from the jobsite, to use a single-copy padded purchase order designed to be handwritten and duplicated.

Offset printing will be most economical for this type of form and it will be just as useful as if a more costly process were used. Of course, offset is useful on many other types of forms as well, but this application is typical.

Mixed system

If a system is partly centralized and partly decentralized, the safest and most economical procedure, in the long run, is to treat it as a decentralized system for the purpose of designing and printing forms. However, with a little additional planning it is usually possible to obtain the advantages of centralized buying.

Camp, Mess Hall, Commissary, and Protective Functions

The first step in setting up a camp, mess hall, or commissary is to get from the manager a requisition for necessary purchases. Most experienced camp managers will have their own checklists, and these will vary a great deal according to the size and location of the jobs the manager has worked on in the past. Schedules 14-2 to 14-5 provide checklists of equipment and supplies for kitchen, mess hall, and camp.

Once the requisition is completed, purchase orders (Chapter 2) are issued for the basic initial requirements.

Forms and procedures

The principles followed in controlling materials and supplies (Chapter 4) apply to camp, mess hall, and commissary inventories, but with special emphasis on protection from theft. If these inventories are not kept under lock and key, and access is not carefully controlled, losses can become serious very quickly.

Whenever a camp, mess hall, or commissary is justified, a reasonable expenditure should be made to provide each activity with its own adequate, locked storage room. Once each storeroom is established, a basic inventory list should be prepared. Usually it is a good idea to design a special sheet (Figure 14-3) for each type of inventory.

Inventory. With sheets prepared in advance showing what should be on hand, frequent and accurate inventories of camp, mess hall, and

SCHEDULE 14-2 CHECKLIST OF KITCHEN AND MESS HALL EQUIPMENT

Bread rack
Butcher block
Can opener, hotel model with spare knives and gear
Coffee maker, urn
Coffee pots, 3 gal
Dishwasher with necessary tables
Doughnut cutters
Egg slicer
Fans, blower type for ranges
Fans, electric for kitchen and mess hall
Flour sifter
Food storage containers
Freezer, deep freeze
Freezer, walk-in type
French fry cutter
French fryer
Fryer, electric
Garbage cans, 10, 20, and 36 gal
Grater, 4-sided, $9 \times 4 \times 4$ in
Griddle, automatic
Hand truck
Hot food tables
Ice cream freezer
Ice machine with storage bin and spare parts
Ice pick
Juice extractor
Knives, butcher, 12 and 14 in
Knives, French forged bolster, 10, 12, and 14 in
Lamb splitter, 12 in
Meat and bone saw
Meat slicer
Mixing spoons
Mixers, 5 and 30 qt
Muffin pan, 12 cup, $14 \times 10 \times 1\frac{1}{2}$ in
Oven, bake
Pans, baking
Pans, long-handled
Pans, loop-handled
Pastry tube
Pie server
Pot racks
Potato peeler
Pots, brazier, heavy aluminum, with 2 loop handles
Pots, sauce, with cover, various sizes
Proof box, 16-gauge galvanized, on casters
Ranges, fuel to suit available supply
Refrigerators, walk-in
Refrigerators, electric or gas
Refrigerator truck
Roast beef slicer
Scale, kitchen
Scale, bakers'
Scoops, sugar and flour
Sinks
Spatulas, bakers' 10 and 12 in
Steak cuber
Steak hammer
Steel, butchers'
Stove, electric for coffee maker
Tables, cooks' work
Tables, bakers'
Tenderizer, meat, with cleaning and sharpening tools
Toasters, electric
Tongs, stainless steel utility
Trays, serving 14×18 in plastic
Waste baskets

commissary supplies and equipment are both economical and practical. However, when the totals to be accounted for are already on the sheets, someone who is not under the camp manager's control should occasionally observe the count. Inventories of food and supplies are best controlled by perpetual inventories kept on visible-record cards (Figure 14-4). Fresh meat and vegetables in the freezer may be inventoried, but unless freezer facilities are available, these items must be bought from day to day and are best charged to cost immediately upon purchase. Withdrawals should be made on a storeroom requisition (Figure 14-5).

Anything charged out of stores is taken into cost at once. If physical counts are taken frequently, adjustments will usually be small, so

SCHEDULE 14-3 CHECKLIST OF KITCHEN AND MESS HALL SUPPLIES

Aprons, cooks' and waiters'
Ash trays
Bowls, 5 and 6 in
Bowls, grapefruit
Bowls, gravy
Bowls, serving
Brooms
Brushes, bottle
Brushes, cleaning and scrub
Caps, cooks' and waiters'
Cruets, oil and vinegar
Cups
Dishtowels
Dustpans
Foil, aluminum, 18 and 24 in
Griddle bricks
Ice cream dishes
Ice cream scoops
Ice cream spade
Mats, door
Mats, hot
Mops, heads and handles
Mustard pots with covers
Napkins, paper, with dispensers
Platters, china, 15½ × 10, 13¾ × 8, 12 × 6, and 11 × 5½ in
Plates, china, 10 and 8 in
Pot lifters
Plumbers' helpers, rubber suction cups
Salt and pepper shakers
Saucers
Scouring pads
Silverware, knives, forks, spoons, soup spoons, serving spoons, steak knives, etc.
Serving dishes
Soap, bar
Soap, powder
Soup plates
Sponges
Spoons and ladles, cooking
Sugar dispensers
Syrup dispensers
Trays, celery
Uniforms, cooks' and waiters'
Vegetable bowls
Water glasses
Wooden bowls

that there will be little or no distortion on the days when they are taken up in the accounts. Some camp, mess hall, and commissary reports make a daily allowance for loss and spoilage so that unless the loss exceeds the reserve thus created, the distortion caused by adjustments is eliminated.

Purchases. If purchases during the course of the job are made on purchase orders through the job office, it is usually necessary to let the disbursements follow the procedures described in Chapter 4. However, because of the large number of day-to-day purchases of milk, fresh vegetables, and fresh meat, many contractors give the camp manager a revolving fund and let him buy directly from the suppliers. When this procedure is followed, there is little to be gained

SCHEDULE 14-4 CHECKLIST OF CAMP EQUIPMENT

Beds, single, steel
Bedsprings, single, steel
Chairs
Coffee tables (for recreation hall)
Lamps, floor and table
Mattresses, innerspring
Motion picture equipment (recreation hall)
Side tables
Upholstered chairs and sofas (recreation hall)
Wardrobes
Washing machine and laundry equipment
Writing desks

SCHEDULE 14-5 CHECKLIST OF CAMP SUPPLIES

Bathmats	Mattress covers
Bathtowels and hand towels	Mops
Bedpads, single	Padlocks and extra keys
Bedspreads, single	Pails
Blankets, single, woolen and cotton	Pillows and pillowcases
Bleach (Clorox or equivalent)	Rat poison
Brooms	Saniflush
Clothesline and clothespins	Sheets, contour and plain
Disinfectant (Lysol or equivalent)	Soap, hand and bath
Dust mops and supplies	Soap, liquid
Dust pans	Soap, laundry, bar and powder
Floor brushes	Toilet paper
Insect spray	Washcloths
Lye	

from using purchase orders. The control comes from delivery tickets attached to the daily report. When the bill is paid the number of the check issued to pay it is shown on the delivery ticket itself. If these day-to-day purchases are accumulated and paid monthly, then a better procedure is to issue a blanket purchase order to the suppliers for a month's supply, and to let the payments go through the regular routine for voucher disbursements.

Income

Income from camp, mess hall, and commissary operations comes from four sources. First, and probably the largest, is from payroll deductions for meal tickets, camp accommodations, and sometimes commissary chits. Second is from over-the-counter sales or collections. Third is from meals and accommodations provided to subcontractors and others on the job. Fourth is from sales of surplus inventory.

Income from payroll deductions is controlled by the payroll department, and is initiated by various forms (Figures 3-9, 3-10, and 7-2). When a meal ticket is sold, the camp manager writes a charge slip which goes to the payroll department to support the charge. This form (Figure 14-6) is designed for a counter billing machine.

Over-the-counter sales may be handled either by a counter biller alone, a counter biller with an attached cash drawer, or by a cash register. Of course, they can be handled through an ordinary cash drawer or out of the manager's pocket too, but these last two methods lack some of the more desirable control features found in the others. The counter biller form (Figure 14-6) is designed to handle a variety of transactions, including income from the third and fourth sources mentioned above.

INVENTORY OF CAMP SUPPLIES AND EQUIPMENT

Sheet No. ....

Date taken.............................By............................Mgr...............................

Items	Storeroom location	Total to account for	Location				Explanation
			Storeroom	Bunkhouse	Laundry and repair	Short	
Bedding Blankets, single, cotton	Sec. 1	100	19	40	40	1	1 burned by cigarette
" double "	"	50	10	20	20		
" single, wool	"	40	18	20	2		
Mattress covers	"	40	18	20	2		

Figure 14-3 Camp Supplies and Equipment Inventory.

Stock Card					CAMP, MESS HALL AND COMMISSARY							Foodstuffs
	ON ORDER			RECEIPTS			ISSUES			BALANCE		
Date	Req. No.	P. O. No.	Quan-tity	P. O. No.	Quan-tity	Invoice cost	Issue ticket	Quan-tity	Charge-out price	On hand	Physical count	Date counted

Item	Unit	Stock No.	Loca-tion	Reorder at

Figure 14-4 Inventory Stock Card. This card is designed for use in a visible-index perpetual inventory record. Note that no effort is made to extend a value for the items on hand after each transaction. Note, too, that "charge-out price" can be so set that a margin for loss and spoilage can be included.

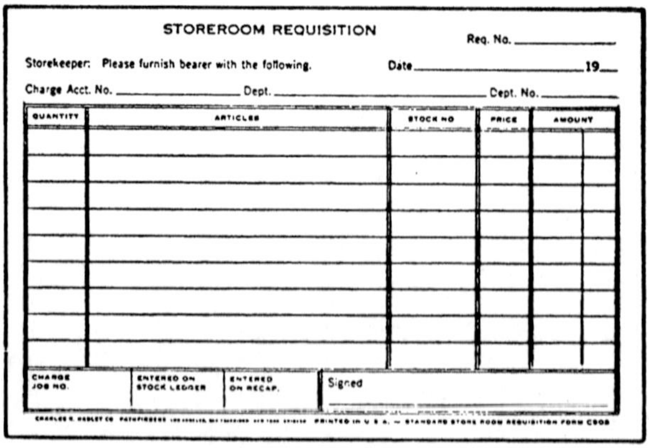

STOREROOM REQUISITION

Req. No. ________

Storekeeper: Please furnish bearer with the following. Date ________ 19__

Charge Acct. No. ________ Dept. ________ Dept. No. ________

QUANTITY	ARTICLES	STOCK NO	PRICE	AMOUNT

CHARGE JOB NO.	ENTERED ON STOCK LEDGER	ENTERED ON RECAP.	Signed

Figure 14-5 **Storeroom Requisition.**

If there is a large, high-activity mess hall, a cash register has many advantages, particularly if some of the meals are sold for cash. One of the most useful features of a cash register is the fact that it can be used to obtain an automatic meal count, as well as a segregation of meal tickets collected (or punched), cash taken in for meals, cash taken in for meal ticket sales, and other cash taken in. However, even though a register is used in the mess hall, it would not replace the counter biller in the camp manager's office.

A cash register (or a counter biller with a cash drawer attached) is also useful in any commissary where a substantial number of sales are for cash. Receipts from sales through automatic vending machines in the commissary are usually recorded as a single sale at the time the cash is removed from each machine. Commissary sales, unlike those in the camp and mess hall, should show a profit, which is controlled by the markup. Whenever the average gross profit on commissary sales falls too far or too sharply, there is a good chance that all the sales are not being recorded.

One of the problems facing any company that makes a policy of charging meals, lodging, and commissary purchases to employees' accounts is how to minimize the number of entries required. Flat daily or weekly charges for room or bunkhouse or other living quarters are usually satisfactory, but they breed dissatisfaction if applied to meals, and are entirely impractical when applied to commissary purchases. A form of meal ticket (Figure 14-7), a meal book (Figure 14-8), and a book of scrip (Figure 14-9) have been used satisfactorily.

This discussion has thus far been chiefly concerned with recording the data from the operations of camp, mess hall, and commissary. The actual control of these operations through accounting lies in

ECKS AND WYE

CONTRACT 51 MOUNTAIN CITY

CAMP AND MESS TRANSACTIONS

Name.................................Date.................

Badge No................Barracks..............................

Cash sale	Charge sale	Sub-contractor charge	Paid on account	Return sale	Memo

Quantity	Description	Price	Amount

No. S- 5024

Figure 14-6 Charge Ticket. This slip is used in the camp manager's counter billing machine.

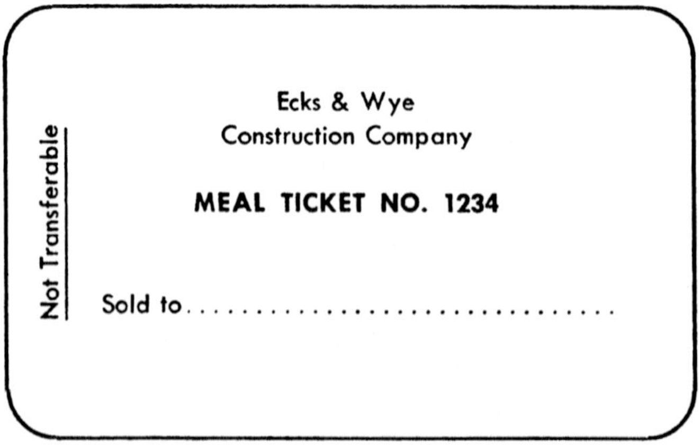
Not Transferable

Ecks & Wye
Construction Company

MEAL TICKET NO. 1234

Sold to. . . . .

Figure 14-7 **Meal Ticket.** This meal ticket is good for 21 meals (one week). After it is punched out, the ticket itself is turned in for the last meal. It is wallet-sized and printed on durable card stock.

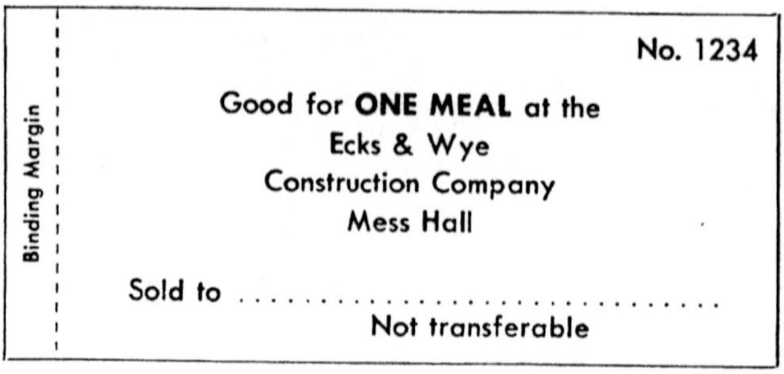
Binding Margin

No. 1234

Good for **ONE MEAL** at the
Ecks & Wye
Construction Company
Mess Hall

Sold to . . . .
Not transferable

Figure 14-8 **Meal Book.** This coupon is sold in books of 20. The cover is turned in for the last meal in the week.

prompt and accurate reporting. Schedule 10-4 showed a monthly report of camp and mess operations, and this is usually adequate for a report from a job which decentralized accounting to the head office. However, for jobsite control, a daily report (Figure 14-10) from the camp manager is needed.

Note that in commissary operations, stock replenishment is treated as the cost of sales. This is done on the principle that the commissary starts out with a basic stock, and that at the end of each day the person in charge of the commissary replenishes that stock from the camp manager's general stores. If this is done daily the cost can be kept quite accurately with a minimum of work. The plant (temporary structures) write-off and the depreciation are fixed amounts which remain constant in the absence of substantial changes in the plant

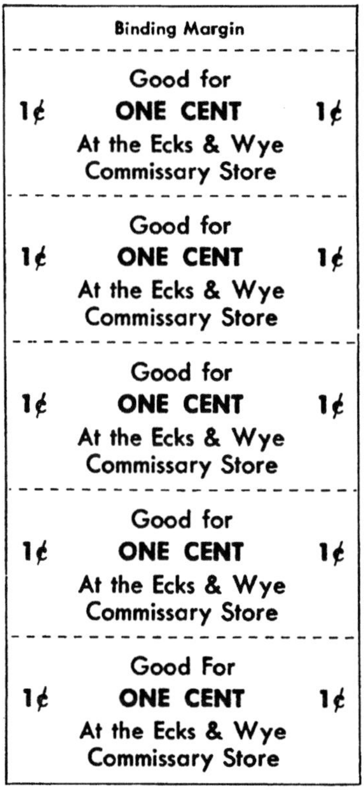

Binding Margin

Good for
1¢ ONE CENT 1¢
At the Ecks & Wye
Commissary Store

Good for
1¢ ONE CENT 1¢
At the Ecks & Wye
Commissary Store

Good for
1¢ ONE CENT 1¢
At the Ecks & Wye
Commissary Store

Good for
1¢ ONE CENT 1¢
At the Ecks & Wye
Commissary Store

Good For
1¢ ONE CENT 1¢
At the Ecks & Wye
Commissary Store

Figure 14-9 **Scrip.** Scrip for commissary purchases is often sold in books of $5 or $10.

and equipment accounts. The treatment of such items as linen, silver, and dishes varies, but the tendency is to treat them as supplies and to spread their cost over the expected life of the job.

One of the problems on any job operating a mess hall is how to account for "free" meals for visiting executives and guests. To those jobsite employees (if any) who are allowed free mess hall privileges, free meals are normally treated as compensation. The number of such meals would appear as the difference between meals served and meals paid for, and if charges were to be made at "cost," they would be made at the unit cost of meals served and the amount included in "other revenue." Various forms (Figures 14-11 to 14-13) are applicable.

Loss and spoilage are most easily accounted for by adding an appropriate percentage (usually from 1 to 5 percent) to food costs reported. However, some companies prefer to set the allowance out as a separate line in the daily report.

ECKS & WYE
Contract No. 51 Mountain City
CAMP MANAGER'S DAILY REPORT Date. . . . . .

Men in Camp		**Meals Served**	
Last night	. . . . . .	Breakfast	. . . . . .
Arrivals	. . . . . .	Lunches (mess hall)	. . . . . .
*Departures	. . . . . .	Lunches (put up)	. . . . . .
Tonight	. . . . . .	Dinners	. . . . . .
* Clearance slips attached		Total meals	. . . . . .

CASH RECEIPTS AND DISBURSEMENTS

Change fund on hand this morning		$. . . . .
Cash received (receipts attached)		. . . .
Total to be accounted for		$. . . . .
Less:		
Cash disbursed — supporting documents attached	$.	
Tomorrow's change fund	. . . .	
Cash (over) or short	. . . .	. . . .
Cash accompanying this report		$. . . . .

Net sales on account — sales slips attached $. . . . .

Purchases on account — delivery memoranda attached $. . . . .

Checks written:

Payee	Purpose	Amount
. . . .	. . . .	$. . . . .
. . . .	. . . .	. . . .
. . . .	. . . .	. . . .
. . . .	. . . .	. . . .

Figure 14-10 **Camp Manager's Daily Report.**

Nº 0160

NON-RESIDENT
MEAL TICKET
Good For 1 Meal

Figure 14-11 **Nonresident Meal Ticket.**

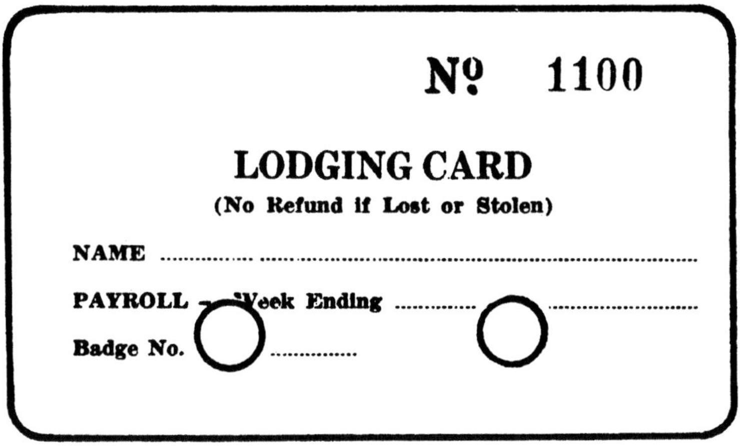
Nº 1100

LODGING CARD

(No Refund if Lost or Stolen)

NAME ....

PAYROLL — Week Ending ....

Badge No. ....

Figure 14-12 Lodging Card.

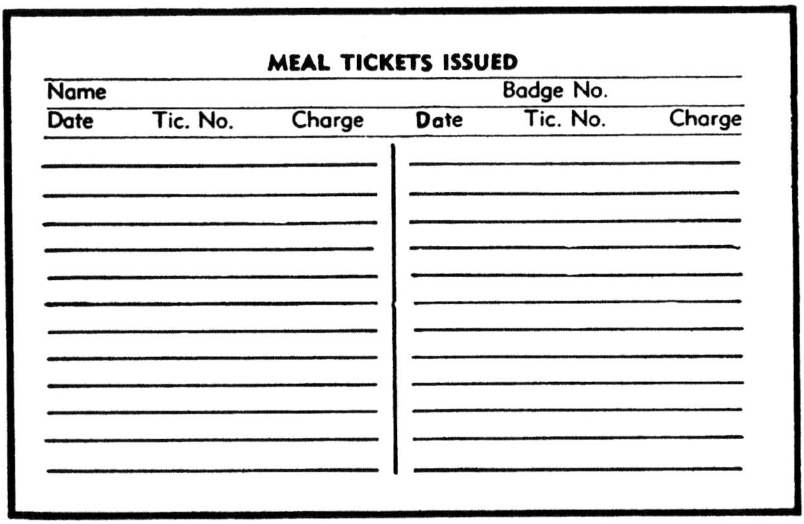
MEAL TICKETS ISSUED

Name				Badge No.	
Date	Tic. No.	Charge	Date	Tic. No.	Charge

Figure 14-13 Summary of Meal Tickets Issued.

Policies

It is rarely good policy to make a profit on a camp or mess hall. The only reason for having such facilities is that the job is remote and local accommodations are inadequate. The very remoteness that creates the problem of feeding and housing employees also creates the problem of hiring them in the first place. One way to get and keep capable people on remote jobs is to have good camp and mess hall facilities at minimum prices. One fairly common rule of thumb is that the company should equip and stock the camp, mess hall, and commissary, and provide the temporary buildings. Thereafter, these activities should charge enough to pay their own way.

Subcontractors often wish to use the camp and mess hall facilities of the prime contractor, and a question arises as to how much they should be charged for the service. Certainly the prime contractor should not assume additional out-of-pocket losses to provide such service, but it is easy to price such facilities so high that their use by the subcontractor is not practicable. Therefore, it is a good idea for a prime contractor to take into account any savings the additional volume will make possible, and to ask the outside users in addition to the direct charges to pay only for their proportionate share of fixed charges.

The net losses which normally result from operating a camp and mess hall will usually be offset, to some extent, by gains from the commissary and (when one is operated) the trailer camp. After offsetting any gains from such related sources, the net loss becomes a part of the job overhead and should be so distributed. If, as provided in some of the account classifications in Chapter 9, job overhead is distributed to the job cost accounts, the rate applied should include these losses.

Some companies try to hold down the losses on camp and mess hall operations by subcontracting them, and some even try to profit by selling the concession. Any apparent saving or gain is usually very expensive. A subcontractor or concessionaire would be forced to hold down quality, variety, and service so that he could operate at a profit, and so would tend to work against the best interests of the contractor. The result is that very few concession or subcontract arrangements for camp management operate satisfactorily from the contractor's viewpoint; they should be considered only after very careful analysis.

Sometimes an unscrupulous concessionaire will start bootlegging liquor, prostitutes, and gambling in the camp. Moral aspects aside, such practices can very quickly cut into the working efficiency of the employees and thus into the contractor's profits. Moreover, once started, they are difficult to stop, even with the help of the local authorities. If any concessionaires or subcontractors are used in the camp, mess hall, or commissary, their contracts should provide for cancellation on their participation in any illegal or improper activities.

Some companies control the normal loss on camp, mess hall, and commissary by turning over the whole operation, accounting included, to the camp manager. Their approach is to start the camp manager with a basic inventory, a revolving fund, and a relatively small salary. Then they let him know that he is expected to complete the job with what he has plus the amounts collected from the workers, and leave the rest strictly to him. The theory is that with the temporary structures, the basic inventory, and the revolving fund representing the amount the company expects to lose on the operation of its camp, mess hall, and commissary, a good camp manager should be able to uphold the quality of meals and accommodations, and still be able to supplement his salary by "commissions" on buying, sales of "surplus" items, and similar methods. However, if he allows any substantial stealing or inefficiency, the margin within which he can operate is reduced, or the quality of the food and service goes down, and the job management can take steps to remedy the situation. In other words, he is still an employee and, as such, is easier to control than a subcontractor or a concessionaire.

Admittedly such a philosophy is hard-boiled. On the other hand, it generally gets better results than concessions or subcontracting the camp, mess hall, and commissary, and it does have the sanction of wide acceptance in the industry. Of course, if the workers pay nothing for their rooms or meals, this system will not work.

A contractor whose operations require him to provide a mess hall will inevitably be faced with the proposal of raising his own beef and pork. The arrangement is risky. If trouble arises in the mess hall, it is usually better to have served only government inspected meat.

Most of the larger and better-managed construction companies are finding it more profitable, if camp and mess facilities must be provided, to run them as a part of the job, and to control their operations through physical inspection and accounting.

First aid, guards, and fire protection

On most large construction jobs, first aid, guards, and the fire department are under the supervision of the job office manager or the project controller. Their cost is a part of overhead and should be distributed as such. Purchases for these activities should originate with requisitions sent to the person responsible for job purchasing, and the actual purchases should be made through the usual channels.

In controlling first aid, the person responsible needs to be aware of two possible trouble spots. First, in contracting to furnish first aid service to subcontractors and others it is possible to place the company in a position of agreeing to practice medicine, and it is possible to subject it to liability for alleged negligence on the part of its first aid personnel in handling accidents. Any such agreements should be cleared through the company's attorney. Second, job dispensaries have been known to be distributing points for narcotics on the job. Foremen should report all requests for time to visit the dispensary and repeaters should be investigated. So, also, should all dispensary purchases and deliveries.

Guards tend to become careless, and there is a possibility of collusive deals if they are in the same place too long. Therefore, guards' duties should be rotated at frequent intervals on an irregular schedule. Attempts to unionize guards and fire fighters should usually be vigorously resisted. First, a strike would leave the company property unprotected, and might well place the company in violation of contractual duties under its contract with the owner, and also with the insurance carriers. Second, a union guard may find himself faced with the job of physically resisting attempts by fellow union members to invade company property.

Occupational Safety and Health Act

No discussion of safety would be complete without a discussion of OSHA (Williams-Steiger Occupational Safety and Health Act of 1970). Essentially, safety is a management function and in larger construction firms administration of OSHA would be the job of the safety engineer. Nevertheless, OSHA requires extensive and detailed record keeping and reporting. Where there is no safety engineer this record keeping and reporting fall on the accounting personnel.

In states which have amended their industrial safety laws to meet the requirements of OSHA and to assume its administration, the law is administered by the state. In any state which has not assumed responsibility, administration is by the U.S. Department of Labor.

Record-keeping requirements are obtained from any office of the Bureau of Labor Statistics under the predictable title *Record Keeping Requirements Under the Williams-Steiger Occupational Safety and Health Act of 1970.*

Essentially, the accounting records consist of the following:

1. A *Log of Occupational Injuries and Illnesses* kept on an official form known as OSHA No. 100. This form has twelve columns and there are detailed instructions as to what information goes in each column.
2. A *Supplemental Record of Industrial Injuries and Illnesses* kept on an official form known as OSHA No. 101. This is an 8½ by 11 inch sheet containing nineteen questions. A separate supplemental record must be prepared for each injury or illness.
3. A *Summary (of) Occupational Injuries and Illnesses* kept on an official form known as OSHA No. 102. This is an 8½ by 11 inch sheet with eight columns summarizing the individual (supplemental record) sickness and injury reports on an annual basis.
4. An official notice which must be posted "in a prominent place in each establishment to which your employees normally report for

work." For construction companies this would probably mean both the head office and each jobsite.

Even where there is a safety engineer, a construction company controller or office manager would do well to be familiar with OSHA and its requirements.

Chapter

15

Contract Claims and Litigation

Contract disputes place big demands on the time and financial resources of contractors. Although contract claims and litigation have become a constant aggravation for many companies, surprisingly little has been published to explain the claims process and, in particular, the quantification of damages.

This chapter discusses such matters as the contract claims environment, types of contract claims, organizing a successful claims team, government contracts, techniques for pricing damages, and other important topics.

Obviously, there is much more to say about contract claims than space allows here, but this chapter will give interested readers some valuable background information.

Basic to any claims procedure is a routine for identifying potential claims and notification of the owner or his representatives of a factual situation that has arisen and the fact that it might be the basis for a future claim. One of the best sources for such factual information is the foreman's or superintendent's daily report. Another good source is a set of detailed notes or minutes of job meetings. Another useful source is written memorandums originating on the job at the time some oral or written order is given by the owner or his representative. The communication need not be formal and it often is a handwritten memo, such as the field work order illustrated in Figure 2-7. In any event, a chronological file of such memorandums constitutes the essential raw material for most successful claims.

Also, an essential preliminary to a successful claim is early legal advice. It is of primary importance, when considering the filing of a claim, to prepare the claim with the assistance of an experienced claims attorney. An ineptly prepared claim may well damage or destroy a useful legal position that might not be immediately apparent to the person preparing the claim.

The Contract Claims Environment

For many construction companies, long-term contracts have become increasingly complex. As work progresses, contracts become susceptible to disputes, and often the result is litigation. In addition, since large sums of money are frequently involved in these disputes, effective resolution is critically important.

Virtually every contract has change orders and requests for additional compensation. This is a routine process, and in most cases changes in the contract price are made without dispute. But when requests for additional compensation are rejected, the next step is usually a formal claim. If the amount of money in question is significant, legal counsel and other consultants may be solicited to ensure proper preparation, documentation, and presentation of the claim. If this process fails, the dispute generally moves into the litigation phase, which is costly to all parties.

Obviously, it is best to resolve a claim without resorting to litigation. Inevitably, however, many disputes do wind up in court. When this occurs, the attorney and client need help not only in developing the best documentation and strategy but also in presenting expert testimony and evaluating the soundness of testimony submitted by others.

Quantifying damages depends in part on the type of dispute. The following section reviews the most common categories of contract claims.

Types of Contract Claims

Claims arise for various reasons. Usually, they involve such matters as contract scope, timing of contract effort, and other actions that have an impact on contract performance and cost. Claims may involve many diverse parties—contractors, subcontractors, customers, owners, sureties, banks, suppliers, and others.

Claims can generally be grouped into five broad categories:

- Delay
- Disruption
- Changed conditions
- Changes in scope
- Termination

Delay. Claims for delay occur because the contractor has not kept to a predetermined schedule. Delays can be caused by problems such as late drawings, defective materials, lack of access, and lack of right-of-way. Delays create additional time-related costs such as contract management, equipment, and facilities. Delay may also affect labor productivity (disruption), cost of materials, and equipment. A delay claim may also give rise to a claim for acceleration if the contractor has added personnel or additional shifts to mitigate the damages caused by the delay.

Disruption. Claims for disruption arise because of the contractor's inability to perform the work in a manner planned. For example, the contractor may have to stop work frequently because of change orders or because of other contractors. This type of claim is based on low productivity, resulting in higher labor costs. These claims are often linked with delay claims or arise from situations in which acceleration occurs when the owner insists upon meeting the original schedule.

Changed conditions. These claims generally occur when data furnished by the owner to the contractor for use in preparing a bid differ from the conditions that the contractor actually encountered. For example, during excavation, the bedrock actually encountered may be different from that indicated by the test borings used in bidding the job. Another example is a situation in which change orders are so numerous that the final product bears little resemblance to what was planned.

Changes in scope. When there are increased costs due to changes in scope, this type of claim often occurs. These changes are made after the contract has been signed and generally after work has begun. For example, after construction has begun, the owners may find that regulations require installation of ramps for the handicapped.

The argument generally centers on whether the alleged change is part of the original contract scope, and it frequently involves the cost of the change rather than the fact that the change occurred.

Termination. These claims result when the contract is terminated before the work is completed. Termination may occur for several reasons: the project being built is no longer needed or is technologically obsolete, the customer does not have sufficient funds, or environmentalists and other intervenors have successfully stopped the project.

Other types of claims. Other types of claims include those made by the United States government for defective pricing, failure to comply with cost accounting standards, and recovery of costs considered unallowable for government contracts. These claims cover various issues concerning cost accounting practices.

While not all-inclusive, Schedule 15-1 provides additional information on events that cause various types of claims.

SCHEDULE 15-1 EVENTS THAT CAUSE CLAIMS

	Type of claim caused				
Event	*Delay*	*Disruption*	*Changed conditions*	*Scope changes*	*Termination*
Late drawings	X	X			
Changes in sequence	X	X	X		
Lack of decision	X	X			
Acceleration		X	X	X	
Deceleration	X	X			
Lack of access	X	X			
Lack of right-of-way	X	X	X		
Delays	X	X			
Interference	X	X			
Errors	X	X			
Differing site conditions	X	X	X		
Suspensions	X	X			X
Bankruptcy					X
Lack of funds					X
Impossibility			X	X	X
Weather	X	X			
Strikes	X	X			X
Lack of approvals	X	X		X	
Erroneous bid specs	X	X	X	X	

Organizing the Litigation Team

How a contractor approaches a contract dispute will depend on many factors, such as the nature of the claim, the ability of the contractor to prepare the claim, the proof required for entitlement, the size of the claim in relation to projected preparation and litigation costs, and the strategies of the defendant. In all cases, experience and teamwork are key factors and can make the difference between winning and losing.

The organization of the team should be considered early in the litigation process. Although many team members may not be needed until later, getting the case under way will be difficult unless the administrative and planning groundwork has already been laid. Many long-term contract disputes require a formal master plan. In relatively simple contract litigation, it may be possible to do without a plan, but in highly complex cases playing it by ear invites disaster.

In large, complex disputes, the team is composed of outside experts and key company personnel. The team is managed by a litigation team leader, who usually is, and should be, the legal counsel. Schedule 15-2 illustrates the team structure for a large contract dispute involving both entitlement and damages considerations. Depending on the nature of the dispute, the contractor, at the legal counsel's suggestion, may hire one or more of these outside experts.

SCHEDULE 15-2 TEAM STRUCTURE

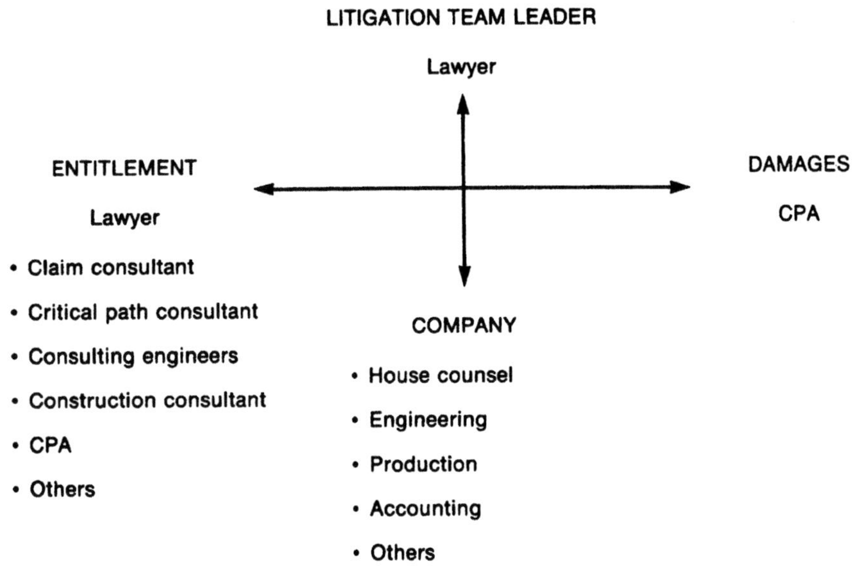

The team leader. In complex construction litigation, the lawyer who serves as the team leader should be a highly skilled manager and administrator and must have considerable legal expertise as well as business expertise. A good team leader keeps costs under control and coordinates the work of the team members. The team leader also serves as the final decision maker in resolving disagreements among team members.

The certified public accountant. The primary role of the certified public accountant (CPA) in a contract dispute is to serve as a consultant and expert witness. The CPA computes and presents the "damages" aspect of the claim. To do this effectively, the CPA should have documented experience in accounting for long-term contracts and significant experience in working with the construction industry. It is important, from a credibility point of view, for expert witnesses to be able to speak the language of the industry. In addition, previous experience in contract claims can make the CPA invaluable to the team leader in helping to develop sound strategies. In particular, he can help by interpreting the relevant accounting and cost accounting principles and practices and certain regulatory requirements that may have an impact on the success of the claim.

Typically, the CPA will:

- Assist the attorney in identifying key issues, information needs, and claim or litigation strategies.
- Advise on appropriate use of automated data retrieval systems.
- Calculate damages and present alternative approaches for the presentation of damages.
- Provide assistance to attorneys in preparing briefs, interrogatories, depositions, and rebuttals of opposing attorneys' arguments.
- Present trial testimony, which usually includes the use of visual aids and other materials in support of the plaintiff's case.

The claim consultant. This team member develops the basic claim documents, including a narrative providing an overview of the project, the parties involved, and the activities that occurred during the project. These documents highlight the problems that were encountered. For example, they may describe how the contractor was delayed numerous times because the owner asked him to stop until he could decide on a change order. Each problem is discussed in detail (e.g., the date of and the names of the people involved in each delay).

The critical path consultant. This expert is most useful in delay and disruption claims. The critical path consultant develops two graphic presentations (networks) of the project: the "as-planned network" (how the project was planned) and the "as-built network" (how the project was actually performed). He then compares the two networks and isolates the delays, disruptions, and other events that caused the deviation from the planned sequence and timing of events. He also assists in preparing the documentation needed to fix the blame for the indicated events.

The consulting engineer. The consulting engineer analyzes the owner's original plans on which the contractor prepared his bid. Then he testifies as to whether the plans would enable the contractor to do what he said he would do, or whether the materials furnished by the customer were defective (e.g., the steel girders were the wrong length, they had hairline fractures, or the holes didn't line up properly).

The construction consultant. In construction claims for delay and disruption, the litigation team often includes a construction consultant. This expert witness would testify that the actual work was performed in such a way that construction had to be disrupted or delayed because of other activities. The construction consultant is often a well-known retired construction executive with extensive experience in supervising the type of project being litigated.

All the expert witnesses must work closely with their counterparts and with people in the contractor's organization. These people are most familiar with the facts and can generate much of the basic data. For example, the critical path consultant must interview the estimators who planned the job to develop the "as-planned network." Then he must interview the project supervisors to determine actual progress so that he can develop the "as-built network." Similarly, the CPA must work closely with the company's accounting personnel in developing the necessary accounting information.

Again, it cannot be stressed strongly enough that the team, particularly the lawyer serving as team leader, is often the key to success. In addition to legal, business, and management skills, the team leader should have a successful track record and personal knowledge of the people who will make the best team members to win the dispute. The legal counsel must also be skilled in the coordination and management of the outside experts to ensure a fit with the legal theories he has decided to follow.

Government Contract Claims

Most disputes take the forms discussed previously. But if a government contract is in dispute, it is subject to many special rules, regulations, and public laws that further complicate both the entitlement and damages aspects of the case. Some of these laws and regulations include:

- Cost Accounting Standards
- Contracts Disputes Act
- Truth in Negotiations Act
- Extraordinary Contractual Action
- Federal Procurement Regulations
- Federal Acquisition Regulations
- Defense Acquisition Regulations

For many government contract claim situations, a knowledgeable CPA can be involved in entitlement issues concerning these rules and regulations. He provides opinions and expert testimony on accounting-related government contract issues, and the work he performs on the damages aspect of the case invariably produces financial information bearing on entitlement.

Pricing the Damages

There are two overall approaches to pricing damages in a contract dispute. These are usually referred to as the total cost approach and the discrete approach.

The total cost approach. This is the simplest approach. In effect, it converts a fixed-price contract into a cost-plus contract. It begins with a determination of the contractor's excess costs ("out-of-pocket" costs), plus overhead and profit, plus interest on the claim amount. It assumes that the events proved on the entitlement side can be shown to have resulted in excess costs or a loss. This approach is not allowed in many jurisdictions and in certain government contracts. But it may be allowed if the plaintiff can show that because of the nature of the entitlement no other approach is feasible. For example, a changed conditions claim may require the total cost approach.

Schedule 15-3 presents a typical claim summary using the total cost approach.

To use the total cost approach successfully, a "four-way" test must be met:

- Total costs incurred on the contract are correct. Costs charged to the contract are in accordance with generally accepted accounting principles or other applicable requirements, and where allocations

SCHEDULE 15-3 TOTAL COST APPROACH

Claim Summary

Contract costs incurred		
Direct costs		
Labor	$200,000	
Materials	100,000	
Subcontracts	900,000	
Equipment	200,000	
Other	100,000	
		$1,500,000
Indirect costs		160,000
Total cost		1,660,000
Total revenue (contract price and change orders)		1,440,000
Cost overrun		220,000
Add:		
Unrecovered home office general and administrative costs (10% of total cost)		166,000
Profit (15% of total cost plus unrecovered overhead)		273,900
Cost of capital		40,000
Total claim		699,900
Claim preparation costs		100,000
Interest from date of notice through (current date of claim)		30,000
Total due at (date of claim)		$ 829,900

of costs are required, the method and base used for such allocations are appropriate.

- The contractor's price was reasonable and the bid was properly prepared.
- The total contract overrun is the result of the other party's acts, and costs leading to portions of or all of the overrun are not due to the contractor's own inefficiencies or mismanagement.
- The nature of the dispute is such that it is impossible or impractical to use any other approach.

In summary, the total cost approach is not favored by most attorneys and/or courts. It requires the plaintiff to be fault-free, and that is rarely the case. It also requires that the contractor not be asserting the impossibility or impracticability of the discrete approach because of a failure to keep the necessary books and records.

The discrete approach. Although this approach is more difficult and more complex than the total cost approach, it is usually the best approach. Each of the events that caused the loss or excess costs is priced separately (discretely). For example, the increased cost of materials as the result of a delay is presented separately from other discrete events. Schedule 15-4 shows a typical discrete claim summary.

Cost schedules. One of the first things that many attorneys have the contractor prepare is a series of schedules that show the final results of the contract under dispute. This provides a better understanding of the situation and serves as a starting point for determining the damages.

It is usually best to prepare this package using the total cost approach even though ultimately the team leader will decide upon a discrete approach to the claim. It will give the attorney an idea of how much recovery is needed to make his "client whole" and give him a fair profit. These schedules also serve as the starting point for the damages witness and are usually susceptible to an audit.

A sample of the financial data that should be prepared is set forth in Schedules 15-4 through 15-15.

The contract audit. Once the preliminary schedules have been prepared, it is usually best to have the damage specialist (CPA) conduct an audit of the contract costs, revenues, resulting cost overrun, and home office overhead. This is one of the key aspects of the CPA's role on the litigation team, particularly if he is likely to give testimony during the trial. A contract audit provides significant benefits in the litigation effort. For example, the audit can:

SCHEDULE 15-4 DISCRETE APPROACH

Claim Summary

Delay	
Extended field support due to late drawings (10 weeks at $24,000 per week)	$240,000
Escalation of costs due to delay (material)	16,000
Disruption	
Decline of 12% in labor productivity due to late drawings	166,000
Unpaid change orders	
Extended sewer excavation	26,000
Acceleration (premium on overtime to attempt to meet original schedule)	16,000
Cost of capital (negative cash flow due to delay, disruption, and unpaid change orders)	8,000
	472,000
Unrecovered home office general and administrative cost at 10% of costs excluding cost of capital	46,400
	518,400
Profit at 15% excluding cost of capital	51,040
Total claim	569,440
Interest from date of notice through (current date)	48,000
Total due at (date of claim)	$617,440

- Identify any "embarrassing" costs and "contractor mistakes" that make the claim overstated or indicate failure to mitigate damages.
- Develop an important database for the costs of labor, material, equipment, miscellaneous, and overhead. This information is necessary to price damages using the discrete approach. Such a documented database also allows for last-minute changes in approaches to quantifying or presenting damages.
- Provide the CPA with knowledge and understanding of the contractor's cost system, history of the contract, and related disputes so that he can provide valuable advice to the attorneys before interrogatories and depositions. He can also assist by suggesting effective approaches to presenting the damages.
- Develop information that has an important impact on the entitlement aspects of the case. This information often complements the work of the critical path consultant and other entitlement consultants.
- Provide a cost-effective way of obtaining information. Actually, the audit usually pays for itself through the benefits cited above.

SCHEDULE 15-5 DAMAGES CHECKLIST

	Additional Cost	
	Yes	*No*
Delay		
Increase in time-related costs		
Escalation in material costs		
Escalation in labor costs		
Additional equipment time		
Effect on other contracts		
Rental versus owned equipment		
Ripple effect		
Increased bond premium		
Increased workmen's compensation premiums		
Other insurance costs		
Subcontractor claims		
Lower labor productivity		
Lost alternative income		
Disruption		
Lower productivity/increased labor cost		
Costs of materials increased by mistakes		
Additional move-in/move-out		
Delay		
Higher workmen's compensation premiums		
Higher performance bond premiums		
Changed conditions		
Additional labor head count		
Change in mix of craft laborers		

SCHEDULE 15-5 DAMAGES CHECKLIST *(Cont.)*

	Additional Cost	
	Yes	*No*
Additional/different equipment		
Additional materials		
Extended schedule		
Additional supervision		
Change orders		
Labor		
Materials		
Equipment		
Subcontracts		
Schedule delay		
Ripple effect		
Other		
Termination		
Demobilization		
Labor		
Equipment		
Materials		
Subcontracts		
Loss on forced sale of equipment		
Supervision		
Jobsite facilities		
Lost alternative income		

There obviously are additional factors that can have an impact on damage calculations. Those listed here will be helpful in providing general guidelines.

- Help to meet the four-way test if the total cost approach is to be used.

During the audit phase, the legal counsel will have decided on the overall approach to presenting and proving the damages. Under either the total cost or discrete approach, the audit will be extremely useful in proving the damages. For the discrete approach, the audit can be used to set out key cost elements being used to price specific events. For example:

- Labor rates being used to price disruption can be reconciled to audited figures.
- Time-related delay costs can be reconciled to the audit for use in pricing delay.
- Material costs that are higher than planned can be shown to have been incurred and matched with planned material costs to identify the resulting overrun.

As mentioned previously, the discrete approach to pricing damages is the most effective way to price a claim and is the most favored method by the course and legal counsel.

Presenting the Damages

Once the audit is completed, the team leader will have decided which approach would be the best one to use to present damages—the total cost approach or the discrete approach. Other than questions of overhead recovery, profit, and other costs outside of the direct cost of the contract overrun, little additional discussion is required, unless the team decides to use the discrete approach. In that case, further discussion would be needed because of the many different ways of pricing the discrete events in the damage claim. The following discussion covers approaches to pricing the major categories of discrete events in a typical claim situation.

Delay. During the audit phase, the CPA should identify all time-related costs. These costs generally include such things as project management salaries, jobsite office, guards, accountants, timekeepers, and clerks, and other costs that are incurred or paid for on a fixed-charge basis for a specific period of time. Once these costs have been identified (they are usually scheduled on an actual per-month basis during the life of the job), they can be applied to the period of delay to determine the dollar impact.

Disruption. As previously mentioned, disruption is the decline in labor productivity resulting from acts of the customer or owner. Entitlement for disruption can be proven by several techniques. But it is one of the most difficult aspects of the discrete approach to a claim. Usually, past productivity for similar work in a nondisrupted environment is the best way to prove a decline in labor productivity on the contract in dispute. Some contractors have detailed productivity figures for various portions of work and for the geographic area in which the work is being performed. These figures are usually based on past empirical data, and in most cases it is only necessary to prove that certain productivity figures are achievable on nondisrupted work.

When past empirical data or productivity figures are not available, the problem of proving the decline in labor productivity becomes more complex. In some situations where repetitive work is being performed, the learning curve technique can be applied to show the productivity that should have been achieved.

Another approach is to apply studies by trade and business associations, such as the National Association of Electrical Contractors, the Business Roundtable, and the Corps of Engineers.

At any rate, once the decline in productivity as a percentage can be proven, it is the CPA's task to apply the dollars to the claim. During the audit phase, the CPA should schedule out all actual labor rates for every type of craft labor during the life of the contract. In summarizing the results of the audit during trial, supplemental schedules reconcilable to the audit should be presented so that when the claim itself is presented, these labor rates can be shown to have been audited and applied properly to the disruption portion of the claim.

Changed conditions. Pricing for changed conditions usually involves presenting the actual cost, less the planned cost, for the phase or phases where the change occurred. This is similar to the total cost approach, except that it may be restricted to a portion of the contract rather than the total contract. Close coordination by the CPA, critical path consultants, and claim consultants is required to prove that the actual costs and planned costs are valid amounts.

Changes in scope. Two types of cost are presented in pricing changes in scope: the actual direct cost of the change (labor, materials, equipment, subcontracts, etc.) and the cost of the impact on the schedule of the balance of the work (ripple effect). The additional overhead relating to these change orders is often presented along with the change order portion of the claim. But it is usually best to present overhead as a separate item after the direct cost of all the individual discrete events has been determined (see Schedule 15-4).

Termination. Pricing in this category involves identifying and presenting actual costs not compensated by progress payments through the date of termination. These costs include such items as demobilization, plant closing, extra equipment time, and supervision.

Claim preparation costs. Upon initiation of the claim process, the contractor should set up cost accounts to separate and keep track of all costs related to preparation of the claim, litigation, and outside witnesses. Costs for the contractor's own employees should also be segregated during this process so that they can be presented as part of the damages.

Interest and cost of capital. The cost of capital caused by negative cash flow during the progress of the work, as well as interest applied to the total claim from date of submission, should be presented. The company's borrowing rate with its commercial bank should be the rate for computing the interest on the negative cash flow, and credit should be given for periods when the contractor is in a positive cash flow using the company's normal investment rate during the period (Treasury bill rate, certificates of deposit, etc.).

Overhead

After presenting the direct costs of the discrete events, the additional overhead incurred by the contractor should be presented. This is one of the more complex and difficult parts of the claim, both to prove and to price, and it usually involves significant amounts of money. The following sections describe a number of various methods used in practice.

The Eichleay formula. The "Eichleay" method got its name from a successful court case involving the Eichleay Corporation. In the Eichleay case, a construction contractor incurred a work suspension (delay claim) while waiting for change orders to the contract. Work could not be abandoned and replacement work was not feasible. To recover the cost of unabsorbed overhead (general and administrative expenses) applicable to the delay, Eichleay applied a daily rate method that was successful in the trial judgment.

The Eichleay method of computing a "daily contract overhead rate" is shown in Schedule 15-6.

There are obvious weaknesses in the Eichleay method, and on occasion it has fallen into disfavor.

First, all fixed overhead is allocated to the delayed contract based on contract billings. As in Schedule 15-6, this relationship may not be meaningful because billings do not generate overhead; costs do. In Schedule 15-6, the disputed contract gets one-third of the overhead since its billings are one-third of total billings. But if you applied the Eichleay method using cost, then the disputed contract would get approximately 43 percent of overhead ($24,000 divided by $56,000), as opposed to 33 percent under the Eichleay method using billings.

Second, the "days of contract performance" used to compute the "daily contract overhead rate" includes the "days of contract delay,"

SCHEDULE 15-6 THE EICHLEAY FORMULA

$$\frac{\text{Disputed contract billings}}{\text{Total contract billings*}} \times \text{total overhead} = \text{overhead applicable to disputed contract}$$

$$\frac{\text{Overhead applicable to disputed contract}}{\text{Days of contract performance}} = \text{daily overhead rate}$$

$$\text{Daily contract overhead rate} \times \text{days of contract delay} = \text{claimed overhead}$$

The following example shows how the Eichleay method works.

Project Results

	Contract		
	Disputed	*Other*	*Total contract*
Billings	$20,000	$40,000	$60,000
Cost	24,000	32,000	56,000
Gross profit (loss)	$(4,000)	$ 8,000	$ 4,000
General and administrative expense			$ 6,000
Days of delay			120
Contract days			600

Overhead Calculation

Overhead applicable to disputed contract

$$\frac{\$20{,}000}{\$60{,}000} \times \$6{,}000 = \$2{,}000$$

Daily overhead rate

$$\$2{,}000 \div 600 \text{ contract days} = \$3.33 \text{ per day}$$

Claimed overhead

$$120 \text{ days of delay} \times \$3.33 = \$399.60$$

*During period of contract in dispute.

which understates the rate and the resulting unabsorbed fixed overhead applicable to the delay.

The advantages of the Eichleay method are its simplicity and its acceptance in previous court cases. The method also enables experienced financial personnel to determine unabsorbed fixed overhead relatively easily.

In addition, the Eichleay method can be easily explained to financially unsophisticated people, and its prior acceptance by the court makes the attorney and contractor more comfortable in using it to compute damages.

Total direct cost allocation. Another widely used approach to applying overhead in contract claims is based on total direct costs. Although this method has drawbacks under certain circumstances, it is relatively easy to use and may be especially valuable when presenting the merits of a claim to a judge or jury.

Generally, this method assigns overhead costs to the direct cost of the contract compared with the total direct costs of all contracts. Schedule 15-7 illustrates the application of this method.

Schedule 15-7 demonstrates how easy it is to compute an overhead rate and apply it to a disputed contract or to any other contract. The overhead rate ($0.10 in this example) is computed by dividing total overhead ($200,000) by total direct costs for all contracts ($2,000,000). Overhead is applied to each contract by multiplying the overhead rate ($0. 10) by total direct costs for each contract ($400,000 for the disputed contract). Direct costs for contracts other than the disputed contract should include only costs incurred during the performance period of the disputed contract.

In theory, this method may not be suitable for many contractors because it does not allow for differences between various types of contracts. For example, some accountants may argue that contracts that consist mostly of subcontractor billings or material costs should not be burdened with the same percentage of overhead costs as contracts that consist primarily of labor costs of the contractor's own work force.

In practice, this method is most applicable to contractors that use approximately the same mix of material costs, labor costs, and subcontract costs for most contracts. If the cost of major components changes significantly from job to job, direct cost allocation probably is not a good method to use.

SCHEDULE 15-7 DIRECT COST ALLOCATION METHOD

	Disputed contract	*All other contracts during period of performance*	*Total*
Direct costs	$400,000	$1,600,000	$2,000,000
Total general and administrative/overhead expenses			200,000
Overhead rate per dollar of direct cost:			

$$\frac{\$200{,}000}{\$2{,}000{,}000} = \$0.10$$

Overhead applied to disputed contract:

$$(\$0.10) \times (\$400{,}000) = \underline{\underline{\$40{,}000}}$$

Other allocation methods may be more useful when a significant time delay is experienced on a long-term contract. If a delay results in significantly reducing the amount of work performed in a year (thereby reducing actual direct costs), direct cost allocation will not provide a contractor with a reasonable allocation of indirect overhead costs for contract claims purposes. In other words, if no direct job costs are incurred during the delay period, overhead will not be allocated to the delayed job. This problem also happens when overhead allocation is based upon direct labor hours or direct labor cost.

If replacement work is not available to use the excess capacity created by a delay, the total allocation base used to distribute indirect overhead costs will be reduced. This, in turn, will increase the overhead rate used on active contracts. Overhead that should be allocated to the delayed contract will then be allocated to other active jobs, thus reducing profits on those jobs.

In summary, the total direct cost allocation method of assigning overhead costs to individual contracts is best suited to companies that (1) typically maintain a relatively constant mix of inputs (e.g., material costs, labor costs, subcontract costs) in relation to total contract costs, and (2) have not experienced significant shutdown periods on their contracts. The primary benefit of using this method is its simplicity. Judges and juries normally are able to understand its logic and application, and this may help the credibility of a claim.

Specific base allocation method. The specific base allocation method (SBAM) is one of the most accurate systems available for distributing overhead costs among specific contracts and/or work items within a contract. It also provides management with specific cost information that can be used to control costs on current jobs and help in estimating costs on future jobs. This method allocates overhead costs based on specific characteristics of each job and each overhead cost element.

SBAM requires that each overhead cost element be analyzed to determine the most appropriate method of allocating it to a specific job. A different means of assigning a particular cost on different jobs may result in a more appropriate allocation base for a specific job. For example, on a labor-intensive job, direct labor hours may be the most appropriate method of allocating utility costs. But on a capital-intensive job that requires an unusually large amount of electricity, equipment hours may be a better allocation base for the job. See Chapter 6 for a more comprehensive discussion of this method.

Schedule 15-8 lists examples of indirect overhead cost pools and the bases used for allocating them by a company building a major dam.

SBAM alleviates many of the problems associated with arbitrary allocation methods, such as assigning overhead costs based on direct

SCHEDULE 15-8 BASES FOR ALLOCATING OVERHEAD

Pool account	*Allocation base*
Equipment, fuel, and supplies	Equipment hours by bid item
Superintendents	Direct labor hours
Utilities	Direct labor hours
Procurement and purchasing	Direct material dollars
Accounting	Total direct cost
Interest income and expense	Total direct cost

SCHEDULE 15-9 DISTORTION CAUSED BY DIRECT LABOR ALLOCATION

	Job A	*Job B*
Contract revenue	$ 3,000,000	$3,000,000
Direct costs		
Direct labor	600,000	200,000
Subcontracts	200,000	600,000
Materials	1,800,000	1,800,000
Equipment	200,000	200,000
	2,800,000	2,800,000
Variable margin	200,000	200,000
Overhead @ $0.50 per labor dollar	300,000	100,000
Operating profit (loss)	$ (100,000)	$ 100,000

labor dollars or labor hours of the total company. To illustrate the distortion of income reported for two jobs using an arbitrary allocation method, consider Schedule 15-9, in which overhead is allocated to two jobs on the basis of direct labor dollars. Both projects contribute the same amount toward overhead and operating profit ($200,000 per job), but in using direct labor dollars as a base, Job A, which used more direct labor than subcontract labor, is shown as a loss job. If the specific base allocation method had been used, comparable profits would have been reported.

Although SBAM is, in theory, one of the soundest approaches to allocating overhead, practical drawbacks limit its use. Few contractors use cost information to the extent that the cost of allocating overhead on multiple bases is justified. Second, wide variations in the type and mix of jobs would require that formulas for allocating overhead cost elements be adjusted for each job. While this task is not impossible, the cost of doing it is seldom justified because of high administration costs.

SBAM is suitable primarily to large, high-risk projects where there is high potential for litigation of a major claim. In certain claim situations, especially when distributable costs are high, professionals believe that the more specific the allocation method is, the more defensible the financial aspects of a claim will be. This consideration should be weighed against the increased complexity (and cost) of the allocation method, and the likelihood that a judge and jury will be able to understand the implications of the method.

Occupancy and other methods. In some contracts overhead cost is more a function of occupancy rather than of contract dollars. If overhead and general and administrative costs are made up primarily of depreciation of premises and equipment, then a standard method for determining contract occupancy should be determined. For example, assume that a contract to build a turbine for a nuclear power plant has been delayed and that most of the overhead is made up of shop facilities. In this case the following method might be used:

$$\frac{\text{Disputed contract floor space}}{\text{Total shop floor space}} \times \text{total overhead} = \text{overhead allocation to disputed contract}$$

In summary, selection of a method for applying overhead in a claim situation should be based on the specific characteristics of the company, the disputed contract, and the components of overhead. Other valid methods (usually variations of the methods presented here) have been used successfully. Different methods should be tested in each claim situation in order to select the most appropriate method for a particular set of circumstances.

Profit

Once the contractor has been able to determine his additional costs as the result of claim entitlement, including recovery of additional home office overhead, it would seem obvious that the contractor should be entitled to a fair profit. But this is not always the case. Questions always seem to arise about equity, profit percentages, and the base to which profit percentages should be applied. There have been cases in arbitrations in which the contractor has clearly proved his entitlement and the costs of the entitlement (damages), but in attempting to please both parties the arbitrators have not given the contractor overhead, profit, and interest. Also, the availability of profit may well depend on the legal theory which counsel advances—damages-for-breach versus quantum merit versus reliance on the contract's change order provisions. The latter of these is the simplest way to get profit in many cases, since the change order provisions usually include a built-in profit and overhead factor.

Whatever the subjective aspects of the claims process and the litigation strategies, a request for profit should be included in the claim, and it should be supported in the best possible manner.

To support the percentage for arriving at the dollar amount of profit, the best approach is usually to present a schedule of profitable contracts completed in the past and a computation of the average profit percentage achieved. Loss jobs should be excluded from the computation since a contract in the claim situation we are talking about would have been profitable, except for acts of the customer or owner. The schedule of profitable jobs can either be a listing of jobs or a summary of the contractor's income statements for several years preceding the disputed job, adjusted for loss contracts. Schedule 15-10 is an example of the latter approach.

SCHEDULE 15-10 HISTORICAL PROFIT ANALYSIS

Based on Audited Income Statement Data

	Years				
	1984	*1985*	*1986*	*1987*	*Total*
Total cost (all profitable jobs)	$1,600,000	$1,400,000	$1,800,000	$2,000,000	$6,800,000
Gross profit (all profitable jobs)	$ 170,000	$ 176,000	$ 178,000	$ 190,000	$ 714,000
Percentage	10.6%	12.6%	9.9%	9.5%	10.5%

SCHEDULE 15-11 CLAIM SUMMARY

Total Cost Approach

Contract costs		
Direct	$1,500,000	
Indirect	160,000	
Total cost		$1,660,000 (A)
Contract revenue		1,440,000
Cost overrun		220,000 (B)
Overhead at 10% of (A)		166,000 (C)
Profit at 15% of (A) and (C)		273,900
Total claim		$ 659,900

SCHEDULE 15-12 COMPARISON OF AS-BID VERSUS AS-BUILD SUMMARY

	As bid	*As built*	*Difference*
Contract revenue	$1,350,000	$1,440,000	$ 90,000
Contract costs			
Direct	1,130,000	1,500,000	(370,000)
Indirect	120,000	160,000	(40,000)
	1,250,000	1,660,000	(410,000)
Profit (loss)	$ 100,000	$ (220,000)	$(320,000)

SCHEDULE 15-13 COMPARISON OF BID VERSUS ACTUAL BY CONTRACT WORK ITEM

Work item	*Cost* *Bid*	*Actual*	*Over (under)*
1	$ 100,000	$ 80,000	$ 20,000)
2	170,000	200,000	30,000
3	170,000	220,000	50,000
4	320,000	400,000	80,000
5	370,000	600,000	230,000
	$1,130,000	$1,500,000	$370,000

SCHEDULE 15-14 RECONCILIATION OF ORIGINAL CONTRACT PRICE WITH FINAL REVENUE

Original contract dated May 1, 19XX		$1,350,000
Change orders		
No. 1, June 4, 19XX	$ 10,000	
No. 2, August 8, 19XX	15,000	
No. 3, December 3, 19XX	40,000	
No. 4, December 12, 19XX	(20,000)	
No. 5, December 20, 19XX	45,000	
		90,000
Total revenue		$1,440,000

There are also questions as to what base should be used for applying the profit percentage. A schedule should be presented showing the contractor's actual final costs and overhead in the same format in which the job was bid. The profit percentage should be applied in the same way that it would have been at the bid stage had the contractor known what the total scope of costs was going to be.

SCHEDULE 15-15 DETAIL OF COST AND REVENUES BY CONTRACT WORK ITEM AND COST FUNCTION

Work item	Costs						
	Labor	*Materials*	*Subcontract*	*Equipment*	*Total*	*Revenue*	*Profit (loss)*
Actual							
1	$ 60,000	$ 20,000	$ —	$ 20,000	$ 100,000	$ 130,000	$ 30,000
2	80,000	30,000	80,000	10,000	200,000	250,000	50,000
3	$ —	20,000	160,000	20,000	200,000	200,000	$ —
4	200,000	100,000	$ —	100,000	400,000	360,000	(40,000)
5	200,000	200,000	160,000	40,000	600,000	500,000	(100,000)
	540,000	370,000	400,000	190,000	1,500,000	1,440,000	(60,000)
Budget	440,000	350,000	400,000	140,000	1,120,000	1,350,000	220,000
Over (under)	$100,000	$ 20,000	$ —	$ 50,000	$ 370,000	$ 90,000	$ 280,000

Appendix A

Contractor's Manual Job Cost System

PUBLISHED FOR THE ACCOUNTING PROFESSION BY BURROUGHS CORPORATION, OFFICE PRODUCTS GROUP, BUSINESS FORMS DIVISION[1]

Reprinted from "The Accountants' Service Bulletin"

The big problem faced by all contractors, whether they be residential builders, general contractors doing all types of construction work, or highway contractors, is the recording of costs. Note that we are speaking of recording costs rather than determining costs.

Today most contractors determine their costs with great accuracy, if only to protect themselves in case of a tax audit. Their great difficulty is that these costs are determined after the job is finished. What they need, and know they need, is an accumulation of costs day by day which they know is accurate.

---

[1]The system illustrated was originated by the Charles R. Hadley Company. It is now marketed by the Office Products Group, Business Forms Division of the Burroughs Corporation. In addition to the system illustrated in this Appendix, the Burroughs Corporation has a similar system of contractor accounting designed for use with a different type of pegboard. Anyone contemplating installation of a pegboard system should compare the two and select the one best suited to the specific business. A suggested chart of accounts is shown in Schedule A-1. While by no means exclusive, it has proved satisfactory for a large number of contractors whose operations are small enough to be adequately served by a pegboard system.

SCHEDULE A-1 SUGGESTED CHART OF ACCOUNTS FOR CONTRACTORS

Assets

Current Assets
100 —Petty Cash
105 —Cash in Bank—General
106 —Cash in Bank—Payroll
115 —Accounts Receivable
120 —Contracts Receivable
130 —Inventory—Materials
135 —Inventory—Work in Progress
140 —Prepaid Expenses

Property, Plant, and Equipment
150 —Real Estate
155 —Buildings
155.1—Accumulated Depreciation—Buildings
160 —Office Furniture and Fixtures
160.1—Accumulated Depreciation—Furn. and Fix.
*165 —Mobile Equip. (bulldozers, earthmoving, etc.)
*165.1—Accumulated Depreciation—Mobile Equip.
170 —Motor Vehicles
170.1—Accum. Depreciation—Motor Vehicles
Deferred Charges
180 —Organization Expenses
185 —Prepaid Expenses (noncurrent)

Revenues

300 —Contract Sales
310 —Discounts Taken
350 —Other Income

Costs

400 —Cost of Contract Sales

Equipment Operating Expense

500 —Fuel
501 —Repairs
502 —Tires
503 —Depreciation—Mobile Equipment
504 —Depreciation—Motor Vehicles
505 —Depreciation—Other Equipment
506 —Property Taxes and Fees
507 —Insurance
550 —Misc. Equipment Operating Expense
599 —Equipment Operating Expense Absorbed

Liabilities

Current Liabilities

201 —Accounts Payable
205 —Notes Payable
210 —Contracts Payable (1 year and less)
215 —Sales Taxes Payable
220 —Employee Payroll Deductions
220.1 —Income Tax Withheld
220.2 —FICA Tax
220.3 —State Disability (or Unemp.) Tax
225 —Employer Payroll Taxes Payable
225.1 —FICA Tax
225.2 —Federal Unemployment Tax
225.3 —State Unemployment Tax
230 —Accrued Payroll
235 —Other Accrued Liabilities

Long-Term Debt

250 —Mortgage Payable
255 —Contracts Payable (over one year)

Capital

270 —Capital Stock Authorized
271 —Capital Stock Unissued
280 —Retained Earnings
290 —Profit and Loss

Indirect Expenses

600 —Supervision
601 —Indirect Labor
605 —Tool Replacements
610 —Officers' Salaries
611 —Office and Clerical Salaries
612 —Stationery and Office Supplies
613 —Utilities
614 —Telephone and Telegrams
615 —Insurance—General
616 —Insurance—Compensation
617 —Interest Paid
618 —Advertising
619 —Taxes and Licenses—General
620 —Payroll Taxes
621 —Dues and Subscriptions
622 —Travel and Entertainment
623 —Professional Fees
624 —Building Repair and Maintenance
625 —Depreciation—Building
626 —Depreciation—Office Furn. and Fix.
630 —Donations
650 —Miscellaneous Indirect Expenses
699 —Indirect Expense Absorbed

---

*Note: Building contractors, and many other contractors, not having heavy equipment, could omit these accounts and include equipment expenses under the indirect expenses.

The accurate recording of costs is essential to support year-end tax returns whether the contractor uses the percentage-of-completion method or the completed-contract method. What is more important to the contractor, however, is the determination of whether costs are running higher or lower than estimated. The most astute contractor can do nothing about excess costs after the job is finished. That same individual may, however, be able to prevent a financial disaster if he learns early enough that costs are piling up faster than the progress on the job warrants.

System standardized

Prior to standardizing the "write it once" accounts payable system with job cost ledger for contractors, the Charles R. Hadley Company had, for several years, been designing and installing special systems of this nature for different types of contractors. From the variety of problems presented, and from the experience gained in solving these problems, our standard system was developed.

Anyone who has ever worked in a contractor's office, either as an accountant or as an auditor, will immediately recognize the truth of the following statements: (1) the job cost ledgers are not kept strictly up to date, and (2) seldom, if ever, does a trial balance of the job cost ledgers agree with the balance in the general ledger account "Work in progress."

With the "write it once" system any number of jobs can be operated at one time through a single set of books and any one of these jobs can be split up into as many subsections as desired. All records can be kept up to date with the least amount of effort and the job cost ledgers on every individual job, and every subsection, can be up to date and in balance at all times.

The heart of the system is the record of invoices and work in progress, which is illustrated with job cost ledger, and accounts payable ledger, superimposed (Figure A-1).

The journal is used not only to record materials or services purchased from outside vendors, but acts as a general journal when transferring internal charges to the individual job cost ledgers. These internal charges may consist of the transfer of materials from inventory to specific jobs, the transfer of materials from one job to another or back to inventory, the distribution of accrued payroll to specific jobs, the allocation of overhead charges to specific jobs, or any other transaction that affects the cost of one or more jobs.

Directly connected with the system through the accounts payable ledger is the check record and the accounts payable check.

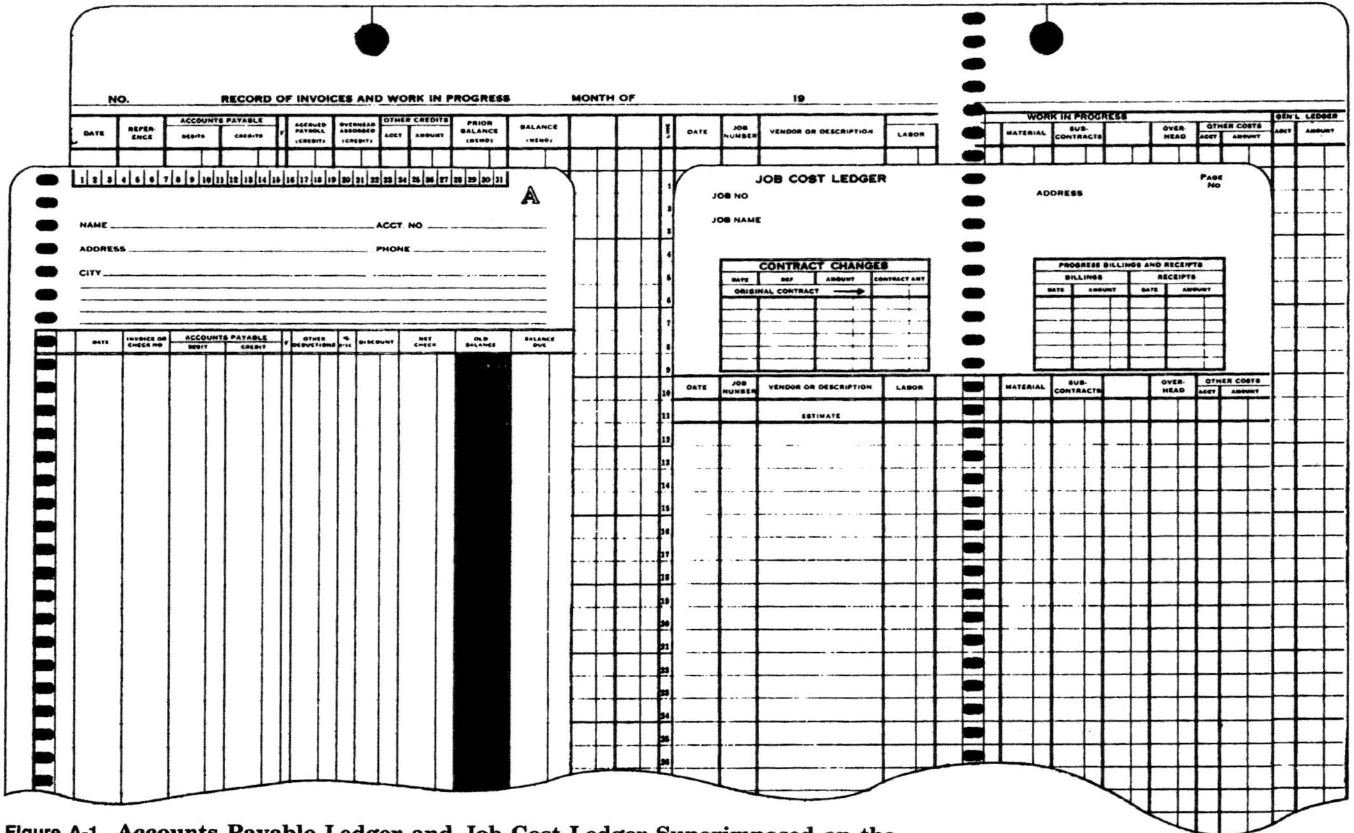

Figure A-1 **Accounts Payable Ledger and Job Cost Ledger Superimposed on the Record of Invoices and Work in Progress.**

Mechanics of the job cost system

The object of the system is not only to keep job costs as up to date as the latest charges available, but to establish absolute control over accounts payable, labor costs, material costs, subcontract costs, and all others which might be charged to "Work in progress."

Some bookkeepers distribute indirect expense items such as rent, utilities, and the like on the check record at the time the items are paid. Such a procedure is not recommended because accounts payable control over these items is lost. It is much better, for the sake of control, to set up an accounts payable ledger for each vendor, including the utility companies and others of this nature, and to record these payables on the record of invoices.

Therefore, every entry on the record of invoices will affect an accounts payable ledger, a job cost ledger, or both.

It follows that no entry can be made on the record of invoices unless the left half of the journal is covered by an accounts payable ledger, the right half covered by the appropriate job cost ledger, or both.

Just as no entry can be made on the journal without the presence of one or both ledgers, neither can any entry be made on a ledger unless it is superimposed upon a journal.

Journal control

To see how control of job cost ledgers is obtained through the journal, a clear understanding of the use of each column is necessary. At the risk of seeming too elementary, we present the following outline: (Columns have been numbered consecutively, from left to right, on the illustration, Figure A-2.)

Column 1. Enter date from original media, such as date of invoice.

Column 2. This is to identify the source of the entry, such as an invoice number, credit memo number, material requisition, or the page number of the payroll journal when recording labor charges.

Column 3. All invoices recorded are a credit to accounts payable, and could be debited to inventory, column 9, directly to a job, columns 15, 16, 17, or 19, or to any appropriate general ledger account, column 20.

The debit column is used to record credit memoranda issued by vendors. They offset credit entries which may be to jobs, to inventory, or to some other general ledger account. If the contractor follows the procedure of issuing a debit memorandum to the vendor upon the return of unacceptable material, the debit memo will be entered here. The vendor's credit memo, when received, becomes, in effect, a confirmation of the debit memo previously recorded.

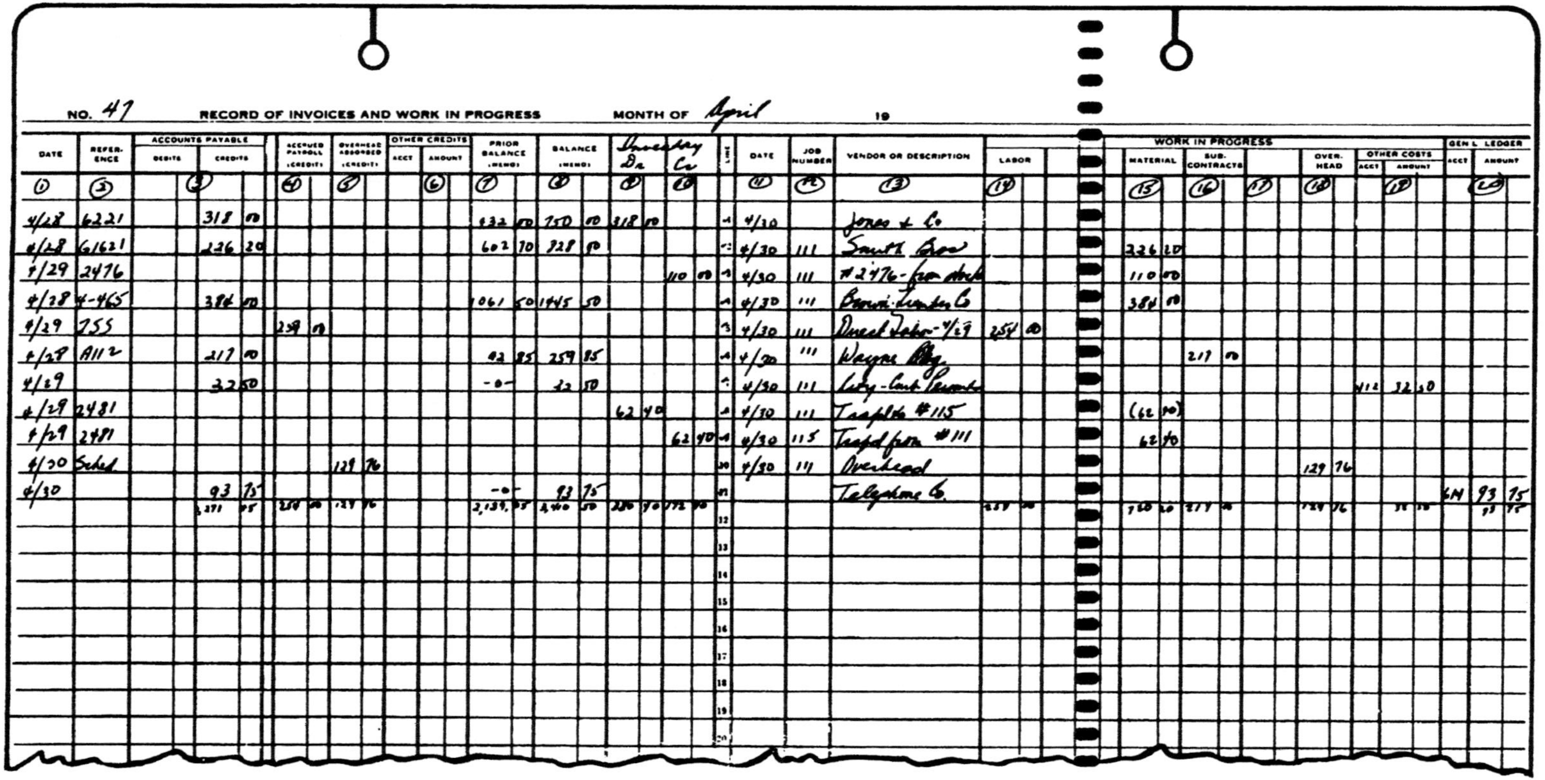

Figure A-2 Illustrative Journal Entries. These entries are shown as they would appear after the removal of ledgers and carbon. The journal is proved after each day's posting is completed.

Column 4. This credit to "Accrued payroll" is to offset labor charged directly to jobs in column 14. The two entries are always equal.

Column 5. As under "Accrued payroll," an offsetting entry is made in this column whenever overhead is charged to a job in column 18.

Column 6. This column adds flexibility. Its frequency of use will depend upon how the blank columns 9, 10, and 17, are used (see below). Its essential purpose is described in the heading "Other credits." When an offsetting credit entry is required and it cannot be assigned to any other column, it goes in column 6.

For example: If a mistake were made, charging architects' fees or permits to the wrong job ("Other charges"), it could be corrected through the use of column 6. The mechanics of such a correction are the same as described under columns 9 and 10. Without column 6, such a correction would have to be made through the general journal.

Columns 7 and 8. These two columns provide posting control for the accounts payable. The balance shown in column 7 on the ledger is entered in column 8. The gross amount of the invoice being recorded is added to the balance to obtain the new balance.

Columns 9 and 10. These have also been left blank to make the form more versatile. In the illustration they are used as debit and credit columns for inventory. If considerable inventory is carried, this is the obvious use. Column 9 would handle the debits for all new materials received or for excess materials returned to stock from a particular job. The offsetting credits for materials taken from inventory and charged to a job, under column 15, would be entered in column 10.

If materials were transferred from one job to another, columns 9 and 10 would act as a clearing account. The procedure would be: debit inventory, column 9, and credit the job, column 15 (an encircled entry), from which the material was taken; then credit inventory, column 10, for the same amount and debit the job, column 15, to which the material was transferred.

Some contractors charge each job periodically with a rental figure for the equipment used on the job. The figure used is approximately what would have to be paid if the equipment were rented rather than owned by the contractor. In such cases, one of the blank columns (9 or 10) could be labeled "Other income" or "Equipment rental income" to receive the offsetting credits when charges are made to the jobs.

If column 10 were used for income from equipment rental, column 9 could still be used for inventory. Both debits and credits would then be entered in the same column. Credit entries would be encircled to distinguish them from the debit entries. The column total would be a net figure, either debit or credit.

Column 11. This is the date the entry is made on the job cost ledger.

Column 12. The job number is entered each time as a control figure. Thus every entry on the journal which has been charged to a job can be traced to the particular job by its number. The repetition of the job number of the job cost ledger is visual proof that the charges have been made to the proper job.

Column 13. Where a vendor's name is not appropriate, a description, such as "overhead charges" or "material from stock," should be used.

We cannot emphasize too strongly that no entries can be made in cost columns 14 through 19 until the appropriate job cost ledger has been placed on the board.

Column 14. In this column will be entered the amount of direct labor to be charged to each job. This will probably be done weekly, although, if UniSort job time cards are used, it can be done daily. The total labor charged to all jobs must, of course, equal the total direct labor shown on the payroll journal. The corresponding credit is to "Accrued payroll," column 4.

Column 15. Charges for all material, whether taken from inventory or purchased specifically for the job, must be entered in this column.

Column 16. Entries in this column will represent invoices from subcontractors at gross. The amounts withheld pending completion and the filing of labor releases, etc., will be reflected in the "Balance" column on the accounts payable ledger for the subcontractor and, of course, in the accounts payable control account in the general ledger.

Column 17. This column was also left blank for flexibility. It may be used for any specific recurring cost, such as equipment rental as explained in connection with column 9.

Column 18. Overhead is usually a monthly charge, prorated to the jobs on some equitable basis. The offsetting credit is to "Overhead absorbed," column 5. Not all contractors charge overhead to the jobs. The practice is sufficiently common, however, to warrant printing the column heading.

Column 19. All costs of the job which are not provided for under columns 14 through 18 are entered here as "Other costs." Such charges as architects' fees, cost of blueprints, surveyors' fees, building permits, and inspection fees are typical.

Column 20. The amount of any item not chargeable to a job (or to inventory) and the account number to which it is to be charged are entered in this column.

Typical entries

To clarify the procedure, ten typical entries are shown on the illustration of the journal (Figure A-2). Ledgers and carbons have been omitted in order to show the entries in full. It is assumed in all cases that the original media, from which these entries were made, have been checked for receipt of materials, prices, and extensions and that the indicated distribution has been approved by the proper authority.

1. Invoice from Jones and Company for $318.00 chargeable to inventory. Accounts payable ledger used.
2. Invoice from Smith Brothers for $226.20 chargeable to materials delivered to job 111. Accounts payable ledger for Smith Brothers and job cost ledger for job 111 used.
3. Materials worth $110.00 were withdrawn from inventory on requisition No. 2476 and delivered to job 111. Job cost ledger used.
4. Brown Lumber Company invoice for $384.00 for materials delivered to job 111. Accounts payable ledger and job cost ledger used.
5. Direct labor amounting to $254.00 charged to job 111 from daily labor distribution sheet for January 29. Job cost ledger used.
6. Invoice from Wayne Plumbing Company for $217.00 for completing the rough plumbing on job 111. Accounts payable and job cost ledger used.
7. Invoice from the city for curb-breaking permit on job 111. Accounts payable and job cost ledger used.
8. Materials costing $62.40 were transferred from job 111 to job 115. This entry requires two lines to complete; on line 8, a debit is made to inventory and a credit to job 111; on line 9, a credit is made to inventory and a debit to job 115. Job cost ledgers for job 111 and job 115 are used.
9. End-of-month entry distributing $129.76, its share of overhead to job 111. Job cost ledger used.
10. Telephone Company bill for $93.75. Accounts payable ledger used.

Any entry, unless it is a transfer of costs from one job to another, seldom affects more than one job. Vendors and subcontractors generally submit separate invoices for different jobs. Should an invoice be received, however, covering more than one job, one line on the left of the journal would be used to record the invoice, but as many lines as there were jobs concerned would be used on the right in the work in progress section. For example: assume that in entry No. 4, above, the invoice from Brown Lumber Company had been for material chargeable half to job 111 and half to job 115. The total amount of the invoice, $384.00, would be recorded on the accounts payable ledger as illustrated on line 4. The charge to job 111 on line 4, however, would be only $192.00 with the other $192.00 being charged to job 115 on line 5.

The line number on the journal should be checked off as each line is used. This is necessary because carbons, when mounted over the journal, cover everything except the two blank columns (which we have headed inventory) and the line number column and make it difficult to check the next open writing line. If the person who posts the invoices is taught to check off the line numbers, it will become automatic and no difficulty will be encountered.

Control

Examination of the illustrative entries shows clearly how costs on each job are classified. Since the journal is self-balancing (total debits equal total credits) and since the journal will be proved at least once each day, control over the accounts payable and job cost ledgers is absolute. Remember these points:

1. No entry can be made on a ledger unless it is on the board, and
2. No entry can be made in columns 3 and 14 through 19 except as a carbon copy of an entry on an accounts payable or job cost ledger.
3. Therefore, proving the journal proves the accounts payable and job cost ledgers.

Daily posting

Daily posting of all available costs should be insisted upon even though on some days only a single posting is made. The major purpose of the system is to gain control of costs. The key to its success is that all related records be kept up to date and in balance at all times. Daily posting is, therefore, an absolute necessity.

That the user gains other benefits such as: (1) more even distribution of the workload in the office, (2) actual savings in clerical time, (3) elimination of month-end bottlenecks, and (4) complete confidence in his costs, is self-evident. But these are by-products. The major goal is up-to-the-minute cost information. Up-to-the-minute cost information means up-to-the-minute records. And up-to-the-minute records mean daily posting with daily proof.

Posting invoices

As stated before, it is assumed that before invoices are posted they will have been checked for receipt of materials, prices, and extensions and that the indicated cost distribution has been approved by the proper authority. Since posting cannot begin until invoices have been properly processed and approved, all personnel concerned with this phase of the operation should be cautioned to process the invoices promptly.

Invoices should be arranged in job order. Usually there will be several invoices, or other original media, for one job. Arranging them in job order will speed the posting operation since the job cost ledger for that job may be left on the board until all the invoices that pertain to it have been posted. If the volume of invoices for any one job is great, those invoices may be arranged alphabetically to facilitate their selection from the tray. Ordinarily, however, this will not be necessary.

While it is unlikely to happen, if an invoice is received which affects more than one job, it should be saved until last. Thus the maximum posting speed will be maintained on all of the invoices.

Assuming that the first invoice affects a job in progress, the steps in posting are as follows:

1. Run an adding machine tape on all of the invoices to be posted. This total will later be used as a control, or proof, total.
2. Select the first job cost ledger affected by the first group of invoices and mount it on the pegs on the right side. Select the accounts payable ledger for the first invoice and mount it on the pegs on the left side, making certain that the next open writing line on each ledger is aligned with the first unchecked line number on the journal.
3. Start the entry by copying the "Balance due" (from the last column on the accounts payable ledger) into the "Old balance" column. Always make this transfer of the "Balance due" to the "Old balance" column first. Complete entering the invoice, starting with the date and entering the gross amount in the "Credit" column under accounts payable as you would on any invoice register (except that the entry here is made on the ledger, of course).
4. On the adding machine, pick up the credit just entered and the last amount shown in the "Balance due" column on the ledger and post the resulting total as the new "Balance due." Note particularly that we did not use the amount we had previously entered in the "Old balance" column. By using the last balance figure in both instances, we prevent the possibility of a repetitive error.
5. Check off the line number.
6. Complete the entry on the job cost ledger distributing the amount of the invoice to the indicated cost classification. Do not neglect to enter the date, job number and description or vendor. The job

number is the number shown on the job cost ledger and should agree with the job number to which distribution was indicated on the invoice. This serves to identify the entry on the journal and assures anyone looking at the ledger that no charges applicable to other jobs have been made to this job.

7. Remove the accounts payable ledger, file it, and select the ledger needed to record the next invoice. Continue as above until all invoices affecting the job cost ledger on the board have been posted.
8. Remove and file the job cost ledger and select the one affected by the next group of invoices.

Invoices covering materials, supplies, or services not chargeable to a job are handled as above except that the job cost ledger is omitted from the board. Where it is desired to enter the vendor's name or a description in column 13, the carbon is lifted and the entry made directly on the journal. The same procedure is followed when posting costs to the job from media other than invoices, where the offsetting credit is to a column covered by carbon. In such a case, of course, the job cost ledger only would be on the board.

Proof

Posting should be proved (a) whenever a journal is completed, and (b) when posting for the day is completed. The various parts of the posting operation are proved as follows:

1. Pencil-foot all columns on the journal.
2. The net total of the accounts payable (credit totals less debit totals) must equal the prelist total of the invoices less any credit memos received from vendors or debit memos issued by the contractor. This will prove not only that we have entered the correct amounts on the journal, but also on each individual accounts payable ledger.
3. The grand total of all debits on the journal must equal the grand total of all credits. If not, check each line to locate the error. This step proves that for all credits (including those to accounts payable, already proved) there have been made offsetting debits in equal amounts.
4. The total of the "Prior balance" column, plus total accounts payable "Credits" less "Debits" (if any) must equal the total of the "Balance" column. Again, if out of balance, check each line to locate the error. Any error of this nature corrected on the journal must also be corrected on the proper ledger. This proves that the

balances have been extended correctly on each accounts payable ledger. (Errorless work will result if the recommended routine is followed exactly; pick up from the "Balance due" column when extending the new balance.)

The proof of the postings of the illustrative entries in Figure A-2 is shown in steps numbered to correspond with those in the foregoing discussion.

1. All columns have been pencil-footed.

2. Prelist total of Invoices	$1,271.45
Total Accounts Payable "Credits"	$1,271.45

3.

Debits		*Credits*	
Inventory	$ 380.40	Accounts payable	$1,271.45
Work in Progress:		Accrued payroll	254.00
Labor	254.00	Overhead absorbed	129.76
Material	720.20	Other credits	—0—
Subcontracts	217.00	Inventory	172.40
Overhead	129.76	Total	$1,827.61
Other costs	32.50		
Tel. & tel.	93.75		
Total	$1,827.61		

4. Prior Balance	$2,139.05
Accounts Payable Credits	1,271.45
Total	$3,410.50
Balance	$3,410.50

Payroll entries

Posting labor costs daily to job cost ledgers, as in illustrative entry No. 5, is the preferred method. If a UniSort, or some other type of daily job time card is used, this method can readily be followed by sorting the cards by job number, totaling, and recording the totals on a labor distribution sheet from which posting can be made to the job cost ledgers.

At the end of the week the only undistributed labor charges are those for indirect labor, supervision, maintenance, office salaries, etc. These are charged to the proper accounts in column 20 and credited to "Accrued payroll" in column 4.

One further entry is needed on the record of invoices. This entry is as follows:

Dr. Accrued payroll (column 4)	Gross amount of payroll
Cr. Accounts payable (column 3)	Net amount of payroll
Cr. Income tax withheld (column 20)	
Cr. F.I.C.A. tax (column 20)	
Cr. State D. or U.I. tax if required (column 20)	

The debit to "Accrued payroll" in the entry should, of course, be equal to the total of all the credits posted to the same column during the week.

The final entry is made on the check record when a check is written on general funds to reimburse the payroll bank account for the net amount of the payroll. The entry would be:

Dr. Accounts payable	Net amount of payroll
Cr. General bank account	Net amount of payroll

If labor costs are posted to the job cost ledgers only once a week, the credit to "Accrued payroll" (for the gross amount of the payroll) is made in total, and that and succeeding lines on the journal are used for posting to the individual jobs and to the indirect labor and expense accounts. The other two entries, given above, would then follow.

Disbursements

Integrally related to the record of invoices and work in progress, through the accounts payable ledger, is the record of checks drawn (Figure A-3). A brief description of the disbursement procedure is, therefore, necessary to complete the accounts payable picture.

No distribution, with the possible exception of petty cash refund checks, is necessary, or recommended, on the check record. The act of filling in the stub of the check automatically debits "Accounts payable," credits "Discounts earned" (if a discount is involved), and credits the general bank account.

Petty cash reimbursements are an allowable exception to the rule that all expenses should be distributed on the record of invoices since the accounts payable control is not affected and the amounts expended from petty cash are generally small and do not materially affect the expense accounts.

If, however, it is the practice to take money from petty cash to pay for some small items purchased for jobs, it will be necessary, then, to set up an accounts payable ledger for petty cash and make all distributions, to jobs and expense, on the record of invoices at the time the fund is to be reimbursed. The debit to "Accounts payable" on the check record, when the reimbursement check is written, offsets the

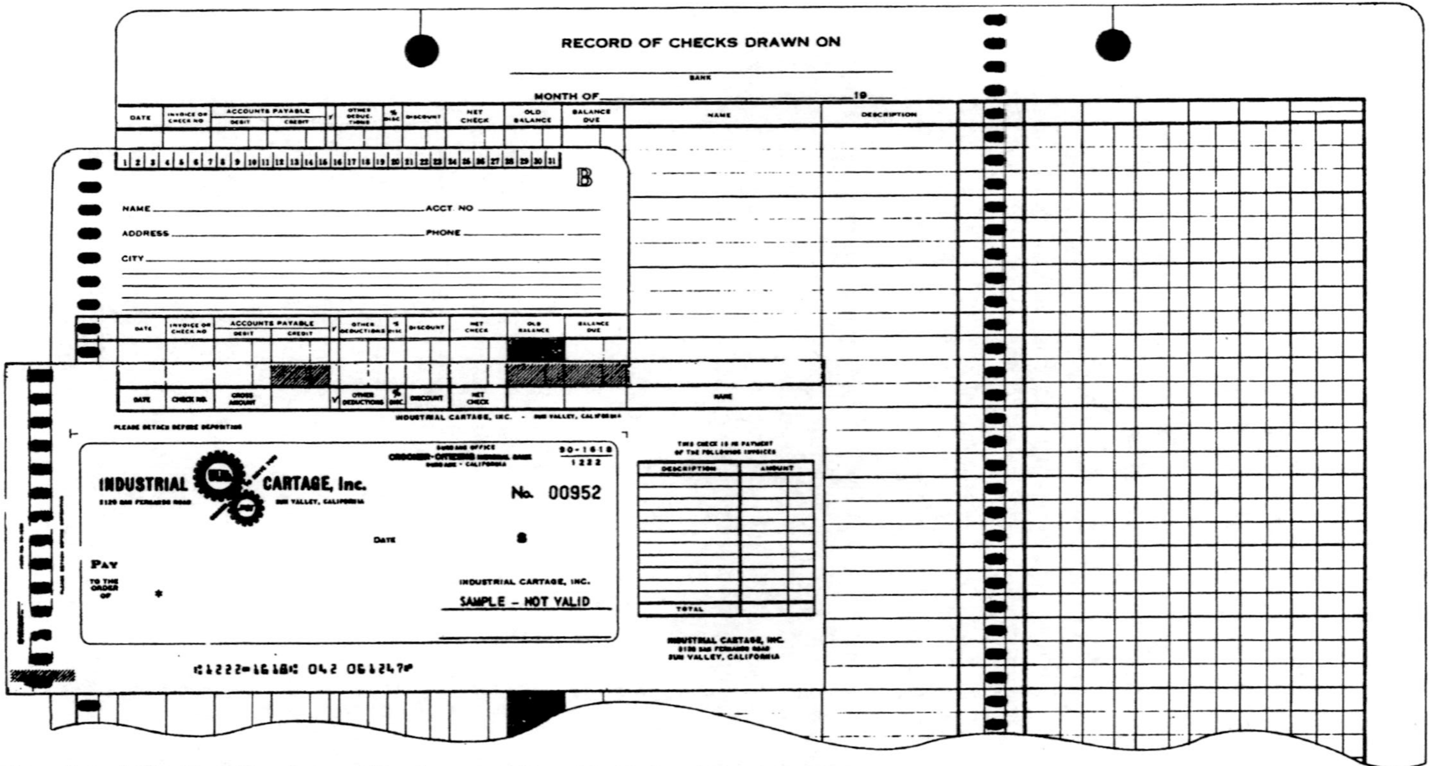

Figure A-3 A Bank of Checks and the Accounts Payable Ledger Superimposed on the Check Record. The check stub has carbon on its reverse side to carry the impression to the ledger and, through carbon, to the journal. The three forms are thus written simultaneously.

credit to "Accounts payable" on the record of invoices at the time of distribution.

Procedure

With a check record and a sheet of carbon on the board, it is only necessary to place a bank of 25 checks on the pegs and adjust the rail until the writing line on the stub of the top check is aligned with the first open line on the journal. Once this is done, all succeeding checks are automatically in the correct position with respect to succeeding lines on the journal.

It is assumed, of course, that the invoices to be paid have been removed from the file and any discounts to be taken have been computed. If a group of invoices from one vendor are being paid, a tape should be attached showing the total gross amount, total discount, and total net amount of the invoices in the group. The invoices to be paid, when removed from the file, will be in alphabetical order; this order should be maintained. A tape should be run on the gross amounts of the entire group of invoices to be paid in order to establish a control total.

The proper accounts payable ledger for the first invoice or group of invoices is selected from the posting tray, slipped under the checks (under one, or any number of checks), slid as far to the left as the pegs on the rail will permit it to go, and then slid up or down until the next open writing line is immediately under the writing line on the stub of the top check. (With a little practice this can be done almost instantaneously.) The first step is the one emphasized in the posting of invoices, picking up the amount shown in the "Balance due" column and entering it in the "Old balance" column. On the check these two column captions do not appear; the blank sections are used, however, just as though the captions were there, although they are shaded.

After picking up the old "balance," the rest of the check stub is completed. The date is entered, the check number (as it appears in the body of the check), the gross amount of the invoice or invoices, the discount percentage and amount, the net amount of the check, and the name of the payee. The check may then be removed from the board or flipped back to the left to expose the stub of the next check. When removed from the board, the checks are completed in the usual manner, and then presented for signature accompanied by the supporting invoices.

The column headed "Other deductions" is rarely used. It might be used, for example, where by mutual agreement the gross amount of the invoice was to be reduced without the formality of either a debit or a credit memo. This would be more likely to happen between a small general contracting firm and one of its small suppliers.

Proof

The same steps would be followed in proving the check record as were followed in proving the record of invoices and work in progress. All columns must first be totaled. Then the total debits to "Accounts payable" must equal the sum of the totals of the three credit columns, "Other deductions" (if any), "Discounts," and "Net check."

As proof of the posting to the ledgers, the total of the "Old balance" column should equal the sum of the totals of the "Accounts payable" debit and "Balance due" columns.

When checks have been written for all of the invoices, the grand total of the "Accounts payable" debit column on all journals (if more than one is used) must equal the control total previously established.

In recent years, since small contractors have, to some extent, begun to use computer service centers, procedures have been devised to utilize the "pegboard" records as a source of input data for the computers. Where analysis proves this to be practical, it has proved much more satisfactory than having the computer center work from "raw" data such as invoices, time cards, and check stubs.

Appendix

B

Computerization in the Construction Industry

Douglas A. Hughes,
Financial Consultant to the Construction Industry, Scottsdale, Arizona

Cost Center or Profit Center?

With the advent of the computer, the construction industry has the tool necessary to be more productive and profitable. The increased capacity and lower costs of acquiring computer power have made this tool one which is available to all contractors, not just the large ones. Its ease of use has made it a tool which is available to even the unsophisticated.

Traditionally, contractors have looked upon the computer budget as an overhead item. It is always easier to cost justify the budget for a piece of construction equipment than it is for the financial person or estimator to get the cost of a computer system accepted. This is because the owners of construction businesses can see the immediate benefits of using a piece of construction equipment to generate revenues. It is more difficult to realize the benefit of accurate and timely data. The lack of accurate and timely data is likely to cause any business to no longer be in business. Bidding decisions based upon erroneous historical data can be fatal to an organization. The inability to prepare the payroll properly and on time can result in penalties which are not only costly but can cause serious morale problems with employees.

The appropriate system enhances the ability to obtain negotiated work by having the capacity to meet customer's unique cost and billing format that they are seeking. If a contractor cannot efficiently provide customers with the data they seek, it will be difficult to be the acceptable low bidder. Furthermore, it might preclude the contractor from being awarded the work in the first place because of the inability to meet the terms and conditions of the contract. Also, it can help impress a contractor's customers, employees, subcontractors, bankers, and bonding companies with the ability and sophistication that exists in the contractor's company. While difficult to put a value to this ability, it can definitely improve the capacity to obtain financing because the records are timely and accurate, thus providing the information necessary to give comfort to a lending institution.

A good data-processing system can enhance the ability to recover costs for changed conditions. By keeping records in such a fashion as to isolate and properly identify costs relating to changed conditions and having records which are easily auditable, the contractor will be able to more readily recover all of his costs that were not anticipated at bid time.

Many times customers are shown around the office as well as the yard or plant by a contractor prior to obtaining a contract. The fact that a computer system (accounting or estimating) exists, and that it can be demonstrated to a potential customer that the contractor uses the system in successfully completing contracts, will reassure the customer that the contractor is capable. Having a good system can help attract employees who are capable and sophisticated. This will further help the contractor to succeed in making profits by providing the talent necessary to achieve the contractor's goals.

Historical Perspective

Data-processing systems consist of two components. The first component is the hardware, or the machinery that does the numbers crunching. The second component is the software, which is the instructions written to allow the hardware to manipulate the numbers and to put them into the format desired. From a hardware standpoint, there are three types of computers: mainframe computers, minicomputers, and microcomputers. The mainframe computer is generally the largest, fastest, most powerful, and most costly in terms of absolute dollars. The minicomputer is smaller and slower than a mainframe computer, but larger and more versatile than a microcomputer. A microcomputer, or "personal computer," is a small desktop machine which is generally used by one person. Some of the more powerful PCs have been networked together to form a basis for providing so-called "client server" systems which can perform the tasks which used to be performed on the larger systems.

As recently as five to six years ago the way in which contractors purchased data-processing systems was to get custom-designed software. This usually meant that the cost was very high and each time that an improvement was made to the base system, additional services were required by the contractor to modify the changes to fit the customized needs of the contractor. Usually this happened each year if for no other reason than that the government changed the tax laws affecting payroll or other areas of the law. The only other alternative was to make the business fit the software. This was obviously not a good situation.

Currently software available to contractors is more flexible, adaptable, portable, and designed to fit various sizes of companies. This allows the contractor to easily provide the customization needed to be unique in the way the individual company does business yet take advantage of the lower cost of so-called "off-the-shelf" programs. The contractor is no longer dependent upon the software manufacturer or programmer to modify and keep the programs current. In many cases, the programs can be run on different machines with very little modification. They are often designed in such a way that when the volume of transactions increases, the result is a need for only a little more powerful hardware and some additional personnel for data entry.

Because the cost of hardware has come down dramatically in the last five years, contractors now have the option of purchasing a minicomputer for about the cost of a few microcomputers. Minicomputer systems have more reliability and are faster to run than a microcomputer network. Minicomputers also offer a platform on which to run some of the most sophisticated programs.

With the development of database programs to be run on a minicomputer system, the software aimed today at the contractor is similar in power and capability to that of the mainframe computers of a few years ago. These programs, when coupled with report writer programs and such other tools as query commands, can provide the user with just about every imaginable piece of information or data that could ever be possibly thought about.

Microcomputers, which were invented in the late 1970s and have been improved upon ever since, provide the basis for the smaller and less sophisticated contractor to get started. The software designed to be run on a microcomputer today can provide information comparable to that which was available only from a minicomputer or larger system a few years ago. The initial cost of the hardware is less than that for a minicomputer and therefore provides a means for the smaller contractor to get started. However, there are often hidden or future costs to running these systems which are not apparent until some time has passed. Generally, minicomputers are less reliable and will have downtime more frequently than the microcomputer. There are always trade-offs which have to be weighed when determining the platform to be chosen.

The challenge of the 1990s

In today's business environment, the financial executive is faced with unprecedented change. This change is creating challenges never thought about in the past due to increased globalization, intense competition, shorter product life cycles, faster delivery to market requirements, rapid production cycles, and extreme cost cutting measures being imposed from higher management. Technology is rapidly being outdated. It becomes inflexible, with limited access, and costly to run in relation to new technology which is being introduced. So what is the financial executive supposed to do in this situation? The remainder of this chapter will be devoted to defining what a system should be capable of achieving and the method for implementing a new system.

Overview of a Typical Contractor Information System

Depending upon the type of contractor and the work being performed, some or all of the following applications are typically found in a contractor's system:

1. Job cost reporting including weekly, monthly, and job to date cost, budget information, projected cost at completion, and unit cost information.
2. General ledger reporting including trial balance, detail ledger data, prior period information (prior one or two years).
3. Payroll including special reporting such as certified payroll reports, paychecks, insurance reporting, fringe benefit reporting, union reporting, payroll tax return reporting, labor distribution reports, and many other human resource type reports such as 401(k) or vacation accrual reporting.
4. Accounts payable including check writing (including joint check capability), aging reports, cash flow reports, and subcontractor reporting and control.
5. Accounts receivable including aging reports, cash flow reports.
6. Equipment costing including analysis by hour, by piece of equipment, by type of cost, by major or minor cost type breakdown, with revenue breakdowns by piece of equipment and including hours of idle as well as productive time.
7. Billing including special forms such as AIA and various revenue reports for multiple contact types such as cost plus or time and materials.

8. Inventory, requisitions, and purchase order applications.
9. Fixed assets including various types of depreciation.
10. Subcontract management including methods of tracking the performance on individual contracts as well as providing information to the estimating department for future bids.
11. Property management including vacancy and renewal tracking as well as reporting on scheduled maintenance.
12. Loan and draw including tracking loan budget and loan provisions as well as requests and approval tracking.
13. Unit sales including tracking of construction by stage as well as providing extensive buyer profiles.
14. Estimating including direct entry of budgets for successful bid work directly into the accounting system, generation of subcontracts and purchase orders, and interfacing into scheduling.
15. Scheduling including manpower and equipment utilization reporting.

Most systems now have on-line inquiry as well as some form of report writing or statement generator capability which will work across each application if the system is set up with a database design. This allows for the flexibility to tailor information reporting to meet the unique requirements of each contractor.

The flowchart (page 552) shows how a typical system might be configured. In this system data is entered through the individual application module such as payroll, accounts payable, equipment, or billing and go directly to the job cost database which in turn feeds data into the general ledger application. The beauty of this system is that it keeps the data integrity between the job cost application and the general ledger application in balance. Furthermore, it requires that data be entered only one time.

Project costing application

The benefits of this application are:

1. Tracks costs and revenues for standard or unique tasks.
2. Instant access to timely and accurate:
   a. Revenues and costs.
   b. Commitments.
   c. Change orders.
   d. Forecasted costs and revenues.
3. Provides the tool necessary to improve project performance by identifying problem areas in time to take corrective action.

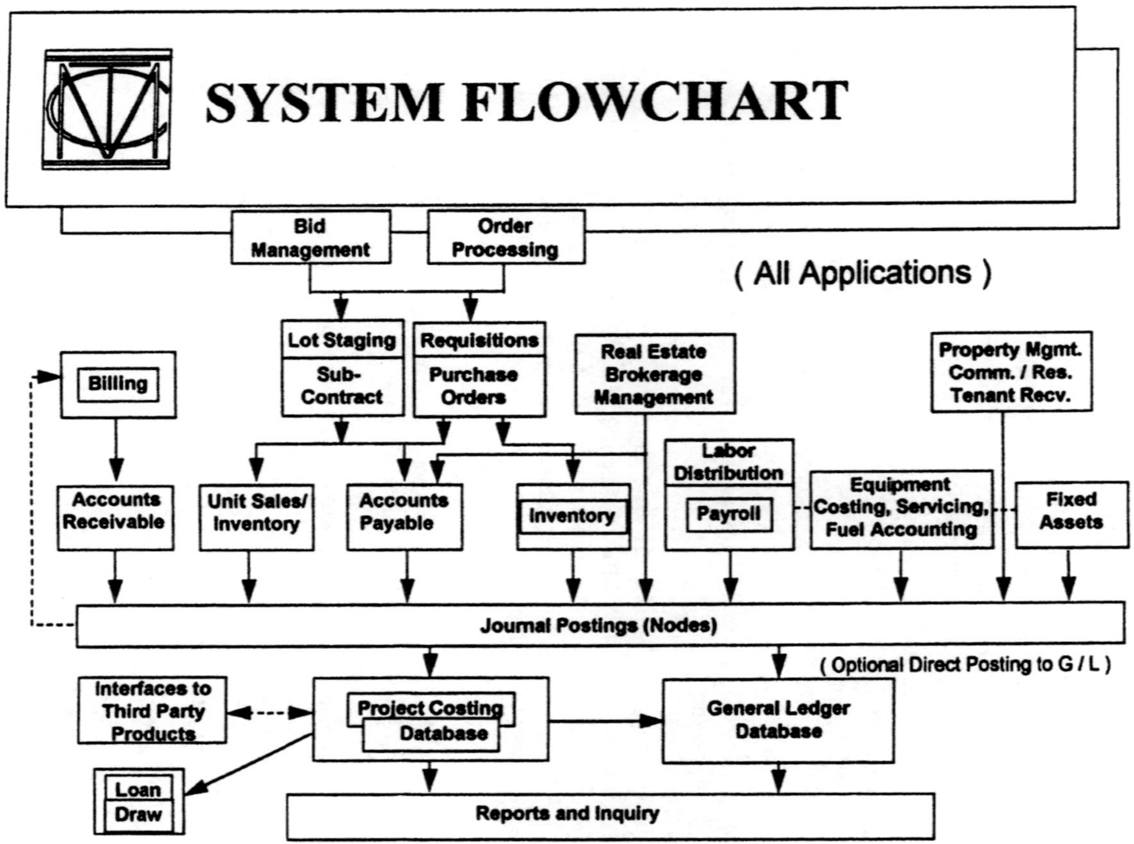
SYSTEM FLOWCHART
(All Applications)
Bid Management
Order Processing
Lot Staging
Sub-Contract
Requisitions
Purchase Orders
Real Estate Brokerage Management
Property Mgmt. Comm. / Res. Tenant Recv.
Billing
Accounts Receivable
Unit Sales/ Inventory
Accounts Payable
Inventory
Labor Distribution
Payroll
Equipment Costing, Servicing, Fuel Accounting
Fixed Assets
Journal Postings (Nodes)
(Optional Direct Posting to G / L)
Interfaces to Third Party Products
Project Costing
Database
General Ledger Database
Loan
Draw
Reports and Inquiry

Some of the standard features of this application include:

1. Multiple user-defined project and cost types.
2. User-defined cost coding structure.
3. Project master used to link multiple jobs.
4. Multiple forecasting methods for cost and revenue at completion.
5. Multiple revenue recognition methods.
6. Powerful cost allocation capability.
7. On-line access to life of job detail and summary information.
8. Comprehensive change event management system.
9. Comprehensive reporting capability.
   a. Labor productivity.
   b. Unit price reporting.
   c. Commitment control.
   d. Over and under billings calculations.
   e. Forecasted cost at completion.
   f. User-defined reference fields.
   g. Flexible report writer.
10. Comprehensive system of requesting, collecting, and analyzing bid information.
11. Enables bid comparison and selection with extensive inquiry and reporting.
12. Awarded bids can automatically generate budgets, subcontracts, and purchase orders.
13. Historical information can be used in the analysis and bidding of future projects.

General ledger application

The benefits of this application are:

1. It provides sound financial management and borrowing capacity with up-to-the-minute inquiry and financial reporting.
2. It will easily adapt to changes in organizational structure.
3. It will easily meet changes in reporting requirements.

Some of the standard features of this application include:

1. Flexible account structures.
2. Universal chart of accounts.
3. Supports income statements and balance sheets at multiple organization levels.
4. Automatic intercompany journals.
5. Extensive consolidation roll up-down capability.

6. Maintains multiple sets of books with a single transaction.
7. Tax, GAAP, and state rule books.
8. Cash and/or accrual accounting sets of books.
9. On-line trial balance allows "drill down" to source documentation.
10. Comprehensive financial statement generator.

Payroll application

The benefits of this application are:

1. Effective control of labor costs.
2. Automate multistate payrolls.
3. Simplify union benefits reporting.
4. Simplify workers' compensation reporting.
5. Manage human resources requirements.

Some of the standard features of this application include:

1. Fully integrated with general ledger and job cost.
2. Flexible default schemes for derivation of codes and wage rates.
3. Certified payroll reporting.
4. Supports employee and employer 401(k) reporting.
5. Prepay generator for calculation of prepay and layoff checks.
6. Project-oriented workers' compensation schemes.
7. Entitlement tracking for accrual purposes.

Accounts payable application

The benefits of this application are:

1. Manage vendor relations through on-line vendor information.
2. Simplify daily processing with streamlined voucher entry and recurring invoices.
3. Simplify cash management through on-line maintenance of target pay dates.

Some of the standard features of this application include:

1. Provides system-generated manual and joint checks.
2. Automatically calculates vendor payment due dates and discounts based upon vendor terms.

3. Purchases by job, cash flow analysis, and summary and detail aging reports.
4. Laser MICR check printing.
5. Supports multiple bank accounts, automatic check reconciliation (bank tape or manual entry), and year-end 1099 processing.
6. Interface with purchase order module so that detail can be automatically retrieved during voucher entry.
7. On-line inquiry.

Accounts receivable application

The benefits of this application are:

1. Help manage customer relationships with up-to-the-minute status of billings and cash receipts.
2. Simplify daily processing routines with billing interface and on-line cash receipts.
3. Simplify cash management through on-line maintenance of expected payment dates.

Some of the standard features of this application include:

1. Supports billings, retention, discounts, miscellaneous sales, and miscellaneous cash receipts.
2. Detail and summary on-line cash and billings inquiry by customer or job.
3. Supports multiple bank accounts.
4. Supports multiple period processing for cash and sales.
5. Automatic retention rebilling.
6. On-demand statements.
7. Comprehensive reporting including:
   a. Cash flow projections.
   b. Summary and detail aging by customer or job.
   c. Contract status.

Equipment costing, servicing, and fuel application

The benefits of this application are:

1. Extend the productive life of equipment through preventive maintenance management.
2. Measure profitability of each piece of equipment.
3. Simplify and automate equipment utilization billing.

4. Gain control over shop costs.

Some of the standard features of this application include:

1. Provides preventive maintenance scheduling and compliance reporting.
2. Monitors the inventory of fuel and tracks fuel purchase issues.
3. Supports extensive on-line cost vs. revenue analysis.
4. Provides multiple rental and recovery rates based upon imposed factors at multiple levels.
   a. Group
   b. Series
   c. Equipment
   d. Job
5. Tracks revenue and cost within ten user-defined rate components.
6. Enables on-line equipment location inquiry.

Billing application

The benefits of this application are:

1. Improve cash flow by generating billings faster and more efficiently.
2. Enhance contract negotiation capabilities by:
   a. Providing the ability to meet owner's reporting requirements.
   b. Supports complex rate and fee commercial terms.
3. Support owner's coding structures without corrupting internal cost coding conventions.

Some of the standard features of this application include:

1. Supports multiple contract types:
   a. Lump sum.
   b. Cost plus.
   c. Time and materials.
   d. Combinations within a single project.
2. Supports owner's coding structures.
3. Provides predefined and user-defined billing formats.
   a. AIA format.
   b. Time and material format.
   c. Detail backup reports.
4. Provides comprehensive date-sensitive commercial terms table:
   a. Multiple cost plus multiplier.
   b. Markups on straight time, overtime, premium time.
   c. Markups on various employee, equipment, and vendor groups.
   d. Supports "all in" rates.
   e. Supports unit price billings.

Purchasing application

The benefits of this application are:

1. Simplify requisition and purchase order generation with standard clauses and reserved lines.
2. Manage purchase order changes through on-line change order feature.
3. Obtain volume discounts by filing multiple requisitions with one purchase order.
4. Prevent purchase order overpayments.

Some of the standard features of this application include:

1. Provides approved and pending commitment information to job cost.
2. Automates requisition, purchasing, receiving, and invoice entry functions.
3. Fully integrated with job cost and accounts payable.
4. Allows grouping of like items from requisitions to optimize purchasing.
5. Prints user-tailored requisitions and purchase orders on demand.
6. Eliminates redundant data entry with extensive replication features.
7. Manage inventory by item and warehouse.
8. Simplify receipt and issue processing.
9. Issue materials to a job, piece of equipment, or another warehouse.

Fixed assets application

The benefits to this application are:

1. Automates the management of property and plant.
2. Allows for forecasting of future expense.
3. Simplifies tracking of location.

Some of the standard features of this application include:

1. Multiple predefined accounting classes or books.
2. Automatic or manual setup of depreciation schedules.
3. Automatic or manual switch to straight-line depreciation.

4. Six different depreciation methods.
5. Four depreciation calculation conventions.
6. Linked to general ledger, job cost, equipment, and purchase order applications.
7. Ability to split an asset after it has been set up for transfer of location purposes or partial disposal situations.

Subcontract application

The benefits of this application are:

1. Easily and accurately maintains change orders.
2. Simplifies tracking of insurance and other submittal requirements.
3. Automates retention tracking and processing.
4. Prevents subcontract overpayment.
5. Simplifies daily processing with flexible entry processes and replication features.

Some of the standard features of this application include:

1. Provides commitment information (approved and pending) to job-cost-based or user-defined milestone.
2. Comprehensive reporting on subcontract status, submittal requirements, turnaround document, and insurance requirements.
3. Automatically withholds retention.
4. Fully integrated with job cost and accounts payable.
5. Allows for company standard and contract unique submittal requirements.
6. Provides extensive replication features using master and standard contracts.
7. Comprehensive joint check processing.

Property management

Some of the standard features of this application include:

1. Each property is owned by a financial entity.
2. Supports multiple entities, but properties can be individually managed.
3. Manages property level expenses.

4. Maintains detail tenant ledger, receivable, and rent roll information.
5. Automatically generates recurring charges and late charges.
6. Maintains comprehensive data-sensitive tickler information at all levels.
7. Provides potential rent reporting and vacancy tracking.
8. Extensive on-line inquiry and reporting capability.
9. Provides flexible calculations for pass-through billings.
10. Escalates recurring charges based on flat percentage or change in CPI.

Loan and draw

Some of the benefits of this application include:

1. Ability to access data already captured through normal work flow, thus eliminating redundant data entry.
2. Manage loan totals, balances, and individual loan draw amounts.
3. Prevents items from being requested multiple times and prevents items from being excluded.

Some of the standard features of this application include:

1. Supports several jobs in one loan, or several loans for one job.
2. Tracks loan budget and budget provisions.
3. Allows items to be "split" to include or exclude portions of a transaction.

Unit sales

Some of the benefits of this application include:

1. Allows for tracking of construction stage by unit.
2. Automates management of sales closings, cancellations, and reservations.
3. Provides for timely reporting of sales and inventory information.
4. Manages and analyzes the sale and order process of buyer options and upgrades.

Some of the standard features of this application include:

1. Extensive buyer demographic profiles available.

2. Allows for tracking of sales contract price components.
3. Automates escrow journals for each closing.
4. Automates cost of sales journals.

When Should a Contractor Computerize the Accounting?

With the advent of the microcomputer, there probably is no reason that any contractor should not have a computerized accounting system. Certainly the time to do this is when the volume of information being processed is too large to be handled manually. A precursor to setting up an automated system is an adequate system of controls. If these are not in place, all that will exist is organized chaos.

One of the principal considerations in assessing whether to convert to an automated accounting system should be the perceived benefit to be derived by the operation of the business in the near and longer term. Will there be better estimating capability, improved management reporting, or better cost controls? Any and all of the benefits of each of the applications described above should also be a factor to consider when looking at installing a computer system.

An option which should be considered by the contractor is time-sharing. Because the cost of a dedicated telephone line has come down dramatically, there is no reason that this should not be considered. The first hurdle to overcome would be to find someone who offers this service. There is at least one software vendor of construction systems that offers this service to contractors. Probably less desirable but yet another source might be to arrange with another contractor to use part of his idle capacity. The initial cost of using a time-sharing service is significantly less than that of purchasing a system. The ongoing cost can be demonstrated to be less also when the personnel cost involved is taken into account. From an operating point of view, the problems associated with keeping a system up and running would then fall on another company. The only downside is that a contractor cannot point to the in-house computer when giving tours to potential customers. With a dedicated telephone line, a contractor can be on-line to the time-share from anywhere in the country and yet have the feel of an in-house system.

Converting Systems

Converting from a familiar to an unfamiliar accounting system is very disruptive to the entire organization. It is time-consuming and can be costly. It is not a decision that is made every day and not one to be made on the spur of the moment. The decision to make the change can be brought about by external factors as well as internal ones. Sometimes the hardware and software are supplied by vendors that no longer continue to be available. On other occasions, the company may outgrow the ability of the software to handle either the volume or type of transactions that occur.

The Process of Selecting and Implementing a Computer System

Lots of written materials have been presented on how to select a computer system. Generally there are seven steps to be followed in this process. First, determine what is needed and what is desired. (See attached Exhibit I for a sample listing to be considered.) Second, determine what software and hardware exists that will allow you to meet these needs and desires. Third, from the list of available suppliers, prepare a request for proposal. (See attached Exhibit II for an outline of an RFP.) Fourth, evaluate the responses to the RFP. (See attached Exhibit III for a checklist for evaluation.) Fifth, arrange a full demonstration of the software that best fits the needs as defined in step one and three. This step can be repeated as many times as necessary until all potential suppliers have been reviewed. This is a very time-consuming effort and is best done by selecting the most appropriate vendor first so that if time runs out, the most effective solutions will have been demonstrated. Sixth, investigate the references of the supplier. From this source is gained some of the most beneficial data. Vendors are obviously going to put their best foot forward. Existing customers will tell you the way it really is. Seventh, select the system that is best for the company.

This all sounds simple, and it can be fairly easily organized. Presented in the exhibits are some helpful guides to completing the process. However, with the advances that have taken place in the last few years, some companies have hired independent consultants to help them evaluate hardware and software proposals. This process can add about 10 to 15 percent of the cost to the project but might make for a better and more complete decision.

XYZ COMPANY
Exhibit I
NEEDS and DESIRES

Elements	NEED	LIKE
FINANCIAL REPORTING/GENERAL		
Company structure		
Multi company	☐	☐
Multi division	☐	☐
Multi department	☐	☐
Chart of accounts		
User defined lengths	☐	☐
Account, subaccount	☐	☐
Statistical accounts	☐	☐
Data entry		
Single period accounting only	☐	☐
Enter to any open period	☐	☐
Enter to future fiscal years	☐	☐
Enter to closed periods	☐	☐
Reporting periods		
Limited to 12	☐	☐
13 to 24	☐	☐
More than 24	☐	☐
Customized financial statements		
User defined	☐	☐
Available on-line	☐	☐
Ratio analysis	☐	☐
Inquire on-line into accounts and subsystems	☐	☐
Enter and maintain budgets:		
General ledger	☐	☐
Project/Job cost application	☐	☐
Work orders	☐	☐
Capital purchases	☐	☐
Process transactions real time	☐	☐
Batch transactions and post only after review	☐	☐
Hold system generated transactions for review	☐	☐
Validate data during entry to fields that cross reference another application	☐	☐
Prevent out of balance between G/L and sub ledger	☐	☐
Secure data access and processing at various levels	☐	☐
Different processing cutoff for each application	☐	☐

XYZ COMPANY
Exhibit I
NEEDS and DESIRES

Elements	NEED	LIKE
PROJECT/JOB COST APPLICATION		
Interface to:		
Scheduling packages	☐	☐
Estimating packages	☐	☐
Categorize jobs in different ways to drive reporting:		
Market	☐	☐
Location	☐	☐
Work type	☐	☐
Customer	☐	☐
Project manager	☐	☐
Estimator	☐	☐
Certified payroll	☐	☐
Other	☐	☐
Data entry – quantities		
Enter daily	☐	☐
Accumulate in weekly qty for review & change	☐	☐
Budgets		
Lock in original budget & prohibit changes	☐	☐
Maintain a revised budget	☐	☐
On–line revision capability	☐	☐
Labor burden		
Report separate from labor	☐	☐
Report in the labor number	☐	☐
Report for a user defined group of items	☐	☐
Standard rate at start of job & change later	☐	☐
Change orders		
Track history by amount	☐	☐
Track history by date	☐	☐
Maintain pending change orders	☐	☐
Force account rate for billing equipment	☐	☐
Ability to reorganize or reclassify dollars within a job for presentation of a claim without altering historical data	☐	☐
Reporting		
Weekly labor & equipment as soon as data is entered	☐	☐
Job performance report for user defined date range	☐	☐
Branch performance report	☐	☐
Executive summary	☐	☐
Monthly detail materials on hand	☐	☐
Trend analysis	☐	☐

XYZ COMPANY
Exhibit I
NEEDS and DESIRES

Elements	NEED	LIKE
PROJECT/JOB COST APPLICATION		
Percentage of completion reporting		
Report	☐	☐
Screen	☐	☐
Calculate as % complete (current cost/est. final cost)	☐	☐
Calculate on other options such as hours	☐	☐
Calculate backlog (est final rev – recognized rev)	☐	☐
Calculate by month to date	☐	☐
Calculate by year to date	☐	☐
Calculate by job to date	☐	☐
Calculate by quarter to date	☐	☐
Recognize entire est final loss in current period	☐	☐
Dates		
Maintain award date	☐	☐
Maintain estimated completion date	☐	☐
Maintain physical completion date	☐	☐
Contract forecasting		
Input forecasted final cost and quantity	☐	☐
Project cost to complete by using:		
Actual unit cost to date	☐	☐
Estimated unit cost	☐	☐
Override unit price	☐	☐
Final cost	☐	☐
Cost to complete	☐	☐
Project revenue at completion:		
Last revenue forecast	☐	☐
Actual revenue to date	☐	☐
Final quantity times unit price	☐	☐
Change unit price	☐	☐
Ability to retain forecast options from last time	☐	☐
Generate reports by item	☐	☐

XYZ COMPANY
Exhibit I
NEEDS and DESIRES

Elements	NEED	LIKE
GENERAL LEDGER		
Fully integrated with all modules		
Maintain standardized chart of accounts		
Ability to process multiple periods concurrently		
Ability to run two or more years		
Ability to maintain both cash and accrual basis		
Budgets		
Report budget vs actual		
Multiple budget versions		
Consolidate or break out budgets		
Access on–line		
Print graphically		
Transaction entry		
Inquire, retrive prior, current and future year data		
Review unposted transactions		
Maintain journal entry transaction description		
Maintain and process recurring entries		
Maintain and process reversing entries		
Automatically post from subsidiary ledger		
Detect out of balance entries		
Validate all account numbers up front from all app.		
Prior period/year adjustments		
Transaction edit reports		
Post real time or by batch at user option		
Post to new year while prior year open		
Automatic inter/intra company journals		
Maintenance		
Ability to retain history from orginal number when account numbers change		
Ability to inactivate an account but not to delete from balance history summary		
Universal chart of accounts		
Consolidate a % of a company		
Reports		
Trial balance		
Detail account analysis		
Income statement		
Balance sheet		
Financial ratios		
Custom reports		

XYZ COMPANY
Exhibit I
NEEDS and DESIRES

Elements	NEED	LIKE
PAYROLL/HUMAN RESOURCES		
Employee information		
Access by name, social security, or emp number		
Check for duplicate employee by ss# or employee #		
Maintain history with:		
Name		
Where worked		
Worker's compensation history		
History of earnings		
Reason for leaving		
Comment text section		
Type of work experience		
Track & record hire and rehire dates		
Track on the job training apprentice programs		
Ability to assign multiple crafts to each employee		
Employee status (active, terminated, temporary, etc.)		
Classify by type (hourly, salary, retired, etc.)		
Maintain negative accumulators		
User defined fields		
Labor rates		
Maintain, control and pay labor rates by job tied to a master file of rates by job		
Allow subsistence in rate file		
Override subsistence rules at time of data entry		
Apprentice wage increases based on length of service		
Multiple labor rate schedules based upon date		
Interim rate changes outside contract changes		
Multiple pay deduction/fringes based upon day, hours, flat rate, lump sum, and user defined		
Mass update for cost of living adjustments		
Across the board salary or hourly rate changes by:		
Percent of base		
Flat amount		
Job grade		
Job classification		
Location		
Time entry		
Batch timecards		
Enter for next pay period before current period closed		
Enter on–line		
Enter off–line		
Automatic validation on screen		
Validate pay type eligibility		
Flag equipment not assigned to job		
Flag is employees assigned vehicle not reported		

XYZ COMPANY
Exhibit I
NEEDS and DESIRES

Elements	NEED	LIKE
PAYROLL/HUMAN RESOURCES		
Time entry (continued)		
Ability to edit/correct time entered and ready for pymt	☐	☐
Enter charge out and hours for crew sheets	☐	☐
Automatically assign overtime based upon multiple overtime schemes	☐	☐
Hash total on data entry	☐	☐
Standard labor rate by equipment number	☐	☐
Support auto-pay processing for non-time card emp	☐	☐
Unlimited number of time card entries per pay period	☐	☐
Work temporarily in a higher class or different cost ctr	☐	☐
Work on multiple jobs in any pay period	☐	☐
User defined holiday table, auto time card entry	☐	☐
Warning for entry of hours for terminated employee	☐	☐
Salary and wage administration		
Manitain wage and salary survey information	☐	☐
Maintain pay grade structures and salary ranges	☐	☐
Employee pay hisyory	☐	☐
On-line review	☐	☐
Multiple salary pay plans	☐	☐
Ability to analyze pay by location	☐	☐
Perform "what if" salary and incentive projections	☐	☐
Historical position compensation data	☐	☐
EEOC designations for each class	☐	☐
Accept FLSA codes	☐	☐
Pay types		
Bonus	☐	☐
Vacation	☐	☐
Sick	☐	☐
Safety	☐	☐
Vehicle	☐	☐
Disability	☐	☐
Moving	☐	☐
Overtime	☐	☐
Living allowances	☐	☐
Subsistence	☐	☐
Shift differential	☐	☐
Severance	☐	☐
Jury	☐	☐
Funeral	☐	☐
User defined	☐	☐
Other compensation such as education	☐	☐
Ability to define as taxable or non-taxable	☐	☐
Increase gross by taxed amount	☐	☐
Maintain leave information	☐	☐

XYZ COMPANY
Exhibit I
NEEDS and DESIRES

Elements	NEED	LIKE
PAYROLL/HUMAN RESOURCES		
Deductions		
Create unlimited number of user defined codes		
Garnish wages in accordance with federal regulations		
Multiple deduction formulas		
Deduct all to a predetermined net, per IRS regulations		
Deduct flat amount per day worked		
Track deductions by week, month, year to date		
Prioritize deductions		
Enter effective date and expiration date		
Distinguish and process pre-tax deductions		
Refund deductions		
Arrears balances deducted over user specified period		
Method for variable employer percentage matching of employee contributions based upon employee election		
Ability to track a declining balance for end to withhold		
Standard deductions:		
Savings bonds		
United Way		
Union dues		
Benefits		
Payroll savings		
Merchandise		
Insurances		
Garnishements		
Credit union		
Ability to override or stop standard deductions		
Process non-cash fringe benefits		
Calculate and track employer paid benefits		
Handle deferred comp and other pre-tax items		
Flexible spending account		
Taxes		
Multiple states/locations for an employee within the same pay period		
Tax all jurisdictions/states/countries		
Override taxes for exempt employees		
Unlimited states worked per employee per year		
Accumulate tax liability monthly, quarterly, and yearly		
Elect tax treatment of earnings categories at company organization level		
Balance total wages to federal, state, and local wages		
Refund taxes on a paycheck as needed		

XYZ COMPANY
Exhibit I
NEEDS and DESIRES

Elements	NEED	LIKE
PAYROLL/HUMAN RESOURCES		
Burden		
Ability to calculate tax burden on actual labor dollars		
Ability to analyze and track burden history by location		
Ability to track workmen's comp for insurance year		
Ability to change mod and excess limits rates		
Ability to track pay types by state		
Ability to charge jobs, items for PL & PD and generate invoice to insurance company		
User defined parameters for builders risk insurance		
Ability to track by:		
Job		
Month		
State		
Wage reporting		
Generate all regulatory compliance reporting on magnetic media		
Generate quarterly reports for workers comp, unemployment ins, and 941s		
Produce W-2's including year end adjustments in multi-copy format.		
Ability to report on form 1099		
Certified payroll		
Generate for any pay period for any job		
Designate as a "c p/r" job when job is set up		
Generate according to agency requirements		
Selectively exclude employees from report		
Shoe total earnings and job related earnings for wk		
Retroactive certified payroll for prior periods		
General for Human Resources		
Accommodate multiple plans		
Accommodate multiple insurance companies		
Capture, store, and search on-line the SPD produce by the insurance companies		
Generate required reports in insurance co format		
Generate reports on annual benefits coverage and use by individual employee		
Ability to perform discriminatiom testing on a different basis for each plan		
COBRA		
Accommodate both employees and dependents		
Generate required letters within legal time limits		
Generate billings to participants		

XYZ COMPANY
Exhibit I
NEEDS and DESIRES

Elements	NEED	LIKE
PAYROLL/HUMAN RESOURCES		
EQUAL EMPLOYMENT OPPORTUNITY		
Ability to maintain the following employee data:		
Race code		
Gender		
Disabled per ADA		
Job classification		
Veteran		
Vietnam Veteran		
Date of separation		
Reason for separation code		
Journeyman/ apprentice status		
Department		
Job		
Date of Hire		
Date of Birth		
Ability to generate standard reports		
EEO−1		
Monthly Employment Utilization Report (257)		
Identify jobs by area using SMSA and EA		
EEO reporting requirements		
Track EEO complaints		
Track subcontractors by disadvantaged business category		
Report dollar volumes from accounts payable for DBE, etc		
Maintain DBE goals by jobs and report actuals		
Performance Review		
Maintain unlimited review history including dates		
User defined form letters regarding review process		
Enter, categorize and report comments and feedback		
Employment/Applicant Profiles		
Access employee data re compensation & benefits		
Unlimited employment history including assignments		
Unlimited training and development history		
Unlimited record of education including degrees, dates, licenses, and institutions		
Unlimited skills inventory		
Associate supervisor with employee		
Emergency notification information		
Dependent information		
Enter into applicant info and roll into employee record		
Applicant Processing		
Build a job order		
Track interviews related to job order		

XYZ COMPANY
Exhibit I
NEEDS and DESIRES

Elements	NEED	LIKE
ACCOUNTS PAYABLE/PURCHASE ORDERS		
Vendor maintenance		
Name, address, telephone #, contact	☐	☐
Federal ID #	☐	☐
EEO status	☐	☐
Multiple locations	☐	☐
Terms	☐	☐
Flag for discounts	☐	☐
Comment section	☐	☐
Flag for tax levy, liens, and bankruptcy	☐	☐
1099 codes	☐	☐
Add new vendors & check for duplication	☐	☐
Invoice transaction entry		
Input from field or remote location on–line or batch	☐	☐
Batch control system	☐	☐
Integrate with purchase order application	☐	☐
Ability to pay alternate payees	☐	☐
Scan invoices on–line	☐	☐
Prorate tax, freight or other misc charges	☐	☐
Ability to enter text comment information	☐	☐
Automatic calculation of pay date/ override pay date	☐	☐
Calculate discount and make distribution	☐	☐
Accrue use tax if appropriate	☐	☐
Payments		
Handle use of manual checks	☐	☐
Multiple bank accounts	☐	☐
Hold payment awaiting ID# for 1099 purpose	☐	☐
Handle multiple joint check situations	☐	☐
Compare A/R balance with A/P balance for same party	☐	☐
Ability to set up recurring payments	☐	☐
Ability to produce MICR encoded checks on laser	☐	☐
Void and regenerate printer damaged checks	☐	☐
Print checks in following order:		
Vendor name	☐	☐
Descending check amount	☐	☐
Job	☐	☐
Special handling required	☐	☐
Zip code	☐	☐
Any combination	☐	☐
Miscellaneous		
Automatic bank reconciliation	☐	☐
Reverse all charge outs when check is voided	☐	☐
Capture stop pymt info for automatic reissue later	☐	☐

XYZ COMPANY
Exhibit I
NEEDS and DESIRES

Elements	NEED	LIKE
ACCOUNTS PAYABLE / PURCHASE ORDERS		
Reports		
Accounts payable cash requirements	☐	☐
List future pay dates and amounts	☐	☐
List vendors with credit balances	☐	☐
List invoices on hold	☐	☐
List invoices processed beyond billing date	☐	☐
Vendor list	☐	☐
Check register	☐	☐
Subcontract liability report	☐	☐
Accounts Payable liability report	☐	☐
Vendor aging report	☐	☐
1099 report	☐	☐
Detail purchases and commitments	☐	☐
Purchase order master list	☐	☐
Requisition list	☐	☐
Purchases by vendor	☐	☐
Purchase orders		
Include special conditions from user-defined options	☐	☐
Generate blanket purchase orders	☐	☐
Warn that equipment is under warranty	☐	☐
Validate charge out codes	☐	☐
Generate control and purchase order number	☐	☐
Enter requisitions on-line and track thru purchase order and account payable phase	☐	☐
Calculate proper sales tax for multi-jurisdictions	☐	☐
Document and track competitive bid situations	☐	☐
Charge to multiple jobs	☐	☐
Charge to multiple items	☐	☐

XYZ COMPANY
Exhibit I
NEEDS and DESIRES

Elements	NEED	LIKE
ACCOUNTS RECEIVABLE / BILLINGS		
Contract billings		
Enter quantities by item # for billing purposes		
Identify job cost items as contract items for billing		
Identify billing items by owner's codes		
Tie billing quantities to cost system quantities		
Billing by:		
Lump sum		
Item of work		
Cost plus		
Percentages		
Quantity driven		
Multiple billing formats		
Distinguish extra work from contract billings		
Different contractual terms for types of work		
Multiple customers on one contract		
Generate credits and rebills		
Ability to overrun at different price		
Batch billings for approval and posting		
Ability to set up retention at time of billing based on contract requirements (vary by item/% complete)		
Ability to set up retention at time of payment		
Account for liquidated damages, penalties, etc.		
Ability to view billings on-line for life of job		
Ability to analyze revenue by type of work		
Interface with subcontact system for payment when paid by owner for both current and retention		
Statements by request		
Differentiate between taxable and non-taxable rev		
Customer master		
Ability to maintain multiple addresses		
Maintain customer as a subsidiary to master customer		
Track customer's resale number and expiration date		
Flag account as subject to interest penalties or not		
Maintain credit limits		
Code customer credit requirements:		
Cash sales only		
COD only		
Deposit required		
Ability to search by name or partial name		
Flag account which is to receive statement		
Access job info from customer account maintenance		
Cash processing		
Ability to view invoices on-line		
Ability to track deposits		

XYZ COMPANY
Exhibit I
NEEDS and DESIRES

Elements	NEED	LIKE
ACCOUNTS RECEIVABLE / BILLINGS		
Cash processing (continued)		
Apply cash to:		
Specific invoices		
Partial payment of invoice		
Aged balance		
Oldest items		
Other		
One check to apply to different customers		
Split a check among several invoices		
Adjust invoices to match payment for small differences		
Set up retention at time of cash entry		
Interface with accounts payable		
Reports		
Cash projection report		
Aging		
Retention receivable		
Billing register		
Accounts receivable summary		
Cash receipts		
Customer Listing		
Customer payment history		
Reports called in detail or summary		
Reports called by customer or for all		

XYZ COMPANY
Exhibit I
NEEDS and DESIRES

Elements	NEED	LIKE
EQUIPMENT		
Master Information Record		
Ability to identify equipment by:		
Number		
Series (e.g. track equipment)		
Category (e.g dozers)		
Association with another piece		
Description of equipment including:		
Equipment identification number		
Capacity		
Horsepower		
Attachments		
Manufacturer		
Model		
New/Used flag		
Fuel code		
Year of manufacture		
Make		
Model		
Warranty information		
Date of acquisition		
Revenue rate		
Revenue unit of measure		
Multiple revenue rates		
User-defined rates		
Department assignment		
Location history		
Division assignment		
Serial number (digits > 20)		
Multiple purchase order numbers		
Work order number		
Number of axles		
Unladen weight		
Gross vehicle weight		
License number, expiration date		
Acquisition		
Enter capital asset requisitions into database		
Create draft budget of capital equipment purchases		
Generate purchase order		
Maintenance		
Interface to equipment mgt system such as ELKE		
Ability to interface with payroll		
Abilty to interface with accounts payable		
Ability to generate a maintenance schedule		
Ability to generate a non-compliance with sch. maint.		

XYZ COMPANY
Exhibit I
NEEDS and DESIRES

Elements	NEED	LIKE
EQUIPMENT		
Repairs		
Link life – to – date revenue units to repair transaction	☐	☐
Associate repair to the job the unit was assigned to	☐	☐
Ability to override job association	☐	☐
Code repair transaction with:		
Class of repair (major, minor, gets, etc.)	☐	☐
Component repaired	☐	☐
Type of repair (overhaul, rebuild, etc.)	☐	☐
Display images of documents scanned into system	☐	☐
Create a repair work order	☐	☐
Collect repair history by component	☐	☐
Components		
Track multiple mechanical components on each		
Type	☐	☐
Warranty description	☐	☐
Location	☐	☐
Other	☐	☐
Accumulate multiple measures of use on each		
Timecard hours	☐	☐
Engine service meters	☐	☐
Fuel consumption	☐	☐
Ability to "reset" measures of use on component	☐	☐
Tires		
Ability to interface with tire management system such as Goodyear's Tyrec or ELKE	☐	☐
Tire data to flow through accounts payable	☐	☐
Track tire location by machine	☐	☐
Assign brand/serial number to tire	☐	☐
Record periodic measurements of tread	☐	☐
Retrieve units (by hours, yds, miles) to determine wear	☐	☐
Enter junk cost to determine if credit is due	☐	☐
Track repair/retread information	☐	☐
Track cost of disposal	☐	☐
Usage		
Interface with payroll for units (miles, hrs, paver tons)	☐	☐
Differentiate between operating and standby hours	☐	☐
Collect engine service meter readings	☐	☐
Assess minimum revenue charges if hours fall below established minimum without affecting units of usage	☐	☐

XYZ COMPANY
Exhibit I
NEEDS and DESIRES

Elements	NEED	LIKE
EQUIPMENT		
Revenues		
Maintain standard revenue rate for each piece of equip		
Twelve components:		
Depreciation		
Rent		
Tax and insurance		
Fuel		
Tires		
Major labor		
Minor labor		
Major parts		
Minor parts		
GETS labor		
GETS parts		
Maintain three exception rates		
Escalate rates by a specified percentage		
Access to cost and fuel consumption history		
Special minimum use rates		
Restate cost results using alternate equipment rates		
Operating Costs		
Allocate costs for labor burden, insurance, lubes, oil, and other		
Reallocate costs monthly based on annualized usage		
Maintain cost–to–operate by:		
Equipment		
Division		
Job–to–date		
Year–to–date		
Month–to–date		
Life–to–date		
Specified period		
Reports costs and use:		
Repair parts cost		
Repair labor costs		
Repair labor hours		
Scheduling		
Ability to look at multi–year future equipment usage		
Division		
Branch		
Job		
Distinguish between working commitments, reserved equipment, and bid commitments		
Ability to allow for multiple bid commitments		
On–line requests for reservation and cancellation		

XYZ COMPANY
Exhibit I
NEEDS and DESIRES

Elements	NEED	LIKE
EQUIPMENT		
Taxes		
Identify equipment as real estate or personal property	☐	☐
Ability to show state/agency taxes	☐	☐
Location history for tax reporting requirements	☐	☐
Associate idle equipment with a location	☐	☐
Input market rate for calculation of assessed value	☐	☐
Input tax rate for auto calculation of tax due	☐	☐
Automatic posting of charge for tax to equipment	☐	☐
Multiple tax jurisdiction tax pd history – sales & use	☐	☐
Track permit, renewal, certification requirements	☐	☐
Disposition		
Maintain:		
Sale date	☐	☐
Sale amount	☐	☐
Sales & Use tax	☐	☐
To whom sold	☐	☐
Text field for comments	☐	☐
State sold in	☐	☐
Generate profit/loss entries	☐	☐
Code as disposed and flag if future charges ok	☐	☐
Fuel		
Multiple inventories by location	☐	☐
Maintain multiple fuel types	☐	☐
Capture gallons purchased	☐	☐
Identify if purchase includes tax	☐	☐
Distinguish base price from taxes and freight	☐	☐
Track storage tank locations & capacity	☐	☐
Enter physical inventories	☐	☐
Calculate book inventory	☐	☐
Calculate gallon variance between book and physical	☐	☐
Ability to transfer between locations	☐	☐
Interface with computerized fuel dispenser	☐	☐
Remote site on–line consumption entry	☐	☐
Code consumption as on– or off–highway	☐	☐
Ability to validate consumption data entry for:		
Job number	☐	☐
General ledger number	☐	☐
Equipment number	☐	☐
Fuel type matched to equipment	☐	☐
Charge out code matched to tank location	☐	☐

XYZ COMPANY
Exhibit I
NEEDS and DESIRES

Elements	NEED	LIKE
FIXED ASSETS		
Ability to track additions and deletions		
Monthly	☐	☐
Quarterly	☐	☐
Annually	☐	☐
User defined	☐	☐
Track location of rolling stock	☐	☐
Maintain multiple sets of books	☐	☐
Reports sorted by user defined parameters	☐	☐
Multiple methods of depreciation	☐	☐
Interface with all subsystems	☐	☐
Ability to split asset if part is disposed	☐	☐
Ability to add cost after acquisition	☐	☐
Track assets with 0 values such as storage vans	☐	☐

XYZ COMPANY
Exhibit I
NEEDS and DESIRES

Elements	NEED	LIKE
SUBCONTRACTS		
Master record to include:		
Name, address, phone number, fax number		
Federal ID number		
Social Security number		
Doing Business as		
Contractors license number and expiration date		
Special ID number by agency		
Contact person/position		
Bonding company name contact & phone number		
Insurance company and type of coverage		
Type of work		
Geographic areas		
MBE, DBE status		
Type of ownership		
Text area		
Maintain standard clauses/retrieve to print a subcontract		
Pay file to include:		
Subcontract number		
Job number		
Period ending date		
Payment number		
Agency bid item numbers		
Item/subitem number		
Item description		
Unit of measure		
Subcontract quantity		
Unit price or lump sum		
Previous quantity		
Quantity this period		
Contract to date quantity		
Material on hand		
Field for adjustments		
Field for discount percentages		
Retention percent		
Joint check payee		
Original contract amount		
This period to contract amount		
Non–retention pay items		
Extra work items		
Withhold, contract, backcharge amounts		
Owners pay estimate number		
Certificate of insurance expiration date		

XYZ COMPANY
Exhibit I
NEEDS and DESIRES

Elements	NEED	LIKE
SUBCONTRACTS		
Payments		
Have single vendor # for s/c who is also a vendor		
Hold payment and show who and why held		
Show discount on pay voucher		
Handle retention – fixed or flexible percentage		
Show multiple locations for each subcontractor		
Generate multiple and joint checks for each estimate		
Handle materials on hand payments		
Generate pay est by entering either quantity to date or quantity this period		
Pull subcontract items from bid items, job cost		
Ability to hold credit balances		
Flag potential duplicates between invoice and pay voucher created by system		
Reports		
List all subcontracts		
Show subcontract liability (pymt, retent, withholding)		
Track subcontractor's insurance expiration		
Generate insurance requirment letter		
Generate letter to sub's surety when payment is in excess of bonded amount		
Create tickler file for contracts and other documents		
Track preliminary notices		
Flag subcontractors listed on prime contract		
Data collection for various agency requirements		
Amount paid to date for MBE/DBE reporting		
Underrun contracts, generate refund request to subcontractor's surety		
Generate request for Federal ID # on first time entry		
Trigger final checklist for all subs when job is 100%		
Show payment history by subcontract		

XYZ COMPANY
Exhibit I
NEEDS and DESIRES

Elements	NEED	LIKE
Safety		
Accumulate incidence reports on–line including:		
Case number		
Insurer's number		
Location code		
Name, address, city of accident location		
Claim number		
Social security number of claimant		
Type of work assigned at time of incident		
Type of injury/illness		
Part of body		
Source (e.g. hit by vehicle, etc.)		
ANSI standard code for type		
Date of injury/illness		
Part of body		
Date employee returned to work		
Date of death		
Total days lost		
Total days of restricted duty		
OSHA recordable		
Foreman		
Superintendednt		
Estimator/Project Manager		
Probable cost		
Ability to associate hours worked for crew assigned to foreman and superintendents		
Generate report by month, year–to–date, job–to–date and most recent 12 months including:		
Total hours worked		
Total injuries		
OSHA injuries		
Lost–time injuries		
Days lost		
Days restricted		
Cost based on average cost/claim		
Ability to produce the OSHA 200 report by:		
Company		
County		
Division		
Department		
State		
Project		
Job		
Ability to calculate average cost per claim		
Produce geographical presentation of accident statistics		
Report incidents/hours of exposure by claim type		

XYZ COMPANY
Exhibit I
NEEDS and DESIRES

Elements	NEED	LIKE
Safety		
Report incidents/hours of exposure grouped as:		
Injuries to employees		
Damage to company–owned equipment		
Personal loss/property damage to third parties		
Report safety information by:		
Foreman		
Superintendent		
Estimator/Project Manager		
Track training records by:		
Name		
Social Security number		
Job number		
Date		
Topic		
Report hours worked by plant/job for MSHA for:		
Salaried employees		
Craft employees		
Where work was performed		
Track exposure by employee to include:		
Employee identifiers		
Date of exposure		
Chemical/agent exposed to		
Concentration of exposure		
Personal protection required		
Length of exposure		
Ability to retain training and exposure records for duration of employement + 30 years		
Track subcontractor safety performance		
Environmental Miscellaneous		
Track permits (air, water quality, storage tanks, use permits, etc.) to include:		
License number		
Fees (permits, application)		
Expiration date		
Maintain manifests of waste disposal by:		
Location		
Political subdivision		
Owner		
Ability to report licenses/permits due to expire		

XYZ COMPANY
Exhibit I
NEEDS and DESIRES

Elements	NEED	LIKE
INSURANCE		
Claims Tracking		
Ability for field to enter claim report on–line	☐	☐
On–line validation and update of all claim info	☐	☐
Report asset losses, liability, and or both	☐	☐
Interface to other applications	☐	☐
Abilty to reference and report claim info including:		
Asset type code	☐	☐
Claim type code	☐	☐
Location code	☐	☐
Loss date	☐	☐
Submit date	☐	☐
Close date	☐	☐
Amount fields	☐	☐
Additional liability field	☐	☐
Text field for describing loss in detail	☐	☐
Insurance company reference	☐	☐
Responsible entity	☐	☐
Ability to accumulate data by policy year	☐	☐
Write off		
Ability to write–off claim based upon multiple user–defined formulas	☐	☐
Ability to store claim history including closed date	☐	☐
Reports		
Report claims by location and total company	☐	☐
Detailed report show current period & cumulative cost	☐	☐
Monthly report to list all write–offs	☐	☐
Insured claims by category in current policy year	☐	☐
Workman's comp forms for various states	☐	☐

XYZ COMPANY
Exhibit I
NEEDS and DESIRES

Elements	NEED	LIKE
MIS		
Development Issues:		
Can we view the source code?		
What development tools were involved?		
Can we obtain the same look and feel in modifications from our developed software as in your software?		
Will menu driver allow us to access other programs?		
Do you offer rapid application development tools?		
Do you offer source–level debuggers?		
Can we fully access database from our applications?		
Vendor support		
Customer support available 7AM to 7PM in all time zones?		
Can we expect problem resolution in 30 minutes or less?		
Will customer support contacts be knowledgeable of the construction industry?		
Do you support a Users Group?		
Is there access to a bulletin board for problem resolution and download solution?		
Can software be upgraded or remove an upgrade without restoring the whole system?		
Can we install your update and preserve our developed features?		
Can we review fixes and features as they become available?		
Is training available for:		
Introduction overview		
Train the trainer		
Technical training		
On–site training		
Are modular training documentation and tools available?		
User interface		
Can we customize field labels displayed on screens, menus, and reports?		
Can we re–word menu choices to be more meaningful to our users?		
Is system consistent in screen layout?		
Is system consistent in key functions?		
Documentation		
Is a comprehensive end–user reference manual available?		
Is help–text displayed on–line?		

XYZ COMPANY
Exhibit I
NEEDS and DESIRES

Elements	NEED	LIKE
MIS		
Implementation		
Do you offer data conversion assistance		
Do you offer application setup assistance?		
Do you help us develop a test suite?		
Can you provide certified benchmarks test results for response times in each of your modules?		
Security		
Is security provided by:		
Value in a field		
User		
Location		
Read/write/update function		
File		
Application		
System		
Access over modem		
Company		
Time of day		
Day of week		
Can security profile be copied from one to another?		
Can we easily change passwords?		
Test environment		
Can we block software developers from seeing production data?		
Can we test new software without affecting the integrity of the production data?		
Do you offer a test suite to test upgrades?		

XYZ Company
EXHIBIT II
General Outline for a Request for Proposal

Introduction

- [ ] Briefly describe your business indicating aproximate sales, number of employees, type of construction performed, number of locations, and current system if any.

- [ ] Indicate the dates when you anticipate wanting to get started and complete the process. Request that a schedule be given based upon start date and end date if it appears that either too much or too little time was provided.

- [ ] Indicate that all responses to the RFP will be incorporated into any contract that is signed for purchase of software license.

- [ ] Indicate the number of users anticipated.

Application software requirements

- [ ] Provide a detail list of needs and wants similar to that provided in Exhibit I. Request that respondents provide information relating to needs and wants by indicating if included in base application, by customization, or not avaialble.

Vendor background

- [ ] Request information about the vendor and its background.
  Number of years in software business?
  Number of employees.
  Number of employees in software development.
  Number of employees in software support.
  Number of active customers.
  Number of customers using current software release.
  Number of customers in specific construction field.

- [ ] Ask for three references specific to type of construction your company performs.

- [ ] Ask for other construction industry references.

- [ ] Request information about financial stability.

Vendor support and Maintenance Policy

- [ ] Request that a description be provided for the types of support that can be expected.

- [ ] Request that a copy of the standard maintenance contract be provided.

- [ ] Request a training and implementation program be included in the response.

Pricing

- [ ] Base package license fees.

- [ ] Custom modifications

- [ ] Installation and training costs

- [ ] Support and maintenance costs

XYZ Company
EXHIBIT III
Checklist for Evaluating Software Alternatives

Functionality:

- ☐ Percentage of critical requirements met.
- ☐ Percentage of all requirements met.
- ☐ Alternative to features not offered.
- ☐ Integration between modules and other vendor's software.
- ☐ Integration with current hardware

Modification required:

- ☐ Will the modifications become part of the basic system?
- ☐ Will modifications be protected with new software releases?
- ☐ Who will make the modifications?
- ☐ Who will support the modification?

Technology direction of software developer:

- ☐ Proprietary or computer industry–standard technology for:
  - ☐ Programming language
  - ☐ Database
  - ☐ Hardware/technology
  - ☐ User interface
  - ☐ Report and inquiry writers
  - ☐ Modification tools
- ☐ Investment in new technology
- ☐ Integration with other software

Installation and training effort required:

- ☐ Vendor support before during and after installation
- ☐ Documentation
- ☐ Number of support personnel and programmers
- ☐ Long term support of the product
- ☐ User groups

Vendor stability and references:

- ☐ Financial stability
- ☐ Installed base in the construction industry
- ☐ Other users of similar size
- ☐ Telephone references and on–site visits
- ☐ Quality of phone support
- ☐ Frequency of enhancement releases

Costs and Terms:

- ☐ What are initial and ongoing costs for base package and modifications?
- ☐ What changes or additions to existing hardware are required?
- ☐ What are the terms and is there any flexibility to this arrangement?
- ☐ On what basis do you pay for maintenance and when does it start?

Appendix

C

Valuing a Construction Company

One of the various tasks which may be assigned to the accountant for a construction company is the valuation of a construction business. The first type of valuation which comes to mind is valuation for credit or bonding purposes. However, this is not really a valuation of the business as much as it is an evaluation of the earning power of the business and the strength of its secondary position with respect to a specific extension of credit. This, in turn, depends in part on what assets can be sold or used as collateral to obtain funds to finish a specific job or pay a specific loan. Thus, while involving some of the same considerations, evaluation for credit and bonding purposes is not the same as valuing the business standing alone.

Liquidating value versus "going concern" value

Stripped to the bare bones, the method of estimating the liquidating value of a construction company may be summarized in the following four parts:

1. The net cash that can be realized from the sale of the physical assets,
2. Plus the net cash that can be realized from collection of the receivables and liquidation of other miscellaneous assets,
3. Plus the cash profits which can be realized from completion of the jobs in progress,
4. Minus the liabilities and the costs of liquidation.

Obviously, this is not the same as book value, although the valuation process often starts with the current balance sheet since the balance sheet purports to show all the existing assets and liabilities. The difference lies in assigning to each asset a valuation which represents its current net realizable value in terms of cash and reducing the total value of the assets by the liabilities at the amount which would have to be paid off if the company were to go out of existence.

In contrast with this liquidating value, the so-called "going concern" value emphasizes the value of the business, a continuing and operating entity, as distinguished from the mere sale of the individual assets and payment of the liabilities. A valuation for this purpose would also consider the ability of the existing organization to bid construction work and perform construction jobs at a profit. Management skills and industry reputation of the individuals in charge of the business are factors. So also is the fact that all the miscellaneous expenses of assembling an organization, coordinating its operation, and equipping it have been paid and the business has the continuing advantage of those expenses. In short, it is the valuation of the business as a continuing profit-making entity.

Goodwill

There is a tendency on the part of lawyers, accountants and tax agents to classify the excess of going concern value over liquidating under the catchall term "goodwill." That term has been variously defined by the courts and authoritative writers. As applied to the construction industry generally it means the ability of the company, given its existing assets and organization, to earn profits in excess of an amount which would be "normal" for a construction business of that size and type. Generally speaking, the ability of a construction company to produce profits, whether greater or less than normal, depends on two things:

1. The ability of the company's organization, on a continuing basis, to obtain profitable construction jobs and complete them at a profit
2. The company's ability to finance as much work as its staff and equipment can handle

This means that in the construction industry, goodwill tends to be intrinsic to the individuals who make up the company's management, including their personal reputations and relationships. The credit and bonding capacity of a construction company depends to a substantial degree (though not exclusively) on the evaluation which the credit grantors place on the ability and integrity of the individuals in management. Labor stability and its contribution to profits depends, to an important degree, on the respect of union officials for the individuals making up the construction company's management team. Subcontractors and suppliers tend to give more favorable prices to contractors who manage their jobs well, pay their bills promptly, and settle claims promptly and fairly.

Factors such as these tend to inhere in the individuals who make up the contractor's organization rather than in the company itself. In addition, a construction company may have advantages due to location, or to its control of such material resources as gravel pits, or to the exclusive rights it has to various types of structures or equipment. In valuing a construction company as a going concern, it may be the existence of advantages such as these that account for the ability of a particular company to earn profits in excess of normal. If so, the various items should be separately valued.

Utilization of resources

Closely related to the management skills of a construction company's management team is the company's ability to utilize its resources to the greatest advantage. However, in placing a going concern value on a construction company at any particular time, the point of asset utilization is important enough for separate consideration.

Typical examples of companies which do not use their resources wisely are those which accumulate unneeded and unused equipment. Some contractors will tie up badly needed working capital to take advantage of what they consider to be bargain purchases of materials, supplies, or equipment for which they have no immediate use. In arriving at the going concern value of a construction company, such nonproductive assets must be heavily discounted and so must the ability of the management which ties up its resources in such a manner.

On the other side of the coin—the wise utilization of assets—is the company that salvages useful materials, supplies, and small tools which have already been written off against operations. Strictly correct accounting would place such assets on the books at their fair value. However, it is typical in valuing construction companies to find in their warehouses valuable inventories of materials, supplies, and small tools which do not appear on the books. Obviously, in valuing a construction company, these assets need to be considered.

Special problems in equipment valuation

No valuation of a construction company's equipment can be complete without consideration of two likelihoods. The first is that fully depreciated equipment may still be in use. It is common practice for construction accountants to remove from the equipment records individual pieces of equipment which have been fully depreciated, even though they are still in use. Second, it is common to find that special attachments may have been stripped off equipment to be junked or sold and either stored or put on other equipment without any adjustments on the company's books.

Obviously, it is essential to have someone familiar with construction equipment take a physical inventory of equipment in order to value any company whose assets include a substantial investment in construction equipment. Such an inventory must obviously take into account the condition of the individual equipment items and any special attachments (bulldozer blades, power control units, shovel buckets, etc.) which are attached to the individual units or held in stock.

Special problems in valuation of current assets

In valuing cash shown on a contractor's balance sheet it is necessary to consider whether any of the bank accounts have been frozen or pledged or for some other reason do not represent cash currently available. Special attention must be paid to bank accounts in foreign countries for possible foreign exchange problems and foreign currency restrictions. Care must be taken to see that accounts receivable are collectible and are not tied up in disputes or lawsuits. If cash in bid deposits or other special deposits is material in amount, the existence and collectability of the deposits needs to be checked.

Contracts in progress is probably the most difficult asset to value at any given date even though a reasonably satisfactory percentage-of-completion accounting system is in effect. It is, of course, standard practice to adjust the value of the balance sheet asset in successive accounting periods to reflect changed conditions.

In valuing a construction company at a given date, those periodic adjustments can only reflect facts known at the valuation date. For that reason greater care needs to be exercised in estimating the cost to complete each job. Predicting the ultimate profit on jobs in progress with any degree of accuracy requires a combination of engineering and accounting skills coupled with a high degree of practical construction know-how. Essentially the method best calculated to produce an accurate result is to have two experienced estimators, who are familiar with the jobs, prepare separate estimates of cost-to-complete just as if they were bidding the job to completion and then compare notes to reconcile their bids. Unfortunately, this degree of time and care is not ordinarily available to the person making the valuation. Often he must make do with whatever estimates of cost-to-complete are available to him.

Contingent assets and contingent liabilities

When a construction company has contracts which, if completed ahead of schedule or below estimated cost, result in bonus payments, the amount of such payments is better treated in an appraisal as a part of estimated profit on the jobs rather than as a contingent asset.

One of the most prolific sources of both contingent assets and contingent liabilities is lawsuits. Whenever possible it is better to exclude these items from the estimate of valuation of the company itself. The same is true of potential tax liabilities and refunds which are still contingent. If they must be valued, it is better that the estimates be made by attorneys and tax specialists other than those who are handling the case.

Past profits and losses

A major element in forming a final judgment on the value of a going concern, and particularly a construction company, is a history of past operations. Usually a five-year history is enough to get an overview of how effective management has been and how well the available assets have been used. Also, five years is usually long enough to evaluate the effect of conditions which seem to constitute an advantage or a disadvantage. For most construction companies it is long enough to average out the jobs so that one or two jobs which are losers or which are exceptionally profitable will not seriously distort the overall results.

If audited statements are available, a comparison of the balance sheets, the operating statements, and the supporting schedules can provide a good checklist. Also, if prepared by accountants who are knowledgeable about construction accounting, audited statements can give the appraiser the benefit of an independent outside opinion to use as a check on his own thinking.

Index

Index note: The *e.* after a page number refers to an exhibit; the *f.* to a figure; the *n.* to a note; and the *s.* to a schedule.

ABOUT THE AUTHORS

WILLIAM J. PALMER is the past national director of Construction Industry Services for Ernst & Young in San Francisco. He has headed the audit management for one of the largest engineering/construction companies in the world, and was vice president of the American Institute of Certified Public Accountants' committee on construction accounting. He is coauthor of the *McGraw-Hill Construction Management Form Book* and of the Dow Jones *Businessman's Guide to Construction.*

WILLIAM E. COOMBS (deceased) was for many years an attorney specializing in taxation and business law. During his career he had been California sales tax auditor and a state senator, as well as senior accountant, controller, treasurer, and house counsel for various construction and other firms.

MARK A. SMITH, Ph.D., CPA, is associate national director of Construction Industry Services and a partner in the firm of Ernst & Young. He earned his Ph.D. in civil engineering at the University of Texas. He testifies frequently as an expert witness in construction litigation.